INTERNATIONAL CODE OF BOTANICAL NOMENCLATURE

1988

International Code of Botanical Nomenclature

Regnum vegetabile, a series of publications for the use of plant taxonomists
published under the auspices of the International Association for Plant Taxonomy,

edited by Werner Greuter

Volume 118

International Code of Botanical Nomenclature

Adopted by the Fourteenth International Botanical Congress, Berlin, July–August 1987

Prepared and edited by

W. GREUTER, Chairman

H. M. BURDET, W. G. CHALONER, V. DEMOULIN, R. GROLLE, D. L. HAWKS-WORTH, D. H. NICOLSON, P. C. SILVA, F. A. STAFLEU, E. G. VOSS, Members

J. MCNEILL, Secretary

of the Editorial Committee

1988

Koeltz Scientific Books

D-6240 Königstein, Federal Republic of Germany

CONTENTS

Contents

PREFACE

The most striking difference between this new "Berlin Code" and previous editions of the International Code of Botanical Nomenclature is that the text is in English only, lacking the French and German versions that have been a feature of the International Code since the very first, the Vienna Code of 1905. This reflects a recommendation made by the Nomenclature Section of the Berlin Congress, which also encouraged publication of the principal text in other languages. A French text is currently in preparation and a German and a Spanish are being considered. In other respects the "Berlin Code" will look quite familiar to those who have used recent editions of the Code. The same system for numbering paragraphs of Articles, Recommendations and Examples is employed as in the previous "Sydney Code". One new Article and several new Recommendations appear, but, as the Article is the last in the Code (Art. 76), a shift in the numbering of Articles and Recommendations has been avoided. Certain paragraphs have necessarily been given new numbers, when new material was inserted ahead of them. The Berlin Congress accepted or referred to the Editorial Committee considerably more proposals than did the Leningrad or Sydney Congress – over 135 as against fewer than 100. As has become usual in recent Congresses, most of these proposals did little, however, to alter the substance of the Code, merely refining existing provisions.

Perhaps the major change in the new Code is the extension of the provisions for the conservation of the names of species. In addition to "species of major economic importance", conservation of species names is now possible in two other situations. The more general is where a name has been widely and persistently used for a taxon or taxa not including its type (Art. 69 situations); henceforth conservation will be permitted in such cases as an alternative to rejection. The other is merely a logical extension of the long-standing option to conserve a generic name with a particular type, in that it permits conservation of the type of a species name that indicates the type of a conserved generic name. Thus, not only may *Amaryllis* be conserved with *A. belladonna* indicated as type, but the type of *A. belladonna* itself may also being conserved – as is, indeed, happening.

Two Articles (66 and 67) have been deleted from the Code. These dealt with illegitimate names and represented something of a leftover from former Codes in which the circumscription rather than the type method had prevailed. The new Article (76) that appears in the Code is a transformation of the former Recommendation 75A dealing with the gender of generic names. The change is logical in view of the fact that wrong gender terminations are errors to be corrected under Art. 75.3.

In terms of text changes, the portion of the Code which has undergone the greatest revision is that relating to typification (Arts. 7-10). Most strikingly, the "Guide for the Determination of Types", which in previous Codes formed an unnumbered appendix at the end of the text, has been incorporated in the main body of the Code, principally by the addition of Rec. 7B and a new Note and paragraph in Art. 7. Five additions to the rules on types and typification have, however, been introduced. The Code now makes it clear that for a lecto-typification (or neotypification) to have priority, the element chosen must be directly cited including the term "type" or an equivalent, and hence mere exclusion of all other elements does not constitute priorable lectotypification. Neo-typification is now permitted in cases where all the surviving original material can be shown to be taxonomically different from a destroyed holotype or previously designated lectotype. The provision that existed for a type to be a description under certain circumstances – something that many felt amounted to a repudiation of the type method – has been deleted from the Code. From 1 Jan 1990, the place of preservation of the type of the name of a species or infraspecific taxon, including a designated lectotype or neotype, must be specified. Finally, and also from 1 Jan 1990, the mandatory indication of the holotype of a name of a new taxon will require use of the word "typus" or "holotypus" or one of their equivalents in a modern language. Art. 10, dealing with the typification of the names of genera and subdivisions of genera, has also been clarified and the previous exclusion of illustrations as types of names conserved under Art. 10.3 has been removed.

Also relevant to typification is the provision that all nomenclatural actions regulated by the Code must appear in effectively published works in order to be taken into consideration. This applies, not only to lectotypification and neotypification, but also to actions such as the uniting of taxa bearing names of equal priority or choosing between homonyms with equal priority.

Some of the rules on valid publication of names have been clarified. In addition to the new obligation, from 1990, to use the word typus or an equivalent when publishing names of new taxa (referred to above), the requirements prior to that date have been redefined. Other clarifications have been made concerning the association of an epithet with the generic or specific name in a new combination, the degree of precision required when referring to the place of publication of a basionym, and the conditions of validation by means of a *descriptio generico-specifica*.

Other changes of note in the present Code include matters relating to illegitimacy, orthography, and to the naming of hybrids. The definition of what constitutes a nomenclaturally superfluous and hence illegitimate name under Art. 63 has been clarified (and made less all embracing than some had considered it to be) by specifying that a name is superfluous only if the corresponding taxon originally included the previously designated lectotype (or the holotype, or all syntypes) of a name which ought to have been adopted under the rules. In addition to the new Article on the gender of generic names referred to above, a few relatively minor modifications have been made in the rules and recommendations dealing with orthography. It was made clear that the ranks of taxa of hybrids (nothotaxa) are the same as the corresponding ranks of non-hybrid taxa, and that where an inappropriate nothotaxon rank is used the name is not invalid but merely incorrect and could be employed later in a taxonomic context in which the rank was appropriate.

The format of the Appendices listing conserved and rejected names has evolved over the years. A major revision of Appendix III was carried out after the Sydney Congress and the criteria upon which this was based were described in Taxon (33: 310-316. 1984). The same criteria have been in general applied in the present edition. Moreover, those entries which involve orthographic variants have been brought into line with Art. 75 which specifies that only one variant is a validly published name. Consequently those entries involving rejected original spellings, formerly prefixed by "(V)", have been deleted. Instead, such cases in which a later spelling is conserved are indicated by the phrase "*orth. cons.*" Another change in the Appendices is the splitting of Appendix II (*nomina familiarum conservanda*) into two, IIA dealing with Fungi and Pteridophyta, in which family names are now conserved only over listed rejected names in the same manner as has always applied to generic names. Appendix IIB covering Bryophyta and Spermatophyta continues in the format of the previous Appendix II with the listed names conserved over all synonyms, and with no rejected names listed. For the first time *nomina specifica conservanda* are listed, forming the new Appendix IIIB (whereby the former Appendix III, *nomina generica conservanda*, becomes Appendix IIIA).

The Code is printed in two different type sizes, the Recommendations, Notes, and Examples (as well as the text of the Appendices and Indices) being set in smaller type. This reflects, so far as the main text of the Code is concerned, the distinction between the rules (primarily the Articles) and the material that explains the implications of the rules (the Notes and Examples) or provides advice on good nomenclatural practice (the Recommendations). The nature of

Articles, Recommendations and Examples seems to be well understood, but, from experience in reviewing proposals to amend the Code, it appears that the role of Notes is less clear. Like an Article, a Note in the Code states something that is mandatory; it differs from an Article, however, in that a Note does not introduce any new rule or concept, but merely spells out something that may not be evident to the user but is provided for elsewhere in the Code.

The procedures for producing this edition of the Code have followed the pattern that has been well established since the Stockholm Congress of 1950. Published proposals for amendment, with technical comments by the Rapporteurs and various relevant reports, were assembled in a Synopsis of Proposals (Taxon 36: 174-281. 1987). Results of the Preliminary Mail Vote on these proposals, a strictly advisory but very helpful expression of opinion, were made available at registration for the Nomenclature Section of the XIV International Botanical Congress, in the "Dachgarten" of the International Congress Centre in Berlin (West). The Section met from July 20-24, just before the regular sessions of the Congress, and made decisions on the 336 proposals before it, accepting some 72 and referring another 63 to the Editorial Committee for modification of the Code. These decisions were adopted by resolution of the closing plenary session of the Congress on August 1 and became official at that time. A list of them appeared in Taxon (36: 858-868. 1987), along with the results of the preliminary mail vote. The full Report of the Section, including the essence of the debates and comments made during the deliberations, is being published in the journal Englera. A preliminary transcript of the complete tape records of the nomenclature sessions, prepared by Dan Nicolson and John McNeill, was available to all members of the Editorial Committee at their meeting in January 1988.

It is the duty of the Editorial Committee, elected by the Section (and, by tradition, from among those present for the discussions), to incorporate the decisions of the Congress into the Code and to make in it whatever strictly editorial changes are desirable for smooth, consistent, accurate, and clear reading. The composition of the Editorial Committee usually changes slightly at each Congress, and this was also the case on this occasion. Ed Voss, Rapporteur-général for the Sydney Congress and hence Chairman of the previous Editorial Committee, did not wish to continue in this capacity at Berlin, and Werner Greuter, vice-rapporteur at Sydney, was there appointed Rapporteur-général for the Berlin Congress. As such he served as Chairman of this Editorial Committee. The new Vice-rapporteur and Secretary of the Committee was John McNeill, a member of the previous Editorial Committee. The Committee was fortunate in being able to continue to benefit from Ed Voss's experience in that he agreed to serve as a member. The Committee benefited enormously by being able to draw on the wisdom and experience of Frans Stafleu, Vice-rapporteur and Secretary of the Editorial Committee from 1954 to 1964, Rapporteur-général and Chairman of the Committee from 1964 to 1979, and President of the Nomenclature Section at Berlin. Returning as a full member of the Committee, and with the insight of having chaired the Nomenclature Section meetings in almost their entirety, Frans Stafleu provided sage advice on very many occasions, particularly when it proved necessary to interpret somewhat conflicting decisions made by the Section. The two new members, Riclef Grolle and David Hawksworth, joined most effectively in the work of the Committee.

After circulation of a draft of the text of the new Code, the Editorial Committee met at the "Botanischer Garten und Botanisches Museum Berlin-Dahlem" from January 3-7, 1988. All 11 members of the Committee were present for this essential meeting that thrashes out the clearest and most concise way to express in the Code the decisions of the Nomenclature Section and deals with a myriad of editorial details, essential in seeking clarity and consistency in an international work, which should not necessarily follow the editorial practices of any particular country or language group. In this context, it should be explained that this edition of the Code, like its predecessors, follows, in cases of discrepancy, British rather than American spelling of English words. In this edition the Committee has adopted the practice of italicizing all scientific names up to the rank of family (the ranks for which priority is mandatory), but not those at higher ranks. The Editorial Committee recognizes that complete clarity and consistency are hardly achievable and is aware of several instances of imprecise, conflicting, or otherwise unsatisfactory wording in the Code. Resolution of these is not always in its power as an editorial matter. Any effort to resolve certain points could be interpreted as extending the operation of the Code or as restricting it, depending upon one's reading of the present text. Failure, therefore, to make a change at a point of ambiguity does not necessarily mean that we consider the present text to be satisfactory. Some matters must await the next Congress!

We have continued our effort, to which major attention was given in the Leningrad and Sydney editions, to be consistent within the Code regarding bibliographic style and details, in a manner which is agreeable and non-confusable to all users. Many aids are now available which did not exist years ago. While none of them provides all the answers to achieving standardization in a document which deals with all groups of plants and an incredible diversity of literature, we have made heavy use of "TL-2" (Taxonomic Literature, ed. 2); "B-P-H" (Botanico-Periodicum-Huntianum); the Draft Index of Author Abbreviations compiled at The Herbarium, Royal Botanic Gardens, Kew; and the Catalogue des Périodiques de la Bibliothèque des Conservatoire et Jardin botaniques de la Ville de Genève.

In addition to the preparation of the principal text of the Code carried out by all members of the Committee at their meeting in Berlin and subsequently by reviewing two revised drafts, other portions of the Code have been produced by particular members of the Committee with help from others. The new Index, which represents a substantial revision of, and we hope improvement on, previous versions, was largely the work of Ed Voss; Appendices II and III were updated by the following: Paul Silva (Algae), Vincent Demoulin (Fungi), Gea Zijlstra (Utrecht, Bryophyta), Bill Chaloner (Fossils), and Dan Nicolson and John McNeill (other groups). The Committee is particularly grateful to Gea Zijlstra, Secretary of the Committee for Bryophyta, for providing the entire update of the Bryophyte entries, to R. K. Brummitt (Kew) who checked the whole Spermatophyta entries and provided substantial comments, and also to the Secretaries of the Committees for Fungi and Lichens and for Pteridophyta (R. Korf and R. E. G. Pichi-Sermolli) for their great help with the entries for their respective groups. The final editing of the whole text, including the appendices and indices, was undertaken by Werner Greuter in close contact with John McNeill and Dan Nicolson.

The speedy publication of the Berlin Code and its relatively modest price are due to a large extent to the fact that the text of the previous edition was available on magnetic tape, in the original phototype format, and that it was possible to transfer it to a personal computer at the IAPT headquarters in Berlin for further editing. Some of the additional matter, in particular many of the new entries to the Appendices, was prepared in Edinburgh, Utrecht and Washington and transmitted to Berlin on magnetic support (floppy disks). Offset-ready copy of the present Code was produced directly in Berlin by means of a laser printer after suitable reformatting. Although a highly sophisticated device, a laser printer is not quite as versatile as professional phototype-setting equipment, which accounts for some minor awkwardness such as the replacement of single by double quotation marks throughout the text and the less than perfect representation of the identity sign used in the Appendices. Reformating the text of the Appendices with their semi-tabular layout seemed at first to pose insoluble problems, and we are much indebted to Mr. Rolf König, Berlin, who devised and wrote the required software at no cost.

The Berlin Code is the first volume of the Regnum Vegetabile series to be published by Koeltz Scientific Books, Königstein/Taunus. It is our pleasure to acknowledge the helpful and amiable attitude of Mr. Sven Koeltz in what we trust will prove the start of a long and fruitful collaboration.

In addition to those who have helped to make possible this new edition of the Code, botanical nomenclature depends on the scores of botanists who serve on the Permanent Nomenclature Committees that work continuously between Congresses, dealing principally with proposals for conservation or rejection, and those others who are members of Special Committees, reviewing and seeking solutions to the problems assigned to them by the Nomenclature Section of the Congress. Botanical Nomenclature is remarkable for the large number of taxonomists who voluntarily work so effectively and for such long hours, to the immeasurable benefit of all their colleagues who must use plant names and on whose behalf this word of sincere thanks is expressed.

The International Code of Botanical Nomenclature is published under the ultimate authority of the International Botanical Congresses. Provisions for modifications of it are detailed in Division III of the Code and are described above. An account of the international organization of botanical nomenclature appears in J. McNeill & W. Greuter, Botanical nomenclature (IUBS Monogr. Ser. 2: 3-26. 1987). The various permanent nomenclature committees listed in Division III operate under the auspices of the International Association for Plant Taxonomy (IAPT), which is itself a Section of the International Union of Biological Sciences (IUBS). The secretaries of these committees, along with additional ex officio and elected members, constitute the General Committee, which represents botanical nomenclature between Congresses and serves also as the Commission on Nomenclature of Plants of IUBS.

The truly international and cooperative nature which characterizes the nomenclature committees, the broad democratic way in which the Code is subject to modification, and the common consent by which its provisions are accepted throughout the world, all make it a pleasure as well as a privilege for each of us to serve in these endeavours.

May 1988

Werner Greuter
John McNeill

IMPORTANT DATES IN THE CODE

DATES UPON WHICH PARTICULAR PROVISIONS OF THE CODE BECOME EFFECTIVE

1 May 1753	Art. 13.1(a), (c), (d), (e)
after 1753	Art. 7.14
1 Jan 1801	Art. 13.1(b)
31 Dec 1801	Art. 13.1(d)
31 Dec 1820	Art. 13.1(f)
1 Jan 1821	Art. 13.1(d)
1 Jan 1848	Art. 13.1(e)
1 Jan 1886	Art. 13.1(e)
1 Jan 1890	Art. 35.3
1 Jan 1892	Art. 13.1(e)
1 Jan 1900	Art. 13.1(e)
1 Jan 1908	Art. 42.2; 44.1
1 Jan 1912	Art. 20.2; 38.1
1 Jan 1935	Art. 36.1
1 Jan 1953	Art. 29.2; 29.4; 31.1; 32.3; 33.2; 34.3; 35.1; 35.2
1 Jan 1958	Art. 36.2; 37.1; 39.1
1 Jan 1959	Art. 28 Note 2
1 Jan 1973	Art. 29.4; 45.1
1 Jan 1990	Art. 8.4; 37.4; 37.5

ARTICLES INVOLVING DATES APPLICABLE TO THE MAIN TAXONOMIC GROUPS

All groups	Art. 8.4; 20.2; 29.2; 29.4; 31.1; 33.2; 34.3; 35.1; 35.2; 35.3; 37.1; 37.4; 37.5; 42.2; 44.1; 45.1
All groups except algae and all fossils	Art. 36.1
Spermatophyta	Art. 13.1(a)
Pteridophyta	Art. 13.1(a)
Bryophyta	Art. 7.14; 13.1(b), (c)
Fungi (incl. Myxomycetes and lichen-forming fungi)	Art. 13.1(d)
Algae	Art. 7.14; 13.1(f); 38.1
Fossil plants	Art. 7.14; 13.1(f); 38.1
Cultivated plants	Art. 28 Note 2

ARTICLES DEFINING THE DATES OF CERTAIN WORKS

Art. 13.1(e), (f); 13.4; 13.5; 23.6(c); Rec. 32A.1; Art. 41 Note 1; Intr. to App. II

INTERNATIONAL CODE OF BOTANICAL NOMENCLATURE

PREAMBLE

1. Botany requires a precise and simple system of nomenclature used by botanists in all countries, dealing on the one hand with the terms which denote the
ranks of taxonomic groups or units, and on the other hand with the scientific
names which are applied to the individual taxonomic groups of plants. The
purpose of giving a name to a taxonomic group is not to indicate its characters
or history, but to supply a means of referring to it and to indicate its taxonomic
rank. This Code aims at the provision of a stable method of naming taxonomic groups, avoiding and rejecting the use of names which may cause error
or ambiguity or throw science into confusion. Next in importance is the avoid-
ance of the useless creation of names. Other considerations, such as absolute
grammatical correctness, regularity or euphony of names, more or less pre-
vailing custom, regard for persons, etc., notwithstanding their undeniable
importance, are relatively accessory.

2. The Principles form the basis of the system of botanical nomenclature.

3. The detailed Provisions are divided into Rules, set out in the Articles, and
Recommendations. Examples (Ex.) are added to the rules and recommendations to illustrate them.

4. The object of the Rules is to put the nomenclature of the past into order
and to provide for that of the future; names contrary to a rule cannot be maintained.

5. The Recommendations deal with subsidiary points, their object being to
bring about greater uniformity and clearness, especially in future nomenclature; names contrary to a recommendation cannot, on that account, be rejected, but they are not examples to be followed.

6. The provisions regulating the modification of this Code form its last division.

7. The rules and recommendations apply to all organisms treated as plants (including fungi and blue-green algae but excluding other prokaryotic groups[1]), whether fossil or non-fossil[2]. Special provisions are needed for certain groups of plants: The International Code of Nomenclature for Cultivated Plants-1980 was adopted by the International Commission for the Nomenclature of Cultivated Plants; provisions for the names of hybrids appear in Appendix I.

8. The only proper reasons for changing a name are either a more profound knowledge of the facts resulting from adequate taxonomic study or the necessity of giving up a nomenclature that is contrary to the rules.

9. In the absence of a relevant rule or where the consequences of rules are doubtful, established custom is followed.

10. This edition of the Code supersedes all previous editions.

---

1) For the nomenclature of other prokaryotic groups, see the International Code of Nomenclature of Bacteria.

2) In this Code, the term "fossil" is applied to a taxon when its name is based on a fossil type and the term "non-fossil" is applied to a taxon when its name is based on a non-fossil type (see Art. 13.3).

DIVISION I. PRINCIPLES

Principle I

Botanical nomenclature is independent of zoological nomenclature.
The Code applies equally to names of taxonomic groups treated as plants whether or not these groups were originally so treated (see Pre. 7).

Principle II

The application of names of taxonomic groups is determined by means of nomenclatural types.

Principle III

The nomenclature of a taxonomic group is based upon priority of publication.

Principle IV

Each taxonomic group with a particular circumscription, position, and rank can bear only one correct name, the earliest that is in accordance with the Rules, except in specified cases.

Principle V

Scientific names of taxonomic groups are treated as Latin regardless of their derivation.

Principle VI

The Rules of nomenclature are retroactive unless expressly limited.

DIVISION II. RULES AND RECOMMENDATIONS

CHAPTER I. RANKS OF TAXA,
AND THE TERMS DENOTING THEM

Article 1

1.1. Taxonomic groups of any rank will, in this Code, be referred to as *taxa* (singular: *taxon*).

Article 2

2.1. Every individual plant is treated as belonging to a number of taxa of consecutively subordinate rank, among which the rank of species (*species*) is basal.

Article 3

3.1. The principal ranks of taxa in ascending sequence are: species (*species*), genus (*genus*), family (*familia*), order (*ordo*), class (*classis*), division (*divisio*), and kingdom (*regnum*). Thus, except for some fossil plants (see Art. 3.2), each species is assignable to a genus, each genus to a family, etc.

3.2. The principal ranks of nothotaxa (hybrid taxa) are nothospecies and nothogenus. These are the same rank as species and genus, only the terms denoting the ranks differing in order to indicate the hybrid character (see Appendix I).

3.3. Because of the fragmentary nature of the specimens on which the species of some fossil plants are based, the genera to which they are assigned are not assignable to a family, although they may be referable to a taxon of higher rank. Such genera are known as form-genera (*forma-genera*).

Ex. 1. Form-genera: *Dadoxylon* Endl. (Coniferopsida), *Pecopteris* (Brongn.) Sternb. (Pteropsida), *Stigmaria* Brongn. (Lepidodendrales), *Spermatites* Miner (seed-bearing plants).

4

Ex. 2. The following are, however, not form-genera: *Lepidocarpon* D. Scott (*Lepidocarpaceae*), *Mazocarpon* M. Benson (*Sigillariaceae*), *Siltaria* Traverse (*Fagaceae*).

Note 1. Art. 59 provides for form-taxa for asexual forms (anamorphs) of certain pleomorphic fungi, at any rank.

3.4. As in the case of certain pleomorphic fungi, the provisions of this Code do not prevent the publication and use of names of form-genera of fossils.

Article 4

4.1. If a greater number of ranks of taxa is required, the terms for these are made either by adding the prefix *sub-* to the terms denoting the ranks or by the introduction of supplementary terms. A plant may thus be assigned to taxa of the following ranks (in descending sequence): *regnum, subregnum, divisio, subdivisio, classis, subclassis, ordo, subordo, familia, subfamilia, tribus, subtribus, genus, subgenus, sectio, subsectio, series, subseries, species, subspecies, varietas, subvarietas, forma, subforma.*

4.2. Further supplementary ranks may be intercalated or added, provided that confusion or error is not thereby introduced.

4.3. The subordinate ranks of nothotaxa are the same as the subordinate ranks of non-hybrid taxa, except that nothogenus is the highest rank permitted (see Appendix I).

Note 1. Throughout this Code the phrase "subdivision of a family" refers only to taxa of a rank between family and genus and "subdivision of a genus" refers only to taxa of a rank between genus and species.

Note 2. For the designation of certain variants of species in cultivation, see Art. 28 Notes 1 and 2.

Note 3. In classifying parasites, especially fungi, authors who do not give specific, subspecific, or varietal value to taxa characterized from a physiological standpoint but scarcely or not at all from a morphological standpoint may distinguish within the species special forms (*formae speciales*) characterized by their adaptation to different hosts, but the nomenclature of special forms is not governed by the provisions of this Code.

Article 5

5.1. The relative order of the ranks specified in Arts. 3 and 4 must not be altered (see Arts. 33.4 and 33.5).

CHAPTER II. NAMES OF TAXA (GENERAL PROVISIONS)

SECTION 1. DEFINITIONS

Article 6

6.1. Effective publication is publication in accordance with Arts. 29-31.

6.2. Valid publication of names is publication in accordance with Arts. 32-45 or H.9 (see also Art. 75).

6.3. A legitimate name is one that is in accordance with the rules.

6.4. An illegitimate name is one that is designated as such in Arts. 18.3 or 63-65 (see also Art. 21 Note 1 and Art. 24 Note 1). A name which according to this Code was illegitimate when published cannot become legitimate later unless it is conserved or sanctioned.

6.5. The correct name of a taxon with a particular circumscription, position, and rank is the legitimate name which must be adopted for it under the rules (see Art. 11).

Ex. 1. The generic name *Vexillifera* Ducke (1922), based on the single species *V. micranthera*, is legitimate because it is in accordance with the rules. The same is true of the generic name *Dussia* Krug & Urban ex Taubert (1892), based on the single species *D. martinicensis.* Both generic names are correct when the genera are thought to be separate. Harms (Repert. Spec. Nov. Regni Veg. 19: 291. 1924), however, united *Vexillifera* Ducke and *Dussia* Krug & Urban ex Taubert in a single genus; when this treatment is accepted the latter name is the only correct one for the genus with this particular circumscription. The legitimate name *Vexillifera* may therefore be correct or incorrect according to different concepts of the taxa.

6.6. In this Code, unless otherwise indicated, the word "name" means a name that has been validly published, whether it is legitimate or illegitimate (see Art. 12).

6.7. The name of a taxon below the rank of genus, consisting of the name of a genus combined with one or two epithets, is termed a combination (see Arts. 21, 23, and 24).

Ex. 2. Combinations: *Gentiana lutea, Gentiana tenella* var. *occidentalis, Equisetum palustre* var. *americanum, Equisetum palustre* f. *fluitans, Mouriri* subg. *Pericrene, Arytera* sect. *Mischarytera.*

6.8. Autonyms are such names as can be established automatically under Arts. 19.4, 22.2, and 26.2, whether they were formally created or not.

SECTION 2. TYPIFICATION

Article 7

7.1. The application of names of taxa of the rank of family or below is determined by means of nomenclatural types (types of names of taxa). The application of names of taxa in the higher ranks is also determined by types when the names are ultimately based on generic names (see Art. 10.5).

7.2. A nomenclatural type (*typus*) is that element to which the name of a taxon is permanently attached, whether as a correct name or as a synonym. The nomenclatural type is not necessarily the most typical or representative element of a taxon.

7.3. A holotype is the one specimen or illustration used by the author or designated by him as the nomenclatural type. As long as a holotype is extant, it automatically fixes the application of the name concerned.

Note 1. Any designation made by the original author, if definitely expressed at the time of the original publication of the name of the taxon, is final (but see Art. 7.4). If the author included only one element, that one must be accepted as the holotype. If a new name is based on a previously published description of the taxon, the same considerations apply to material included by the earlier author (see Arts. 7.14-7.16).

7.4. If no holotype was indicated by the author of a name, or when the holotype has been lost or destroyed, or when the material designated as type is found to belong to more than one taxon, a lectotype or, if permissible (Art. 7.9), a neotype as a substitute for it may be designated. A lectotype always takes precedence over a neotype, except as provided by Art. 7.10. An isotype, if such exists, must be chosen as the lectotype. If no isotype exists, the lectotype must be chosen from among the syntypes, if such exist. If neither an isotype nor a syntype nor any of the original material[1] is extant, a neotype may be selected.

---

1) For the purposes of this Code, "original material" includes illustrations examined by an author prior to publication of a name and associated by the author with the concept of the named taxon.

7.5. A lectotype is a specimen or illustration selected from the original material to serve as a nomenclatural type when no holotype was indicated at the time of publication or as long as it is missing. When two or more specimens have been designated as types by the author of a specific or infraspecific name (e.g. male and female, flowering and fruiting, etc.), the lectotype must be chosen from among them.

7.6. An isotype is any duplicate[1] of the holotype; it is always a specimen.

7.7. A syntype is any one of two or more specimens cited by the author when no holotype was designated, or any one of two or more specimens simultaneously designated as types.

7.8. A paratype is a specimen or illustration cited in the protologue that is neither the holotype nor an isotype, nor one of the syntypes if two or more specimens were simultaneously designated as types.

7.9. A neotype is a specimen or illustration selected to serve as nomenclatural type as long as all of the material on which the name of the taxon was based is missing (see also Art. 7.10).

7.10. When a holotype or a previously designated lectotype has been lost or destroyed and it can be shown that all the other original material differs taxonomically from the destroyed type, a neotype may be selected to preserve the usage established by the previous typification (see also Art. 8.5).

7.11. A new name published as an avowed substitute (*nomen novum*) for an older name is typified by the type of the older name (see Art. 33.2; but see Art. 33 Note 1).

Ex. 1. Myrcia lucida McVaugh (1969) was published as a nomen novum for *M. laevis* O. Berg (1862), an illegitimate homonym of *M. laevis* G. Don (1832). The type of *M. lucida* is therefore the type of *M. laevis* O. Berg (non G. Don), namely, Spruce 3502.

7.12. A new name formed from a previously published legitimate name (*stat. nov., comb. nov.*) is, in all circumstances, typified by the type of the basionym (see Art. 55.2).

Ex. 2. Iridaea splendens (Setch. & Gardner) Papenf., *I. cordata* var. *splendens* (Setch. & Gardner) Abbott, and *Gigartina cordata* var. *splendens* (Setch. & Gardner) Kim all have the same type as their basionym, *Iridophycus splendens* Setch. & Gardner, namely, Gardner 7781 (UC 539565).

---

1) Here and elsewhere, the word duplicate is given its usual meaning in herbarium curatorial practice. It is part of a single gathering made by a collector at one time. However, the possibility of a mixed gathering must always be considered by an author choosing a lectotype, and corresponding caution used.

7.13. A name which was nomenclaturally superfluous when published (see Art. 63) is automatically typified by the type of the name which ought to have been adopted under the rules, unless the author of the superfluous name has definitely indicated a different type. Automatic typification does not apply to names sanctioned under Art. 13.1(d).

7.14. The type of a name of a taxon assigned to a group with a nomenclatural starting-point later than 1753 (see Art. 13) is to be determined in accordance with the indication or description and other matter accompanying its valid publication (see Arts. 32-45).

7.15. When valid publication is by reference to a pre-starting-point description, the latter must be used for purposes of typification.

7.16. A name validly published by reference to a previously and effectively published description or diagnosis (Art. 32.3) is to be typified by an element selected from the context of the validating description or diagnosis, unless the validating author has definitely designated a different type.

Ex. 3. Since the name *Adenanthera bicolor* Moon (1824) is validated solely by reference to Rumphius, Herbarium Amboinense 3: t. 112, the type of the name, in the absence of the specimen from which it was figured, is the illustration referred to. It is not the specimen, at Kew, collected by Moon and labelled "*Adenanthera bicolor*", since Moon did not definitely designate the latter as the type.

Ex. 4. *Echium lycopsis* L. (Fl. Angl. 12. 1754) was published without a description but with reference to Ray (Syn. Meth. Stirp. Brit. ed. 3. 227. 1724), in which a "*Lycopsis*" species was discussed with citation of earlier references, including Bauhin (Pinax 255. 1623), but also with no description. The accepted validating description of *E. lycopsis* is that of Bauhin, and the type must be chosen from the context of his work. Consequently the Sherard specimen in the Morison herbarium (OXF), selected by Klotz (Wiss. Z. Martin-Luther-Univ. Halle-Wittenberg Math.-Naturwiss. Reihe 9: 375-376. 1960), although probably consulted by Ray, is not eligible as type. The first acceptable choice is that of the illustration, cited by both Ray and Bauhin, of "*Echii altera species*" in Dodonaeus (Stirp. Hist. Pempt. 620. 1583), suggested by P. E. Gibbs (Lagascalia 1: 60-61. 1971) and formally made by W. T. Stearn (Ray Soc. Publ. 149, Introd. 65. 1973).

7.17. A change of the listed type of a conserved generic name (see Art. 14 and App. III) can be effected only by a procedure similar to that adopted for the conservation of generic names.

Ex. 5. Bullock and Killick (Taxon 6: 239. 1957) published a proposal that the type of *Plectranthus* L'Hér. be changed from *P. punctatus* (L.f.) L'Hér. to *P. fruticosus* L'Hér. This proposal was approved by the appropriate Committees and by an International Botanical Congress.

7.18. The type of the name of a taxon of fossil plants of the rank of species or below is the specimen whose figure accompanies or is cited in the valid publication of the name (see Art. 38). If figures of more than one specimen were given or cited when the name was validly published, one of those specimens must be chosen as the type.

7.19. The typification of names of form-genera of plant fossils (Art. 3.3), of fungal anamorphs (Art. 59), and of any other analogous genera or lower taxa does not differ from that indicated above.

Note 2. See also Art. 59 for details regarding typification of names in certain pleomorphic fungi.

7.20. Typification of names adopted in one of the works specified in Art. 13.1(d), and thereby sanctioned, may be effected in the light of anything associated with the name in that work.

7.21. The type of an autonym is the same as that of the name from which it is derived.

Recommendation 7A

7A.1. It is strongly recommended that the material on which the name of a taxon is based, especially the holotype, be deposited in a public herbarium or other public collection and that it be scrupulously conserved.

Recommendation 7B

7B.1. If no holotype was indicated by the original author and if no syntypes are extant, the lectotype should be chosen from among duplicates of the syntypes (isosyntypes), if such exist. If neither an isotype, nor a syntype, nor an isosyntype is extant, a paratype, if such exists, may be chosen as lectotype.

7B.2. Typification of names for which no holotype was designated should only be carried out with an understanding of the author's method of working; in particular it should be realized that some of the material used by the author in describing the taxon may not be in the author's own herbarium or may not even have survived, and conversely, that not all the material surviving in the author's herbarium was necessarily used in describing the taxon.

7B.3. Designation of a lectotype should be undertaken only in the light of an understanding of the group concerned. In choosing a lectotype, all aspects of the protologue should be considered as a basic guide. Mechanical methods, such as the automatic selection of the first species or specimen cited or of a specimen collected by the person after whom a species is named, should be avoided as unscientific and productive of possible future confusion and further changes.

7B.4. In choosing a lectotype, any indication of intent by the author of a name should be given preference unless such indication is contrary to the protologue. Such indications are manuscript notes, annotations on herbarium sheets, recognizable figures, and epithets such as *typicus, genuinus,* etc.

7B.5. In cases where two or more heterogeneous elements were included in or cited with the original description, the lectotype should be so selected as to preserve current usage. In particular, if another author has already segregated one or more elements as other taxa, the residue or part of it should be designated as the lectotype provided that this element is not in conflict with the original description or diagnosis (see Art. 8.1).

7B.6. For the name of a fossil species, the lectotype, when one is needed, should, if possible, be a specimen illustrated at the time of the valid publication of the name (Art. 7.18).

7B.7. When a combination in a rank of subdivision of a genus has been published under a generic name that has not yet been typified, the lectotype of the generic name should be selected from the subdivision of the genus that was designated as nomenclaturally typical, if that is apparent.

Recommendation 7C

7C.1. In selecting a neotype particular care and critical knowledge should be exercised, because the reviewer usually has no guide except personal judgement as to what best fits the protologue, and if this selection proves to be faulty, it will inevitably result in further change.

Article 8

8.1. The author who first designates a lectotype or a neotype must be followed, but his choice is superseded if *(a)* the holotype or, in the case of a neotype, any of the original material is rediscovered; it may also be superseded if *(b)* it can be shown that it is in serious conflict with the protologue[1] and another element is available which is not in conflict with the protologue, or *(c)* that it was based on a largely mechanical method of selection, or *(d)* that it is contrary to Art. 9.2.

Ex. 1. Authors following the American Code of Botanical Nomenclature, Canon 15 (Bull. Torrey Bot. Club 34: 172. 1907), designated as the type "the first binomial species in order" eligible under certain provisions. This method of selection has been considered to be largely mechanical. Thus the first lectotypification of *Delphinium* L., by Britton (in Britton & Brown, Ill. Fl. N. U.S. ed. 2, 2: 93. 1913), who followed the American Code and chose *D. consolida* L., has been superseded by the choice of *D. peregrinum* L. by Hitchcock & Green (Nomencl. Prop. Brit. Botanists 162. 1929). As Linnaeus described *Delphinium* as having "germina tria vel unum", the unicarpellate *D. consolida* is not in "serious conflict with the protologue". It could not otherwise be displaced as the type, even though the tricarpellate *D. peregrinum* would seem a better choice for the type of the name of a genus assigned by its author to "Polyandria Trigynia".

8.2. For purposes of priority under Art. 8.1, designation of a type is achieved only by effective publication (Arts. 29-31).

8.3. For purposes of priority under Art. 8.1, designation of a type is achieved only if the type is definitely accepted as such by the typifying author, and if the type element is clearly indicated by direct citation including the term "type" or an equivalent.

Ex. 2. The phrase "standard species" as used by Hitchcock & Green (Nomencl. Prop. Brit. Botanists 110-199. 1929) and by the same authors in the Cambridge Rules (1935) (in which it is subordinate to the words "species lectotypicae propositae") is now regarded as equivalent to "type", and hence lectotypifications in these works are acceptable.

---

1) Protologue (from the Greek *protos*, first; *logos*, discourse): everything associated with a name at its valid publication, i.e., diagnosis, description, illustrations, references, synonymy, geographical data, citation of specimens, discussion, and comments.

Ex. 3. When originally described, *Stapelia* L. included two species, *S. variegata* and *S. hirsuta.* Haworth (Syn. Pl. Succ. 19, 40. 1812) transferred the former to his new genus *Orbea*, retaining the latter in *Stapelia.* As he did not use the term "type" or an equivalent, his action does not constitute lectotypification under Art. 8.1. The first lectotypification of *Stapelia* acceptable under Art. 8.1 appears to be that by Hitchcock & Green (Nomencl. Prop. Brit. Botanists 137. 1929), who chose *S. variegata.*

8.4. On or after 1 Jan. 1990, lectotypification or neotypification of a name of a species or infraspecific taxon by a specimen or unpublished illustration is not effected unless the herbarium or institution in which the type is conserved is specified.

8.5. A neotype selected under Art. 7.10 may be superseded if it can be shown to differ taxonomically from the holotype or lectotype that it replaced.

Article 9

9.1. The type (holotype, lectotype, or neotype) of a name of a species or infraspecific taxon is a single specimen or illustration except in the following case: for small herbaceous plants and for most non-vascular plants, the type may consist of more than one individual, which ought to be conserved permanently on one herbarium sheet or in one equivalent preparation (e.g., box, packet, jar, microscope slide).

9.2. If it is later proved that such a type herbarium sheet or preparation contains parts belonging to more than one taxon (Art. 7.4), the name must remain attached to that part (lectotype) which corresponds most nearly with the original description.

Ex. 1. The holotype of the name *Rheedia kappleri* Eyma, which applies to a polygamous species, is a male specimen collected by Kappler (593a in U). The author designated a hermaphroditic specimen collected by the Forestry Service of Surinam as a paratype (B. W. 1618 in U).

Ex. 2. The type of the name *Tillandsia bryoides* Griseb. ex Baker (1878) is Lorentz 128 in BM; this, however, proved to be a mixture. L. B. Smith (Proc. Amer. Acad. Arts 70: 192. 1935) acted in accordance with this rule in designating one part of Lorentz's gathering as the lectotype.

9.3. If it is impossible to preserve a specimen as the type of a name of a species or infraspecific taxon of non-fossil plants, or if such a name is without a type specimen, the type may be an illustration.

9.4. One whole specimen used in establishing a taxon of fossil plants is to be considered the nomenclatural type. If this specimen is cut into pieces (sections of fossil wood, pieces of coal-ball plants, etc.), all parts originally used in establishing the diagnosis ought to be clearly marked.

9.5. Type specimens of names of taxa must be preserved permanently and cannot be living plants or cultures.

9A.1. Whenever practicable a living culture should be prepared from the holotype material of the name of a newly described taxon of fungi or algae and deposited in a reputable culture collection. (Such action does not obviate the requirement for a holotype specimen under Art. 9.5.).

Article 10

10.1. The type of a name of a genus or of any subdivision of a genus is the type of a name of a species (except as provided by Art. 10.3). For purposes of designation or citation of a type, the species name alone suffices, i.e., it is considered as the full equivalent of its type.

10.2. If in the protologue of the name of a genus or of any subdivision of a genus reference is made to the name(s) of one or more definitely included species, the type must be chosen from among the types of these names. If such a reference is lacking, a type must be otherwise chosen, but the choice is to be superseded if it can be demonstrated that the selected type is not conspecific with any of the material associated with the protologue.

10.3. By conservation (Art. 14.8), the type of the name of a genus can be a specimen or illustration used by the author in the preparation of the protologue, other than the type of a name of an included species.

Ex. 1. The General Committee has approved conservation of *Physconia* Poelt with the specimen "Germania, Lipsia in *Tilia*, 1767, Schreber sub "*Lichen pulverulentus*" (**M**)" as the type.

10.4. The type of a name of a family or of any subdivision of a family is the same as that of the generic name on which it is based (see Art. 18.1). For purposes of designation or citation of a type, the generic name alone suffices. The type of a name of a family or subfamily not based on a generic name is the same as that of the corresponding alternative name (Arts. 18.5 and 19.7).

10.5. The principle of typification does not apply to names of taxa above the rank of family, except for names that are automatically typified by being based on generic names (see Art. 16). The type of such a name is the same as that of the generic name on which it is based.

Note 1. For the typification of some names of subdivisions of genera see Art. 22.

10A.1. If the element selected under Art. 10.3 is the type of a species name, that name may be cited as the type of the generic name. If the element selected is not the type of a species name the type element should be cited and, optionally, a parenthetical reference to its correct name may be given.

SECTION 3. PRIORITY

Article 11

11.1. Each family or taxon of lower rank with a particular circumscription, position, and rank can bear only one correct name, special exceptions being made for 9 families and 1 subfamily for which alternative names are permitted (see Arts. 18.5 and 19.7). However, the use of separate names for the form-taxa of fungi and for form-genera of fossil plants is allowed under Arts. 3.3 and 59.5.

11.2. For any taxon from family to genus inclusive, the correct name is the earliest legitimate one with the same rank, except in cases of limitation of priority by conservation (see Art. 14) or where Arts. 13.1(d), 19.3, 58, or 59 apply.

11.3. For any taxon below the rank of genus, the correct name is the combination of the final epithet[1] of the earliest legitimate name of the taxon in the same rank, with the correct name of the genus or species to which it is assigned, except *(a)* in cases of limitation of priority under Arts. 13.1(d) and 14, or *(b)* if the resulting combination would be invalid under Art. 32.1(b) or illegitimate under Art. 64, or *(c)* if Arts. 22.1, 26.1, 58, or 59 rule that a different combination is to be used.

11.4. The principle of priority is not mandatory for names of taxa above the rank of family (but see Rec. 16B).

Article 12

12.1. A name of a taxon has no status under this Code unless it is validly published (see Arts. 32-45).

SECTION 4. LIMITATION OF THE PRINCIPLE OF PRIORITY

Article 13

13.1. Valid publication of names for plants of the different groups is treated as beginning at the following dates (for each group a work is mentioned which is treated as having been published on the date given for that group):

---

1) Here and elsewhere in this Code, the phrase "final epithet" refers to the last epithet in sequence in any particular combination, whether that of a subdivision of a genus, or of a species, or of an infraspecific taxon.

Non-fossil plants:

(a) SPERMATOPHYTA and PTERIDOPHYTA, 1 May 1753 (Linnaeus, Species Plantarum ed. 1).

(b) MUSCI (the *Sphagnaceae* excepted), 1 Jan. 1801 (Hedwig, Species Muscorum).

(c) SPHAGNACEAE and HEPATICAE, 1 May 1753 (Linnaeus, Species Plantarum ed. 1).

(d) FUNGI (including Myxomycetes and lichen-forming fungi), 1 May 1753 (Linnaeus, Species Plantarum ed. 1). Names in the Uredinales, Ustilaginales, and Gasteromycetes adopted by Persoon (Synopsis Methodica Fungorum, 31 Dec. 1801) and names of other fungi (excluding Myxomycetes) adopted by Fries (Systema Mycologicum, vols. 1 (1 Jan. 1821) to 3, with additional Index (1832), and Elenchus Fungorum, vols. 1-2), are sanctioned, i.e., are treated as if conserved against earlier homonyms and competing synonyms. For nomenclatural purposes names given to lichens shall be considered as applying to their fungal component.

(e) ALGAE, 1 May 1753 (Linnaeus, Species Plantarum ed. 1). Exceptions:

NOSTOCACEAE HOMOCYSTEAE, 1 Jan. 1892 (Gomont, Monographie des Oscillariées, Ann. Sci. Nat. Bot. ser. 7, 15: 263-368; 16: 91-264). The two parts of Gomont's "Monographie", which appeared in 1892 and 1893 respectively, are treated as having been published simultaneously on 1 Jan. 1892.

NOSTOCACEAE HETEROCYSTEAE, 1 Jan. 1886 (Bornet & Flahault, Révision des Nostocacées hétérocystées, Ann. Sci. Nat. Bot. ser. 7, 3: 323-381; 4: 343-373; 5: 51-129; 7: 177-262). The four parts of the "Révision", which appeared in 1886, 1886, 1887, and 1888 respectively, are treated as having been published simultaneously on 1 Jan. 1886.

DESMIDIACEAE, 1 Jan. 1848 (Ralfs, British Desmidieae).

OEDOGONIACEAE, 1 Jan. 1900 (Hirn, Monographie und Iconographie der Oedogoniaceen, Acta Soc. Sci. Fenn. 27(1)).

Fossil plants:

(f) ALL GROUPS, 31 Dec. 1820 (Sternberg, Flora der Vorwelt, Versuch 1: 1-24. t. 1-13). Schlotheim, Petrefactenkunde, 1820, is regarded as published before 31 Dec. 1820.

13.2. The group to which a name is assigned for the purposes of this Article is determined by the accepted taxonomic position of the type of the name.

Ex. 1. The genus *Porella* and its single species, *P. pinnata*, were referred by Linnaeus (1753) to the Musci; if the type specimen of *P. pinnata* is accepted as belonging to the Hepaticae, the names were validly published in 1753.

15

Ex. 2. The lectotype of *Lycopodium* L. (1753) is *L. clavatum* L. (1753) and the type specimen of this is currently accepted as a pteridophyte. Accordingly, although the genus is listed by Linnaeus among the Musci, the generic name and the names of the pteridophyte species included by Linnaeus under it were validly published in 1753.

13.3. For nomenclatural purposes, a name is treated as pertaining to a non-fossil taxon unless its type is fossil in origin. Fossil material is distinguished from non-fossil material by stratigraphic relations at the site of original occurrence. In cases of doubtful stratigraphic relations, provisions for non-fossil taxa apply.

13.4. Generic names which first appear in Linnaeus' Species Plantarum ed. 1 (1753) and ed. 2 (1762-63) are associated with the first subsequent description given under those names in Linnaeus' Genera Plantarum ed. 5 (1754) and ed. 6 (1764) (see Art. 41). The spelling of the generic names included in the Species Plantarum ed. 1 is not to be altered because a different spelling has been used in the Genera Plantarum ed. 5.

13.5. The two volumes of Linnaeus' Species Plantarum ed. 1 (1753), which appeared in May and August, 1753, respectively, are treated as having been published simultaneously on the former date (1 May 1753).

Ex. 3. The generic names *Thea* L. Sp. Pl. 515 (May 1753) and *Camellia* L. Sp. Pl. 698 (Aug. 1753), Gen. Pl. ed. 5. 311 (1754) are treated as having been published simultaneously in May 1753. Under Art. 57 the combined genus bears the name *Camellia*, since Sweet (Hort. Suburb. Lond. 157. 1818), who was the first to unite the two genera, chose that name, citing *Thea* as a synonym.

13.6. Names of anamorphs of fungi with a pleomorphic life cycle do not, irrespective of priority, affect the nomenclatural status of the names of the correlated holomorphs (see Art. 59.4).

Article 14

14.1. In order to avoid disadvantageous changes in the nomenclature of families, genera, and species entailed by the strict application of the rules, and especially of the principle of priority in starting from the dates given in Art. 13, this Code provides, in Appendices II and III, lists of names that are conserved (*nomina conservanda*) and must be retained as useful exceptions.

Note 1. The rules on conserved names also apply to names at any rank sanctioned under Art. 13.1(d).

14.2. Conservation aims at retention of those names which best serve stability of nomenclature (see Rec. 50E). Conservation of specific names is restricted to species of major economic importance and to cases provided for by Arts. 14.3 and 69.3 (see also Art. 13.1(d)).

14.3. The application of both conserved and rejected names is determined by nomenclatural types. When typification of the specific name cited as the type of a conserved generic name is in dispute, the type of the specific name may be conserved, and listed in Appendix IIIA, so that the application of the generic name is not in doubt.

14.4. A conserved name of a family or genus is conserved against all other names in the same rank based on the same type (nomenclatural synonyms, which are to be rejected) whether these are cited in the corresponding list of rejected names or not, and against those names based on different types (taxonomic synonyms) that are cited in that list[1]. A conserved name of a species is conserved against all names listed as rejected, and against all combinations based on the rejected names.

14.5. When a conserved name competes with one or more other names based on different types and against which it is not explicitly conserved, the earliest of the competing names is adopted in accordance with Art. 57.1, except for names sanctioned under Art. 13.1(d) and for some conserved family names (Appendix IIB), which are conserved against unlisted names.

Ex. 1. If the genus *Weihea* Sprengel (1825) is united with *Cassipourea* Aublet (1775), the combined genus will bear the prior name *Cassipourea*, although *Weihea* is conserved and *Cassipourea* is not.

Ex. 2. If *Mahonia* Nutt. (1818) is united with *Berberis* L. (1753), the combined genus will bear the prior name *Berberis*, although *Mahonia* is conserved.

Ex. 3. Nasturtium R. Br. (1812) was conserved only against the homonym *Nasturtium* Miller (1754) and the nomenclatural synonym *Cardaminum* Moench (1794); consequently if reunited with *Rorippa* Scop. (1760) it must bear the name *Rorippa*.

14.6. When a name of a taxon has been conserved against an earlier name based on a different type, the latter is to be restored, subject to Art. 11, if it is considered the name of a taxon at the same rank distinct from that of the nomen conservandum except when the earlier rejected name is a homonym of the conserved name.

Ex. 4. The generic name *Luzuriaga* Ruiz & Pavón (1802) is conserved against the earlier names *Enargea* Banks & Sol. ex Gaertner (1788) and *Callixene* Comm. ex A. L. Juss. (1789). If, however, *Enargea* Banks & Sol. ex Gaertner is considered to be a separate genus, the name *Enargea* is retained for it.

14.7. A rejected name, or a combination based on a rejected name, may not be restored for a taxon which includes the type of the corresponding conserved name.

---

1) The International Code of Zoological Nomenclature and the International Code of Nomenclature of Bacteria use the terms "objective synonym" and "subjective synonym" for nomenclatural and taxonomic synonym, respectively.

Ex. 5. Enallagma Baillon (1888) is conserved against *Dendrosicus* Raf. (1838), but not against *Amphitecna* Miers (1868); if *Enallagma* and *Amphitecna* are united, the combined genus must bear the name *Amphitecna*, although the latter is not explicitly conserved against *Dendrosicus.*

14.8. A name may be conserved with a different type from that designated by the author or determined by application of the Code (see Art. 10.3). A name with a type so conserved (*typ. cons.*) is legitimate even if it would otherwise be illegitimate under Art. 63. When a name is conserved with a type different from that of the original author, the author of the name as conserved, with the new type, must be cited.

Ex. 6. Bulbostylis Kunth (1837), nom. cons. (non *Bulbostylis* Steven 1817). This is not to be cited as *Bulbostylis* Steven emend. Kunth, since the type listed was not included in *Bulbostylis* by Steven in 1817.

14.9. A conserved name, with its corresponding autonyms, is conserved against all earlier homonyms. An earlier homonym of a conserved or sanctioned name is not made illegitimate by that conservation or sanctioning but is unavailable for use; if legitimate, it may serve as basionym of another name or combination based on the same type (see also Art. 68.3).

Ex. 7. The generic name *Smithia* Aiton (1789), conserved against *Damapana* Adanson (1763), is thereby conserved automatically against the earlier homonym *Smithia* Scop. (1777).

14.10. A name may be conserved in order to preserve a particular orthography or gender. A name so conserved is to be attributed without change of priority to the author who validly published it, not to an author who later introduced the conserved spelling or gender.

Ex. 8. The spelling *Rhodymenia*, used by Montagne (1839), has been conserved against the original spelling *Rhodomenia*, used by Greville (1830). The name is to be cited as *Rhodymenia* Grev. (1830).

Note 2. The date of conservation or sanctioning does not affect the nomenclatural status of the conserved or sanctioned name, whose priority depends on its date of valid publication. When two or more conserved or sanctioned names are considered to be synonyms, the first to have been validly published has priority. When two or more homonyms are sanctioned only the earliest of them can be used, the later being illegitimate under Art. 64.

14.11. The lists of conserved names will remain permanently open for additions and changes. Any proposal of an additional name must be accompanied by a detailed statement of the cases both for and against its conservation. Such proposals must be submitted to the General Committee (see Division III), which will refer them for examination to the committees for the various taxonomic groups.

14.12. Entries of conserved names cannot be deleted. Similarly, a name once sanctioned remains sanctioned, even if elsewhere in the sanctioning works the sanctioning author does not recognize it.

Article 15

15.1. When a proposal for the conservation (or rejection under Art. 69) of a name has been approved by the General Committee after study by the Committee for the taxonomic group concerned, retention (or rejection) of that name is authorized subject to the decision of a later International Botanical Congress.

Recommendation 15A

15A.1. When a proposal for the conservation or rejection of a name has been referred to the appropriate Committee for study, authors should follow existing usage as far as possible pending the General Committee's recommendation on the proposal.

CHAPTER III. NOMENCLATURE OF TAXA ACCORDING TO THEIR RANK

SECTION 1. NAMES OF TAXA ABOVE THE RANK OF FAMILY

Article 16

16.1. Names of taxa above the rank of family are automatically typified if they are based on generic names (see Art. 10.5); for such automatically typified names, the name of a subdivision which includes the type of the adopted name of a division, the name of a subclass which includes the type of the adopted name of a class, and the name of a suborder which includes the type of the adopted name of an order, are to be based on the generic name equivalent to that type, but without the citation of an author's name.

16.2. Where one of the word elements *-monado-, -cocco-, -nemato-,* or *-clado-* as the second part of a generic name has been omitted before the termination *-phyceae* or *-phyta,* the shortened class or division name is regarded as based on the generic name in question if such derivation is obvious or is indicated at establishment of the group name.

Ex. 1. Raphidophyceae Chadefaud ex P. C. Silva (1980) was indicated by its author to be based on *Raphidomonas* F. Stein (1878).

Note 1. The principle of priority is not mandatory for names of taxa above the rank of family (Art. 11.4).

Recommendation 16A

16A.1. The name of a division is taken either from distinctive characters of the division (descriptive names) or from a name of an included genus; it should end in *-phyta,* unless it is a division of fungi, in which case it should end in *-mycota.*

16A.2. The name of a subdivision is formed in a similar manner; it is distinguished from a divisional name by an appropriate prefix or suffix or by the termination *-phytina*, unless it is a subdivision of fungi, in which case it should end in *-mycotina.*

16A.3. The name of a class or of a subclass is formed in a similar manner and should end as follows:

(a) In the algae: *-phyceae* (class) and *-phycidae* (subclass);

(b) In the fungi: *-mycetes* (class) and *-mycetidae* (subclass);

(c) In other groups of plants: *-opsida* (class) and *-idae* (subclass).

16A.4. When a name has been published with a Latin termination not agreeing with this recommendation, the termination may be changed to accord with it, without change of author's name or date of publication.

Recommendation 16B

16B.1. In choosing among typified names for a taxon above the rank of family, authors should generally follow the principle of priority.

Article 17

17.1. The name of an order or suborder is taken either from distinctive characters of the taxon (descriptive name) or from a legitimate name of an included family based on a generic name (automatically typified name). An ordinal name of the second category is formed by replacing the termination *-aceae* by *-ales* . A subordinal name of the second category is similarly formed, with the termination *-ineae.*

Ex. 1. Descriptive names of orders: Centrospermae, Parietales, Farinosae; of a suborder: Enantioblastae.

Ex. 2. Automatically typified names: Fucales, Polygonales, Ustilaginales; Bromeliineae, Malvineae.

17.2. Names intended as names of orders, but published with their rank denoted by a term such as "cohors", "nixus", "alliance", or "Reihe" instead of "order", are treated as having been published as names of orders.

17.3. When the name of an order or suborder based on a name of a genus has been published with an improper Latin termination, this termination must be changed to accord with the rule, without change of the author's name or date of publication.

Recommendation 17A

17A.1. Authors should not publish new names of orders for taxa of that rank which include a family from whose name an existing ordinal name is derived.

SECTION 2. NAMES OF FAMILIES AND SUBFAMILIES, TRIBES AND SUBTRIBES

Article 18

18.1. The name of a family is a plural adjective used as a substantive; it is formed from the genitive singular of a legitimate name of an included genus by replacing the genitive singular inflection (Latin *-ae, -i, -us, -is;* transliterated Greek *-ou, -os, -es, -as,* or *-ous,* including the latter's equivalent *-eos*) with the termination *-aceae.* For generic names of non-classical origin, when analogy with classical names is insufficient to determine the genitive singular, *-aceae* is added to the full word. For generic names with alternative genitives the one implicitly used by the original author must be maintained.

Ex. 1. Family names based on a generic name of classical origin: *Rosaceae* (from *Rosa, Rosae*), *Salicaceae* (from *Salix, Salicis*), *Plumbaginaceae* (from *Plumbago, Plumbaginis*), *Rhodophyllaceae* (from *Rhodophyllus, Rhodophylli*), *Rhodophyllidaceae* (from *Rhodophyllis, Rhodophyllidos*), *Sclerodermataceae* (from *Scleroderma, Sclerodermatos*), *Aextoxicaceae* (from *Aextoxicon, Aextoxicou*), *Potamogetonaceae* (from *Potamogeton, Potamogetonos*).

Ex. 2. Family names based on a generic name of non-classical origin: *Nelumbonaceae* (from *Nelumbo, Nelumbonis,* declined by analogy with *umbo, umbonis*), *Ginkgoaceae* (from *Ginkgo,* indeclinable).

18.2. Names intended as names of families, but published with their rank denoted by one of the terms "order" (*ordo*) or "natural order" (*ordo naturalis*) instead of "family", are treated as having been published as names of families.

18.3. A name of a family or subdivision of a family based on an illegitimate generic name is illegitimate unless conserved. Contrary to Art. 32.1(b) such a name is validly published if it complies with the other requirements for valid publication.

Ex. 3. Caryophyllaceae, nom. cons. (from *Caryophyllus* Miller non L.); *Winteraceae,* nom. cons. (from *Wintera* Murray, an illegitimate synonym of *Drimys* Forster & Forster f.).

18.4. When a name of a family has been published with an improper Latin termination, the termination must be changed to conform with the rule, without change of the author's name or date of publication (see Art. 32.5).

Ex. 4. "*Coscinodisceae*" Kütz. is to be accepted as *Coscinodiscaceae* Kütz. and not attributed to De Toni, who first used the correct spelling (Notarisia 5: 915. 1890).

Ex. 5. "*Atherospermeae*" R. Br. is to be accepted as *Atherospermataceae* R. Br. and not attributed to Airy Shaw (in Willis, Dict. Fl. Pl. ed. 7. 104. 1966), who first used the correct spelling, or to Lindley, who used the spelling "*Atherospermaceae*" (Veg. Kingd. 300. 1846).

Ex. 6. However, Tricholomées Roze (Bull. Soc. Bot. France 23: 49. 1876) is not to be accepted as *Tricholomataceae* Roze, because it has a French rather than a Latin termination. The name *Tricholomataceae* was later validated by Pouzar (1983; see App. II).

18.5. The following names, sanctioned by long usage, are treated as validly published: *Palmae* (*Arecaceae*; type, *Areca* L.); *Gramineae* (*Poaceae*; type, *Poa* L.); *Cruciferae* (*Brassicaceae*; type, *Brassica* L.); *Leguminosae* (*Fabaceae*; type, *Faba* Miller (= *Vicia* L. p.p.)); *Guttiferae* (*Clusiaceae*; type, *Clusia* L.); *Umbelliferae* (*Apiaceae*; type, *Apium* L.); *Labiatae* (*Lamiaceae*; type, *Lamium* L.); *Compositae* (*Asteraceae*; type, *Aster* L.). When the *Papilionaceae* (*Fabaceae*; type, *Faba* Miller) are regarded as a family distinct from the remainder of the *Leguminosae,* the name *Papilionaceae* is conserved against *Leguminosae* (see Art. 51.2).

18.6. The use, as alternatives, of the names indicated in parentheses in Art. 18.5 is authorized.

Article 19

19.1. The name of a subfamily is a plural adjective used as a substantive; it is formed in the same manner as the name of a family (Art. 18.1) but by using the termination *-oideae* instead of *-aceae.*

19.2. A tribe is designated in a similar manner, with the termination *-eae,* and a subtribe similarly with the termination *-inae.*

19.3. The name of any subdivision of a family that includes the type of the adopted, legitimate name of the family to which it is assigned is to be based on the generic name equivalent to that type, but not followed by an author's name (see Art. 46). Such names are termed autonyms (Art. 6.8; see also Art. 7.18).

Ex. 1. The type of the family name *Rosaceae* A. L. Juss. is *Rosa* L. and hence the subfamily and tribe which include *Rosa* are to be called *Rosoideae* and *Roseae.*

Ex. 2. The type of the family name *Poaceae* Barnhart (nom. alt., *Gramineae* A. L. Juss. – see Art. 18.5) is *Poa* L. and hence the subfamily and tribe which include *Poa* are to be called *Pooideae* and *Poëae.*

Note 1. This provision applies only to the names of those subordinate taxa that include the type of the adopted name of the family (but see Rec. 19A).

Ex. 3. The subfamily including the type of the family name *Ericaceae* A. L. Juss. (*Erica* L.) is called *Ericoideae,* and the tribe including this type is called *Ericeae.* However, the correct name of the tribe including both *Rhododendron* L., the type of the subfamily name *Rhododendroideae* Endl., and *Rhodora* L. is *Rhodoreae* G. Don (the oldest legitimate name), and not *Rhododendreae.*

Ex. 4. The subfamily of the family *Asteraceae* Dumort. (nom. alt., *Compositae* Giseke) including *Aster* L., the type of the family name, is called *Asteroideae,* and the tribe and subtribe including *Aster* are called *Astereae* and *Asterinae,* respectively. However, the correct name of the tribe including both *Cichorium* L., the type of the subfamily name *Cichorioideae* Kitamura, and *Lactuca* L. is *Lactuceae* Cass., not *Cichorieae,* while that of the subtribe including both *Cichorium* and *Hyoseris* L. is *Hyoseridinae* Less., not *Cichoriinae* (unless the *Cichoriaceae* A. L. Juss. are accepted as a family distinct from *Compositae*).

23

19.4. The first valid publication of a name of a subdivision of a family that does not include the type of the adopted, legitimate name of the family automatically establishes the corresponding autonym (see also Arts. 32.6 and 57.3).

19.5. The name of a subdivision of a family may not be based on the same generic name as is the name of the family or of any subdivision of the same family unless it has the same type as that name.

19.6. When a name of a taxon assigned to one of the above categories has been published with an improper Latin termination, such as *-eae* for a subfamily or *-oideae* for a tribe, the termination must be changed to accord with the rule, without change of the author's name or date of publication (see Art. 32.5).

Ex. 5. The subfamily name "*Climacieae*" Grout (Moss Fl. N. Amer. 3: 4. 1928) is to be changed to *Climacioideae* with rank and author's name unchanged.

19.7. When the *Papilionaceae* are included in the family *Leguminosae* (nom.-alt., *Fabaceae*; see Art. 18.5) as a subfamily, the name *Papilionoideae* may be used as an alternative to *Faboideae*.

Recommendation 19A

19A.1. If a legitimate name is not available for a subdivision of a family which includes the type of the correct name of another taxon of higher or lower rank (e.g., subfamily, tribe, or subtribe), but not of the family to which it is assigned, the new name of that taxon should be based on the same generic name as the name of the higher or lower taxon.

Ex. 1. Three tribes of the family *Ericaceae*, none of which includes the type of that family name (*Erica* L.), are *Pyroleae* D. Don, *Monotropeae* D. Don, and *Vaccinieae* D. Don. The names of the later-described subfamilies *Pyroloideae* (D. Don) A. Gray, *Monotropoideae* (D. Don) A. Gray, and *Vaccinioideae* (D. Don) Endl. are based on the same generic names.

SECTION 3. NAMES OF GENERA AND SUBDIVISIONS OF GENERA

Article 20

20.1. The name of a genus is a substantive in the singular number, or a word treated as such, and is written with a capital initial letter (see Art. 73.2). It may be taken from any source whatever, and may even be composed in an absolutely arbitrary manner.

Ex. 1. Rosa, Convolvulus, Hedysarum, Bartramia, Liquidambar, Gloriosa, Impatiens, Rhododendron, Manihot, Ifloga (an anagram of *Filago*).

20.2. The name of a genus may not coincide with a technical term currently used in morphology unless it was published before 1 Jan. 1912 and accompanied by a specific name published in accordance with the binary system of Linnaeus.

Ex. 2. The generic name *Radicula* Hill (1756) coincides with the technical term "radicula" (radicle) and was not accompanied by a specific name in accordance with the binary system of Linnaeus. The name is correctly attributed to Moench (1794), who first combined it with specific epithets, but at that time he included in the genus the type of the generic name *Rorippa* Scop. (1760). *Radicula* Moench is therefore rejected in favour of *Rorippa*.

Ex. 3. *Tuber* Wigg. : Fr., when published in 1780, was accompanied by a binary specific name (*Tuber gulosorum* Wigg.) and is therefore validly published.

Ex. 4. The generic names *Lanceolatus* Plumstead (1952) and *Lobata* V. J. Chapman (1952) coincide with technical terms and are therefore not validly published.

Ex. 5. Names such as *Radix, Caulis, Folium, Spina,* etc., cannot now be validly published as generic names.

20.3. The name of a genus may not consist of two words, unless these words are joined by a hyphen.

Ex. 6. The generic name *Uva ursi* Miller (1754) as originally published consisted of two separate words unconnected by a hyphen, and is therefore rejected; the name is correctly attributed to Duhamel (1755) as *Uva-ursi* (hyphened when published).

Ex. 7. However, names such as *Quisqualis* (formed by combining two words into one when originally published), *Sebastiano-schaueria,* and *Neves-armondia* (both hyphenated when originally published) are validly published.

Note 1. The names of intergeneric hybrids are formed according to the provisions of Art. H.6.

20.4. The following are not to be regarded as generic names:

(a) Words not intended as names.

Ex. 8. *Anonymos* Walter (Fl. Carol. 2, 4, 9, etc. 1788) is rejected as being a word applied to 28 °ifferent genera by Walter to indicate that they were without names.

Ex. 9. *Schaenoides* and *Scirpoides,* as used by Rottbøll (Descr. Pl. Rar. Progr. 14, 27. 1772) to indicate unnamed genera resembling *Schoenus* and *Scirpus* which he stated (on page 7) he intended to name later, are token words and not generic names. *Kyllinga* Rottb. and *Fuirena* Rottb. (1773) are the first legitimate names of these genera.

(b) Unitary designations of species.

Ex. 10. Ehrhart (Phytophylacium 1780, and Beitr. 4: 145-150. 1789) proposed unitary names for various species known at that time under binary names, e.g. *Phaeocephalum* for *Schoenus fuscus,* and *Leptostachys* for *Carex leptostachys.* These names, which resemble generic names, should not be confused with them and are to be rejected, unless they have been published as generic names by a subsequent author; for example, the name *Baeothryon,* employed as a unitary name of a species by Ehrhart, was subsequently published as a generic name by A. Dietrich.

Ex. 11. Necker in his Elementa Botanica, 1790, proposed unitary designations for his "species naturales". These names, which resemble generic names, are not to be treated as such, unless they have been published as generic names by a subsequent author; for example *Anthopogon,* employed by Necker for one of his "species naturales", was published as a generic name by Rafinesque: *Anthopogon* Raf. non Nutt.

20A.1. Authors forming generic names should comply with the following suggestions:

(a) To use Latin terminations insofar as possible.

(b) To avoid names not readily adaptable to the Latin language.

(c) Not to make names which are very long or difficult to pronounce in Latin.

(d) Not to make names by combining words from different languages.

(e) To indicate, if possible, by the formation or ending of the name the affinities or analogies of the genus.

(f) To avoid adjectives used as nouns.

(g) Not to use a name similar to or derived from the epithet of one of the species of the genus.

(h) Not to dedicate genera to persons quite unconnected with botany or at least with natural science.

(i) To give a feminine form to all personal generic names, whether they commemorate a man or a woman (see Rec. 73B).

(j) Not to form generic names by combining parts of two existing generic names, e.g. *Hordelymus* from *Hordeum* and *Elymus*, because such names are likely to be confused with nothogeneric names (see Art. H.6).

Article 21

21.1. The name of a subdivision of a genus is a combination of a generic name and a subdivisional epithet connected by a term (subgenus, sectio, series, etc.) denoting its rank.

21.2. The epithet is either of the same form as a generic name, or a plural adjective agreeing in gender with the generic name and written with a capital initial letter (see Art. 32.5).

21.3. The epithet in the name of a subdivision of a genus is not to be formed from the name of the genus to which it belongs by adding the prefix *Eu-*.

Ex. 1. Costus subg. *Metacostus; Ricinocarpos* sect. *Anomodiscus; Sapium* subsect. *Patentinervia; Valeriana* sect. *Valerianopsis; Euphorbia* sect. *Tithymalus; Euphorbia* subsect. *Tenellae; Arenaria* ser. *Anomalae;* but not *Carex* sect. *Eucarex.*

Note 1. The use within the same genus of the same epithet in names of subdivisions of the genus, even in different ranks, based on different types is illegitimate under Art. 64.

Note 2. The names of hybrids with the rank of a subdivision of a genus are formed according to the provisions of Art. H.7.

21A.1. When it is desired to indicate the name of a subdivision of the genus to which a particular species belongs in connection with the generic name and specific epithet, its epithet should be placed in parentheses between the two; when desirable, its rank may also be indicated.

Ex. 1. Astragalus (Cycloglottis) contortuplicatus; Astragalus (Phaca) umbellatus; Loranthus (sect. Ischnanthus) gabonensis.

Recommendation 21B

21B.1. The epithet of a subgenus or section is preferably a substantive, that of a subsection or lower subdivision of a genus preferably a plural adjective.

21B.2. Authors, when proposing new epithets for subdivisions of genera, should avoid those in the form of a substantive when other co-ordinate subdivisions of the same genus have them in the form of a plural adjective, and vice-versa. They should also avoid, when proposing an epithet for a subdivision of a genus, one already used for a subdivision of a closely related genus, or one which is identical with the name of such a genus.

Article 22

22.1. The name of any subdivision of a genus that includes the type of the adopted, legitimate name of the genus to which it is assigned is to repeat that generic name unaltered as its epithet, but not followed by an author's name (see Art. 46). Such names are termed autonyms (Art. 6.8; see also Art. 7.21).

Note 1. This provision applies only to the names of those subordinate taxa that include the type of the adopted name of the genus (but see Rec. 22A).

22.2. The first valid publication of a name of a subdivision of a genus that does not include the type of the adopted, legitimate name of the genus automatically establishes the corresponding autonym (see also Arts. 32.6 and 57.3).

Ex. 1. The subgenus of *Malpighia* L. which includes the lectotype of the generic name (*M. glabra* L.) is called *Malpighia* subg. *Malpighia,* and not *Malpighia* subg. *Homoiostylis* Niedenzu.

Ex. 2. The section of *Malpighia* L. including the lectotype of the generic name is called *Malpighia* sect. *Malpighia,* and not *Malpighia* sect. *Apyrae* DC.

Ex. 3. However, the correct name of the section of the genus *Rhododendron* L. which includes *Rhododendron luteum* Sweet, the type of *Rhododendron* subg. *Anthodendron* (Reichenb.) Rehder, is *Rhododendron* sect. *Pentanthera* G. Don, the oldest legitimate name for that section, and not *Rhododendron* sect. *Anthodendron.*

22.3. The epithet in the name of a subdivision of a genus may not repeat unchanged the correct name of the genus, except when the two names have the same type.

22.4. When the epithet of a subdivision of a genus is identical with or derived from the epithet of one of its constituent species, the type of the name of the subdivision of the genus is the same as that of the species name, unless the original author of the subdivisional name designated another type.

Ex. 4. The type of *Euphorbia* subg. *Esula* Pers. is *E. esula* L.; the designation of *E. peplus* L. as lectotype by Croizat (Revista Sudamer. Bot. 6: 13. 1939) is rejected.

Ex. 5. The type of *Lobelia* sect. *Eutupa* Wimmer is *L. tupa* L.

22.5. When the epithet of a subdivision of a genus is identical with or derived from the epithet of a specific name that is a later homonym, it is the type of that later homonym, whose correct name necessarily has a different epithet, that is the nomenclatural type.

Recommendation 22A

22A.1. A section including the type of the correct name of a subgenus, but not including the type of the correct name of the genus, should, where there is no obstacle under the rules, be given a name with the same epithet and type as the subgeneric name.

22A.2. A subgenus not including the type of the correct name of the genus should, where there is no obstacle under the rules, be given a name with the same epithet and type as a name of one of its subordinate sections.

Ex. 1. Instead of using a new name at the subgeneric level, Brizicky raised *Rhamnus* sect. *Pseudofrangula* Grubov to the rank of subgenus as *Rhamnus* subg. *Pseudofrangula* (Grubov) Briz. The type of both names is the same, *R. alnifolia* L'Hér.

SECTION 4. NAMES OF SPECIES

Article 23

23.1. The name of a species is a binary combination consisting of the name of the genus followed by a single specific epithet in the form of an adjective, a noun in the genitive, or a word in apposition, but not a phrase in the ablative (see Art. 23.6(c)). If an epithet consists of two or more words, these are to be united or hyphenated. An epithet not so joined when originally published is not to be rejected but, when used, is to be united or hyphenated (see Art. 73.9).

23.2. The epithet in the name of a species may be taken from any source whatever, and may even be composed arbitrarily (but see Art. 73.1).

Ex. 1. *Cornus sanguinea, Dianthus monspessulanus, Papaver rhoeas, Uromyces fabae, Fumaria gussonei, Geranium robertianum, Embelia sarasiniorum, Atropa bella-donna, Impatiens noli-tangere, Adiantum capillus-veneris, Spondias mombin* (an indeclinable epithet).

23.3. Symbols forming part of specific epithets proposed by Linnaeus do not invalidate the relevant names but must be transcribed.

Ex. 2. *Scandix pecten* ♀ L. is to be transcribed as *Scandix pecten-veneris; Veronica anagallis* ▽ L. is to be transcribed as *Veronica anagallis-aquatica.*

23.4. The specific epithet may not exactly repeat the generic name with or without the addition of a transcribed symbol (tautonym).

Ex. 3. Linaria linaria, Nasturtium nasturtium-aquaticum.

23.5. The specific epithet, when adjectival in form and not used as a substantive, agrees grammatically with the generic name (see Art. 32.5).

Ex. 4. Helleborus niger, Brassica nigra, Verbascum nigrum; Vinca major, Tropaeolum majus; Rubus amnicola ("amnicolus"), the specific epithet being a Latin substantive; *Peridermium balsameum* Peck, but also *Gloeosporium balsameae* J. J. Davis, both derived from the epithet of *Abies balsamea,* the specific epithet of which is treated as a substantive in the second example.

23.6. The following are not to be regarded as specific epithets:

(a) Words not intended as epithets.

Ex. 5. Viola "qualis" Krocker (Fl. Siles. 2: 512, 517. 1790); *Urtica "dubia?"* Forsskål (Fl. Aegypt.-Arab. cxxi. 1775), the word "dubia?" being repeatedly used in that work for species which could not be reliably identified.

Ex. 6. Atriplex "nova" Winterl (Index Horti Bot. Univ. Pest. fol. A. 8, recto et verso. 1788), the word "nova" being here used in connection with four different species of *Atriplex.*

Ex. 7. However, in *Artemisia nova* A. Nelson (Bull. Torrey Bot. Club 27: 274. 1900), *nova* was intended as a specific epithet, the species having been newly distinguished from others.

(b) Ordinal adjectives used for enumeration.

Ex. 8. Boletus vicesimus sextus, Agaricus octogesimus nonus.

(c) Epithets published in works in which the Linnaean system of binary nomenclature for species is not consistently employed. Linnaeus is regarded as having used binary nomenclature for species consistently from 1753 onwards, although there are exceptions, e.g. *Apocynum fol. androsaemi* L. (Sp. Pl. 213. 1753 = *Apocynum androsaemifolium* L. Syst. Nat. ed. 10: 946. 1759).

Ex. 9. Abutilon album Hill (Brit. Herb. 49. 1756) is a descriptive phrase reduced to two words, not a binary name in accordance with the Linnaean system, and is to be rejected: Hill's other species was *Abutilon flore flavo.*

Ex. 10. Secretan (Mycographie Suisse. 1833) introduced a large number of new specific names, more than half of them not binomials, e.g. *Agaricus albus corticis, Boletus testaceus scaber, Boletus aereus carne lutea.* He is therefore considered not to have consistently used the Linnaean system of binary nomenclature and none of the specific names, even those with a single epithet, in this work are validly published.

Ex. 11. Other works in which the Linnaean system of binary nomenclature is not consistently employed: Gilibert, Fl. Lit. Inch. 1781; Gilibert, Exerc. Phyt. 1792; Miller, Gard. Dict. Abr. ed. 4. 1754; W. Kramer, Elench. Veg. 1756.

(d) Formulae designating hybrids (see Art. H.10.3).

Recommendation 23A

23A.1. Names of men and women and also of countries and localities used as specific epithets should be in the form of substantives in the genitive (*clusii, porsildiorum, saharae*) or of adjectives (*clusianus, dahuricus*) (see also Art. 73, Recs. 73C and D).

23A.2. The use of the genitive and the adjectival form of the same word to designate two different species of the same genus should be avoided (e.g. *Lysimachia hemsleyana* Oliver and *L. hemsleyi* Franchet).

Recommendation 23B

23B.1. In forming specific epithets, authors should comply also with the following suggestions:

(a) To use Latin terminations insofar as possible.

(b) To avoid epithets which are very long and difficult to pronounce in Latin.

(c) Not to make epithets by combining words from different languages.

(d) To avoid those formed of two or more hyphenated words.

(e) To avoid those which have the same meaning as the generic name (pleonasm).

(f) To avoid those which express a character common to all or nearly all the species of a genus.

(g) To avoid in the same genus those which are very much alike, especially those which differ only in their last letters or in the arrangement of two letters.

(h) To avoid those which have been used before in any closely allied genus.

(i) Not to adopt epithets from unpublished names found in correspondence, travellers' notes, herbarium labels, or similar sources, attributing them to their authors, unless these authors have approved publication.

(j) To avoid using the names of little-known or very restricted localities, unless the species is quite local.

SECTION 5. NAMES OF TAXA BELOW THE RANK OF SPECIES
(INFRASPECIFIC TAXA)

Article 24

24.1. The name of an infraspecific taxon is a combination of the name of a species and an infraspecific epithet connected by a term denoting its rank.

Ex. 1. Saxifraga aizoon subf. *surculosa* Engler & Irmscher. This can also be cited as *Saxifraga aizoon* var. *aizoon* subvar. *brevifolia* f. *multicaulis* subf. *surculosa* Engler & Irmscher; in this way a full classification of the subforma within the species is given.

24.2. Infraspecific epithets are formed as those of species and, when adjectival in form and not used as substantives, they agree grammatically with the generic name (see Art. 32.5).

Ex. 2. Trifolium stellatum forma *nanum* (not *nana*).

24.3. Infraspecific epithets such as *typicus, originalis, originarius, genuinus, verus,* and *veridicus,* purporting to indicate the taxon containing the type of the name of the next higher taxon, are not validly published unless they repeat the specific epithet because Art. 26 requires their use.

24.4. The use of a binary combination instead of an infraspecific epithet is not admissible. Contrary to Art. 32.1(b), names so constructed are validly published but are to be altered to the proper form without change of the author's name or date of publication.

Ex. 3. "*Salvia grandiflora* subsp. *S. willeana*" Holmboe is to be cited as *Salvia grandiflora* subsp. *willeana* Holmboe.

Ex. 4. "*Phyllerpa prolifera* var. *Ph. firma*" Kütz. is to be altered to *Phyllerpa prolifera* var. *firma* Kütz.

24.5. Infraspecific taxa within different species may bear the same epithets; those within one species may bear the same epithets as other species (but see Rec. 24B).

Ex. 5. Rosa jundzillii var. *leioclada* and *Rosa glutinosa* var. *leioclada; Viola tricolor* var. *hirta* in spite of the previous existence of a different species named *Viola hirta.*

Note 1. The use within the same species of the same epithet for infraspecific taxa, even if they are of different rank, based on different types is illegitimate under Art. 64.3.

Recommendation 24A

24A.1. Recommendations made for specific epithets (Recs. 23A, B) apply equally to infraspecific epithets.

Recommendation 24B

24B.1. Authors proposing new infraspecific epithets should avoid those previously used for species in the same genus.

Article 25

25.1. For nomenclatural purposes, a species or any taxon below the rank of species is regarded as the sum of its subordinate taxa, if any. In fungi, a holomorph (see Art. 59.4) also includes its correlated form-taxa.

Ex. 1. When *Montia parvifolia* (DC.) Greene is treated as containing two subspecies, the name *M. parvifolia* applies to the sum of these subordinate taxa. Under this taxonomic treatment, one must write *M. parvifolia* (DC.) Greene subsp. *parvifolia* if only that part of *M. parvifolia* which includes its nomenclatural type and excludes the type of the name of the other subspecies (*M. parvifolia* subsp. *flagellaris* (Bong.) Ferris) is meant.

Article 26

26.1. The name of any infraspecific taxon that includes the type of the adopted, legitimate name of the species to which it is assigned is to repeat the specific epithet unaltered as its final epithet, but not followed by an author's name (see Art. 46). Such names are termed autonyms (Art. 6.8; see also Art. 7.21).

Ex. 1. The combination *Lobelia spicata* var. *originalis* McVaugh, applying to a taxon which includes the type of the name *Lobelia spicata* Lam., is to be replaced by *Lobelia spicata* Lam. var. *spicata.*

Note 1. This provision applies only to the names of those subordinate taxa that include the type of the adopted name of the species (but see Rec. 26A).

26.2. The first valid publication of a name of an infraspecific taxon that does not include the type of the adopted, legitimate name of the species automatically establishes the corresponding autonym (see also Arts. 32.6 and 57.3).

Ex. 2. The publication of the name *Lycopodium inundatum* var. *bigelovii* Tuckerman (1843) automatically established the name of another variety, *Lycopodium inundatum* L. var. *inundatum,* the type of which is that of the name *Lycopodium inundatum* L.

Ex. 3. Utricularia stellaris L. f. (1781) includes *U. stellaris* var. *coromandeliana* A. DC. (1844) and *U. stellaris* L. f. var. *stellaris* (1844) automatically established at the same time. When *U. stellaris* is included in *U. inflexa* Forsskål (1775) as a variety the correct name of the variety, under Art. 57.3, is *U. inflexa* var. *stellaris* (L. f.) P. Taylor (1961).

Recommendation 26A

26A.1. A variety including the type of the correct name of a subspecies, but not including the type of the correct name of the species, should, where there is no obstacle under the rules, be given a name with the same epithet and type as the subspecies name.

26A.2. A subspecies not including the type of the correct name of the species should, where there is no obstacle under the rules, be given a name with the same epithet and type as a name of one of its subordinate varieties.

26A.3. A taxon of lower rank than variety which includes the type of the correct name of a subspecies or variety, but not the type of the correct name of the species, should, where there is no obstacle under the rules, be given a name with the same epithet and type as the name of the subspecies or variety. On the other hand, a subspecies or variety which does not include the type of the correct name of the species should not be given a name with the same epithet as the name of one of its subordinate taxa below the rank of variety.

Ex. 1. Fernald treated *Stachys palustris* subsp. *pilosa* (Nutt.) Epling as composed of five varieties, for one of which (that including the type of *S. palustris* subsp. *pilosa*) he made the combination *S. palustris* var. *pilosa* (Nutt.) Fern., there being no legitimate varietal name available.

Ex. 2. There being no legitimate name available at the rank of subspecies, Bonaparte made the combination *Pteridium aquilinum* subsp. *caudatum* (L.) Bonap., using the same epithet that Sadebeck had used earlier in the combination *P. aquilinum* var. *caudatum* (L.) Sadeb. (both names based on *Pteris caudata* L.). Each name is legitimate, and both can be used, as by Tryon, who treated *P. aquilinum* var. *caudatum* as one of four varieties under subsp. *caudatum.*

Article 27

27.1. The final epithet in the name of an infraspecific taxon may not repeat unchanged the epithet of the correct name of the species to which the taxon is assigned unless the two names have the same type.

SECTION 6. NAMES OF PLANTS IN CULTIVATION

Article 28

28.1. Plants brought from the wild into cultivation retain the names that are applied to the same taxa growing in nature.

28.2. Hybrids, including those arising in cultivation, may receive names as provided in Appendix I (see also Arts. 40 and 50).

Note 1. Additional, independent designations for plants used in agriculture, forestry, and horticulture (and arising either in nature or cultivation) are dealt with in the International Code of Nomenclature for Cultivated Plants, where regulations are provided for their formation and use. However, nothing precludes the use for cultivated plants of names published in accordance with the requirements of the International Code of Botanical Nomenclature.

Note 2. Epithets published in conformity with the International Code of Botanical Nomenclature may be used as cultivar epithets under the rules of the International Code of Nomenclature for Cultivated Plants, when this is considered to be the appropriate status for the groups concerned. Otherwise, cultivar epithets published on or after 1 January 1959 in conformity with the International Code of Nomenclature for Cultivated Plants are required to be fancy names markedly different from epithets of names in Latin form governed by the International Code of Botanical Nomenclature (see that Code, Art. 27).

Ex. 1. Cultivar names: *Taxus baccata* 'Variegata' or *Taxus baccata* cv. Variegata (based on *T. baccata* var. *variegata* Weston), *Phlox drummondii* 'Sternenzauber', *Viburnum* x*bodnantense* "Dawn".

CHAPTER IV. EFFECTIVE AND VALID PUBLICATION

SECTION 1. CONDITIONS AND DATES OF EFFECTIVE PUBLICATION

Article 29

29.1. Publication is effected, under this Code, only by distribution of printed matter (through sale, exchange, or gift) to the general public or at least to botanical institutions with libraries accessible to botanists generally. It is not effected by communication of new names at a public meeting, by the placing of names in collections or gardens open to the public, or by the issue of microfilm made from manuscripts, type-scripts or other unpublished material.

Ex. 1. Cusson announced his establishment of the genus *Physospermum* in a memoir read at the Société des Sciences de Montpellier in 1770, and later in 1782 or 1783 at the Société de Médecine de Paris, but its effective publication dates from 1787 in the Mémoires de la Société Royale de Médecine de Paris 5(1): 279.

29.2. Publication by indelible autograph before 1 Jan. 1953 is effective.

Ex. 2. Salvia oxyodon Webb & Heldr. was effectively published in July 1850 in an autograph catalogue placed on sale (Webb & Heldreich, Catalogus Plantarum Hispanicarum ... ab A. Blanco lectarum. Paris, July 1850, folio).

Ex. 3. H. Léveillé, Flore du Kouy Tchéou (1914-1915), a work lithographed from the handwritten manuscript, is effectively published.

29.3. For the purpose of this Article, handwritten material, even though reproduced by some mechanical or graphic process (such as lithography, offset, or metallic etching), is still considered as autographic.

29.4. Publication on or after 1 Jan. 1953 in tradesmen's catalogues or non-scientific newspapers, and on or after 1 Jan. 1973 in seed-exchange lists, does not constitute effective publication.

29A.1. It is strongly recommended that authors avoid publishing new names and descriptions of new taxa in ephemeral printed matter of any kind, in particular that which is multiplied in restricted and uncertain numbers, where the permanence of the text may be limited, where the effective publication in terms of number of copies is not obvious, or where the printed matter is unlikely to reach the general public. Authors should also avoid publishing new names and descriptions in popular periodicals, in abstracting journals, or on correction slips.

Article 30

30.1. The date of effective publication is the date on which the printed matter became available as defined in Art. 29. In the absence of proof establishing some other date, the one appearing in the printed matter must be accepted as correct.

Ex. 1. Individual parts of Willdenow's Species Plantarum were published as follows: 1(1), 1797; 1(2), 1798; 2(1), 1799; 2(2), 1799 or January 1800; 3(1) (to page 850), 1800; 3(2) (to page 1470), 1802; 3(3) (to page 2409), 1803 (and later than Michaux's Flora Boreali-Americana); 4(1) (to page 630), 1805; 4(2), 1806; these dates, which are partly in disagreement with those on the title-pages of the volumes, are accepted as the correct dates of effective publication.

30.2. When separates from periodicals or other works placed on sale are issued in advance, the date on the separate is accepted as the date of effective publication unless there is evidence that it is erroneous.

Ex. 2. Publication in separates issued in advance: the names of the *Selaginella* species published by Hieronymus in Hedwigia 51: 241-272 (1912) were effectively published on 15 Oct. 1911, since the volume in which the paper appeared states (p. ii) that the separate appeared on that date.

30A.1. The date on which the publisher or his agent delivers printed matter to one of the usual carriers for distribution to the public should be accepted as its date of effective publication.

Article 31

31.1. The distribution on or after 1 Jan. 1953 of printed matter accompanying exsiccata does not constitute effective publication.

Note 1. If the printed matter is also distributed independently of the exsiccata, this constitutes effective publication.

Ex. 1. Works such as Schedae operis . . . plantae finlandiae exsiccatae, Helsingfors 1. 1906, 2. 1916, 3. 1933, 1944, or Lundell & Nannfeldt, Fungi exsiccati suecici etc., Uppsala 1-. . ., 1934-. . ., distributed independently of the exsiccata, whether published before or after 1 Jan. 1953, are effectively published.

SECTION 2. CONDITIONS AND DATES OF VALID PUBLICATION OF NAMES

Article 32

32.1. In order to be validly published, a name of a taxon (autonyms excepted) must *(a)* be effectively published (see Art. 29) on or after the starting-point date of the respective group (Art. 13.1); *(b)* have a form which complies with the provisions of Arts. 16-27 and Arts. H.6-7; *(c)* be accompanied by a description or diagnosis or by a reference to a previously and effectively published description or diagnosis (except as provided in Art. H.9); and *(d)* comply with the special provisions of Arts. 33-45.

Ex. 1. Egeria Néraud (in Gaudichaud, Voy. Uranie, Bot. 25, 28. 1826), published without a description or a diagnosis or a reference to a former one, was not validly published.

Ex. 2. The name *Loranthus macrosolen* Steudel originally appeared without a description or diagnosis on the printed labels issued about the year 1843 with Sect. II. no. 529, 1288, of Schimper's herbarium specimens of Abyssinian plants; it was not validly published, however, until A. Richard (Tent. Fl. Abyss. 1: 340. 1847) supplied a description.

Ex. 3. In Sweet's Hortus Britannicus, ed. 3 (1839), for each listed species the flower colour, the duration of the plant, and a translation into English of the specific epithet are given in tabular form. In many genera the flower colour and duration may be identical for all species and clearly their mention is not intended as a validating description. New names appearing in that work are therefore not validly published, except in some cases where reference is made to earlier descriptions or to validly published basionyms.

32.2. A diagnosis of a taxon is a statement of that which in the opinion of its author distinguishes the taxon from others.

32.3. For the purpose of valid publication of a name, reference to a previously and effectively published description or diagnosis may be direct or indirect (Art. 32.4). For names published on or after 1 Jan. 1953 it must, however, be full and direct as specified in Art. 33.2.

32.4. An indirect reference is a clear indication, by the citation of the author's name or in some other way, that a previously and effectively published description or diagnosis applies.

Ex. 4. Kratzmannia Opiz (in Berchtold & Opiz, Oekon.-Techn. Fl. Böhm. 1: 398. 1836) is published with a diagnosis, but it was not definitely accepted by the author and is therefore not validly published. It is accepted definitely in Opiz (Seznam 56. 1852), but without any description. or diagnosis. The citation of "*Kratzmannia* O." includes an indirect reference to the previously published diagnosis in 1836.

Ex. 5. Opiz published the name of the genus *Hemisphace* (Bentham) Opiz (1852) without a description or diagnosis, but as he wrote "*Hemisphace* Benth." he indirectly referred to the previously effectively published description by Bentham (Labiat. Gen. Spec. 193. 1833) of *Salvia* sect. *Hemisphace.*

Ex. 6. The new combination *Cymbopogon martini* (Roxb.) W. Watson (1882) is validated by the addition of the number "309", which, as explained at the top of the same page, is the running-number of the species (*Andropogon martini* Roxb.) in Steudel (Syn. Pl. Glum. 1: 388. 1854). Although the reference to the basionym *Andropogon martini* is indirect, it is perfectly unambiguous.

32.5. Names published with an incorrect Latin termination but otherwise in accordance with this Code are regarded as validly published; they are to be changed to accord with Arts. 17-19, 21, 23, and 24, without change of the author's name or date of publication (see also Art. 73.10).

32.6. Autonyms (Art. 6.8) are accepted as validly published names, dating from the publication in which they were established (see Arts. 19.4, 22.2, 26.2), whether or not they appear in print in that publication.

Note 1. In certain circumstances an illustration with analysis is accepted as equivalent to a description (see Arts. 42 and 44).

Note 2. For names of plant taxa that were originally not treated as plants, see Art. 45.4.

Recommendation 32A

32A.1. A name should not be validated solely by a reference to a description or diagnosis published before 1753.

Recommendation 32B

32B.1. The description or diagnosis of any new taxon should mention the points in which the taxon differs from its allies.

Recommendation 32C

32C.1. Authors should avoid adoption of a name which has been previously but not validly published for a different taxon.

Recommendation 32D

32D.1. In describing new taxa, authors should, when possible, supply figures with details of structure as an aid to identification.

32D.2. In the explanation of the figures, it is valuable to indicate the specimen(s) on which they are based.

32D.3. Authors should indicate clearly and precisely the scale of the figures which they publish.

Recommendation 32E

32E.1. The description or diagnosis of parasitic plants should always be followed by an indication of the hosts, especially those of parasitic fungi. The hosts should be designated by their scientific names and not solely by names in modern languages, the applications of which are often doubtful.

Article 33

33.1. A combination (autonyms excepted) is not validly published unless the author definitely associates the final epithet with the name of the genus or species, or with its abbreviation.

Ex. 1. Combinations validly published: In Linnaeus's Species Plantarum the placing of the epithet in the margin opposite the name of the genus clearly associates the epithet with the name of the genus. The same result is attained in Miller's Gardeners Dictionary, ed. 8, by the inclusion of the epithet in parentheses immediately after the name of the genus, in Steudel's Nomenclator Botanicus by the arrangement of the epithets in a list headed by the name of the genus, and in general by any typographical device which associates an epithet with a particular generic or specific name.

Ex. 2. Combinations not validly published: Rafinesque's statement under *Blephilia* that "Le type de ce genre est la *Monarda ciliata* Linn." (J. Phys. Chim. Hist. Nat. Arts 89: 98. 1819) does not constitute valid publication of the combination *Blephilia ciliata*, since he did not definitely associate the epithet *ciliata* with the generic name *Blephilia*. Similarly, the combination *Eulophus peucedanoides* is not to be ascribed to Bentham on the basis of the listing of "*Cnidium peucedanoides*, H. B. et K." under *Eulophus* (in Bentham & Hooker, Gen. Pl. 1: 885. 1867).

33.2. A new combination, or an avowed substitute (*nomen novum*), published on or after 1 Jan. 1953, for a previously and validly published name is not validly published unless its basionym (name-bringing or epithet-bringing synonym) or the replaced synonym (when a new name is proposed) is clearly indicated and a full and direct reference given to its author and place of valid publication with page or plate reference[1] and date. Errors of bibliographic citation and incorrect forms of author citation (see Art. 46) do not invalidate publication of a new combination or nomen novum.

Ex. 3. In transferring *Ectocarpus mucronatus* Saund. to *Giffordia*, Kjeldsen & Phinney (Madroño 22: 90. 27 Apr. 1973) cited the basionym and its author but without reference to its place of valid publication. They later (Madroño 22: 154. 2 Jul. 1973) validated the binomial *Giffordia mucronata* (Saund.) Kjeldsen & Phinney by giving a full and direct reference to the place of valid publication of the basionym.

Ex. 4. *Aronia arbutifolia* var. *nigra* (Willd.) Seymour (1969) was published as a new combination "Based on *Mespilus arbutifolia* L. var. *nigra* Willd., in Sp. Pl. 2: 1013. 1800." Willdenow treated these plants in the genus *Pyrus*, not *Mespilus*, and publication was in 1799, not 1800; these errors are treated as bibliographic errors of citation and do not invalidate the new combination.

Ex. 5. The combination *Trichipteris kalbreyeri* was proposed by Tryon (Contr. Gray Herb. 200: 45. 1970) with a full and direct reference to *Alsophila kalbreyeri* C. Chr. (Index Filic. 44. 1905). This, however, was not the place of valid publication of the basionym, which had previously been published, with the same type, by Baker (Summ. New Ferns 9. 1892). Tryon's bibliographic error of citation does not invalidate this new combination, which is to be cited as *Trichipteris kalbreyeri* (Baker) Tryon.

---

1) A page reference (for publications with a consecutive pagination) is here understood to mean a reference to the page or pages on which the basionym was validly published or on which the protologue is printed, but not to the pagination of the whole publication unless it is coextensive with that of the protologue.

Ex. 6. The combination *Lasiobelonium corticale* was proposed by Raitviir (1980) with a full and direct reference to *Peziza corticalis* Fr. (Syst. Mycol. 2: 96. 1822). This, however, was not the place of valid publication of the basionym, which, under the Code operating in 1980, was in Mérat (Nouv. Fl. Env. Paris ed. 2, 1: 22. 1821), and under the present Code is in Persoon (Obs. Mycol. 1: 28. 1796). Raitviir's bibliographic error of citation does not invalidate the new combination, which is to be cited as *Lasiobelonium corticale* (Pers.) Raitviir.

33.3. Mere reference to the Index Kewensis, the Index of Fungi, or any work other than that in which the name was validly published does not constitute a full and direct reference to the original publication of a name.

Ex. 7. Ciferri (Mycopath. Mycol. Appl. 7: 86-89. 1954), in proposing 142 new combinations in *Meliola*, omitted references to places of publication of basionyms, stating that they could be found in Petrak's lists or in the Index of Fungi; none of these combinations was validly published. Similarly, Grummann (Cat. Lich. Germ. 18. 1963) introduced a new combination in the form *Lecanora campestris* f. "*pseudistera* (Nyl.) Grumm. c.n. – *L. p.* Nyl., Z 5: 521", in which "Z 5" referred to Zahlbruckner (Cat. Lich. Univ., vol. 5: 521. 1928), who gave the full citation of the basionym, *Lecanora pseudistera* Nyl.; Grummann's combination is not validly published.

Note 1. The publication of a name for a taxon previously known under a misapplied name must be valid under Arts. 32-45. This procedure is not the same as publishing an avowed substitute (nomen novum) for a validly published but illegitimate name (Art. 72.1(b)), the type of which is necessarily the same as that of the name which it replaced (Art. 7.11).

Ex. 8. *Sadleria hillebrandii* Robinson (1913) was introduced as a "nom. nov." for "*Sadleria pallida* Hilleb. Fl. Haw. Is. 582. 1888. Not Hook. & Arn. Bot. Beech. 75. 1832." Since the requirements of Arts. 32-45 were satisfied (for valid publication prior to 1935, simple reference to a previous description in any language is sufficient), the name is validly published. It is, however, to be considered the name of a new species, validated by the citation of the misapplication of *S. pallida* Hooker & Arn. by Hillebrand, and not a nomen novum as stated; hence, Art. 7.11 does not apply.

Ex. 9. *Juncus bufonius* var. *occidentalis* F. J. Herm. (U.S. Forest Serv. Techn. Rep. RM-18: 14. 1975) was published as a "nom. et stat. nov." for *J. sphaerocarpus* "auct. Am., non Nees". Since there is no Latin diagnosis, designation of type, or reference to any previous publication providing these requirements, the name is not validly published.

33.4. A name given to a taxon whose rank is at the same time denoted by a misplaced term (one contrary to Art. 5) is treated as not validly published, examples of such misplacement being a form divided into varieties, a species containing genera, or a genus containing families or tribes.

Ex. 10. The name sectio *Orontiaceae* was not validly published by R. Brown (Prodr. 337. 1810) since he misapplied the term "sectio" to taxa of a rank higher than genus.

Ex. 11. The names tribus *Involuta* Huth and tribus *Brevipedunculata* Huth (Bot. Jahrb. Syst. 20: 365, 368. 1895) are not validly published, since Huth misapplied the term "tribus" to a taxon of a rank lower than section, within the genus *Delphinium*.

Ex. 12. Gandoger, in his Flora Europae (1883-1891), applied the term species ("espèce") and used binary nomenclature for two categories of taxa of consecutive rank, the higher rank being equivalent to that of species in contemporary literature. He misapplied the term species to the lower rank and the names of these taxa ("Gandoger's microspecies") are not validly published.

33.5. An exception to Art. 33.4 is made for names of the subdivisions of genera termed tribes (*tribus*) in Fries's Systema Mycologicum, which are treated as validly published names of subdivisions of genera.

Ex. 13. Agaricus tribus *Pholiota* Fr. (1821) is a validly published basionym for the generic name *Pholiota* (Fr.) P. Kummer (1871).

Recommendation 33A

33A.1. The full and direct reference to the place of publication of the basionym or replaced synonym should immediately follow a proposed new combination or nomen novum. It should not be provided by mere cross-reference to a bibliography at the end of the publication or to other parts of the same publication, e.g. by use of the abbreviations "*loc. cit.*" or "*op. cit.*"

Article 34

34.1. A name is not validly published *(a)* when it is not accepted by the author in the original publication; *(b)* when it is merely proposed in anticipation of the future acceptance of the group concerned, or of a particular circumscription, position, or rank of the group (so-called provisional name); *(c)* when it is merely cited as a synonym; *(d)* by the mere mention of the subordinate taxa included in the taxon concerned.

34.2. Art. 34.1(a) does not apply to names published with a question mark or other indication of taxonomic doubt, yet published and accepted by the author. Art. 34.1(b) does not apply to names for anamorphs of fungi published in holomorphic genera in anticipation of the discovery of a particular kind of teleomorph (see Art. 59, Ex. 2).

Ex. 1. (a) The name of the monotypic genus *Sebertia* Pierre (ms.) was not validly published by Baillon (Bull. Mens. Soc. Linn. Paris 2: 945. 1891) because he did not accept it. Although he gave a description of the taxon, he referred its only species *Sebertia acuminata* Pierre (ms.) to the genus *Sersalisia* R. Br. as *Sersalisia ? acuminata*; under the provision of Art. 34.2 this combination is validly published. The name *Sebertia* Pierre (ms.) was later validly published by Engler (1897).

Ex. 2. (a) The names listed in the left-hand column of the Linnaean thesis Herbarium Amboinense defended by Stickman (1754) were not accepted by Linnaeus upon publication and are not validly published.

Ex. 3. (a) (b) The generic name *Conophyton* Haw., suggested by Haworth (Rev. Pl. Succ. 82. 1821) for *Mesembryanthemum* sect. *Minima* Haw. (Rev. Pl. Succ. 81. 1821) in the words "If this section proves to be a genus, the name of *Conophyton* would be apt", was not validly published, since Haworth did not adopt that generic name or accept that genus. The correct name for the genus is *Conophytum* N. E. Br. (1922).

Ex. 4. (c) *Acosmus* Desv. (in Desf., Cat. Pl. Horti Paris. 233. 1829), cited as a synonym of the generic name *Aspicarpa* Rich., was not validly published thereby.

Ex. 5. (c) *Ornithogalum undulatum* hort. Bouch. (in Kunth, Enum. Pl. 4: 348. 1843), cited as a synonym under *Myogalum boucheanum* Kunth, was not validly published thereby; when transferred to *Ornithogalum*, this species is to be called *O. boucheanum* (Kunth) Ascherson (1866).

Ex. 6. (c) *Erythrina micropteryx* Poeppig was not validly published by being cited as a synonym of *Micropteryx poeppigiana* Walp. (1850); the species concerned, when placed under *Erythrina*, is to be called *E. poeppigiana* (Walp.) Cook (1901).

Ex. 7. (d) The family name *Rhaptopetalaceae* Pierre (Bull. Mens. Soc. Linn. Paris 2: 1296. May 1897), which was accompanied merely by mention of constituent genera, *Brazzeia*, *Scytopetalum*, and *Rhaptopetalum*, was not validly published, as Pierre gave no description or diagnosis; the family bears the later name *Scytopetalaceae* Engler (Oct. 1897), which was accompanied by a description.

Ex. 8. (d) The generic name *Ibidium* Salisb. (Trans. Hort. Soc. London 1: 291. 1812) was published merely with the mention of four included species. As Salisbury supplied no generic description or diagnosis, his *Ibidium* is not validly published.

34.3. When, on or after 1 Jan. 1953, two or more different names (so-called alternative names) are proposed simultaneously for the same taxon by the same author, none of them is validly published. This rule does not apply in those cases where the same combination is simultaneously used at different ranks, either for infraspecific taxa within a species or for subdivisions of a genus within a genus (see Recs. 22A.1-2, 26A.1-3).

Ex. 9. The species of *Brosimum* described by Ducke (Arch. Jard. Bot. Rio de Janeiro 3: 23-29. 1922) were published with alternative names under *Piratinera* added in a footnote (pp. 23-24). The publication of these names, being effected before 1 Jan. 1953, is valid.

Ex. 10. *Euphorbia jaroslavii* Polj. (Bot. Mater. Gerb. Bot. Inst. Komarova Akad. Nauk SSSR 15: 155. tab. 1953) was published with an alternative name, *Tithymalus jaroslavii*. Neither name was validly published. However, one of the names, *Euphorbia yaroslavii* (with a different transliteration of the initial letter), was validly published by Poljakov (1961), who effectively published it with a new reference to the earlier publication and simultaneously rejected the other name.

Ex. 11. Description of "*Malvastrum bicuspidatum* subsp. *tumidum* S. R. Hill var. *tumidum*, subsp. et var. nov." (Brittonia 32: 474. 1980) simultaneously validated both *M. bicuspidatum* subsp. *tumidum* S. R. Hill and *M. bicuspidatum* var. *tumidum* S. R. Hill.

Note 1. The name of a fungal holomorph and that of a correlated anamorph (see Art. 59), even if validated simultaneously, are not alternative names in the sense of Art. 34.3. They have different types and do not pertain to the same taxon: the circumscription of the holomorph is considered to include the anamorph, but not vice versa.

Ex. 12. *Lasiosphaeria elinorae* Linder (1929), the name of a fungal holomorph, and the simultaneously published name of a correlated anamorph, *Helicosporium elinorae* Linder, are both valid, and both can be used under Art. 59.5.

Recommendation 34A

34A.1. Authors should avoid publishing or mentioning in their publications unpublished names which they do not accept, especially if the persons responsible for these unpublished names have not formally authorized their publication (see Rec. 23B.1(i)).

Article 35

35.1. A new name or combination published on or after 1 Jan. 1953 without a clear indication of the rank of the taxon concerned is not validly published.

35.2. A new name or combination published before 1 Jan. 1953 without a clear indication of rank is validly published provided that all other requirements for valid publication are fulfilled; it is, however, inoperative in questions of priority except for homonymy (see Art. 64.4). If it is a new name, it may serve as a basionym or replaced synonym for subsequent combinations or avowed substitutes in definite ranks.

Ex. 1. The groups *Soldanellae, Sepincoli, Occidentales*, etc., were published without any indication of rank under the genus *Convolvulus* by House (Muhlenbergia 4: 50. 1908). These names are validly published but they are not in any definite rank and have no status in questions of priority except that they may act as homonyms.

Ex. 2. In the genus *Carex*, the epithet *Scirpinae* was published for an infrageneric taxon of no stated rank by Tuckerman (Enum. Caric. 8. 1843); this was assigned sectional rank by Kükenthal (in Engler, Pflanzenr. 38 (IV.20): 81. 1909) and if recognized at this rank is to be cited as *Carex* sect. *Scirpinae* (Tuckerman) Kükenthal.

35.3. If in a given publication prior to 1 Jan. 1890 only one infraspecific rank is admitted it is considered to be that of variety unless this would be contrary to the statements of the author himself in the same publication.

35.4. In questions of indication of rank, all publications appearing under the same title and by the same author, such as different parts of a Flora issued at different times (but not different editions of the same work), must be considered as a whole, and any statement made therein designating the rank of taxa included in the work must be considered as if it had been published together with the first instalment.

Article 36

36.1. In order to be validly published, a name of a new taxon of plants, the algae and all fossils excepted, published on or after 1 Jan. 1935 must be accompanied by a Latin description or diagnosis or by a reference to a previously and effectively published Latin description or diagnosis (but see Art. H.9).

Ex. 1. The names *Schiedea gregoriana* Degener (Fl. Hawaiiensis, fam. 119. 1936, Apr. 9) and *S. kealiae* Caum & Hosaka (Occas. Pap. Bernice Pauahi Bishop Mus. 11(23): 3. 1936, Apr. 10) were proposed for the same plant; the type of the former is a part of the original material of the latter. Since the name *S. gregoriana* is not accompanied by a Latin description or diagnosis it is not validly published; the later *S. kealiae* is legitimate.

36.2. In order to be validly published, a name of a new taxon of non-fossil algae published on or after 1 Jan. 1958 must be accompanied by a Latin description or diagnosis or by a reference to a previously and effectively published Latin description or diagnosis.

36A.1. Authors publishing names of new taxa of non-fossil plants should give or cite a full description in Latin in addition to the diagnosis.

Article 37

37.1. Publication on or after 1 Jan. 1958 of the name of a new taxon of the rank of genus or below is valid only when the holotype of the name is indicated (see Arts. 7-10; but see Art. H.9, Note 1 for the names of certain hybrids).

37.2. For the name of a new genus or subdivision of a genus, inclusion of reference (direct or indirect) to a single type of a name of a species is acceptable as indication of the holotype (see also Art. 22.4; but see Art. 37.4).

37.3. For the name of a new species or infraspecific taxon, citation of a single element is acceptable as indication of the holotype (but see Art. 37.4). Mere citation of a locality without concrete reference to a specimen does not however constitute indication of a holotype. Citation of the collector's name and/or collecting number and/or date of collection and/or reference to any other detail of the type specimen or illustration is required.

37.4. For the name of a new taxon published on or after 1 Jan. 1990, indication of the holotype must include one of the words "typus" or "holotypus", or its abbreviation, or its equivalent in a modern language.

37.5. For the name of a new species or infraspecific taxon published on or after 1 Jan. 1990 whose type is a specimen or unpublished illustration, the herbarium or institution in which the type is conserved must be specified.

Note 1. Specification of the herbarium or institution may be made in an abbreviated form, e.g. as given in the Index Herbariorum.

37A.1. The indication of the nomenclatural type should immediately follow the description or diagnosis and should use the Latin word "typus" or "holotypus".

Article 38

38.1. In order to be validly published, a name of a new taxon of fossil plants of specific or lower rank published on or after 1 Jan. 1912 must be accompanied by an illustration or figure showing the essential characters, in addition to the description or diagnosis, or by a reference to a previously and effectively published illustration or figure.

Article 39

39.1. In order to be validly published, a name of a new taxon of non-fossil algae of specific or lower rank published on or after 1 Jan. 1958 must be accompanied by an illustration or figure showing the distinctive morphological features, in addition to the Latin description or diagnosis, or by a reference to a previously and effectively published illustration or figure.

Recommendation 39A

39A.1. The illustration or figure required by Art. 39 should be prepared from actual specimens, preferably including the holotype.

Article 40

40.1. In order to be validly published, names of hybrids of specific or lower rank with Latin epithets must comply with the same rules as names of non-hybrid taxa of the same rank.

Ex. 1. The name *Nepeta* ×*faassenii* Bergmans (Vaste Pl. ed. 2. 544. 1939) with a description in Dutch, and in Gentes Herb. 8: 64 (1949) with a description in English, is not validly published, not being accompanied by or associated with a Latin description or diagnosis. The name *Nepeta* ×*faassenii* Bergmans ex Stearn (1950) is validly published, being accompanied by a Latin description with designation of type.

Ex. 2. The name *Rheum* ×*cultorum* Thorsrud & Reis. (Norske Plantenavr. 95. 1948), being there a nomen nudum, is not validly published.

Ex. 3. The name *Fumaria* ×*salmonii* Druce (List Brit. Pl. 4. 1908) is not validly published, because only its presumed parentage *F. densiflora* × *F. officinalis* is stated.

Note 1. For names of hybrids of the rank of genus or subdivision of a genus, see Art. H.9.

40.2. For purposes of priority, names in Latin form given to hybrids are subject to the same rules as are those of non-hybrid taxa of equivalent rank.

Ex. 4. The name ×*Solidaster* Wehrh. (1932) antedates the name ×*Asterago* Everett (1937) for the hybrid *Aster* × *Solidago.*

Ex. 5. The name ×*Gaulnettya* W. J. Marchant (1937) antedates the name ×*Gaulthettya* Camp (1939) for the hybrid *Gaultheria* × *Pernettya.*

Ex. 6. *Anemone* ×*hybrida* Paxton (1848) antedates *A.* ×*elegans* Decne. (1852), pro sp., as the binomial for the hybrids derived from *A. hupehensis* × *A. vitifolia.*

Ex. 7. In 1927, Aimée Camus (Bull. Mus. Hist. Nat. (Paris) 33: 538) published the name ×*Agroelymus* as the name of a nothogenus, without a Latin diagnosis or description, mentioning only the names of the parents involved (*Agropyron* and *Elymus*). Since this name was not validly published under the Code then in force (Stockholm 1950), Jacques Rousseau, in 1952 (Mém. Jard. Bot. Montréal 29: 10-11), published a Latin diagnosis. However, the date of valid publication of the name ×*Agroelymus* under this Code (Art. H.9) is 1927, not 1952, and the name also antedates ×*Elymopyrum* Cugnac (Bull. Soc. Hist. Nat. Ardennes 33: 14. 1938) which is accompanied by a statement of parentage and a description in French but not Latin.

Article 41

41.1. In order to be validly published, a name of a family must be accompanied *(a)* by a description or diagnosis of the family, or *(b)* by a reference (direct or indirect) to a previously and effectively published description or diagnosis of a family or subdivision of a family.

Ex. 1. The name "Pseudoditrichaceae fam. nov." was not validly published by Steere and Iwatsuki (Canad. J. Bot. 52: 701. 1974) as there was no Latin diagnosis, description, or reference to either, but only mention of the single included genus and species (see Art. 34.1(e)), "*Pseudoditrichum mirabile* gen. et sp. nov.", for both of which the name was validated under Art. 42 by a single Latin diagnosis.

41.2. In order to be validly published, a name of a genus must be accompanied *(a)* by a description or diagnosis of the genus (but see Art. 42), or *(b)* by a reference (direct or indirect) to a previously and effectively published description or diagnosis of a genus or subdivision of a genus.

Ex. 2. Validly published generic names: *Carphalea* A. L. Juss., accompanied by a generic description; *Thuspeinanta* T. Durand, accompanied by a reference to the previously described genus *Tapeinanthus* Boiss. (non Herbert); *Aspalathoides* (DC.) K. Koch, based on a previously described section, *Anthyllis* sect. *Aspalathoides* DC.; *Scirpoides* Scheuchzer ex Séguier (Pl. Veron. Suppl. 73. 1754), accepted there but without a generic description, validated by indirect reference (through the title of the book and a general statement in the preface) to the generic diagnosis and further direct references in Séguier (Pl. Veron. 1: 117. 1745).

Note 1. An exception to Art. 41.2 is made for the generic names first published by Linnaeus in Species Plantarum ed. 1 (1753) and ed. 2 (1762-1763), which are treated as having been validly published on those dates (see Art. 13.4).

Note 2. In certain circumstances, an illustration with analysis is accepted as equivalent to a generic description (see Art. 42.2).

41.3. In order to be validly published, a name of a species must be accompanied *(a)* by a description or diagnosis of the species (but see Arts. 42 and 44), or *(b)* by a reference to a previously and effectively published description or diagnosis of a species or infraspecific taxon, or *(c),* under certain circumstances, by reference to a genus whose name was previously and validly published simultaneously with its description or diagnosis. A reference as mentioned under *(c)* is acceptable only if neither the author of the name of the genus nor the author of the name of the species indicate that more than one species belongs to the genus in question.

Ex. 3. Trilepisium Thouars (1806) was validated by a generic description but without mention of a name of a species. *Trilepisium madagascariense* DC. (1828) was subsequently proposed without a description of the species. Neither author gave any indication that there was more than one species in the genus. Augustin-Pyramus de Candolle's specific name is therefore validly published.

Article 42

42.1. The names of a genus and a species may be simultaneously validated by provision of a single description (*descriptio generico-specifica*) or diagnosis, even though this may have been intended as only generic or specific, if all of the following conditions obtain: *(a)* the genus is at that time monotypic; *(b)* no other names (at any rank) have previously been validly published based on the same type; and *(c)* the names of the genus and species otherwise fulfil the requirements for valid publication. Reference to an earlier description or diagnosis is not accepted as provision of such a description or diagnosis.

Note 1. In this context a monotypic genus is one for which a single binomial is validly published, even though the author may indicate that other species are attributable to the genus.

Ex. 1. The names *Kedarnatha* Mukherjee & Constance (Brittonia 38: 147. 1986) and *Kedarnatha sanctuarii* Mukherjee & Constance, the latter designating the only species in the new genus, are both validly published although a Latin description is provided only under the generic name.

Ex. 2. Piptolepis phillyreoides Bentham is a new species assigned to the monotypic new genus *Piptolepis* published with a combined generic and specific description.

Ex. 3. In publishing *Phaelypea* without a generic description, P. Browne (Civ. Nat. Hist. Jamaica 269. 1756) included and described a single species, but he gave the species a phrase-name and did not provide a valid binomial. Art. 42 does not therefore apply and *Phaelypea* is not validly published.

42.2. Prior to 1 Jan. 1908 an illustration with analysis, or for non-vascular plants a single figure showing details aiding identification, is acceptable, for the purpose of this Article, in place of a written description or diagnosis.

Note 2. An analysis in this context is a figure or group of figures, commonly separate from the main illustration of the plant (though usually on the same page or plate), showing details aiding identification, with or without a separate caption.

Ex. 4. The generic name *Philgamia* Baillon (1894) was validly published, as it appeared on a plate with analysis of the only included species, *P. hibbertioides* Baillon, and was published before 1 Jan. 1908.

Article 43

43.1. A name of a taxon below the rank of genus is not validly published unless the name of the genus or species to which it is assigned is validly published at the same time or was validly published previously.

Ex. 1. Binary designations for six species of *Suaeda*, including *Suaeda baccata* and *S. vera*, were published with diagnoses and descriptions by Forsskål (Fl. Aegypt.-Arab. 69-71. 1775), but he provided no diagnosis or description for the genus: these specific names were therefore, like the generic name, not validly published by him.

Ex. 2. In 1880, Müller Argoviensis (Flora 63: 286) published the new genus *Phlyctidia* with the species *P. hampeana* n. sp., *P. boliviensis* (= *Phlyctis boliviensis* Nyl.), *P. sorediiformis* (= *Phlyctis sorediiformis* Kremp.), *P. brasiliensis* (= *Phlyctis brasiliensis* Nyl.), and *P. andensis* (= *Phlyctis andensis* Nyl.). These specific names are, however, not validly published in this place, because the generic name *Phlyctidia* was not validly published; Müller gave no generic description or diagnosis but only a description and a diagnosis of the new species *P. hampeana.* This description and diagnosis cannot validate the generic name as a descriptio generico-specifica under Art. 42 since the new genus was not monotypic. Valid publication of the name *Phlyctidia* was by Müller (1895), who provided a short generic diagnosis. The only species mentioned here were *P. ludoviciensis* n. sp. and *P. boliviensis* (Nyl.). The latter combination was validly published in 1895 by the reference to the basionym.

Note 1. This Article applies also to specific and other epithets published under words not to be regarded as generic names (see Art. 20.4).

Ex. 3. The binary combination *Anonymos aquatica* Walter (Fl. Carol. 230. 1788) is not validly published. The correct name for the species concerned is *Planera aquatica* J. F. Gmelin (1791), and the date of the name, for purposes of priority, is 1791. The species must not be cited as *Planera aquatica* (Walter) J. F. Gmelin.

Ex. 4. The binary combination *Scirpoides paradoxus* Rottb. (Descr. Pl. Rar. Progr. 27. 1772) is not validly published since *Scirpoides* in this context is a word not intended as a generic name. The first validly published name for this species is *Fuirena umbellata* Rottb. (1773).

Article 44

44.1. The name of a species or of an infraspecific taxon published before 1 Jan. 1908 is validly published if it is accompanied only by an illustration with analysis (see Art. 42, Note 2).

Ex. 1. Panax nossibiensis Drake (1896) was validly published on a plate with analysis.

44.2. Single figures of non-vascular plants showing details aiding identification are considered as illustrations with analysis (see Art. 42, Note 2).

Ex. 2. Eunotia gibbosa Grunow (1881), a name of a diatom, was validly published by provision of a single figure of the valve.

Article 45

45.1. The date of a name is that of its valid publication. When the various conditions for valid publication are not simultaneously fulfilled, the date is that on which the last is fulfilled. However, the name must always be explicitly accepted in the place of its validation. A name published on or after 1 Jan. 1973 for which the various conditions for valid publication are not simultaneously fulfilled is not validly published unless a full and direct reference (Art. 33.2) is given to the places where these requirements were previously fulfilled.

Ex. 1. Clypeola minor first appeared in the Linnaean thesis Flora Monspeliensis (1756), in a list of names preceded by numerals but without an explanation of the meaning of these numerals and without any other descriptive matter; when the thesis was reprinted in vol. 4 of the Amoenitates Academicae (1759), a statement was added explaining that the numbers referred to earlier descriptions published in Magnol's Botanicon Monspeliense. However, *Clypeola minor* was absent from the reprint, being no longer accepted by Linnaeus, and the name is not therefore validly published.

Ex. 2. When proposing *Graphis meridionalis* as a new species, in 1966, Nakanishi (J. Sci. Hiroshima Univ., ser. B (2), 11: 75) provided a Latin description but failed to designate a holotype. *Graphis meridionalis* Nakanishi was validly published only in 1967 (J. Sci. Hiroshima Univ., ser. B (2), 11: 265) when he designated the holotype of the name and provided a full and direct reference to the previous publication.

45.2. A correction of the original spelling of a name (see Art. 73) does not affect its date of valid publication.

Ex. 3. The correction of the orthographic error in *Gluta benghas* L. (Mant. 293. 1771) to *Gluta renghas* L. does not affect the date of publication of the name even though the correction dates only from 1883 (Engler in A. DC. & C. DC., Monogr. Phan. 4: 225).

45.3. For purposes of priority only legitimate names are taken into consideration (see Arts. 11, 63-65). However, validly published earlier homonyms, whether legitimate or not, shall cause rejection of their later homonyms, unless the latter are conserved or sanctioned (but see Art. 14 Note 2).

45.4. If a taxon originally assigned to a group not covered by this Code is treated as belonging to a group of plants other than algae, the authorship and date of any of its names are determined by the first publication that satisfies the requirements for valid publication under this Code. If the taxon is treated as belonging to the algae, any of its names need satisfy only the requirements of the pertinent non-botanical code for status equivalent to valid publication under the botanical Code (but see Art. 65, regarding homonymy).

Ex. 4. Amphiprora Ehrenb. (1843) is an available[1] name for a genus of animals first treated as belonging to the algae by Kützing (1844). *Amphiprora* has priority in botanical nomenclature from 1843, not 1844.

Ex. 5. Petalodinium J. Cachon & M. Cachon (Protistologica 5: 16. 1969) is available under the International Code of Zoological Nomenclature as the name of a genus of dinoflagellates. When the taxon is treated as belonging to the algae, its name retains its original authorship and date even though the original publication lacked a Latin diagnosis.

Ex. 6. Labyrinthodyction Valkanov (Progr. Protozool. 3: 373. 1969), although available under the International Code of Zoological Nomenclature as the name of a genus of rhizopods, is not valid when the taxon is treated as belonging to the fungi because the original publication lacked a Latin diagnosis.

Ex. 7. Protodiniferidae Kofoid & Swezy (Mem. Univ. Calif. 5: 111. 1921), available under the International Code of Zoological Nomenclature, is validly published as a name of a family of algae with its original authorship and date but with the termination *-idae* changed to *-aceae* (in accordance with Arts. 18.4 and 32.5).

---

1) The word "available" in the International Code of Zoological Nomenclature is equivalent to "validly published" in the International Code of Botanical Nomenclature.

Recommendation 45A

45A.1. Authors using new names in works written in a modern language (floras, catalogues, etc.) should simultaneously comply with the requirements of valid publication.

Recommendation 45B

45B.1. Authors should indicate precisely the dates of publication of their works. In a work appearing in parts the last-published sheet of the volume should indicate the precise dates on which the different fascicles or parts of the volume were published as well as the number of pages and plates in each.

Recommendation 45C

45C.1. On separately printed and issued copies of works published in a periodical, the name of the periodical, the number of its volume or parts, the original pagination, and the date (year, month, and day) should be indicated.

SECTION 3. CITATION OF AUTHORS' NAMES FOR PURPOSES OF PRECISION

Article 46

46.1. For the indication of the name of a taxon to be accurate and complete, and in order that the date may be readily verified, it is necessary to cite the name of the author(s) who validly published the name concerned unless the provisions for autonyms apply (Arts. 19.3, 22.1, and 26.1; see also Art. 16.1).

Ex. 1. Rosaceae A. L. Juss., *Rosa* L., *Rosa gallica* L., *Rosa gallica* var. *eriostyla* R. Keller, *Rosa gallica* L. var. *gallica.*

46.2. When a name of a taxon and its description or diagnosis (or reference to a description or diagnosis) are supplied by one author but published in a work by another author, the word "in" is to be used to connect the names of the two authors. When it is desirable to simplify such a citation, the name of the author who supplied the description or diagnosis is to be retained.

Ex. 2. Viburnum ternatum Rehder in Sargent, Trees and Shrubs 2: 37 (1907), or *V. ternatum* Rehder; *Teucrium charidemi* Sandw. in Lacaita, Cavanillesia 3: 38 (1930), or *T. charidemi* Sandw.

46.3. When an author who validly publishes a name ascribes it to another person, e.g. to an author who failed to fulfil all requirements for valid publication of the name or to an author who published the name prior to the nomenclatural starting point of the group concerned (see Art. 13.1), the correct author citation is the name of the validating author, but the name of the other person, followed by the connecting word "ex", may be inserted before the name of the validating author (see also Rec. 50A.2). The same holds for names of garden origin ascribed to "hort.", meaning "hortulanorum".

Ex. 3. Gossypium tomentosum Seemann or *G. tomentosum* Nutt. ex Seemann; *Lithocarpus polystachyus* (A. DC.) Rehder or *L. polystachyus* (Wall. ex A. DC.) Rehder; *Orchis rotundifolia* Pursh or *O. rotundifolia* Banks ex Pursh; *Carex stipata* Willd. or *C. stipata* Muhlenb. ex Willd.; *Gesneria donklarii* Hooker or *G. donklarii* hort. ex Hooker.

Ex. 4. Lupinus L. or *Lupinus* Tourn. ex L.; *Euastrum binale* Ralfs or *E. binale* Ehrenb. ex Ralfs.

Ex. 5. The name *Lichen debilis,* which was validly published by Smith (1812) with "*Calicium debile.* Turn. and Borr. Mss." cited as a synonym, is not to be attributed to "Turner & Borrer ex Smith" (see also Rec. 50A.2).

Recommendation 46A

46A.1. Authors' names put after names of plants may be abbreviated, unless they are very short. For this purpose, particles should be suppressed unless they are an inseparable part of the name, and the first letters should be given without any omission (Lam. for J. B. P. A. Monet Chevalier de Lamarck, but De Wild. for E. De Wildeman).

46A.2. If a name of one syllable is long enough to make it worth while to abridge it, the first consonants only should be given (Fr. for Elias Magnus Fries); if the name has two or more syllables, the first syllable and the first letter of the following one should be taken, or the two first when both are consonants (Juss. for Jussieu, Rich. for Richard).

46A.3. When it is necessary to give more of a name to avoid confusion between names beginning with the same syllable, the same system should be followed. For instance, two syllables should be given together with the one or two first consonants of the third; or one of the last characteristic consonants of the name be added (Bertol. for Bertoloni, to distinguish it from Bertero; Michx. for Michaux, to distinguish it from Micheli).

46A.4. Given names or accessory designations serving to distinguish two botanists of the same name should be abridged in the same way (Adr. Juss. for Adrien de Jussieu, Gaertner f. for Gaertner filius, J. F. Gmelin for Johann Friedrich Gmelin, J. G. Gmelin for Johann Georg Gmelin, C. C. Gmelin for Carl Christian Gmelin, S. G. Gmelin for Samuel Gottlieb Gmelin, Müll. Arg. for Jean Müller of Aargau).

46A.5. When it is a well-established custom to abridge a name in another manner, it is advisable to conform to it (L. for Linnaeus, DC. for Augustin-Pyramus de Candolle, St.-Hil. for Saint-Hilaire, R. Br. for Robert Brown).

Recommendation 46B

46B.1. In citing the author of the scientific name of a taxon, the romanization of the author's name(s) given in the original publication should normally be accepted. Where an author failed to give a romanization, or where an author has at different times used different romanizations, then the romanization known to be preferred by the author or that most frequently adopted by the author should be accepted. In the absence of such information the author's name should be romanized in accordance with an internationally available standard.

46B.2. Authors of scientific names whose personal names are not written in Roman letters should romanize their names, preferably (but not necessarily) in accordance with an internationally available standard and, as a matter of typographic convenience, without diacritical signs. Once authors have selected the romanization of their personal names, they should use it consistently thereafter. Whenever possible, authors should not permit editors or publishers to change the romanization of their personal names.

Recommendation 46C

46C.1. When a name has been published jointly by two authors, the names of both should be cited, linked by means of the word "et" or by an ampersand (&).

Ex. 1. Didymopanax gleasonii Britton et Wilson (or Britton & Wilson).

46C.2. When a name has been published jointly by more than two authors, the citation should be restricted to that of the first one followed by "et al."

Ex. 2. Lapeirousia erythrantha var. *welwitschii* (Baker) Geerinck, Lisowski, Malaisse & Symoens (Bull. Soc. Roy. Bot. Belgique 105: 336. 1972) should be cited as *L. erythrantha* var. *welwitschii* (Baker) Geerinck et al.

Recommendation 46D

46D.1. Authors should cite their own names after each new name they publish; the expression "nobis" (*nob.*) or a similar reference to themselves should be avoided.

Article 47

47.1. An alteration of the diagnostic characters or of the circumscription of a taxon without the exclusion of the type does not warrant the citation of the name of an author other than the one who first published its name.

Examples: see under Art. 51.

Recommendation 47A

47A.1. When an alteration as mentioned in Art. 47 has been considerable, the nature of the change may be indicated by adding such words, abbreviated where suitable, as "emendavit" (*emend.*) (followed by the name of the author responsible for the change), "mutatis characteribus" (*mut. char.*), "pro parte" (*p. p.*), "excluso genere" or "exclusis generibus" (*excl. gen.*), "exclusa specie" or "exclusis speciebus" (*excl. sp.*), "exclusa varietate" or "exclusis varietatibus" (*excl. var.*), "sensu amplo" (*s. ampl.*), "sensu lato" (*s. l.*), "sensu stricto" (*s. str.*), etc.

Ex. 1. Phyllanthus L. emend. Müll. Arg.; *Globularia cordifolia* L. excl. var. (emend. Lam.).

Article 48

48.1. When an author adopts an existing name but explicitly excludes its original type, he is considered to have published a later homonym that must be ascribed solely to him. Similarly, when an author who adopts a name refers to an apparent basionym but explicitly excludes its type, he is considered to have published a new name that must be ascribed solely to him. Explicit exclusion can be effected by simultaneous explicit inclusion of the type in a different taxon by the same author (see also Art. 59.6).

Ex. 1. Sirodot (1872) placed the type of *Lemanea* Bory (1808) in *Sacheria* Sirodot (1872); hence *Lemanea*, as treated by Sirodot (1872), is to be cited as *Lemanea* Sirodot non Bory and not as *Lemanea* Bory emend. Sirodot.

51

Ex. 2. The name *Amorphophallus campanulatus*, published by Decaisne, was apparently based on *Arum campanulatum* Roxb. However, the type of the latter was explicitly excluded by Decaisne, and the name is to be cited as *Amorphophallus campanulatus* Decne., not as *Amorphophallus campanulatus* (Roxb.) Decne.

Note 1. Misapplication of a new combination to a different taxon, but without explicit exclusion of the type of the basionym, is dealt with under Arts. 55.2 and 56.2.

Note 2. Retention of a name in a sense that excludes the type can be effected only by conservation (see Art. 14.8).

Article 49

49.1. When a genus or a taxon of lower rank is altered in rank but retains its name or epithet, the author of the earlier, epithet-bringing legitimate name (the author of the basionym) must be cited in parentheses, followed by the name of the author who effected the alteration (the author of the new name). The same holds when a taxon of lower rank than genus is transferred to another genus or species, with or without alteration of rank.

Ex. 1. Medicago polymorpha var. *orbicularis* L. when raised to the rank of species becomes *Medicago orbicularis* (L.) Bartal.

Ex. 2. Anthyllis sect. *Aspalathoides* DC. raised to generic rank, retaining the epithet *Aspalathoides* as its name, is cited as *Aspalathoides* (DC.) K. Koch.

Ex. 3. Cineraria sect. *Eriopappus* Dumort. (Fl. Belg. 65. 1827) when transferred to *Tephroseris* (Reichenb.) Reichenb. is cited as *Tephroseris* sect. *Eriopappus* (Dumort.) Holub (Folia Geobot. Phytotax. Bohem. 8: 173. 1973).

Ex. 4. Cistus aegyptiacus L. when transferred to *Helianthemum* Miller is cited as *Helianthemum aegyptiacum* (L.) Miller.

Ex. 5. Fumaria bulbosa var. *solida* L. (1753) was elevated to specific rank as *F. solida* (L.) Miller (1771). The name of this species when transferred to *Corydalis* is to be cited as *C. solida* (L.) Clairv. (1811), not *C. solida* (Miller) Clairv.

Ex. 6. However, *Pulsatilla montana* var. *serbica* W. Zimmerm. (Feddes Repert. Spec. Nov. Regni Veg. 61: 95. 1958), originally placed under *P. montana* subsp. *australis* (Heuffel) Zam., retains the same author citation when placed under *P. montana* subsp. *dacica* Rummelsp. (see Art. 24.1) and is not cited as var. *serbica* (W. Zimmerm.) Rummelsp. (Feddes Repert. 71: 29. 1965).

Ex. 7. Salix subsect. *Myrtilloides* C. Schneider (Ill. Handb. Laubholzk. 1: 63. 1904), originally placed under *S.* sect. *Argenteae* Koch, retains the same author citation when placed under *S.* sect. *Glaucae* Pax and is not cited as *S.* subsect. *Myrtilloides* (C. Schneider) Dorn (Canad. J. Bot. 54: 2777. 1976).

Article 50

50.1. When a taxon at the rank of species or below is transferred from the non-hybrid category to the hybrid category of the same rank (Art. H.10.2), or vice versa, the author citation remains unchanged but may be followed by an indication in parentheses of the original category.

Ex. 1. Stachys ambigua Smith was published as the name of a species. If regarded as applying to a hybrid, it may be cited as *Stachys* × *ambigua* Smith (pro sp.).

Ex. 2. The binary name *Salix* × *glaucops* Andersson was published as the name of a hybrid. Later, Rydberg (Bull. New York Bot. Gard. 1: 270. 1899) considered the taxon to be a species. If this view is accepted, the name may be cited as *Salix glaucops* Andersson (pro hybr.).

SECTION 4. GENERAL RECOMMENDATIONS ON CITATION

Recommendation 50A

50A.1. In the citation of a name published as a synonym, the words "as synonym" or "pro syn." should be added.

50A.2. When an author has published as a synonym a manuscript name of another author, the word "ex" should be used in citations to connect the names of the two authors (see also Art. 46.3).

Ex. 1. Myrtus serratus, a manuscript name of Koenig published by Steudel as a synonym of *Eugenia laurina* Willd., should be cited thus: *Myrtus serratus* Koenig ex Steudel, pro syn.

Recommendation 50B

50B.1. In the citation of a nomen nudum, its status should be indicated by adding the words "nomen nudum" or "nom. nud."

Ex. 1. Carex bebbii Olney (Car. Bor.-Am. 2: 12. 1871), published without a diagnosis or description, should be cited as a nomen nudum.

Recommendation 50C

50C.1. The citation of a later homonym should be followed by the name of the author of the earlier homonym preceded by the word "non", preferably with the date of publication added. In some instances it will be advisable to cite also any other homonyms, preceded by the word "nec".

Ex. 1. Ulmus racemosa Thomas, Amer. J. Sci. Arts 19: 170 (1831), non Borkh. 1800; *Lindera* Thunb., Nov. Gen. Pl. 64 (1783), non Adanson 1763; *Bartlingia* Brongn., Ann. Sci. Nat. (Paris) 10: 373 (1827), non Reichenb. 1824 nec F. Muell. 1882.

Recommendation 50D

50D.1. Misidentifications should not be included in synonymies but added after them. A misapplied name should be indicated by the words "auct. non" followed by the name of the original author and the bibliographic reference of the misidentification.

Ex. 1. Ficus stortophylla Warb. in Warb. & De Wild., Ann. Mus. Congo Belge, B, Bot. ser. 4, 1: 32 (1904). *F. irumuensis* De Wild., Pl. Bequaert. 1: 341 (1922). *F. exasperata* auct. non Vahl: De Wild. & T. Durand, Ann. Mus. Congo Belge, B, Bot. ser. 2, 1: 54 (1899); De Wild., Miss. Em. Laurent 26 (1905); T. Durand & H. Durand, Syll. Fl. Congol. 505 (1909).

Recommendation 50E

50E.1. If a generic or specific name is accepted as a nomen conservandum (see Art. 14 and App. III) the abbreviation "nom. cons." should be added in a full citation.

Ex. 1. Protea L., Mant. Pl. 187 (1771), nom. cons., non L. 1753; *Combretum* Loefl. (1758), nom. cons. (syn. prius *Grislea* L. 1753).

50E.2. If it is desirable to indicate the sanctioned status of the names of fungi adopted by Persoon or Fries (see Art. 13.1(d)), ": Pers." or ": Fr." should be added to the citation.

Ex. 2. Boletus piperatus Bull. : Fr.

Recommendation 50F

50F.1. If a name is cited with alterations from the form as originally published, it is desirable that in full citations the exact original form should be added, preferably between single or double quotation marks.

Ex. 1. Pyrus calleryana Decne. (*Pyrus mairei* H. Léveillé, Repert. Spec. Nov. Regni Veg. 12: 189. 1913, "*Pirus*").

Ex. 2. Zanthoxylum cribrosum Sprengel, Syst. Veg. 1: 946 (1825), "*Xanthoxylon*". (*Zanthoxylum caribaeum* var. *floridanum* (Nutt.) A. Gray, Proc. Amer. Acad. Arts 23: 225. 1888, "*Xanthoxylum*").

Ex. 3. Spathiphyllum solomonense Nicolson, Amer. J. Bot. 54: 496 (1967), "*solomonensis*".

CHAPTER V. RETENTION, CHOICE, AND REJECTION OF NAMES AND EPITHETS

SECTION 1. RETENTION OF NAMES OR EPITHETS WHEN TAXA ARE REMODELLED OR DIVIDED

Article 51

51.1. An alteration of the diagnostic characters or of the circumscription of a taxon does not warrant a change in its name, except as may be required (*a*) by transference of the taxon (Arts. 54-56), or (*b*) by its union with another taxon of the same rank (Arts. 57,58), or (*c*) by a change of its rank (Art. 60).

Ex. 1. The genus *Myosotis* as revised by R. Brown differs from the original genus of Linnaeus, but the generic name has not been changed, nor is a change allowable, since the type of *Myosotis* L. remains in the genus; it is cited as *Myosotis* L. or as *Myosotis* L. emend. R. Br. (see Art. 47, Rec. 47A).

Ex. 2. Various authors have united with *Centaurea jacea* L. one or two species which Linnaeus had kept distinct; the taxon so constituted is called *Centaurea jacea* L. sensu amplo or *Centaurea jacea* L. emend. Cosson & Germ., emend. Vis., or emend. Godron, etc.; any new name for this taxon, such as *Centaurea vulgaris* Godron, is superfluous and illegitimate.

51.2. An exception to Art. 51.1 is made for the family name *Papilionaceae* (see Art. 18.5).

Article 52

52.1. When a genus is divided into two or more genera, the generic name, if correct, must be retained for one of them. If a type was originally designated the generic name must be retained for the genus including that type. If no type has been designated, a type must be chosen (see Rec. 7B).

Ex. 1. The genus *Dicera* Forster & Forster f. was divided by Rafinesque into the two genera *Misipus* and *Skidanthera*. This procedure is contrary to the rules: the name *Dicera* must be kept for one of the genera, and it is now retained for that part of *Dicera* including the lectotype, *D. dentata*.

Ex. 2. Among the sections which have been recognized in the genus *Aesculus* L. are *Aesculus* sect. *Aesculus*, sect. *Pavia* (Miller) Walp., sect. *Macrothyrsus* (Spach) K. Koch, and sect. *Calothyrsus* (Spach) K. Koch, the last three of which were regarded as distinct genera by the authors cited in parentheses. In the event of these four sections being treated as genera, the name *Aesculus* must be kept for the first of them, which includes *Aesculus hippocastanum* L., the type of the generic name.

Article 53

53.1. When a species is divided into two or more species, the specific name, if correct, must be retained for one of them. If a particular specimen, description, or figure was originally designated as the type, the specific name must be retained for the species including that element. If no type has been designated, a type must be chosen (see Rec. 7B).

Ex. 1. Arabis beckwithii S. Watson (1887) was based on specimens which represented at least two species in the opinion of Munz, who based *A. shockleyi* Munz (1932) on one of the specimens cited by Watson, retaining the name *A. beckwithii* for the others (one of which may be designated as lectotype of *A. beckwithii*).

Ex. 2. Hemerocallis lilioasphodelus L. (1753) was originally treated by Linnaeus as consisting of two varieties: var. *flava* ("*flavus*") and var. *fulva* ("*fulvus*"). In 1762 he recognized these as distinct species, calling them *H. flava* and *H. fulva*. The original specific epithet was reinstated for one of these by Farwell (Amer. Midl. Naturalist 11: 51. 1928) and the two species are correctly named *H. lilioasphodelus* L. and *H. fulva* (L.) L.

53.2. The same rule applies to infraspecific taxa, for example, to a subspecies divided into two or more subspecies, or to a variety divided into two or more varieties.

SECTION 2. RETENTION OF EPITHETS OF TAXA BELOW THE RANK OF
GENUS ON TRANSFERENCE TO ANOTHER GENUS OR SPECIES

Article 54

54.1. When a subdivision of a genus is transferred to another genus or placed under another generic name for the same genus without change of rank, the epithet of its formerly correct name must be retained unless one of the following obstacles exists:

(a) The resulting combination has been previously and validly published for a subdivision of a genus based on a different type;

(b) The epithet of an earlier legitimate name of the same rank is available (but see Arts. 13.1(d), 58, 59);

(c) Arts. 21 or 22 provide that another epithet be used.

Ex. 1. Saponaria sect. *Vaccaria* DC. when transferred to *Gypsophila* becomes *Gypsophila* sect. *Vaccaria* (DC.) Godron.

Ex. 2. Primula sect. *Dionysiopsis* Pax (1909) when transferred to the genus *Dionysia* becomes *Dionysia* sect. *Dionysiopsis* (Pax) Melchior (1943); the name *Dionysia* sect. *Ariadne* Wendelbo (1959), based on the same type, is not to be used.

Article 55

55.1. When a species is transferred to another genus or placed under another generic name for the same genus without change of rank, the epithet of its formerly correct name must be retained unless one of the following obstacles exists:

(a) The resulting binary name is a later homonym (Art. 64) or a tautonym (Art. 23.4);

(b) The epithet of an earlier legitimate specific name is available (but see Arts. 13.1(d), 58, 59).

Ex. 1. Antirrhinum spurium L. (1753) when transferred to the genus *Linaria* must be called *Linaria spuria* (L.) Miller (1768).

Ex. 2. Spergula stricta Sw. (1799) when transferred to the genus *Arenaria* must be called *Arenaria uliginosa* Schleicher ex Schlechtendal (1808) because of the existence of the name *Arenaria stricta* Michx. (1803), referring to a different species; but on further transfer to the genus *Minuartia* the epithet *stricta* must be used and the species called *Minuartia stricta* (Sw.) Hiern (1899).

Ex. 3. Conyza candida L. (1753) was illegitimately renamed *Conyza limonifolia* Smith (1813) and *Inula limonifolia* Boiss. (1843). However, the Linnaean epithet must be retained and the correct name of the species, in the genus *Inula*, is *I. candida* (L.) Cass. (1822).

Ex. 4. When transferring *Serratula chamaepeuce* L. (1753) to his new genus *Ptilostemon*, Cassini renamed the species *P. muticus* Cass. (1826, "*muticum*"). Lessing rightly reinstated the original specific epithet, creating the combination *Ptilostemon chamaepeuce* (L.) Less. (1832).

Ex. 5. Spartium biflorum Desf. (1798) when transferred to the genus *Cytisus* by Spach in 1849 could not be called *C. biflorus*, because this name had been previously and validly published for a different species by L'Héritier in 1791; the name *C. fontanesii* given by Spach is therefore legitimate.

Ex. 6. Arum dracunculus L. (1753) when transferred to the genus *Dracunculus* was renamed *Dracunculus vulgaris* Schott (1832), as use of the Linnaean epithet would create a tautonym.

Ex. 7. Melissa calamintha L. (1753) when transferred to the genus *Thymus* becomes *T. calamintha* (L.) Scop. (1772); placed in the genus *Calamintha* it may not be called *C. calamintha* (a tautonym) but has been named *C. officinalis* Moench (1794). However, when *C. officinalis* is transferred to the genus *Satureja*, the Linnaean epithet is again available and the name becomes *S. calamintha* (L.) Scheele (1843).

Ex. 8. Cucubalus behen L. (1753) was legitimately renamed *Behen vulgaris* Moench (1794) to avoid the tautonym *Behen behen*. If the species is transferred to the genus *Silene*, it may not retain its original epithet because of the existence of a *Silene behen* L. (1753). Therefore, the substitute name *Silene cucubalus* Wibel (1799) was created. However, the specific epithet *vulgaris* was still available under *Silene*. It was rightly reinstated in the combination *Silene vulgaris* (Moench) Garcke (1869).

55.2. On transference of a specific epithet under another generic name, the resulting combination must be retained for the species to which the type of the basionym belongs, and attributed to the author who first published it, even though it may have been applied erroneously to a different species (Art. 7.12; but see Arts. 48.1 and 59.6).

Ex. 9. Pinus mertensiana Bong. was transferred to the genus *Tsuga* by Carrière, who, however, as is evident from his description, erroneously applied the new combination *Tsuga mertensiana* to another species of *Tsuga*, namely *T. heterophylla* (Raf.) Sarg. The combination *Tsuga mertensiana* (Bong.) Carrière must not be applied to *T. heterophylla* (Raf.) Sarg. but must be retained for *Pinus mertensiana* Bong. when that species is placed in *Tsuga*; the citation in parentheses (under Art. 49) of the name of the original author, Bongard, indicates the type of the name.

Article 56

56.1. When an infraspecific taxon is transferred without change of rank to another genus or species, the final epithet of its formerly correct name must be retained unless one of the following obstacles exists:

(a) The resulting ternary combination, with a different type, has been previously and validly published for an infraspecific taxon of any rank;

(b) The epithet of an earlier legitimate name at the same rank is available (but see Arts. 13.1(d), 58, 59);

(c) Art. 26 provides that another epithet be used.

Ex. 1. Helianthemum italicum var. *micranthum* Gren. & Godron (Fl. France 1: 171. 1847) when transferred as a variety to *H. penicillatum* Thibaud ex Dunal retains its varietal epithet, becoming *H. penicillatum* var. *micranthum* (Gren. & Godron) Grosser (in Engler, Pflanzenr. 14 (IV.193): 115. 1903).

56.2. On transference of an infraspecific epithet under another specific name, the resulting combination must be retained for the taxon to which the type of the basionym belongs, and attributed to the author who first published it, even though it may have been applied erroneously to a different taxon (Art. 7.12; but see Arts. 48.1 and 59.6).

SECTION 3. CHOICE OF NAMES WHEN TAXA OF THE SAME RANK ARE UNITED

Article 57

57.1. When two or more taxa of the same rank are united, the earliest legitimate name or (for taxa below the rank of genus) the final epithet of the earliest legitimate name is retained, unless another epithet or a later name must be accepted under the provisions of Arts. 13.1(d), 14, 16.1, 19.3, 22.1, 26.1, 27, 55.1, 58, or 59.

Ex. 1. Schumann (in Engler & Prantl, Nat. Pflanzenfam. III, 6: 5. 1890), uniting the three genera *Sloanea* L. (1753), *Echinocarpus* Blume (1825), and *Phoenicosperma* Miq. (1865), rightly adopted the earliest of these three generic names, *Sloanea* L., for the resulting genus.

57.2. The author who first unites taxa bearing names of equal priority must choose one of them, unless an autonym is involved (see Art. 57.3). As soon as that choice is effectively published (Arts. 29-31), the name thus chosen is treated as having priority.

Ex. 2. If the two genera *Dentaria* L. (1 May 1753) and *Cardamine* L. (1 May 1753) are united, the resulting genus must be called *Cardamine* because that name was chosen by Crantz (Cl. Crucif. Emend. 126. 1769), who was the first to unite the two genera.

Ex. 3. R. Brown (in Tuckey, Narr. Exp. Congo 484. 1818) appears to have been the first to unite *Waltheria americana* L. (1 May 1753) and *W. indica* L. (1 May 1753). He adopted the name *W. indica* for the combined species, and this name is accordingly to be retained.

Ex. 4. Baillon (Adansonia 3: 162. 1863), when uniting for the first time *Sclerocroton integerrimus* Hochst. (Flora 28: 85. 1845) and *Sclerocroton reticulatus* Hochst. (Flora 28: 85. 1845), adopted the epithet *integerrimus* in the name of the combined taxon. Consequently this epithet is to be retained irrespective of the generic name (*Sclerocroton, Stillingia, Excoecaria, Sapium*) with which it is combined.

Ex. 5. Linnaeus in 1753 simultaneously published the names *Verbesina alba* and *V. prostrata*. Later (1771), he published *Eclipta erecta*, a superfluous name because *V. alba* is cited in synonymy, and *E. prostrata*, based on *V. prostrata*. The first author to unite these taxa was Roxburgh (Fl. Ind. 3: 438. 1832), who did so under the name *Eclipta prostrata* (L.) L., which therefore is to be used if these taxa are united and placed in the genus *Eclipta*.

Ex. 6. When the genera *Entoloma* (Fr. ex Rabenh.) P. Kummer (1871), *Leptonia* (Fr.) P. Kummer (1871), *Eccilia* (Fr.) P. Kummer (1871), *Nolanea* (Fr.) P. Kummer (1871), and *Claudopus* Gillet (1876) are united, one of the generic names simultaneously published by Kummer must be used for the whole, as was done by Donk (Bull. Jard. Bot. Buitenzorg ser. 3, 18(1): 157. 1949) who selected *Entoloma*. The name *Rhodophyllus* Quélet (1886), introduced to cover these combined genera, is superfluous.

57.3. An autonym is treated as having priority over the name or names of the same date and rank that established it.

Note 1. When the final epithet of an autonym is used in a new combination under the requirements of Art. 57.3, the basionym of that combination is the name from which the autonym is derived.

Ex. 7. *Heracleum sibiricum* L. (1753) includes *H. sibiricum* subsp. *lecokii* (Godron & Gren.) Nyman (1879) and *H. sibiricum* subsp. *sibiricum* (1879) automatically established at the same time. When *H. sibiricum* is included in *H. sphondylium* L. (1753) as a subspecies, the correct name for the taxon is *H. sphondylium* subsp. *sibiricum* (L.) Simonkai (1887), not subsp. *lecokii*, whether or not subsp. *lecokii* is treated as distinct.

Ex. 8. The publication of *Salix tristis* var. *microphylla* Andersson (Salices Bor.-Amer. 21. 1858) created the autonym *S. tristis* Aiton (1789) var. *tristis*. If *S. tristis*, including var. *microphylla*, is recognized as a variety of *S. humilis* Marshall (1785), the correct name is *S. humilis* var. *tristis* (Aiton) Griggs (Proc. Ohio Acad. Sci. 4: 301. 1905). However, if both varieties of *S. tristis* are recognized as varieties of *S. humilis*, then the names *S. humilis* var. *tristis* and *S. humilis* var. *microphylla* (Andersson) Fernald (Rhodora 48: 46. 1946) are both used.

Ex. 9. In the classification adopted by Rollins and Shaw, *Lesquerella lasiocarpa* (Hooker ex A. Gray) S. Watson is composed of two subspecies, subsp. *lasiocarpa* (which includes the type of the name of the species and is cited without an author) and subsp. *berlandieri* (A. Gray) Rollins & E. Shaw. The latter subspecies is composed of two varieties. In this classification the correct name of the variety which includes the type of subsp. *berlandieri* is *L. lasiocarpa* var. *berlandieri* (A. Gray) Payson (1922), not *L. lasiocarpa* var. *berlandieri* (cited without an author) or *L. lasiocarpa* var. *hispida* (S. Watson) Rollins & E. Shaw (1972), based on *Synthlipsis berlandieri* var. *hispida* S. Watson (1882), since publication of the latter name established the autonym *Synthlipsis berlandieri* A. Gray var. *berlandieri* which, at varietal rank, is treated as having priority over var. *hispida*.

Recommendation 57A

57A.1. Authors who have to choose between two generic names should note the following suggestions:

(a) Of two names of the same date, to prefer that which was first accompanied by the description of a species.

(b) Of two names of the same date, both accompanied by descriptions of species, to prefer that which, when the author makes his choice, includes the larger number of species.

(c) In cases of equality from these various points of view, to select the more appropriate name.

Article 58

58.1. When a non-fossil taxon of plants, algae excepted, and a fossil (or subfossil) taxon of the same rank are united, the correct name of the non-fossil taxon is treated as having priority (see Pre.7 and Art. 13.3).

Ex. 1. If *Platycarya* Siebold & Zucc. (1843), a non-fossil genus, and *Petrophiloides* Bowerbank (1840), a fossil genus, are united, the name *Platycarya* is accepted for the combined genus, although it is antedated by *Petrophiloides.*

Ex. 2. The generic name *Metasequoia* Miki (1941) was based on the fossil type of *M. disticha* (Heer) Miki. After discovery of the non-fossil species *M. glyptostroboides* Hu & Cheng, conservation of *Metasequoia* Hu & Cheng (1948) as based on the non-fossil type was approved. Otherwise, any new generic name based on *M. glyptostroboides* would have had to be treated as having priority over *Metasequoia* Miki.

SECTION 4. NAMES OF FUNGI WITH A PLEOMORPHIC LIFE CYCLE

Article 59

59.1. In ascomycetous and basidiomycetous fungi (including Ustilaginales) with mitotic asexual morphs (anamorphs) as well as a meiotic sexual morph (teleomorph), the correct name covering the holomorph (i.e., the species in all its morphs) is – except for lichen-forming fungi – the earliest legitimate name typified by an element representing the teleomorph, i.e. the morph characterized by the production of asci/ascospores, basidia/basidiospores, teliospores, or other basidium-bearing organs.

59.2. For a binary name to qualify as a name of a holomorph, not only must its type specimen be teleomorphic, but also the protologue must include a diagnosis or description of this morph (or be so phrased that the possibility of reference to the teleomorph cannot be excluded).

59.3. If these requirements are not fulfilled, the name is that of a form-taxon and is applicable only to the anamorph represented by its type, as described or referred to in the protologue. The accepted taxonomic disposition of the type of the name determines the application of the name, no matter whether the genus to which a subordinate taxon is assigned by the author(s) is holomorphic or anamorphic.

59.4. The priority of names of holomorphs at any rank is not affected by the earlier publication of names of anamorphs judged to be correlated morphs of the holomorph.

59.5. The provisions of this article shall not be construed as preventing the publication and use of binary names for form-taxa when it is thought necessary or desirable to refer to anamorphs alone.

Note 1. When not already available, specific or infraspecific names for anamorphs may be proposed at the time of publication of the name for the holomorphic fungus or later. The epithets may, if desired, be identical, as long as they are not in homonymous combinations.

59.6. As long as there is direct and unambiguous evidence for the deliberate introduction of a new morph judged by the author(s) to be correlated with the morph typifying a purported basionym, and this evidence is strengthened by fulfilment of all requirements in Arts. 32-45 for valid publication of a name of a new taxon, any indication such as "comb. nov." or "nom. nov." is regarded as a formal error, and the name introduced is treated as that of a new taxon, and attributed solely to the author(s) thereof. When only the requirements for valid publication of a new combination (Arts. 33, 34) have been fulfilled, the name is accepted as such and based, in accordance with Art. 55, on the type of the declared or implicit basionym.

Ex. 1. The name *Penicillium brefeldianum* Dodge, based on teleomorphic and anamorphic material, is a valid and legitimate name of a holomorph, in spite of the attribution of the species to a form-genus. It is legitimately combined in a holomorphic genus as *Eupenicillium brefeldianum* (Dodge) Stolk & Scott. *P. brefeldianum* is not available for use in a restricted sense for the anamorph alone.

Ex. 2. The name *Ravenelia cubensis* Arthur & Johnston, based on a specimen bearing only uredinia (an anamorph), is a valid and legitimate name of an anamorph, in spite of the attribution of the species to a holomorphic genus. It is legitimately combined in a form-genus as *Uredo cubensis* (Arthur & Johnston) Cummins. *R. cubensis* is not available for use inclusive of the teleomorph.

Ex. 3. *Mycosphaerella aleuritidis* was published as "(Miyake) Ou comb. nov., syn. *Cercospora aleuritidis* Miyake" but with a Latin diagnosis of the teleomorph. The indication "comb. nov." is taken as a formal error, and *M. aleuritidis* Ou is accepted as a validly published new specific name for the holomorph, typified by the teleomorphic material described by Ou.

Ex. 4. Corticium microsclerotium was published in 1939 as "(Matz) Weber, comb. nov., syn. *Rhizoctonia microsclerotia* Matz" with a description, only in English, of the teleomorph. Because of Art. 36, this may not be considered as the valid publication of the name of a new species, and so *C. microsclerotium* (Matz) Weber must be considered a validly published and legitimate new combination based on the specimen of the anamorph that typifies its basionym. *C. microsclerotium* Weber, as published in 1951 with a Latin description and a teleomorphic type, is an illegitimate later homonym of the combination *C. microsclerotium* (Matz) Weber (1939), typified by an anamorph.

Ex. 5. Hypomyces chrysospermus Tul. (Ann. Sci. Nat. Bot. ser. 4, 13: 16. 1860), presented as the name of a holomorph without the indication "comb. nov." but with explicit reference to *Mucor chrysospermus* (Bull.) Bull. and *Sepedonium chrysospermum* (Bull.) Fr., which are names of its anamorph, is not to be considered as a new combination but as the name of a newly described species, with a teleomorphic type.

Recommendation 59A

59A.1. When a new morph of a fungus is described, it should be published either as a new taxon (e.g., gen. nov., sp. nov., var. nov.) whose name has a teleomorphic type, or as a new anamorph (anam. nov.) whose name has an anamorphic type.

59A.2. When in naming a new morph of a fungus the epithet of the name of a different, earlier described morph of the same fungus is used, the new name should be designated as the name of a new taxon or anamorph, as the case may be, but not as a new combination based on the earlier name.

SECTION 5. CHOICE OF NAMES WHEN THE RANK OF A TAXON IS CHANGED

Article 60

60.1. In no case does a name have priority outside its own rank (but see Art. 64.4).

Ex. 1. Campanula sect. *Campanopsis* R. Br. (Prodr. 561. 1810) as a genus is called *Wahlenbergia* Roth (1821), a name conserved against the taxonomic synonym *Cervicina* Delile (1813), and not *Campanopsis* (R. Br.) Kuntze (1891).

Ex. 2. Magnolia virginiana var. *foetida* L. (1753) when raised to specific rank is called *Magnolia grandiflora* L. (1759), not *M. foetida* (L.) Sarg. (1889).

Ex. 3. Lythrum intermedium Ledeb. (1822) when treated as a variety of *Lythrum salicaria* L. (1753) has been called *L. salicaria* var. *glabrum* Ledeb. (Fl. Ross. 2: 127. 1843), and hence may not be called *L. salicaria* var. *intermedium* (Ledeb.) Koehne (Bot. Jahrb. Syst. 1: 327. 1881).

Article 61

61.1. When a taxon at the rank of family or below is changed to another such rank, the correct name is the earliest legitimate one available in the new rank.

61A.1. When a family or subdivision of a family is changed in rank and no earlier legitimate name is available in the new rank, the name should be retained, and only its termination (*-aceae, -oideae, -eae, -inae*) altered, unless the resulting name would be a later homonym.

Ex. 1. The subtribe *Drypetinae* Pax (1890) (*Euphorbiaceae*) when raised to the rank of tribe was named *Drypeteae* (Pax) Hurusawa (1954); the subtribe *Antidesmatinae* Pax (1890) (*Euphorbiaceae*) when raised to the rank of subfamily was named *Antidesmatoideae* (Pax) Hurusawa (1954).

61A.2. When a section or a subgenus is raised in rank to a genus, or the inverse change occurs, the original name or epithet should be retained unless the resulting name would be contrary to this Code.

61A.3. When an infraspecific taxon is raised in rank to a species, or the inverse change occurs, the original epithet should be retained unless the resulting combination would be contrary to this Code.

61A.4. When an infraspecific taxon is changed in rank within the species, the original epithet should be retained unless the resulting combination would be contrary to this Code.

SECTION 6. REJECTION OF NAMES AND EPITHETS

Article 62

62.1. An epithet or a legitimate name must not be rejected merely because it is inappropriate or disagreeable, or because another is preferable or better known, or because it has lost its original meaning, or (in pleomorphic fungi with names governed by Art. 59) because the generic name does not accord with the morph represented by its type.

Ex. 1. The following changes are contrary to the rule: *Staphylea* to *Staphylis, Tamus* to *Thamnos, Thamnus,* or *Tamnus, Mentha* to *Minthe, Tillaea* to *Tillia, Vincetoxicum* to *Alexitoxicum;* and *Orobanche rapum* to *O. sarothamnophyta, O. columbariae* to *O. columbarihaerens, O. artemisiae* to *O. artemisiepiphyta.* All these modifications are to be rejected.

Ex. 2. *Ardisia quinquegona* Blume (1825) is not to be changed to *A. pentagona* A. DC. (1834), although the specific epithet *quinquegona* is a hybrid word (Latin and Greek) (see Rec. 23B.1(c)).

Ex. 3. The name *Scilla peruviana* L. is not to be rejected merely because the species does not grow in Peru.

Ex. 4. The name *Petrosimonia oppositifolia* (Pallas) Litv., based on *Polycnemum oppositifolium* Pallas, is not to be rejected merely because the species has leaves only partly opposite, and partly alternate, although there is another closely related species, *Petrosimonia brachiata* (Pallas) Bunge, having all its leaves opposite.

Ex. 5. *Richardia* L. is not to be changed to *Richardsonia,* as was done by Kunth, although the name was originally dedicated to the British botanist, Richardson.

Article 63

63.1. A name, unless conserved (Art. 14) or sanctioned under Art. 13.1(d), is illegitimate and is to be rejected if it was nomenclaturally superfluous when published, i.e. if the taxon to which it was applied, as circumscribed by its author, definitely included the holotype or all syntypes or the previously designated lectotype of a name which ought to have been adopted, or whose epithet ought to have been adopted, under the rules (but see Art. 63.3).

Ex. 1. The generic name *Cainito* Adanson (1763) is illegitimate because it was a superfluous name for *Chrysophyllum* L. (1753) which Adanson cited as a synonym.

Ex. 2. *Chrysophyllum sericeum* Salisb. (1796) is illegitimate, being a superfluous name for *C. cainito* L. (1753), which Salisbury cited as a synonym.

Ex. 3. On the other hand, *Salix myrsinifolia* Salisb. (1796) is legitimate, being explicitly based upon *S. myrsinites* of Hoffmann (Hist. Salic. Ill. 71. 1787), a misapplication of the name *S. myrsinites* L.

Ex. 4. *Picea excelsa* Link is illegitimate because it is based on *Pinus excelsa* Lam. (1778), a superfluous name for *Pinus abies* L. (1753). Under *Picea* the proper name is *Picea abies* (L.) H. Karsten.

Ex. 5. On the other hand, *Cucubalus latifolius* Miller and *C. angustifolius* Miller (1768) are not illegitimate names, although these species are now united with the species previously named *C. behen* L. (1753): *C. latifolius* Miller and *C. angustifolius* Miller as circumscribed by Miller did not include the type of *C. behen* L., which name he adopted for another independent species.

Note 1. The inclusion, with an expression of doubt, of an element in a new taxon, e.g. the citation of a name with a question mark, does not make the name of the new taxon nomenclaturally superfluous.

Ex. 6. The protologue of *Blandfordia grandiflora* R. Br. (1810) includes, in synonymy, "Aletris punicea. *Labill. nov. holl.* 1. *p.* 85. *t.* 111 ?", indicating that the new species might be the same as *Aletris punicea* previously published by Labillardière (1805). *Blandfordia grandiflora* is nevertheless a legitimate name.

Note 2. The inclusion, in a new taxon, of an element that was subsequently designated as the lectotype of a name which, so typified, ought to have been adopted, or whose epithet ought to have been adopted, does not in itself make the name of the new taxon illegitimate.

63.2. The inclusion of a type (see Art. 7) is here understood to mean the citation of the type specimen, the citation of an illustration of the type specimen, the citation of the type of a name, or the citation of the name itself unless the type is at the same time excluded either explicitly or by implication.

Ex. 7. Explicit exclusion of type: When publishing the name *Galium tricornutum*, Dandy (Watsonia 4: 47. 1957) cited *G. tricorne* Stokes (1787) pro parte as a synonym, but explicitly excluded the type of the latter name.

Ex. 8. Exclusion of type by implication: *Cedrus* Duhamel (1755) is a legitimate name even though *Juniperus* L. was cited as a synonym; only some of the species of *Juniperus* L. were included in *Cedrus* by Duhamel, and the differences between the two genera were discussed, *Juniperus* (including the type of its name) being recognized in the same work as an independent genus.

Ex. 9. *Tmesipteris elongata* Dangeard (Botaniste 2: 213. 1891) was published as a new species but *Psilotum truncatum* R. Br. was cited as a synonym. However, on the following page (214), *T. truncata* (R. Br.) Desv. is recognized as a different species and on p. 216 the two are distinguished in a key, thus showing that the meaning of the cited synonym was either "*P. truncatum* R. Br. pro parte" or "*P. truncatum* auct. non R. Br."

Ex. 10. *Solanum torvum* Sw. (Prodr. 47. 1788) was published with a new diagnosis but *S. indicum* L. (1753) was cited as a synonym. In accord with the practice in his Prodromus, Swartz indicated where the species was to be inserted in the latest edition [14, Murray] of the Systema Vegetabilium. *S. torvum* was to be inserted between species 26 (*S. insanum*) and 27 (*S. ferox*); the number of *S. indicum* in this edition of the Systema is 32. *S. torvum* is thus a legitimate name; the type of *S. indicum* is excluded by implication.

63.3. A name that was nomenclaturally superfluous when published is not illegitimate if its basionym is legitimate, or if it is based on the stem of a legitimate generic name. When published it is incorrect, but it may become correct later.

Ex. 11. *Chloris radiata* (L.) Sw. (1788), based on *Agrostis radiata* L. (1759), was nomenclaturally superfluous when published, since Swartz also cited *Andropogon fasciculatus* L. (1753) as a synonym. It is, however, the correct name in the genus *Chloris* for *Agrostis radiata* when *Andropogon fasciculatus* is treated as a different species, as was done by Hackel (in A. DC. & C. DC., Monogr. Phan. 6: 177. 1889).

Ex. 12. The generic name *Hordelymus* (Jessen) Jessen (1885), based on the legitimate *Hordeum* subg. *Hordelymus* Jessen (Deutschl. Gräser 202. 1863), was superfluous when published because its type, *Elymus europaeus* L., is also the type of *Cuviera* Koeler (1802). *Cuviera* Koeler has since been rejected in favour of its later homonym *Cuviera* DC., and *Hordelymus* (Jessen) Jessen can now be used as a correct name for the segregate genus containing *Elymus europaeus* L.

Note 3. In no case does a statement of parentage accompanying the publication of a name for a hybrid make the name superfluous (see Art. H.5).

Ex. 13. The name *Polypodium* ×*shivasiae* Rothm. (1962) was proposed for hybrids between *P. australe* and *P. vulgare* subsp. *prionodes*, while at the same time the author accepted *P.* ×*fontqueri* Rothm. (1936) for hybrids between *P. australe* and *P. vulgare* subsp. *vulgare*. Under Art. H.4.1, *P.* ×*shivasiae* is a synonym of *P.* ×*font-queri*; nevertheless, it is not a superfluous name.

Article 64

64.1. A name, unless conserved (Art. 14) or sanctioned under Art. 13.1(d), is illegitimate if it is a later homonym, that is, if it is spelled exactly like a name based on a different type that was previously and validly published for a taxon of the same rank.

Note 1. Even if the earlier homonym is illegitimate, or is generally treated as a synonym on taxonomic grounds, the later homonym must be rejected.

Ex. 1. The name *Tapeinanthus* Boiss. ex Bentham (1848), given to a genus of *Labiatae*, is a later homonym of *Tapeinanthus* Herbert (1837), a name previously and validly published for a genus of *Amaryllidaceae*. *Tapeinanthus* Boiss. ex Bentham is therefore rejected. It was renamed *Thuspeinanta* by T. Durand (1888).

Ex. 2. The name *Amblyanthera* Müll. Arg. (1860) is a later homonym of the validly published *Amblyanthera* Blume (1849) and is therefore rejected, although *Amblyanthera* Blume is now considered to be a synonym of *Osbeckia* L. (1753).

Ex. 3. The name *Torreya* Arnott (1838) is a nomen conservandum and is therefore not to be rejected because of the existence of the earlier homonym *Torreya* Raf. (1818).

Ex. 4. Astragalus rhizanthus Boiss. (1843) is a later homonym of the validly published name *Astragalus rhizanthus* Royle (1835) and it is therefore rejected, as was done by Boissier in 1849, who renamed it *A. cariensis.*

64.2. A sanctioned name is illegitimate if it is a later homonym of another sanctioned name (see also Art. 14 Note 2).

64.3. When two or more generic, specific, or infraspecific names based on different types are so similar that they are likely to be confused[1] (because they are applied to related taxa or for any other reason) they are to be treated as homonyms.

Ex. 5. Names treated as homonyms: *Astrostemma* Bentham and *Asterostemma* Decne.; *Pleuripetalum* Hooker and *Pleuropetalum* T. Durand; *Eschweilera* DC. and *Eschweileria* Boerl.; *Skytanthus* Meyen and *Scytanthus* Hooker.

Ex. 6. The three generic names *Bradlea* Adanson, *Bradleja* Banks ex Gaertner, and *Braddleya* Vell., all commemorating Richard Bradley, are treated as homonyms because only one can be used without serious risk of confusion.

Ex. 7. Kadalia Raf. and *Kadali* Adanson (both *Melastomataceae*) are treated as homonyms (Taxon 15: 287. 1966); *Acanthoica* Lohmann and *Acanthoeca* W. Ellis (both flagellates) are sufficiently alike to be considered homonyms (Taxon 22: 313. 1973); *Solanum saltiense* S. L. Moore and *S. saltense* (Bitter) C. Morton should be treated as homonyms (Taxon 22: 153. 1973).

Ex. 8. Epithets so similar that they are likely to be confused if combined under the same generic or specific name: *chinensis* and *sinensis; ceylanica* and *zeylanica; napaulensis, nepalensis,* and *nipalensis; polyanthemos* and *polyanthemus; macrostachys* and *macrostachyus; heteropus* and *heteropodus; poikilantha* and *poikilanthes; pteroides* and *pteroideus; trinervis* and *trinervius; macrocarpon* and *macrocarpum; trachycaulum* and *trachycaulon.*

Ex. 9. Names not likely to be confused: *Rubia* L. and *Rubus* L.; *Monochaete* Doell and *Monochaetum* (DC.) Naudin; *Peponia* Grev. and *Peponium* Engler; *Iria* (Pers.) Hedwig and *Iris* L.; *Desmostachys* Miers and *Desmostachya* (Stapf) Stapf; *Symphyostemon* Miers and *Symphostemon* Hiern; *Gerrardina* Oliver and *Gerardiina* Engler; *Durvillaea* Bory and *Urvillea* Kunth; *Peltophorus* Desv. (*Gramineae*) and *Peltophorum* (Vogel) Bentham (*Leguminosae*); *Senecio napaeifolius* (DC.) Schultz-Bip. and *S. napifolius* MacOwan (the epithets being derived respectively from *Napaea* and *Napus*); *Lysimachia hemsleyana* Oliver and *L. hemsleyi* Franchet (see, however, Rec. 23A.2); *Euphorbia peplis* L. and *E. peplus* L.; *Acanthococcus* Lagerh., an alga, and *Acanthococos* Barb. Rodr., a palm (see Taxon 18: 735. 1969).

Ex. 10. Names ruled (by the Berlin Congress, 1987) as not likely to be confused: *Cathayeia* Ohwi (1931) and *Cathaya* Chun & Kuang (1962), for which the General Committee, upon unanimous advice from the Committee for Spermatophyta, noted that *Cathayeia* (*Flacourtiaceae*) is a nomenclatural synonym of *Idesia* Maxim. (1866), nom. cons., and hence cannot be used, that even if used it is unlikely to appear in the same context as *Cathaya* (fossil *Pinaceae*), and that the two names have a different number of syllables (Taxon 36: 429. 1987); *Cristella* Pat. (1887; Fungi) and *Christella* H. Léveillé (1915; Pteridophyta), which were regarded by the Committee for Fungi and Lichens, by the Committee for Pteridophyta and, upon their advice, by the General Committee (Taxon 35: 551. 1986) not to be confusable since the older name is in disuse for taxonomic reasons, since the taxa are not closely related, and since the etymology of the names is different.

---

1) When it is doubtful whether names are sufficiently alike to be confused, a request for a decision may be submitted to the General Committee (see Division III) which will refer it for examination to the committee or committees for the appropriate taxonomic group or groups. A recommendation may then be put forward to an International Botanical Congress, and, if ratified, will become a binding decision (see Ex. 10).

Ex. 11. Names conserved against earlier names treated as homonyms (see App. III): *Lyngbya* Gomont (vs. *Lyngbyea* Sommerf.); *Columellia* Ruiz & Pavón (vs. *Columella* Lour.), both commemorating Columella, the Roman writer on agriculture; *Cephalotus* Labill. (vs. *Cephalotos* Adanson); *Simarouba* Aublet (vs. *Simaruba* Boehmer).

64.4. The names of two subdivisions of the same genus, or of two infraspecific taxa within the same species, even if they are of different rank, are treated as homonyms if they have the same epithet and are not based on the same type. The same epithet may be used for subdivisions of different genera, and for infraspecific taxa within different species.

Ex. 12. *Verbascum* sect. *Aulacosperma* Murb. (1933) is allowed, although there was already a *Celsia* sect. *Aulacospermae* Murb. (1926). This, however, is not an example to be followed, since it is contrary to Rec. 21B.2.

Ex. 13. The names *Andropogon sorghum* subsp. *halepensis* (L.) Hackel and *A. sorghum* var. *halepensis* (L.) Hackel (in A. DC & C.DC., Monogr. Phan. 6: 502. 1889) are legitimate, since both have the same type and the epithet may be repeated under Rec. 26A.1.

Ex. 14. *Anagallis arvensis* var. *caerulea* (L.) Gouan (Fl. Monsp. 30. 1765), based on *A. caerulea* L. (1759), makes illegitimate the combination *A. arvensis* subsp. *caerulea* Hartman (Sv. Norsk Exc.- Fl. 32. 1846), based on the later homonym *A. caerulea* Schreber (1771).

64.5. When two or more homonyms have equal priority, the first of them that is adopted in an effectively published text (Arts. 29-31) by an author who simultaneously rejects the other(s) is treated as having priority. Likewise, if an author in an effectively published text substitutes other names for all but one of these homonyms, the homonym for the taxon that is not renamed is treated as having priority.

Ex. 15. Linnaeus simultaneously published both *Mimosa* 10 *cinerea* (Sp. Pl. 517. 1753) and *Mimosa* 25 *cinerea* (Sp. Pl. 520. 1753). In 1759, he renamed species 10 *Mimosa cineraria* and retained the name *Mimosa cinerea* for species 25; *Mimosa cinerea* is thus a legitimate name for species 25.

Ex. 16. Rouy & Foucaud (Fl. France 2: 30. 1895) published the name *Erysimum hieraciifolium* var. *longisiliquum*, with two different types, for two different taxa under different subspecies. Only one of these names can be maintained.

Article 65

65.1. Consideration of homonymy does not extend to the names of taxa not treated as plants, except as stated below:

(a) Later homonyms of the names of taxa once treated as plants are illegitimate, even though the taxa have been reassigned to a different group of organisms to which this Code does not apply.

(b) A name originally published for a taxon other than a plant, even if validly published under Arts. 32-45 of this Code, is illegitimate if it becomes a homonym of a plant name when the taxon to which it applies is first treated as a plant (see also Art. 45.4).

Note 1. The International Code of Nomenclature of Bacteria provides that a bacterial name is illegitimate if it is a later homonym of a name of a taxon of bacteria, fungi, algae, protozoa, or viruses.

Article 66

[Article 66, dealing with illegitimate names of subdivisions of genera, was deleted by the Berlin Congress, 1987.]

Article 67

[Article 67, dealing with illegitimate specific and infraspecific names, was deleted by the Berlin Congress, 1987.]

Article 68

68.1. A specific name is not illegitimate merely because its epithet was originally combined with an illegitimate generic name, but is to be taken into consideration for purposes of priority if the epithet and the corresponding combination are in other respects in accordance with the rules.

Ex. 1. Agathophyllum A. L. Juss. (1789) is an illegitimate name, being a superfluous substitute for *Ravensara* Sonn. (1782). Nevertheless the name *A. neesianum* Blume (1851) is legitimate. Because Meisner (1864) cited *A. neesianum* as a synonym of his new *Mespilodaphne mauritiana* but did not adopt the epithet *neesiana, M. mauritiana* is a superfluous name and hence illegitimate.

68.2. An infraspecific name, autonyms excepted (Art. 26.1), may be legitimate even if its final epithet was originally placed under an illegitimate name.

68.3. The names of species and of subdivisions of genera assigned to genera whose names are conserved or sanctioned later homonyms, and which had earlier been assigned to the genera under the rejected homonyms, are legitimate under the conserved or sanctioned names without change of authorship or date if there is no other obstacle under the rules.

Ex. 2. Alpinia languas J. F. Gmelin (1791) and *Alpinia galanga* (L.) Willd. (1797) are to be accepted although *Alpinia* L. (1753), to which they were assigned by their authors, is rejected and the genus in which they are now placed is *Alpinia* Roxb. (1810), nom. cons.

Article 69

69.1. A name may be ruled as rejected if it has been widely and persistently used for a taxon or taxa not including its type. A name thus rejected, or its basionym if it has one, is placed on a list of nomina rejicienda (Appendix IV). Along with the listed names, all combinations based on them are similarly rejected, and none is to be used.

69.2. The list of rejected names will remain permanently open for additions and changes. Any proposal of an additional name must be accompanied by a detailed statement of the cases both for and against its rejection. Such proposals must be submitted to the General Committee (see Division III), which will refer them for examination to the committees for the various taxonomic groups (see also Art. 15 and Rec. 15A).

69.3. A name of a genus or species that has been widely and persistently used for a taxon or taxa not including its type and would, but for Art. 69.4, be the correct name of another taxon may also be conserved or rejected under Art. 14^1.

Note 1. The name proposed for conservation can be either the name that has been misapplied or a later homonym or synonym against which the misapplied name is rejected.

69.4. A name that has been widely and persistently used for a taxon or taxa not including its type is not to be used in a sense that conflicts with current usage unless and until a proposal to deal with it under Art. 14.1 or 69.1 has been submitted and rejected.

Article 70

[Article 70, dealing with discordant elements, was deleted by the Leningrad Congress, 1975.]

Article 71

[Article 71, dealing with monstrosities, was deleted by the Leningrad Congress, 1975.]

Article 72

72.1. A name rejected under Arts. 63-65 or 69 is replaced by the name that has priority (Art. 11) in the rank concerned. If none exists in any rank a new name must be chosen: *(a)* the taxon may be treated as new and another name published for it, or *(b)* if the illegitimate name is a later homonym, an avowed substitute (nomen novum) based on the same type as the rejected name may be published for it. If a name is available in another rank, one of the above alternatives may be chosen, or *(c)* a new combination, based on the name in the other rank, may be published.

---

1) The Berlin Congress (1987) ruled that names of genera and species previously rejected, or recommended for rejection, under Art. 69 are to be reconsidered by the Nomenclature Committees concerned which may, when appropriate, recommend conservation of the name that will best serve stability. Such names are to be listed in the appropriate Appendix of the Code.

72.2. Similar action is to be taken if transfer of an epithet of a legitimate name would result in a combination that cannot be validly published under Arts. 21.3 or 23.4.

Ex. 1. *Linum radiola* L. (1753) when transferred to the genus *Radiola* may not be named *Radiola radiola* (L.) H. Karsten (1882), as that combination is invalid (see Arts. 23.4 and 32.1(b)). The next oldest name, *L. multiflorum* Lam. (1779), is illegitimate, being a superfluous name for *L. radiola* L. Under *Radiola*, the species has been given the legitimate name *R. linoides* Roth (1788).

Note 1. When a new epithet is required, an author may adopt an epithet previously given to the taxon in an illegitimate name if there is no obstacle to its employment in the new position or sense; the resultant combination is treated as the name of a new taxon or as a nomen novum, as the case may be.

Ex. 2. The name *Talinum polyandrum* Hooker (1855) is illegitimate, being a later homonym of *T. polyandrum* Ruiz & Pavón (1798). When Bentham, in 1863, transferred *T. polyandrum* Hooker to *Calandrinia*, he called it *Calandrinia polyandra*. This name is treated as having priority from 1863, and should be cited as *Calandrinia polyandra* Bentham, not *C. polyandra* (Hooker) Bentham.

Ex. 3. *Cenomyce ecmocyna* Achar. (1810) is a superfluous name for *Lichen gracilis* L. (1753), and so is *Scyphophora ecmocyna* Gray (1821), the type of *L. gracilis* still being included. However, when proposing the combination *Cladonia ecmocyna*, Leighton (1866) explicitly excluded that type and thereby published a new, legitimate name, *Cladonia ecmocyna* Leighton.

Recommendation 72A

72A.1. Authors should avoid adoption of the epithet of an illegitimate name previously published for the same taxon.

CHAPTER VI. ORTHOGRAPHY OF NAMES AND EPITHETS AND GENDER OF GENERIC NAMES

SECTION 1. ORTHOGRAPHY OF NAMES AND EPITHETS

Article 73

73.1. The original spelling of a name or epithet is to be retained, except for the correction of typographic or orthographic errors and the standardizations imposed by Arts. 73.8 (compounding forms), 73.9 (hyphens), and 73.10 (terminations; see also Art. 32.5).

Ex. 1. Retention of original spelling: The generic names *Mesembryanthemum* L. (1753) and *Amaranthus* L. (1753) were deliberately so spelled by Linnaeus and the spelling is not to be altered to *Mesembrianthemum* and *Amarantus* respectively, although these latter forms are philologically preferable (see Bull. Misc. Inform. 1928: 113, 287). – *Phoradendron* Nutt. is not to be altered to *Phoradendrum*. – *Triaspis mozambica* Adr. Juss. is not to be altered to *T. mossambica*, as in Engler (Pflanzenw. Ost-Afrikas C: 232. 1895). – *Alyxia ceylanica* Wight is not to be altered to *A. zeylanica*, as in Trimen (Handb. Fl. Ceyl. 3: 127. 1895). – *Fagus sylvatica* L. is not to be altered to *F. silvatica*. The classical spelling *silvatica* is recommended for adoption in the case of a new name (Rec. 73E), but the mediaeval spelling *sylvatica* is not treated as an orthographic error. – *Scirpus cespitosus* L. is not to be altered to *S. caespitosus*.

Ex. 2. Typographic errors: *Globba brachycarpa* Baker (1890) and *Hetaeria alba* Ridley (1896) are typographic errors for *Globba trachycarpa* Baker and *Hetaeria alta* Ridley respectively (see J. Bot. 59: 349. 1921). – *Thevetia nereifolia* Adr. Juss. ex Steudel is an obvious typographic error for *T. neriifolia*.

Ex. 3. Orthographic error: *Gluta benghas* L. (1771), being an orthographic error for *G. renghas*, should be cited as *G. renghas* L., as has been done by Engler (in A. DC. & C. DC., Monogr. Phan. 4: 225. 1883); the vernacular name used as a specific epithet by Linnaeus is "Renghas", not "Benghas".

Note 1. Art. 14.10 provides for the conservation of an altered spelling of a generic name.

Ex. 4. *Bougainvillea* (see Appendix IIIA, Spermatophyta, no. 2350).

73.2. The words "original spelling" in this Article mean the spelling employed when the name was validly published. They do not refer to the use of an initial capital or small letter, this being a matter of typography (see Arts. 20.1 and 21.2, Rec. 73F).

73.3. The liberty of correcting a name is to be used with reserve, especially if the change affects the first syllable and, above all, the first letter of the name.

Ex. 5. The spelling of the generic name *Lespedeza* is not to be altered, although it commemorates Vicente Manuel de Céspedes (see Rhodora 36: 130-132, 390-392. 1934). – *Cereus jamacaru* DC. may not be altered to *C. mandacaru*, even if *jamacaru* is believed to be a corruption of the vernacular name "mandacaru".

73.4. The letters *w* and *y*, foreign to classical Latin, and *k*, rare in that language, are permissible in Latin plant names. Other letters and ligatures foreign to classical Latin that may appear in Latin plant names, such as the German *ß* (double *s*), are to be transcribed.

73.5. When a name or epithet has been published in a work where the letters *u, v* or *i, j* are used interchangeably or in any other way incompatible with modern practices (one of those letters is not used or only in capitals), those letters should be transcribed in conformity with modern botanical usage.

Ex. 6. *Uffenbachia* Fabr., not *Vffenbachia; Taraxacum* Zinn, not *Taraxacvm; Curculigo* Gaertner, not *Cvrcvligo.*

Ex. 7. *Geastrvm hygrometricvm* Pers. and *Vredo pvstvlata* Pers. (1801) should be written respectively *Geastrum hygrometricum* and *Uredo pustulata.*

Ex. 8. *Bromus iaponicus* Thunb. (1784) should be written *Bromus japonicus.*

73.6. Diacritical signs are not used in Latin plant names. In names (either new or old) drawn from words in which such signs appear, the signs are to be suppressed with the necessary transcription of the letters so modified; for example *ä, ö, ü* become respectively *ae, oe, ue; é, è, ê* become *e*, or sometimes *ae; ñ* becomes *n; ø* becomes *oe; å* becomes *ao*. The diaeresis, indicating that a vowel is to be pronounced separately from the preceding vowel (as in *Cephaëlis, Isoëtes*), and the ligatures *-æ-* and *-œ-* indicating that the letters are to be pronounced together (*Arisæma, Schœnus*), are permissible.

73.7. When changes made in orthography by earlier authors who adopt personal, geographic, or vernacular names in nomenclature are intentional latinizations, they are to be preserved, except for terminations covered by Art. 73.10.

Ex. 9. *Valantia* L. (1753), *Gleditsia* L. (1753), and *Clutia* L. (1753), commemorating Vaillant, Gleditsch, and Cluyt respectively, are not to be altered to *Vaillantia, Gleditschia,* and *Cluytia;* Linnaeus latinized the names of these botanists deliberately as "Valantius", "Gleditsius", and "Clutius".

Ex. 10. *Zygophyllum billardierii* DC. was named for J. J. H. de Labillardière (de la Billardière). The intended latinization is "Billardierius" (in nominative), but that termination is not acceptable under Art. 73.10 and the name is correctly spelled *Z. billardierei* DC.

73.8. The use of a compounding form contrary to Rec. 73G in an adjectival epithet is treated as an error to be corrected.

Ex. 11. Pereskia opuntiaeflora DC. is to be cited as *P. opuntiiflora* DC. However, in *Andromeda polifolia* L. (1753), the epithet is a pre-Linnean plant name ("*Polifolia*" Buxb.) used in apposition and not an adjective; it is not to be corrected to "*poliifolia*".

Ex. 12. Cacalia napeaefolia DC. and *Senecio napeaefolius* (DC.) Schultz-Bip. are to be cited as *Cacalia napaeifolia* DC. and *Senecio napaeifolius* (DC.) Schultz-Bip. respectively; the specific epithet refers to the resemblance of the leaves to those of the genus *Napaea* (not *Napea*), and the substitute (connecting) vowel *-i* should have been used instead of the genitive singular inflection *-ae*.

73.9. The use of a hyphen in a compound epithet is treated as an error to be corrected by deletion of the hyphen, except if an epithet is formed of words that usually stand independently, when a hyphen is permitted (see Arts. 23.1 and 23.3).

Ex. 13. Deletion of the hyphen: *Acer pseudoplatanus* L., not *A. pseudo-platanus; Ficus neoëbudarum* Summerh., not *F. neo-ebudarum; Lycoperdon atropurpureum* Vitt., not *L. atro-purpureum; Croton ciliatoglandulifer* Ortega, not *C. ciliato-glandulifer; Scirpus* sect. *Pseudoëriophorum* Jurtzer, not *S.* sect. *Pseudo-eriophorum.*

Ex. 14. Hyphen permitted: *Aster novae-angliae* L., *Coix lacryma-jobi* L., *Peperomia san-felipensis* J. D. Smith, *Arctostaphylos uva-ursi* (L.) Sprengel, *Veronica anagallis-aquatica* L. (Art. 23.3).

Note 2. Art. 73.9 refers only to epithets (in combinations), not to names of genera or taxa in higher ranks; a generic name published with a hyphen can be changed only by conservation.

Ex. 15. Pseudo-salvinia Piton (1940).

73.10. The use of a termination (for example *-i, -ii, -ae, -iae, -anus*, or *-ianus*) contrary to Rec. 73C.1 is treated as an error to be corrected (see also Art. 32.5).

Ex. 16. Rosa pissarti Carrière (Rev. Hort. 1880: 314) is a typographic error for *R. pissardi* (see Rev. Hort. 1881: 190), which in its turn is treated as an error for *R. pissardii* (see Rec. 73C.1(b)).

Note 3. If the gender and/or number of a substantival epithet derived from a personal name is inappropriate for the sex and/or number of the person(s) whom the name commemorates, the termination is to be corrected in conformity with Rec. 73C.1.

Ex. 17. Rosa ×*toddii* was named by Wolley-Dod (J. Bot. 69, suppl. 106. 1931) for "Miss E. S. Todd"; the epithet is to be corrected to *toddiae*

Ex. 18. Astragalus matthewsii, dedicated by Podlech and Kirchhoff (Mitt. Bot. Staatssamml. München 11: 432. 1974) to Victoria A. Matthews, is to be corrected to *A. matthewsiae* Podlech & Kirchhoff; it is not therefore a later homonym of *A. matthewsii* S. Watson (see Agerer-Kirchhoff & Podlech in Mitt. Bot. Staatssamml. München 12: 375. 1976).

Ex. 19. Codium geppii O. C. Schmidt (Biblioth. Bot. 23(91): 50. 1923), which commemorates "A. & E. S. Gepp", is to be corrected to *C. geppiorum.*

Recommendation 73A

73A.1. When a new name or epithet is to be derived from Greek, the transliteration to Latin should conform to classical usage.

73A.2. The spiritus asper should be transcribed in Latin as the letter *h.*

Recommendation 73B

73B.1. When a new name for a genus, subgenus, or section is taken from the name of a person, it should be formed as follows:

(*a*) When the name of the person ends in a vowel, the letter *-a* is added (thus *Ottoa* after Otto; *Sloanea* after Sloane), except when the name ends in *-a*, when *-ea* is added (e.g. *Collaea* after Colla), or in *-ea* (as *Correa*), when no letter is added.

(*b*) When the name of the person ends in a consonant, the letters *-ia* are added; when the name ends in *-er*, the terminations *-ia* and *-a* are both in use (e.g. *Sesleria* after Sesler and *Kernera* after Kerner).

(*c*) In latinized personal names ending in *-us* this termination is dropped (e.g. *Dillenia* after Dillenius) before applying the procedure described under (a) and (b).

(*d*) The syllables not modified by these endings retain their original spelling, unless they contain letters foreign to Latin plant names or diacritical signs (see Art. 73.6).

Note 1. Names may be accompanied by a prefix or a suffix, or be modified by anagram or abbreviation. In these cases they count as different words from the original name.

Ex. 1. Durvillaea and *Urvillea; Lapeirousia* and *Peyrousea; Englera, Englerastrum,* and *Englerella; Bouchea* and *Ubochea; Gerardia* and *Graderia; Martia* and *Martiusia.*

Recommendation 73C

73C.1. Modern personal names may be given Latin terminations and used to form specific and infraspecific epithets as follows (but see Rec. 73C.2):

(*a*) If the personal name ends in a vowel or *-er*, substantive epithets are formed by adding the genitive inflection appropriate to the sex and number of the person(s) honoured (e.g., *scopoli-i* for Scopoli (m), *fedtschenko-i* for Fedtschenko (m), *glaziou-i* for Glaziou (m), *lace-ae* for Lace (f), *hooker-orum* for the Hookers), except when the name ends in *-a*, in which case adding *-e* (singular) or *-rum* (plural) is appropriate (e.g. *triana-e* for Triana (m)).

(*b*) If the personal name ends in a consonant (except *-er*), substantive epithets are formed by adding *-i-* (stem augmentation) plus the genitive inflection appropriate to the sex and number of the person(s) honoured (e.g. *lecard-ii* for Lecard (m), *wilson-iae* for Wilson (f), *verlot-iorum* for the Verlot brothers, *braun-iarum* for the Braun sisters).

(*c*) If the personal name ends in a vowel, adjectival epithets are formed by adding *-an-* plus the nominative singular inflection appropriate to the gender of the generic name (e.g., *Cyperus heyne-anus* for Heyne, *Vanda lindley-ana* for Lindley, *Aspidium bertero-anum* for Bertero), except when the personal name ends in *-a* in which case *-n-* plus the appropriate inflection is added (e.g. *balansa-nus* (m), *balansa-na* (f), and *balansa-num* (n) for Balansa).

(*d*) If the personal name ends in a consonant, adjectival epithets are formed by adding *-i-* (stem augmentation) plus *-an-* (stem of adjectival suffix) plus the nominative singular inflection appropriate to the gender of the generic name (e.g. *Rosa webb-iana* for Webb, *Desmodium griffith-ianum* for Griffith, *Verbena hassler-iana* for Hassler).

Note 1. The hyphens in the above examples are used only to set off the total appropriate termination.

73C.2. Personal names already in Greek or Latin, or possessing a well-established latinized form, should be given their appropriate Latin genitive to form substantive epithets (e.g. *alexandri* from Alexander or Alexandre, *augusti* from Augustus or August or Auguste, *linnaei* from Linnaeus, *martii* from Martius, *beatricis* from Beatrix or Béatrice, *hectoris* from Hector). (However, modern personal names are subject to the provisions of Art. 73.10.) Treating modern names as if they were in Third Declension should be avoided (e.g. *munronis* from Munro, *richardsonis* from Richardson).

73C.3. In forming new epithets based on personal names the original spelling of the personal name should not be modified unless it contains letters foreign to Latin plant names or diacritical signs (see Arts. 73.4 and 73.6).

73C.4. Prefixes and particles ought to be treated as follows:

(a) The Scottish patronymic prefix "Mac", "Mc" or "M'", meaning "son of", should be spelled "mac" and united with the rest of the name, e.g. *macfadyenii* after Macfadyen, *macgillivrayi* after MacGillivray, *macnabii* after McNab, *mackenii* after M'Ken.

(b) The Irish patronymic prefix "O" should be united with the rest of the name or omitted, e.g. *obrienii, brienianus* after O'Brien, *okellyi* after O'Kelly.

(c) A prefix consisting of an article, e.g. le, la, l', les, el, il, lo, or containing an article e.g. du, de la, des, del, della, should be united to the name, e.g. *leclercii* after Le Clerc, *dubuyssonii* after DuBuysson, *lafarinae* after La Farina, *logatoi* after Lo Gato.

(d) A prefix to a surname indicating ennoblement or canonization should be omitted, e.g. *candollei* after de Candolle, *jussieui* after de Jussieu, *hilairei* after Saint-Hilaire, *remyi* after St. Rémy; in geographical epithets, however, "St." is rendered as *sanctus* (m) or *sancta* (f), e.g. *sancti-johannis*, of St. John, *sanctae-helenae*, of St. Helena.

(e) A German or Dutch prefix when it is normally treated as part of the family name, as often happens outside its country of origin, e.g. in the United States, may be included in the epithet, e.g. *vonhausenii* after Vonhausen, *vanderhoekii* after Vanderhoek, *vanbruntiae* after Mrs. Van Brunt, but should otherwise be omitted, e.g. *iheringii* after von Ihering, *martii* after von Martius, *steenisii* after van Steenis, *strassenii* after zu Strassen, *vechtii* after van der Vecht.

Recommendation 73D

73D.1. An epithet derived from a geographical name is preferably an adjective and usually takes the termination *-ensis, -(a)nus, -inus,* or *-icus.*

Ex. 1. Rubus quebecensis (from Quebec), *Ostrya virginiana* (from Virginia), *Eryngium amorginum* (from Amorgos), *Polygonum pensylvanicum* (from Pennsylvania).

Recommendation 73E

73E.1. A new epithet should be written in conformity with the original spelling of the word or words from which it is derived and in accordance with the accepted usage of Latin and latinization (see Art. 23.5).

Ex. 1. sinensis (not *chinensis*).

Recommendation 73F

73F.1. All specific and infraspecific epithets should be written with a small initial letter, although authors desiring to use capital initial letters may do so when the epithets are directly derived from the names of persons (whether actual or mythical), or are vernacular (or non-Latin) names, or are former generic names.

Recommendation 73G

73G.1. A compound name or an epithet which combines elements derived from two or more Greek or Latin words should be formed, as far as practicable, in accordance with classical usage (see Art. 73.8). This may be stated as follows:

(a) In a true compound, a noun or adjective in non-final position appears as a compounding form generally obtained by

    (1) removing the case ending of the genitive singular (Latin *-ae, -i, -us, -is;* Greek *-os, -es, -as, -ous* and the latter's equivalent *-eos*) and

    (2) before a consonant, adding a connecting vowel (*-i-* for Latin elements, *-o-* for Greek elements).

    (3) Exceptions are common, and one should review earlier usages of a particular compounding form.

(b) A pseudocompound is a noun or adjectival phrase treated as if it were a single compound word. In a pseudocompound, a noun or adjective in a non-final position appears as a word with a case ending, not as a modified stem. Examples are: *nidus-avis* (nest of bird), *Myos-otis* (ear of mouse), *cannae-folius* (leaf of canna), *albo-marginatus* (margined with white), etc. In epithets where tingeing is expressed, the modifying initial colour often is in the ablative because the preposition *e, ex,* is implicit, e.g., *atropurpureus* (blackish purple) from *ex atro purpureus* (purple tinged with black). Others have been deliberately introduced to reveal etymological differences when different word elements have the same compounding forms, such as *tubi-* from tube (*tubus, tubi,* stem *tubo-*) or from trumpet (*tuba, tubae,* stem *tuba-*) where *tubaeflorus* can only mean trumpet-flowered; also *carici-* is the compounding form from both papaya (*carica, caricae,* stem *carica-*) and sedge (*carex, caricis,* stem *caric-*) where *caricaefolius* can only mean papaya-leaved. The latter use of the genitive singular of the first declension for pseudocompounding is treated as an error to be corrected unless it makes an etymological distinction.

(c) Some common irregular forms are used in compounding. Examples are *hydro-* and *hydr-* (*Hydro-phyllum*) where the regular noun stem is *hydat-; calli-* (*Calli-stemon*) where the regular adjective stem is *calo-;* and *meli-* (*Meli-osma, Meli-lotus*) where the regular noun stem is *melit-*.

Note 1. The hyphens in the above examples are given solely for explanatory reasons. For the use of hyphens in botanical names and epithets see Arts. 20.3, 23.1, and 73.9.

Recommendation 73H

73H.1. Epithets of fungus names derived from the generic name of the host plant should be spelled in accordance with the accepted spelling of this name; other spellings are regarded as orthographic variants to be corrected (see Art. 75).

Ex. 1. Phyllachora anonicola Chardon is to be altered to *P. annonicola,* since the spelling *Annona* is now accepted in preference to *Anona.* – *Meliola albizziae* Hansford & Deighton is to be altered to *M. albiziae,* since the spelling *Albizia* is now accepted in preference to *Albizzia.*

Recommendation 73I

73I.1 The etymology of new names and epithets should be given when the meaning of these is not obvious.

Article 74

[Article 74, dealing with variant spellings of Linnaean generic names, was deleted by the Sydney Congress, 1981 (but see Art. 13.4).]

Article 75

75.1. Only one orthographic variant of any one name is treated as validly published, the form which appears in the original publication except as provided in Art. 73 (orthographic and typographic errors), Art. 14.10 (conserved spellings), and Art. 32.5 (incorrect Latin terminations).

Note 1. Orthographic variants are the various spelling, compounding, and inflectional forms of a name or epithet (including typographic errors), only one type being involved.

75.2. If orthographic variants of a name appear in the original publication, the one that conforms to the rules and best suits the recommendations of Art. 73 is to be retained; otherwise the first author who, in an effectively published text (Arts. 29-30), explicitly adopts one of the variants, rejecting the other(s), must be followed.

75.3. The orthographic variants of a name are to be automatically corrected to the validly published form of that name. Whenever such a variant appears in print, it is to be treated as if it were printed in its corrected form.

Note 2. In full citations it is desirable that the original form of an automatically corrected orthographic variant of a name be added (Rec. 50F).

75.4. Confusingly similar names based on the same type are treated as orthographic variants. (For confusingly similar names based on different types, see Art. 64.3.)

Ex. 1. Geaster Fr. (1829) and *Geastrum* Pers. (1794) : Pers. (1801) are similar names with the same type (Taxon 33: 498. 1984); they are treated as orthographic variants despite the fact that they are derived from two different nouns, *aster* (*asteris*) and *astrum* (*astri*).

SECTION 2. GENDER OF GENERIC NAMES

Article 76

76.1. A generic name retains the gender assigned by its author, unless this is contrary to botanical tradition. The following names must be treated as feminine in accordance with botanical tradition, irrespective of classical usage or the author's original usage: *Adonis, Diospyros, Hemerocallis, Orchis, Stachys,* and *Strychnos. Lotus* and *Melilotus* must be treated as masculine.

Note 1. Botanical tradition usually maintains the classical gender of a Greek or Latin word, when this was well established.

Ex. 1. Although their ending suggests masculine gender, *Cedrus* and *Fagus* are feminine like most other classical tree names; similarly, *Rhamnus* is feminine, despite the fact that Linnaeus gave it masculine gender. *Eucalyptus*, a neologism, is also feminine, retaining the gender assigned by its author. *Phyteuma* (neuter), *Sicyos* (masculine), and *Erigeron* (masculine) are other names for which botanical usage has reestablished the classical gender despite another choice by Linnaeus. The classical gender of *Atriplex* varied (e.g. feminine in Columella, neuter in Pliny) and Linnaeus' choice of feminine gender stands.

76.2. Compound generic names take the gender of the last word in the nominative case in the compound. If the termination is altered, however, the gender is altered accordingly.

(a) Modern compounds ending in *-codon, -myces, -odon, -panax, -pogon, -stemon*, and other masculine words are masculine, irrespective of the fact that the generic names *Andropogon* L. and *Oplopanax* (Torrey & A. Gray) Miq. were originally treated as neuter by their authors.

(b) Similarly, all modern compounds ending in *-achne, -chlamys, -daphne, -mecon, -osma* (the modern transcription of the feminine Greek word osmé) and other feminine words are feminine, irrespective of the fact that *Dendromecon* Bentham and *Hesperomecon* E. Greene were originally ascribed the neuter gender. An exception is made in the case of names ending in *-gaster*, which strictly speaking ought to be feminine, but which are treated as masculine in accordance with botanical tradition.

(c) Similarly, all modern compounds ending in *-ceras, -dendron, -nema, -stigma, -stoma* and other neuter words are neuter, irrespective of the fact that Robert Brown and Bunge respectively made *Aceras* and *Xanthoceras* feminine. An exception is made for names ending in *-anthos* (or *-anthus*) and *-chilos* (*-chilus* or *-cheilos*), which ought to be neuter, since that is the gender of the Greek words anthos and cheilos, but are treated as masculine in accordance with botanical tradition.

Ex. 2. Compound generic names in which the termination of the last word is altered: *Stenocarpus, Dipterocarpus*, and all other modern compounds ending in the Greek masculine *-carpos* (or *-carpus*), e.g. *Hymenocarpos*, are masculine; those in *-carpa* or *-carpaea*, however, are feminine, e.g. *Callicarpa* and *Polycarpaea*; and those in *-carpon, -carpum*, or *-carpium* are neuter, e.g. *Polycarpon, Ormocarpum*, and *Pisocarpium*.

76.3. Arbitrarily formed generic names or vernacular names or adjectives used as generic names, whose gender is not apparent, take the gender assigned to them by their authors. If the original author failed to indicate the gender, the next subsequent author may choose a gender, and his choice, if effectively published (Arts. 29-31), is to be accepted.

Ex. 3. Taonabo Aublet is feminine: Aublet's two species were *T. dentata* and *T. punctata.*

Ex. 4. Agati Adanson was published without indication of gender: the feminine gender was assigned to it by Desvaux (J. Bot. Agric. 1: 120. 1813), who was the first subsequent author to adopt the name in an effectively published text, and his choice is to be accepted.

Ex. 5. Boehmer (in Ludwig, Def. Gen. Pl. ed. 3. 436. 1760) and Adanson (Fam. Pl. 2: 356. 1763) failed to indicate the gender of *Manihot.* Crantz (Inst. Rei Herb. 1: 167. 1766) was the first author who, by publishing the names *Manihot gossypiifolia*, etc., indicated the gender of *Manihot,* and *Manihot* is therefore to be treated as feminine.

76.4. Generic names ending in *-oides* or *-odes* are treated as feminine and those ending in *-ites* as masculine, irrespective of the gender assigned to them by the original author.

Recommendation 76A

76A.1. When a genus is divided into two or more genera, the gender of the new generic name or names should be that of the generic name that is retained.

Ex. 1. When *Boletus* is divided, the gender of the new generic names should be masculine: *Xerocomus, Boletellus,* etc.

DIVISION III. PROVISIONS FOR MODIFICATION OF THE CODE

Div.III.1. Modification of the Code. The Code may be modified only by action of a plenary session of an International Botanical Congress on a resolution moved by the Nomenclature Section of that Congress.[1]

Div.III.2. Nomenclature Committees. Permanent Nomenclature Committees are established under the auspices of the International Association for Plant Taxonomy. Members of these committees are elected by an International Botanical Congress. The Committees have power to co-opt and to establish subcommittees; such officers as may be desired are elected.

(1) General Committee, composed of the secretaries of the other committees, the rapporteur-général, the president and the secretary of the International Association for Plant Taxonomy, and at least 5 members to be appointed by the Nomenclature Section. The rapporteur-général is charged with the presentation of nomenclature proposals to the International Botanical Congress.

(2) Committee for Spermatophyta.

(3) Committee for Pteridophyta.

(4) Committee for Bryophyta.

(5) Committee for Fungi and Lichens.

(6) Committee for Algae.

(7) Committee for Hybrids.

(8) Committee for Fossil Plants.

(9) Editorial Committee, charged with the preparation and publication of the Code in conformity with the decisions adopted by the International Botanical Congress. Chairman: the rapporteur-général of the previous Congress, who is charged with the general duties in connection with the editing of the Code.

---

1) In the event that there should not be another International Botanical Congress, authority for the International Code of Botanical Nomenclature shall be transferred to the International Union of Biological Sciences or to an organization at that time corresponding to it. The General Committee is empowered to define the machinery to achieve this.

Div.III.3. The Bureau of Nomenclature of the International Botanical Congress. Its officers are: *(1)* the president of the Nomenclature Section, elected by the organizing committee of the International Botanical Congress in question; *(2)* the recorder, appointed by the same organizing committee; *(3)* the rapporteur-général, elected by the previous Congress; *(4)* the vice-rapporteur, elected by the organizing committee on the proposal of the rapporteur-général.

Div.III.4. The voting on nomenclature proposals is of two kinds: *(a)* a preliminary guiding mail vote and *(b)* a final and binding vote at the Nomenclature Section of the International Botanical Congress.

Qualifications for voting:

(a) Preliminary mail vote:

> *(1)* The members of the International Association for Plant Taxonomy.

> *(2)* The authors of proposals.

> *(3)* The members of the nomenclature committees.

Note 1. No accumulation or transfer of personal votes is permissible.

(b) Final vote at the sessions of the Nomenclature Section:

> *(1)* All officially enrolled members of the Section. No accumulation or transfer of personal votes is permissible.

> *(2)* Official delegates or vice-delegates of the institutes appearing on a list drawn up by the Bureau of Nomenclature of the International Botanical Congress and submitted to the General Committee for final approval; such institutes are entitled to 1-7 votes, as specified on the list.[1] Transfer of institutional votes to specified vice-delegates is permissible, but no single person will be allowed more than 15 votes, his personal vote included. Institutional votes may be deposited at the Bureau of Nomenclature to be counted in a specified way for specified proposals.

---

1) The Sydney Congress directed that no single institution, even in the wide sense of the term, shall be entitled to more than 7 votes.

APPENDIX I

NAMES OF HYBRIDS

Article H.1

H.1.1. Hybridity is indicated by the use of the multiplication sign x, or by the addition of the prefix "notho-"[1] to the term denoting the rank of the taxon.

Article H.2

H.2.1. A hybrid between named taxa may be indicated by placing the multiplication sign between the names of the taxa; the whole expression is then called a hybrid formula.

Ex. 1. Agrostis L. x *Polypogon* Desf.; *Agrostis stolonifera* L. x *Polypogon monspeliensis* (L.) Desf.; *Salix aurita* L. x *S. caprea* L.; *Mentha aquatica* L. x *M. arvensis* L. x *M. spicata* L.; *Polypodium vulgare* subsp. *prionodes* Rothm. x subsp. *vulgare.*

Recommendation H.2A

H.2A.1. It is usually preferable to place the names or epithets in a formula in alphabetical order. The direction of a cross may be indicated by including the sexual symbols (♀ : female; ♂: male) in the formula, or by placing the female parent first. If a non-alphabetical sequence is used, its basis should be clearly indicated.

Article H.3

H.3.1. Hybrids between representatives of two or more taxa may receive a name. For nomenclatural purposes, the hybrid nature of a taxon is indicated by placing the multiplication sign x before the name of an intergeneric hybrid or before the epithet in the name of an interspecific hybrid, or by prefixing the term "notho-" (optionally abbreviated "n-") to the term denoting the rank of the taxon (see Arts. 3.2 and 4.3). All such taxa are designated nothotaxa.

---

1) From the Greek *nothos,* meaning hybrid.

Ex. 1. (The putative or known parentage is found in Art. H.2, Ex.1.) x*Agropogon* P. Fourn.; x*Agropogon littoralis* (Smith) C. E. Hubb.; *Salix* x*capreola* Kerner ex Andersson; *Mentha* x*smithiana* R. A. Graham; *Polypodium vulgare* nothosubsp. *mantoniae* (Rothm.) Schidlay.

H.3.2. A nothotaxon cannot be designated unless at least one parental taxon is known or can be postulated.

H.3.3. The epithet in the name of a nothospecies is termed a collective epithet.

H.3.4. For purposes of homonymy and synonymy the multiplication sign and the prefix "notho-" are disregarded.

Ex. 2. x*Hordelymus* Bacht. & Darevskaja (1950) (= *Elymus* L. x *Hordeum* L.) is a later homonym of *Hordelymus* (Jessen) Jessen (1885).

Note 1. Taxa which are believed to be of hybrid origin need not be designated as nothotaxa.

Ex. 3. The true-breeding tetraploid raised from the artificial cross *Digitalis grandiflora* L. x *D. purpurea* L. may, if desired, be referred to as *D. mertonensis* Buxton & Darl.; *Triticum aestivum* L. is treated as a species although it is not found in nature and its genome has been shown to be composed of those of *T. monococcum, Aegilops speltoides,* and *A. squarrosa*; the taxon known as *Phlox divaricata* subsp. *laphamii* (Wood) Wherry is believed by Levin (Evolution 21: 92-108. 1967) to be a stabilized product of hybridization between *P. divaricata* L. subsp. *divaricata* and *P. pilosa* subsp. *ozarkana* Wherry; *Rosa canina* L., a polyploid believed to be of ancient hybrid origin, is treated as a species.

Note 2. The term "collective epithet" is used in the International Code of Nomenclature for Cultivated Plants-1980 to include also epithets in modern language.

Recommendation H.3A

H.3A.1. The multiplication sign in the name of a nothotaxon should be placed against the initial letter of the name or epithet. However, if the mathematical symbol is not available and the letter *x* is used instead, a single letter space may be left between it and the epithet if this helps to avoid ambiguity. The letter *x* should be in lower case.

Article H.4

H.4.1. When all the parent taxa can be postulated or are known, a nothotaxon is circumscribed so as to include all individuals (as far as they can be recognized) derived from the crossing of representatives of the stated parent taxa (i.e. not only the F_1 but subsequent filial generations and also back-crosses and combinations of these). There can thus be only one correct name corresponding to a particular hybrid formula; this is the earliest legitimate name (see Art. 6.3) in the appropriate rank (Art. H.5), and other names to which the same hybrid formula applies are synonyms of it.

Ex. 1. The names *Oenothera* x*wienii* Renner ex Rostański (1977) and *O.* x*hoelscheri* Renner ex Rostański (1968) are both considered to apply to the hybrid *O. rubricaulis* x *O. depressa;* the types of the two nothospecific names are known to differ by a whole gene-complex; nevertheless, the later name is treated as a synonym of the earlier.

Note 1. Variation within nothospecies and nothotaxa of lower rank may be treated according to Art. H.12 or, if appropriate, according to the International Code of Nomenclature for Cultivated Plants-1980.

Article H.5

H.5.1. The appropriate rank of a nothotaxon is that of the postulated or known parent taxa.

H.5.2. If the postulated or known parent taxa are of unequal rank the appropriate rank of the nothotaxon is the lowest of these ranks.

Note 1. When a taxon is designated by a name in a rank inappropriate to its hybrid formula, the name is incorrect in relation to that hybrid formula but may nevertheless be correct, or may become correct later (see also Art. 63 Note 3).

Ex. 1. The combination *Elymus* ×*laxus* (Fries) Melderis & D. McClintock, based on *Triticum laxum* Fries, was published for hybrids with the formula *E. farctus* subsp. *boreoatlanticus* (Simonet & Guinochet) Melderis × *E. repens* (L.) Gould, so that the combination is in a rank inappropriate to the hybrid formula. It is, however, the correct name applicable to all hybrids between *E. farctus* (Viv.) Melderis and *E. repens.*

Ex. 2. Radcliffe-Smith incorrectly published the nothospecific name *Euphorbia* ×*cornubiensis* for *E. amygdaloides* L. × *E. characias* subsp. *wulfenii* (Koch) A. R. Sm., although the correct designation for hybrids between *E. amygdaloides* and *E. characias* is *E.* ×*martini* Rouy; later, he remedied his mistake by publishing the combination *E.* ×*martini* nothosubsp. *cornubiensis* (A. R. Sm.) A. R. Sm. However, the name *E.* ×*cornubiensis* is potentially correct for hybrids with the formula *E. amygdaloides* × *E. wulfenii.*

Recommendation H.5A

H.5A.1. When publishing a name of a new nothotaxon at the rank of species or below, authors should provide any available information on the taxonomic identity, at lower ranks, of the known or postulated parent plants of the type of the name.

Article H.6

H.6.1. A nothogeneric name (i.e. the name at generic rank for a hybrid between representatives of two or more genera) is a condensed formula or is equivalent to a condensed formula.

H.6.2. The nothogeneric name of a bigeneric hybrid is a condensed formula in which the names adopted for the parental genera are combined into a single word, using the first part or the whole of one, the last part or the whole of the other (but not the whole of both) and, if desirable, a connecting vowel.

Ex. 1. ×*Agropogon* P. Fourn. (= *Agrostis* × *Polypogon*); ×*Gymnanacamptis* Asch. & Graebner (= *Anacamptis* × *Gymnadenia*); ×*Cupressocyparis* Dallimore (= *Chamaecyparis* × *Cupressus);* ×*Seleniphyllum* Rowley (= *Epiphyllum* × *Selenicereus).*

Ex. 2. x*Amarcrinum* Coutts (1925) is correct for *Amaryllis* L. x *Crinum* L., not x*Crindonna* Ragion. (1921). The latter name was proposed for the same nothogenus, but was formed from the generic name adopted for one parent (*Crinum*) and a synonym (*Belladonna* Sweet) of the generic name adopted for the other (*Amaryllis*). Being contrary to Art. H.6, it is not validly published under Art. 32.1(b).

Ex. 3. The name x*Leucadenia* Schlechter is correct for *Leucorchis* E. Meyer x *Gymnadenia* R. Br., but if the generic name *Pseudorchis* Séguier is adopted instead of *Leucorchis*, x*Pseudadenia* P. Hunt is correct.

Ex. 4. x*Aporophyllum* Johnson when first published was defined as *Aporocactus* x members of the "Orchid Cacti". The latter constitute the epicacti ("epiphyllums" of horticulture) – a complex descended from 4 or 5 separate genera. This name is hence not validly published (Art. 32.1(b)) because it conflicts with Art. H.6.3. For the bigeneric hybrid *Aporocactus* x *Epiphyllum* a different name applies(x*Aporepiphyllum* Rowley).

Ex. 5. Boivin (1967) published x*Maltea* for what he considered to be the intergeneric hybrid *Phippsia* x *Puccinellia*. As this is not a condensed formula, the name cannot be used for that intergeneric hybrid, for which the correct name is x*Pucciphippsia* Tzvelev (1971). Boivin did, however, provide a Latin description and designate a type; consequently, *Maltea* is a validly published generic name and is correct if its type is treated as belonging to a separate genus, not to a nothogenus.

H.6.3. The nothogeneric name of an intergeneric hybrid derived from four or more genera is formed from the name of a person to which is added the termination -*ara*; no such name may exceed eight syllables. Such a name is regarded as a condensed formula.

Ex. 6. x*Potinara* Charlesworth & Co. (= *Brassavola* x *Cattleya* x *Laelia* x *Sophronitis*).

H.6.4. The nothogeneric name of a trigeneric hybrid is either *(a)* a condensed formula in which the three names adopted for the parental genera are combined into a single word not exceeding eight syllables, using the whole or first part of one, followed by the whole or any part of another, followed by the whole or last part of the third (but not the whole of all three) and, if desirable, one or two connecting vowels, or *(b)* a name formed like that of a nothogenus derived from four or more genera, i.e., from a personal name to which is added the termination -*ara*.

Ex. 7. x*Sophrolaeliocattleya* Hurst (= *Cattleya* x *Laelia* x *Sophronitis*); x*Vascostylis* Takakura (= *Ascocentrum* x *Rhynchostylis* x *Vanda*); x*Rodrettiopsis* Moir (= *Comparettia* x *Ionopsis* x *Rodriguezia*); x*Wilsonara* Charlesworth & Co. (= *Cochlioda* x *Odontoglossum* x *Oncidium*).

Recommendation H.6A

H.6A.1. When a nothogeneric name is formed from the name of a person by adding the termination -*ara*, that person should preferably be a collector, grower, or student of the group.

Article H.7

H.7.1. The name of a nothotaxon which is a hybrid between subdivisions of a genus is a combination of an epithet, which is a condensed formula formed in the same way as a nothogeneric name (Art. H.6.2), with the name of the genus.

Ex. 1. Ptilostemon nothosect. *Platon* Greuter (Boissiera 22: 159. 1973), comprising hybrids between *Ptilostemon* sect. *Platyrhaphium* Greuter and *P.* sect. *Ptilostemon; Ptilostemon* nothosect. *Plinia* Greuter (Boissiera 22: 158. 1973), comprising hybrids between *Ptilostemon* sect. *Platyrhaphium* and *P.* sect. *Cassinia* Greuter.

Article H.8

H.8.1. When the name or the epithet in the name of a nothotaxon is a condensed formula (Arts. H.6 and H.7), the parental names used in its formation must be those which are correct for the particular circumscription, position, and rank accepted for the parental taxa.

Ex. 1. If the genus *Triticum* L. is interpreted on taxonomic grounds as including *Triticum* (s. str.) and *Agropyron* Gaertner, and the genus *Hordeum* L. as including *Hordeum* (s. str.) and *Elymus* L., then hybrids between *Agropyron* and *Elymus* as well as between *Triticum* (s. str.) and *Hordeum* (s. str.) are placed in the same nothogenus, x*Tritordeum* Asch. & Graebner (1902). If, however, *Agropyron* is separated generically from *Triticum,* hybrids between *Agropyron* and *Hordeum* (s. str. or s. lat.) are placed in the nothogenus x*Agrohordeum* A. Camus (1927). Similarly, if *Elymus* is separated generically from *Hordeum,* hybrids between *Elymus* and *Triticum* (s. str. or s. lat.) are placed in the nothogenus x*Elymotriticum* P. Fourn. (1935). If both *Agropyron* and *Elymus* are given generic rank, hybrids between them are placed in the nothogenus x*Agroelymus* A. Camus (1927); x*Tritordeum* is then restricted to hybrids between *Hordeum* (s. str.) and *Triticum* (s. str.), and hybrids between *Elymus* and *Hordeum* are placed in x*Elyhordeum* Mansf. ex Tsitsin & Petrova (1955), a substitute name for x*Hordelymus* Bacht. & Darevskaja (1950) non *Hordelymus* (Jessen) Jessen (1885).

H.8.2. Names ending in *-ara* for nothogenera, which are equivalent to condensed formulae (Art. H.6.3-4), are applicable only to plants which are accepted taxonomically as derived from the parents named.

Ex. 2. If *Euanthe* is recognized as a distinct genus, hybrids simultaneously involving its only species, *E. sanderiana,* and the three genera *Arachnis, Renanthera,* and *Vanda* must be placed in x*Cogniauxara* Garay & H. Sweet; if on the other hand *E. sanderiana* is included in *Vanda,* the same hybrids are placed in x*Holttumara* hort. (*Arachnis* x *Renanthera* x *Vanda*).

Article H.9

H.9.1. In order to be validly published, the name of a nothogenus or of a nothotaxon with the rank of subdivision of a genus (Arts. H.6 and H.7) must be effectively published (see Art. 29) with a statement of the names of the parent genera or subdivisions of genera, but no description or diagnosis is necessary, whether in Latin or in any other language.

Ex. 1. Validly published names: x*Philageria* Masters (1872), published with a statement of parentage, *Lapageria* x *Philesia; Eryngium* nothosect. *Alpestria* Burdet & Miège, pro sect. (Candollea 23: 116. 1968), published with a statement of its parentage, *Eryngium* sect. *Alpina* x sect. *Campestria;* x*Agrohordeum* A. Camus (1927) (= *Agropyron* Gaertner x *Hordeum* L.), of which x*Hordeopyron* Simonet (1935, "*Hordeopyrum*") is a later synonym.

Note 1. Since the names of nothogenera and nothotaxa with the rank of a subdivision of a genus are condensed formulae or treated as such, they do not have types.

Ex. 2. The name ×*Ericalluna bealei* Krüssm. (1960) was published for plants which were thought to be variants of the cross *Calluna vulgaris* x *Erica cinerea*. If it is considered that these are not hybrids, but are forms of *Erica cinerea*, the name ×*Ericalluna* Krüssm. remains available for use if and when known or postulated plants of *Calluna* x *Erica* should appear.

Ex. 3. ×*Arabidobrassica* Gleba & Fr. Hoffm. (Naturwissenschaften 66: 548. 1979), a nothogeneric name which was validly published with a statement of parentage for the result of somatic hybridization by protoplast fusion of *Arabidopsis thaliana* with *Brassica campestris*, is also available for intergeneric hybrids resulting from normal crosses between *Arabidopsis* and *Brassica*, should any be produced.

Note 2. However, names published merely in anticipation of the existence of a hybrid are not validly published under Art. 34.1(b).

Article H.10

H.10.1. Names of nothotaxa at the rank of species or below must conform with the provisions *(a)* in the body of the Code applicable to the same ranks and *(b)* in Art. H.3. Infringements of Art. H.3.1. are treated as errors to be corrected.

H.10.2. Taxa previously published as species or infraspecific taxa which are later considered to be nothotaxa may be indicated as such, without change of rank, in conformity with Arts. 3 and 4 and by the application of Art. 50 (which also operates in the reverse direction).

H.10.3. The following are considered to be formulae and not true epithets: designations consisting of the epithets of the names of the parents combined in unaltered form by a hyphen, or with only the termination of one epithet changed, or consisting of the specific epithet of the name of one parent combined with the generic name of the other (with or without change of termination).

Ex. 1. The designation *Potentilla atrosanguinea-pedata* published by Maund (Bot. Gard. 5: no. 385, t. 97. 1833) is considered to be a formula meaning *Potentilla atrosanguinea* Lodd. ex D. Don x *P. pedata* Nestler.

Ex. 2. *Verbascum nigro-lychnitis* Schiede (Pl. Hybr. 40. 1825) is considered to be a formula, *Verbascum lychnitis* L. x *V. nigrum* L.; the correct binary name for this hybrid is *Verbascum* ×*schiedeanum* Koch (1844).

Ex. 3. The following names include true epithets: *Acaena* ×*anserovina* Orch. (1969) (from *anserinifolia* and *ovina*); *Micromeria* ×*benthamineolens* Svent. (1969) (from *benthamii* and *pineolens*).

Note 1. Since the name of a nothotaxon at the rank of species or below has a type, statements of parentage play a secondary part in determining the application of the name.

Ex. 4. *Quercus* ×*deamii* Trel. was described as *Q. alba* L. x *Q. muehlenbergii* Engelm. However, progeny grown from acorns from the type tree led Bartlett to conclude that the parents were in fact *Q. macrocarpa* Michx. and *Q. muehlenbergii*. If this conclusion is accepted, the name *Q.* ×*deamii* applies to *Q. macrocarpa* x *Q. muehlenbergii*, and not to *Q. alba* x *Q. muehlenbergii*.

H.10A.1. In forming epithets for nothotaxa at the rank of species and below, authors should avoid combining parts of the epithets of the names of the parents.

H.10B.1. When contemplating the publication of new names for hybrids between named infraspecific taxa, authors should carefully consider whether they are really needed, bearing in mind that formulae, though more cumbersome, are more informative.

Article H.11

H.11.1. The name of a nothospecies of which the postulated or known parent species belong to different genera is a combination of a nothospecific (collective) epithet with a nothogeneric name.

Ex. 1. x*Heucherella tiarelloides* (Lemoine) Wehrh. ex Stearn (considered to be *Heuchera* x*brizoides* hort. x*Tiarella cordifolia* L., for which *Heuchera* x*tiarelloides* Lemoine is incorrect).

Ex. 2. When *Orchis fuchsii* Druce was renamed *Dactylorhiza fuchsii* (Druce) Soó the name x*Orchicoeloglossum mixtum* Asch. & Graebner (for its hybrid with *Coeloglossum viride* (L.) Hartman) became the basis of the necessary new combination x*Dactyloglossum mixtum* (Asch. & Graebner) Rauschert (1969).

H.11.2. The epithet of an infraspecific nothotaxon, of which the postulated or known parental taxa are assigned to different taxa at a higher rank, may be placed subordinate to the name of a nothotaxon at that higher rank (see Art. 24.1). If this higher-ranking nothotaxon is a nothospecies the name of the subordinate nothotaxon is a combination of its epithet with the nothospecific name (but see Rec. H.10B).

Ex. 3. *Mentha* x*piperita* L. nothosubsp. *piperita* (= *M. aquatica* L. x *M. spicata* L. subsp. *spicata*); *Mentha* x*piperita* nothosubsp. *pyramidalis* (Ten.) R. Harley (= *M. aquatica* L. x *M. spicata* subsp. *tomentosa* (Briq.) R. Harley).

Article H.12

H.12.1. Subordinate taxa within nothotaxa of specific or infraspecific rank may be recognized without an obligation to specify parent taxa at the subordinate rank. In this case non-hybrid infraspecific categories of the appropriate rank are used.

Ex. 1. *Mentha* x*piperita* forma *hirsuta* Sole; *Populus* x*canadensis* var. *serotina* (Hartig) Rehder and *P.* x*canadensis* var. *marilandica* (Poiret) Rehder (see also Art. H.4, Note 1).

Note 1. As there is no statement of parentage at the rank concerned there is no control of circumscription at this rank by parentage (compare Art. H.4.).

Note 2. It is not feasible to treat subdivisions of nothospecies by the methods of both Art. H.10 and H.12.1 at the same rank.

H.12.2. Names published at the rank of nothomorph[1] are treated as having been published as names of varieties (see Art. 50).

---

1) Previous editions of the Code (1978, Art. H.10, and the corresponding article in earlier editions) permitted only one rank under provisions equivalent to H.12. That rank was equivalent to variety and the category was termed "nothomorph".

APPENDIX IIA

NOMINA FAMILIARUM ALGARUM, FUNGORUM ET PTERIDOPHYTORUM
CONSERVANDA ET REJICIENDA

In the following lists the **nomina conservanda** have been inserted in the left column; they have been printed in **bold-face** type. Synonyms and earlier homonyms (*nomina rejicienda*) have been listed in the right column.

T. type.

=   taxonomic synonym(s) to be rejected only in favour of the conserved name. Based on a type different from that of the conserved name.

*   Conservation approved by the General Committee; use authorized under Art. 15 pending final decision by the next Congress.

Some names listed as conserved have no corresponding nomina rejicienda because they were conserved explicitly to conserve a particular type, because evidence after their conservation may have indicated that conservation was unnecessary, or because they were conserved to eliminate doubt about their legitimacy.

CHLOROPHYCEAE

Cladophoraceae Wille in Warming, Haandb. Syst. Bot. ed. 2. 30. 1884.
T.: *Cladophora* Kützing, nom. cons.

(=) *Pithophoraceae* Wittrock, Nova Acta Regiae Soc. Sci. Upsal. ser. 3, vol. extra ord. (19): 47. 1877.
T.: *Pithophora* Wittrock.

Siphonocladaceae Schmitz, Ber. Sitzungen Naturf. Ges. Halle 1878: 20. 1879 ("*Siphonocladiaceae*").
T.: *Siphonocladus* Schmitz.

* **Corticiaceae** Herter in Warnstorf et al., Kryptogamenfl. Mark Brandenburg **6**(1): 70. 1910.
T.: *Corticium* Persoon.

(=) *Cyphellaceae* Lotsy, Vortr. Bot. Stammesgesch. **1**: 695. 1907.
T.: *Cyphella* Fries : Fries.

(=) *Peniophoraceae* Lotsy, Vortr. Bot. Stammesgesch. **1**: 687. 1907.
T.: *Peniophora* Cooke.

(=) *Vuilleminiaceae* Maire ex Lotsy, Vortr. Bot. Stammesgesch. **1**: 678. 1907.
T.: *Vuilleminia* Maire.

* **Cortinariaceae** Heim ex Pouzar, Česká Mykol. **37**: 173. 1983.
T.: *Cortinarius* (Persoon) S. F. Gray.

(=) *Crepidotaceae* (Imai) Singer, Lilloa **22**: 584. 1951 ("1949").
T.: *Crepidotus* (Fries) Staude.

(=) *Galeropsidaceae* Singer, Bol. Soc. Argent. Bot. **10**: 61. 1962.
T.: *Galeropsis* Velenovský.

(=) *Thaxterogasteraceae* Singer, Bol. Soc. Argent. Bot. **10**: 63. 1962 ("*Thaxterogastraceae*").
T.: *Thaxterogaster* Singer.

(=) *Hebelomataceae* Locquin, Flore Mycol. **3**: 146. 1977.
T.: *Hebeloma* (Fries) P. Kummer.

(=) *Inocybaceae* Jülich, Higher Taxa Basid. 374. 1982 ("1981").
T.: *Inocybe* (Fries) Fries.

(=) *Verrucosporaceae* Jülich, Higher Taxa Basid. 393. 1982 ("1981").
T.: *Verrucospora* Horak.

* **Tricholomataceae** Heim ex Pouzar, Česká Mykol. **37**: 175. 1983.
T.: *Tricholoma* (Fries) Staude.

(=) *Hydnangiaceae* Gäumann et Dodge, Compar. Morphol. Fungi 485. 1928.
T.: *Hydnangium* Wallroth.

(=) *Physalacriaceae* Corner, Beih. Nova Hedwigia **33**: 10. 1970.
T.: *Physalacria* Peck.

(=) *Amparoinaceae* Singer, Rev. Mycol. **40**: 58. 1976.
T.: *Amparoina* Singer.

(=) *Dermolomataceae* (Bon) Bon, Docum. Mycol. **9**(35): 43. 1979.
T.: *Dermoloma* (J. E. Lange) Herink.

(=) *Macrocystidiaceae* Kühner, Bull. Mens. Soc. Linn. Lyon **48**: 172. 1979.
T.: *Macrocystidia* Josserand.

(=) *Rhodotaceae* Kühner, Bull. Mens. Soc. Linn. Lyon **49**: 235. 1980.
T.: *Rhodotus* Maire.

(=) *Pleurotaceae* Kühner, Bull. Mens. Soc. Linn. Lyon **49**: 184. 1980.
T.: *Pleurotus* (Fries) P. Kummer.

(=) *Marasmiaceae* Kühner, Bull. Mens. Soc. Linn. Lyon **49**: 76. 1980.
T.: *Marasmius* Fries.

(=) *Hygrophoropsidaceae* Kühner, Hymé-nomyc. Agaric. 900. 1980.
T.: *Hygrophoropsis* (Schroeter) Maire ex Martin-Sans.

(=) *Biannulariaceae* Jülich, Higher Taxa Basid. 356. 1982 ("1981").
T.: *Biannularia* Beck.

(=) *Cyphellopsidaceae* Jülich, Higher Taxa Basid. 362. 1982 ("1981").
T.: *Cyphellopsis* Donk.

(=) *Fayodiaceae* Jülich, Higher Taxa Basid. 367. 1982 ("1981").
T.: *Fayodia* Kühner.

(=) *Laccariaceae* Jülich, Higher Taxa Basid. 374. 1982 ("1981").
T.: *Laccaria* Berkeley et Broome.

(=) *Lentinaceae* Jülich, Higher Taxa Basid. 376. 1982 ("1981").
T.: *Lentinus* Fries : Fries.

(=) *Leucopaxillaceae* (Singer) Jülich, Higher Taxa Basid. 376. 1982 ("1981").
T.: *Leucopaxillus* Boursier.

(=) *Lyophyllaceae* Jülich, Higher Taxa Basid. 378. 1982 ("1981").
T.: *Lyophyllum* P. Karsten.

(=) *Nyctalidaceae* Jülich, Higher Taxa Basid. 381. 1982 ("1981").
T.: *Nyctalis* Fries.

(=) *Panellaceae* Jülich, Higher Taxa Basid. 382. 1982 ("1981").
T.: *Panellus* P. Karsten.

(=) *Resupinataceae* (Singer) Jülich, Higher Taxa Basid. 388. 1982 ("1981").
T.: *Resupinatus* (Nees) S. F. Gray.

(=) *Squamanitaceae* Jülich, Higher Taxa Basid. 390. 1982 ("1981").
T.: *Squamanita* Imbach.

(=) *Termitomycetaceae* Jülich, Higher Taxa Basid. 391. 1982 ("1981").
T.: *Termitomyces* Heim.

(=) *Xerulaceae* Jülich, Higher Taxa Basid. 394. 1982 ("1981").
T.: *Xerula* Maire.

PTERIDOPHYTA

Adiantaceae (K. B. Presl) Ching, Sunyatsenia 5: 229. 1940.
T.: *Adiantum* Linnaeus.

(=) *Parkeriaceae* W. J. Hooker, Exot. Fl. 2(20): *t. 147.* 1825.
T.: *Parkeria* W. J. Hooker.

(=) *Acrostichaceae* Mettenius ex Frank in Leunis, Syn. Pflanzenk. ed. 2, 3: 1458. 1877.
T.: *Acrostichum* Linnaeus.

(=) *Sinopteridaceae* Koidzumi, Acta Phyto-tax. Geobot. 3: 50. 1934.
T.: *Sinopteris* Christensen et Ching.

Dicksoniaceae Bower, Origin Land Fl. 591. 1908 ("*Dicksonieae*").
T.: *Dicksonia* L'Héritier.

(=) *Thyrsopteridaceae* K. B. Presl, Gefäß-bündel Farrn 38. 1847 ("*Thyrsopteri-deae*").
T.: *Thyrsopteris* Kunze.

Dryopteridaceae Ching, Acta Phytotax. Sin. **10**: 1. 1965.
T.: *Dryopteris* Adanson, nom. cons.

(=) *Peranemataceae* (K. B. Presl) Ching, Sunyatsenia **5**: 246. 1940 ("*Perenema-ceae*"), nom. cons.
T.: *Peranema* D. Don.

Peranemataceae (K. B. Presl) Ching, Sunya-tsenia **5**: 246. 1940 ("*Perenemaceae*").
T.: *Peranema* D. Don.

APPENDIX IIB

NOMINA FAMILIARUM BRYOPHYTORUM ET SPERMATOPHYTORUM
CONSERVANDA

The names of families printed in **bold-face** type are to be retained in all cases, with priority over unlisted synonyms (Art. 14.5) and homonyms (Art. 14.9).

When two listed names compete, the earlier must be retained unless the contrary is indicated or one of the competing names is listed in Art. 18.5. For any family including the type of an alternative family name, one or the other of these alternative names is to be used.

For purposes of this list the starting point for Spermatophyta is Jussieu's Genera Plantarum (Jul-Aug 1789); earlier usage of listed names is to be disregarded. For unlisted names the starting point is the same as that for all other taxa of Spermatophyta (1 May 1753).

T.   type.

*   Conservation approved by the General Committee; use authorized under Art. 15 pending final decision by the next Congress.

MUSCI

Bryoxiphiaceae Bescherelle, J. Bot. (Morot) 6: 183. 1892.
T.: *Bryoxiphium* Mitten, nom. cons.

Ditrichaceae Limpricht in Rabenhorst, Deutschl. Krypt.-Fl. ed. 2, 4: 482. 1887.
T.: *Ditrichum* Hampe, nom. cons.

Entodontaceae N. C. Kindberg, Gen. Eur. N.-Amer. Bryin. 7. 1889.
T.: *Entodon* K. Müller Hal.

Eustichiaceae Brotherus in Engler et Prantl, Nat. Pflanzenfam. ed. 2, 10: 420. 1924.
T.: *Eustichia* (Bridel) Bridel.

Pottiaceae Schimper, Coroll. Bryol. Eur. 24. 1856.
T.: *Pottia* (Reichenbach) Fürnrohr.

Sematophyllaceae Brotherus in Engler et Prantl, Nat. Pflanzenfam. 1(3): 1098. 1908.
T.: *Sematophyllum* Mitten.

HEPATICAE

Lejeuneaceae Casares-Gil, Fl. Ibér. Brióf. Hepát. 703. 1919.
T.: *Lejeunea* Libert, nom. cons.

Porellaceae Cavers, New Phytol. 9: 292. 1910.
T.: *Porella* Linnaeus.

GYMNOSPERMAE

Araucariaceae Henkel et W. Hochstetter, Syn. Nadelhölzer xvii, 1. 1865 (*"Araucarieae"*).
T.: *Araucaria* A. L. Jussieu.

Cephalotaxaceae Neger, Nadelhölzer 23, 30. 1907.
T.: *Cephalotaxus* Siebold et Zuccarini ex Endlicher.

Cupressaceae Bartling, Ord. Nat. Pl. 90, 95. 1830 (*"Cupressinae"*).
T.: *Cupressus* Linnaeus.

Cycadaceae Persoon, Syn. Pl. 2: 630. 1807 (*"Cycadeae"*).
T.: *Cycas* Linnaeus.

Ephedraceae Dumortier, Anal. Fam. Pl. 11, 12. 1829.
T.: *Ephedra* Linnaeus.

Ginkgoaceae Engler in Engler et Prantl, Nat. Pflanzenfam. Nachtr. [1]: 19. 1897.
T.: *Ginkgo* Linnaeus.

Gnetaceae Lindley, Bot. Reg. 20: sub *t. 1686*. 1834.
T.: *Gnetum* Linnaeus.

Pinaceae Lindley, Nat. Syst. Bot. ed. 2. 313. 1836.
T.: *Pinus* Linnaeus.

Podocarpaceae Endlicher, Syn. Conif. 203. 1847 (*"Podocarpeae"*).
T.: *Podocarpus* L'Héritier ex Persoon, nom. cons.

Taxaceae S. F. Gray, Nat. Arr. Brit. Pl. 2: 222, 226. 1821 (*"Taxideae"*).
T.: *Taxus* Linnaeus.

Taxodiaceae Warming, Haandb. Syst. Bot. ed. 2. 163. 1884.
T.: *Taxodium* L. C. Richard.

Welwitschiaceae Markgraf in Engler et Prantl, Nat. Pflanzenfam. ed. 2. 13: 419. 1926.
T.: *Welwitschia* J. D. Hooker, nom. cons.

ANGIOSPERMAE

Acanthaceae A. L. Jussieu, Gen. Pl. 102. 1789 (*"Acanthi"*).
T.: *Acanthus* Linnaeus.

Aceraceae A. L. Jussieu, Gen. Pl. 250. 1789 (*"Acera"*).
T.: *Acer* Linnaeus.

Achariaceae Harms in Engler et Prantl, Nat. Pflanzenfam. Nachtr. [1]: 256. 1897.
T.: *Acharia* Thunberg.

Achatocarpaceae Heimerl in Engler et Prantl, Nat. Pflanzenfam. ed. 2. 16c: 174. 1934.
T.: *Achatocarpus* Triana.

Actinidiaceae Hutchinson, Fam. Fl. Pl. 1: 177. 1926.
T.: *Actinidia* Lindley.
Note: If this family is united with *Saurauiaceae*, the name *Actinidiaceae* must be used.

Adoxaceae Trautvetter, Estestv. Istorija Gub. Kievsk. Ucebn. Okr. 35. 1853.
T.: *Adoxa* Linnaeus.

Aextoxicaceae Engler et Gilg, Syllabus ed. 8. 250. 1919.
T.: *Aextoxicon* Ruiz et Pavón.

Agavaceae Endlicher, Ench. Bot. 105. 1841 (*"Agaveae"*).
T.: *Agave* Linnaeus.

Aizoaceae Rudolphi, Syst. Orb. Veg. 53. 1830 (*"Aizoideae"*).
T.: *Aizoon* Linnaeus.

Akaniaceae O. Stapf, Bull. Misc. Inform. 1912: 380. 1912.
T.: *Akania* J. D. Hooker.

Alangiaceae A.-P. de Candolle, Prodr. 3: 203. 1828 (*"Alangieae"*).
T.: *Alangium* Lamarck, nom. cons.

Alismataceae Ventenat, Tabl. Règne Vég. 2: 157. 1799 (*"Alismoideae"*).
T.: *Alisma* Linnaeus.

Alliaceae J. G. Agardh, Theor. Syst. Pl. 32 1858.
T.: *Allium* Linnaeus.

Alsinaceae Bartling in Bartling et Wendland, Beitr. Bot. 2: 159. 1825 (*"Alsineae"*).
T.: *Alsine* Linnaeus.

Alstroemeriaceae Dumortier, Anal. Fam. Pl. 57, 58. 1829.
T.: *Alstroemeria* Linnaeus.

Altingiaceae Lindley, Veg. Kingd. 253. 1846.
T.: *Altingia* Noronha.

Amaranthaceae A. L. Jussieu, Gen. Pl. 87. 1789 ("*Amaranthi*").
T.: *Amaranthus* Linnaeus.

Amaryllidaceae Jaume Saint-Hilaire, Expos. Fam. 1: 134. 1805 ("*Amaryllideae*").
T.: *Amaryllis* Linnaeus.

Amborellaceae Pichon, Bull. Mus. Hist. Nat. Paris ser. 2. 20: 384. 1948.
T.: *Amborella* Baillon.

Ambrosiaceae Dumortier, Anal. Fam. Pl. 15, 16. 1829; Link, Handb. 1: 816. 1829.
T.: *Ambrosia* Linnaeus.

Amygdalaceae D. Don, Prodr. Fl. Nepal. 239. 1825 ("*Amydalinae*").
T.: *Amygdalus* Linnaeus.

Anacardiaceae Lindley, Intr. Nat. Syst. Bot. 127. 1830.
T.: *Anacardium* Linnaeus.

Ancistrocladaceae Walpers, Ann. Bot. 2: 175. 1851 ("*Ancistrocladeae*").
T.: *Ancistrocladus* Wallich, nom. cons.

Annonaceae A. L. Jussieu, Gen. Pl. 283. 1789 ("*Anonae*").
T.: *Annona* Linnaeus.

Apiaceae Lindley, Nat. Syst. Bot. ed. 2. 21. 1836. – Nom. alt.: *Umbelliferae.*
T.: *Apium* Linnaeus.

Apocynaceae A. L. Jussieu, Gen. Pl. 143. 1789 ("*Apocineae*").
T.: *Apocynum* Linnaeus.

Aponogetonaceae J. G. Agardh, Theoria Syst. Pl. 44. 1858 ("*Aponogetaceae*").
T.: *Aponogeton* Linnaeus f., nom. cons.

Apostasiaceae Lindley, Nix. Pl. 22, 1833 ("*Apostasieae*"); Blume, Tijdschr. Nat. Gesch. 1: 137. (Nov. Pl. Fam. Expos. 7). 1833 ("*Apostasieae*").
T.: *Apostasia* Blume.

Aquifoliaceae Bartling, Ord. Nat. Pl. 228, 376. 1830.
T.: *Aquifolium* P. Miller, nom. illeg. (*Ilex* Linnaeus).

Araceae A. L. Jussieu, Gen. Pl. 23. 1789 ("*Aroideae*").
T.: *Arum* Linnaeus.

Araliaceae A. L. Jussieu, Gen. Pl. 217. 1789 ("*Araliae*").
T.: *Aralia* Linnaeus.

Arecaceae C. H. Schultz-Schultzenstein, Nat. Syst. Pflanzenr. 317. 1832. – Nom. alt.: *Palmae.*
T.: *Areca* Linnaeus.

Aristolochiaceae A. L. Jussieu, Gen. Pl. 72. 1789 ("*Aristolochiae*").
T.: *Aristolochia* Linnaeus.

Asclepiadaceae R. Brown, Asclepiadeae 12, 17. 1810 ("*Asclepiadeae*").
T.: *Asclepias* Linnaeus.

Asparagaceae A. L. Jussieu, Gen. Pl. 40. 1789 ("*Asparagi*").
T.: *Asparagus* Linnaeus.

Asteraceae Dumortier, Comment. Bot. 55. 1822 ("*Astereae*"). – Nom. alt.: *Compositae.*
T.: *Aster* Linnaeus.

Asteranthaceae Knuth in Engler, Pflanzenr. iv.219b (Heft 105): 1. 1939.
T.: *Asteranthos* Desfontaines.

Austrobaileyaceae Croizat, Cact. Succ. J. (Los Angeles) 15: 64. 1943.
T.: *Austrobaileya* C. T. White.

Avicenniaceae Endlicher, Ench. Bot. 314. 1841 ("*Avicennieae*").
T.: *Avicennia* Linnaeus.

Balanitaceae Endlicher, Ench. Bot. 547. 1841 ("*Balaniteae*").
T.: *Balanites* Delile, nom. cons.

Balanopaceae Bentham in Bentham et J. D. Hooker, Gen. Pl. 3: v, 341. 1880 ("*Balanopseae*").
T.: *Balanops* Baillon.

Balanophoraceae L. C. Richard et A. Richard, Mém. Mus. Hist. Nat. 8: 429. 1822 ("*Balanophoreae*").
T.: *Balanophora* J. R. Forster et G. Forster.

Balsaminaceae A. Richard, Dict. Class. Hist. Nat. 2: 173. 1822 ("*Balsamineae*").
T.: *Balsamina* P. Miller, nom. illeg. (*Impatiens* Linnaeus).

Barbeyaceae Rendle in Thiselton-Dyer, Fl. Trop. Afr. 6(2): 14. 1916.
T.: *Barbeya* Schweinfurth.

Barringtoniaceae Rudolphi, Syst. Orb. Veg. 56. 1830 ("*Barringtonieae*").
T.: *Barringtonia* J. R. Forster et G. Forster, nom. cons.

96

Basellaceae Moquin-Tandon, Chenopod. Monogr. Enum. x. 1840.
T.: *Basella* Linnaeus.

Bataceae C. F. P. Martius ex Meisner, Pl. Vasc. Gen., Tab. Diagn. 345, Comm. 260. 1842 ("*Batideae*").
T.: *Batis* P. Browne.

Begoniaceae C. A. Agardh, Aphor. Bot. 200. 1825.
T.: *Begonia* Linnaeus.

Berberidaceae A. L. Jussieu, Gen. Pl. 286. 1789 ("*Berberides*").
T.: *Berberis* Linnaeus.

Betulaceae S. F. Gray, Nat. Arr. Brit. Pl. 2: 222, 243. 1821 ("*Betulideae*").
T.: *Betula* Linnaeus.
Note: If this family is united with *Corylaceae*, the name *Betulaceae* must be used.

Bignoniaceae A. L. Jussieu, Gen. Pl. 137. 1789 ("*Bignoniae*").
T.: *Bignonia* Linnaeus.

Bixaceae Link, Handbuch 2: 371. 1831 ("*Bixinae*").
T.: *Bixa* Linnaeus.

Bombacaceae Kunth, Malvac. 5. 1822 ("*Bombaceae*").
T.: *Bombax* Linnaeus.

Boraginaceae A. L. Jussieu, Gen. Pl. 128. 1789 ("*Borragineae*").
T.: *Borago* Linnaeus.

Brassicaceae Burnett, Outl. Bot. 1123. 1835. – Nom. alt.: *Cruciferae*.
T.: *Brassica* Linnaeus.

Bretschneideraceae Engler et Gilg, Syllabus ed. 9.-10. 218. 1924.
T.: *Bretschneidera* Hemsley.

Bromeliaceae A. L. Jussieu, Gen. Pl. 49. 1789 ("*Bromeliae*").
T.: *Bromelia* Linnaeus.

Brunelliaceae Engler in Engler et Prantl, Nat. Pflanzenfam. Nachtr. [1]: 182. 1897.
T.: *Brunellia* Ruiz et Pavón.

Bruniaceae A.-P. de Candolle, Prodr. 2: 43. 1825.
T.: *Brunia* Linnaeus, nom. cons.

Brunoniaceae Dumortier, Anal. Fam. Pl. 19, 21. 1829.
T.: *Brunonia* J. E. Smith.

Buddlejaceae Wilhelm, Samenpfl. 90. 1910 ("*Buddleiaceae*").
T.: *Buddleja* Linnaeus.

Burmanniaceae Blume, Enum. Pl. Javae 1: 27. 1827.
T.: *Burmannia* Linnaeus.

Burseraceae Kunth, Ann. Sci. Nat. (Paris) 2: 346. 1824.
T.: *Bursera* N. J. Jacquin ex Linnaeus, nom. cons.

Butomaceae L. C. Richard, Mém. Mus. Hist. Nat. 1: 366. 1815 vel 1816 prim. ("*Butomeae*").
T.: *Butomus* Linnaeus.

Buxaceae Dumortier, Comment. Bot. 54. 1822.
T.: *Buxus* Linnaeus.

Byblidaceae Domin, Act. Bot. Bohem. 1: 3. 1922.
T.: *Byblis* R. A. Salisbury.

Byttneriaceae R. Brown in Flinders, Voy. Terra Austr. 2: 540. 1814 ("*Buttneriaceae*").
T.: *Byttneria* Loefling, nom. cons.
Note: If this family is united with *Sterculiaceae*, the name *Byttneriaceae* is rejected in favour of *Sterculiaceae*.

Cabombaceae A. Richard, Nouv. Elém. Bot. ed. 4. 420. 1828 ("*Cabombeae*").
T.: *Cabomba* Aublet.

Cactaceae A. L. Jussieu, Gen. Pl. 310. 1789 ("*Cacti*").
T.: *Cactus* Linnaeus (*Mammillaria* Haworth, nom. cons.).

Caesalpiniaceae R. Brown in Flinders, Voy. Terra Austr. 2: 551. 1814 ("*Caesalpineae*").
T.: *Caesalpinia* Linnaeus.

Callitrichaceae Link, Enum. Hort. Berol. Alt. 1: 7. 1821 ("*Callitrichinae*").
T.: *Callitriche* Linnaeus.

Calycanthaceae Lindley, Bot. Reg. 5: sub t. 404. 1819 ("*Calycantheae*").
T.: *Calycanthus* Linnaeus, nom. cons.

Calyceraceae L. C. Richard, Mém. Mus. Hist. Nat. 6: 74. 1820 ("*Calycereae*").
T.: *Calycera* Cavanilles.

Campanulaceae A. L. Jussieu, Gen. Pl. 163. 1789.
T.: *Campanula* Linnaeus.

Canellaceae C. F. P. Martius, Nov. Gen. Sp. Pl. 3: 170. 1832.
T.: *Canella* P. Browne, nom. cons.

Cannabaceae Endlicher, Gen. Pl. 286. 1837 ("*Cannabineae*").
T.: *Cannabis* Linnaeus.

Cannaceae A. L. Jussieu, Gen. Pl. 62. 1789 ("*Cannae*").
T.: *Canna* Linnaeus.

Capparaceae A. L. Jussieu, Gen. Pl. 242. 1789 ("*Capparides*").
T.: *Capparis* Linnaeus.

Caprifoliaceae A. L. Jussieu, Gen. Pl. 210. 1789 ("*Caprifolia*").
T.: *Caprifolium* P. Miller.

Cardiopteridaceae Blume, Rumphia 3: 205. 1849 ("*Cardiopterideae*").
T.: *Cardiopteris* Wallich ex Royle.

Caricaceae Dumortier, Anal. Fam. Pl. 37, 42. 1829.
T.: *Carica* Linnaeus.

Cartonemataceae Pichon, Notul. Syst. (Paris) 12: 219. 1946.
T.: *Cartonema* R. Brown.

Caryocaraceae Szyszylowicz in Engler et Prantl, Nat. Pflanzenfam. 3(6): 153. 1893.
T.: *Caryocar* Allemand ex Linnaeus.

Caryophyllaceae A. L. Jussieu, Gen. Pl. 299. 1789 ("*Caryophylleae*").
T.: *Caryophyllus* P. Miller non Linnaeus, nom. illeg. (*Dianthus* Linnaeus).

Cassythaceae Bartling ex Lindley, Nix. Pl. 15. 1833 ("*Cassytheae*").
T.: *Cassytha* Linnaeus.

Casuarinaceae R. Brown in Flinders, Voy. Terra Austr. 2: 571. 1814 ("*Casuarineae*").
T.: *Casuarina* Adanson.

Celastraceae R. Brown in Flinders, Voy. Terra Austr. 2: 554. 1814 ("*Celastrinae*").
T.: *Celastrus* Linnaeus.
Note: If this family is united with *Hippocrateaceae*, the name *Celastraceae* must be used.

Centrolepidaceae Endlicher, Gen. Pl. 119. 1836 ("*Centrolepideae*").
T.: *Centrolepis* Labillardière.

Cephalotaceae Dumortier, Anal. Fam. Pl. 59, 61. 1829 ("*Cephaloteae*").
T.: *Cephalotus* Labillardière, nom. cons.

Ceratophyllaceae S. F. Gray, Nat. Arr. Brit. Pl. 2: 395, 554. 1821 ("*Ceratophyllae*").
T.: *Ceratophyllum* Linnaeus.

Cercidiphyllaceae Engler, Syllabus ed. 6. 132. 1909.
T.: *Cercidiphyllum* Siebold et Zuccarini.

Chailletiaceae, see *Dichapetalaceae*.

Chenopodiaceae Ventenat, Tabl. Règne Vég. 2: 253. 1799 ("*Chenopodeae*").
T.: *Chenopodium* Linnaeus.

Chloranthaceae R. Brown ex Lindley, Collect. Bot. sub *t. 17.* 1821 ("*Chlorantheae*").
T.: *Chloranthus* Swartz.

Chrysobalanaceae R. Brown in Tuckey, Narr. Exped. Congo 433. 1818 ("*Chrysobalaneae*").
T.: *Chrysobalanus* Linnaeus.

Cichoriaceae A. L. Jussieu, Gen. Pl. 168. 1789 ("*Cichoraceae*").
T.: *Cichorium* Linnaeus.

Circaeasteraceae Hutchinson, Fam. Fl. Pl. 1: 98. 1926.
T.: *Circaeaster* Maximowicz.

Cistaceae A. L. Jussieu, Gen. Pl. 294. 1789 ("*Cisti*").
T.: *Cistus* Linnaeus.

Clethraceae Klotzsch, Linnaea 24: 12. 1851.
T.: *Clethra* Linnaeus.

Clusiaceae Lindley, Nat. Syst. Bot. ed. 2. 74. 1836. – Nom. alt.: *Guttiferae*.
T.: *Clusia* Linnaeus.

Cneoraceae Link, Handbuch 2: 440. 1831 ("*Cneoreae*").
T.: *Cneorum* Linnaeus.

Cochlospermaceae J. E. Planchon in W. J. Hooker, London J. Bot. 6: 305. 1847 ("*Cochlospermeae*").
T.: *Cochlospermum* Kunth ex A.-P. de Candolle, nom. cons.

Colchicaceae A.-P. de Candolle in Lamarck et A.-P. de Candolle, Fl. Franç. ed. 3. 3: 192. 1805.
T.: *Colchicum* Linnaeus.

Columelliaceae D. Don, Edinburgh New Philos. J. 6: 46. 1828 ("*Columellieae*").
T.: *Columellia* Ruiz et Pavón, nom. cons.

Combretaceae R. Brown, Prodr. 351. 1810.
T.: *Combretum* Loefling, nom. cons.

Commelinaceae R. Brown, Prodr. 268. 1810
("*Commelineae*").
T.: *Commelina* Linnaeus.

Compositae Giseke, Prael. Ord. Nat. Pl. 538.
1792. – Nom. alt.: *Asteraceae.*
T.: *Aster* Linnaeus.

Connaraceae R. Brown in Tuckey, Narr. Exped. Congo 431. 1818.
T.: *Connarus* Linnaeus.

Convolvulaceae A. L. Jussieu, Gen. Pl. 132.
1789 ("*Convolvuli*").
T.: *Convolvulus* Linnaeus.

Cordiaceae R. Brown ex Dumortier, Anal.
Fam. Pl. 20, 25. 1829.
T.: *Cordia* Linnaeus.

Coriariaceae A.-P. de Candolle, Prodr. 1: 739.
1824 ("*Coriarieae*").
T.: *Coriaria* Linnaeus.

Cornaceae Dumortier, Anal. Fam. Pl. 33, 34.
1829 ("*Corneae*").
T.: *Cornus* Linnaeus.

Corsiaceae Beccari, Malesia 1: 328. 1878.
T.: *Corsia* Beccari.

Corylaceae Mirbel, Elém. Phys. Vég. Bot. 2:
906. 1815.
T.: *Corylus* Linnaeus.
Note: If this family is united with *Betulaceae*,
the name *Corylaceae* is rejected in favour of
Betulaceae.

Corynocarpaceae Engler in Engler et Prantl,
Nat. Pflanzenfam. Nachtr. [1]: 215. 1897.
T.: *Corynocarpus* J. R. Forster et G. Forster.

Crassulaceae A.-P. de Candolle in Lamarck et
A.-P. de Candolle, Fl. Franç. ed. 3. 4(1): 382.
1805.
T.: *Crassula* Linnaeus.

Crossosomataceae Engler in Engler et Prantl,
Nat. Pflanzenfam. Nachtr. [1]: 185. 1897.
T.: *Crossosoma* Nuttall.

Cruciferae A. L. Jussieu, Gen. Pl. 237. 1789. –
Nom. alt.: *Brassicaceae.*
T.: *Brassica* Linnaeus.

Crypteroniaceae A. de Candolle, Prodr. 16(2):
677. 1868.
T.: *Crypteronia* Blume.

Cucurbitaceae A. L. Jussieu, Gen. Pl. 393.
1789.
T.: *Cucurbita* Linnaeus.

Cunoniaceae R. Brown in Flinders, Voy. Terra
Austr. 2: 548. 1814.
T.: *Cunonia* Linnaeus, nom. cons.

Cuscutaceae Dumortier, Anal. Fam. Pl. 20, 25.
1829.
T.: *Cuscuta* Linnaeus.

Cyanastraceae Engler, Bot. Jahrb. Syst. 28:
357. 1900.
T.: *Cyanastrum* Oliver.

Cyclanthaceae Dumortier, Anal. Fam. Pl. 65,
66. 1829 ("*Cyclanthae*", "*Cyclanteae*").
T.: *Cyclanthus* Poiteau.

Cymodoceaceae N. Taylor, N. Amer. Fl. 17(1):
31. 1909.
T.: *Cymodocea* C. Konig, nom. cons.

Cynomoriaceae Lindley, Nix. Pl. 23. 1833 ("*Cy-
nomorieae*").
T.: *Cynomorium* Linnaeus.

Cyperaceae A. L. Jussieu, Gen. Pl. 26. 1789
("*Cyperoideae*").
T.: *Cyperus* Linnaeus.

Cyrillaceae Endlicher, Ench. Bot. 578. 1841
("*Cyrilleae*").
T.: *Cyrilla* Garden ex Linnaeus.

Cytinaceae, see *Rafflesiaceae.*

Daphniphyllaceae Müller Arg. in A. de Candolle, Prodr. 16(1): 1. 1869.
T.: *Daphniphyllum* Blume.

Datiscaceae Lindley, Intr. Nat. Syst. Bot. 109.
1830 ("*Datisceae*").
T.: *Datisca* Linnaeus.

Degeneriaceae Bailey et A. C. Smith, J. Arnold
Arbor. 23: 357. 1942.
T.: *Degeneria* Bailey et A. C. Smith.

Desfontainiaceae Endlicher, Ench. Bot. 336.
1841 ("*Desfontaineae*").
T.: *Desfontainia* Ruiz et Pavón.

Dialypetalanthaceae Rizzini et Occhioni, Lilloa
17: 253. 1949.
T.: *Dialypetalanthus* Kuhlmann.

Diapensiaceae Lindley, Nat. Syst. Bot. ed. 2.
233. 1836.
T.: *Diapensia* Linnaeus.

Dichapetalaceae Baillon in C. F. P. Martius, Fl. Bras. **12**(1): 365. 1886 ("*Dichapetaleae*").
T.: *Dichapetalum* Thouars.

Dichondraceae Dumortier, Anal. Fam. Pl. 20, 24. 1829.
T.: *Dichondra* J. R. Forster et G. Forster.

Diclidantheraceae J. G. Agardh, Theor. Syst. Pl. 195. 1858 ("*Diclidanthereae*").
T.: *Diclidanthera* C. F. P. Martius.

Didiereaceae Drake del Castillo, Bull. Mus. Hist. Nat. (Paris) **9**: 36. 1903.
T.: *Didierea* Baillon.

Dilleniaceae Salisbury, Parad. Lond. **2**(1): sub *t. 73*. 1807 ("*Dilleneae*").
T.: *Dillenia* Linnaeus.

Dioncophyllaceae Airy Shaw, Kew Bull. **1951**: 33. 1952.
T.: *Dioncophyllum* Baillon.

Dioscoreaceae R. Brown, Prodr. 294. 1810 ("*Dioscoreae*").
T.: *Dioscorea* Linnaeus.

Dipentodontaceae Merrill, Brittonia **4**: 73. 1941 ("*Dipentodonaceae*").
T.: *Dipentodon* Dunn.

Dipsacaceae A. L. Jussieu, Gen. Pl. 194. 1789 ("*Dipsaceae*").
T.: *Dipsacus* Linnaeus.

Dipterocarpaceae Blume, Bijdr. 222. 1825 ("*Dipterocarpeae*").
T.: *Dipterocarpus* C. F. Gaertner.

Dodonaeaceae Link, Handbuch **2**: 441. 1831.
T.: *Dodonaea* P. Miller.

Donatiaceae Takhtajan ex Dostál, Bot. Nomenkl. 204. 1957.
T.: *Donatia* J. R. Forster et G. Forster, nom. cons.

Dracaenaceae R. A. Salisbury, Gen. Pl. ed. J. E. Gray 73. 1866 ("*Dracaeneae*").
T.: *Dracaena* Linnaeus.

Droseraceae R. A. Salisbury, Parad. Lond. sub *t. 95*. 1808 ("*Drosereae*").
T.: *Drosera* Linnaeus.

Dysphaniaceae Pax, Bot. Jahrb. Syst. **61**: 230. 1927.
T.: *Dysphania* R. Brown.

Ebenaceae Gürke in Engler et Prantl, Nat. Pflanzenfam. **4**(1): 153. 1891.
T.: *Ebenus* O. Kuntze non Linnaeus, nom. illeg. (*Maba* J. R. Forster et G. Forster).

Ehretiaceae Lindley, Intr. Nat. Syst. Bot. 242. 1830.
T.: *Ehretia* P. Browne.

Elaeagnaceae A. L. Jussieu, Gen. Pl. 74. 1789 ("*Elaeagni*").
T.: *Elaeagnus* Linnaeus.

Elaeocarpaceae A.-P. de Candolle, Prodr. **1**: 519. 1824 ("*Elaeocarpeae*").
T.: *Elaeocarpus* Linnaeus.

Elatinaceae Dumortier, Anal. Fam. Pl. 44, 49. 1829 ("*Elatinideae*").
T.: *Elatine* Linnaeus.

Empetraceae S. F. Gray, Nat. Arr. Brit. Pl. **2**: 732. 1821 ("*Empetrideae*").
T.: *Empetrum* Linnaeus.

Epacridaceae R. Brown, Prodr. 535. 1810 ("*Epacrideae*").
T.: *Epacris* Cavanilles, nom. cons.

Ericaceae A. L. Jussieu, Gen. Pl. 159. 1789 ("*Ericae*").
T.: *Erica* Linnaeus.

Eriocaulaceae Desvaux, Ann. Sci. Nat. (Paris) **13**: 47. 1828 ("*Eriocauloneae*").
T.: *Eriocaulon* Linnaeus.

Erythropalaceae Sleumer in Engler et Prantl, Nat. Pflanzenfam. ed. 2. **20b**: 401. 1942.
T.: *Erythropalum* Blume.

Erythroxylaceae Kunth in Humbolt, Bonpland et Kunth, Nov. Gen. Sp. **5**: ed. fol. 135; ed. qu. 175. 1822 ("*Erythroxyleae*").
T.: *Erythroxylum* P. Browne.

Escalloniaceae Dumortier, Anal. Fam. Pl. 35, 37. 1829.
T.: *Escallonia* Mutis ex Linnaeus f.

Eucommiaceae Engler, Syllabus ed. 6. 145. 1909.
T.: *Eucommia* Oliver.

Eucryphiaceae Endlicher, Ench. Bot. 528. 1841 ("*Eucryphieae*").
T.: *Eucryphia* Cavanilles.

Euphorbiaceae A. L. Jussieu, Gen. Pl. 384. 1789 ("*Euphorbiae*").
T.: *Euphorbia* Linnaeus.

Eupomatiaceae Endlicher, Ench. Bot. 425. 1841 ("*Eupomatieae*").
T.: *Eupomatia* R. Brown.

Eupteleaceae Wilhelm, Samenpfl. 17. 1910.
T.: *Euptelea* Siebold et Zuccarini.

Fabaceae Lindley, Nat. Syst. Bot. ed. 2. 148. 1836. – Nom. alt.: *Leguminosae* vel *Papilionaceae*.
T.: *Faba* P. Miller.

Fagaceae Dumortier, Anal. Fam. Pl. 11, 12. 1829 ("*Fagineae*").
T.: *Fagus* Linnaeus.

Ficoidaceae, see *Aizoaceae*.

Flacourtiaceae A.-P. de Candolle, Prodr. 1: 255. 1824 ("*Flacourtianeae*").
T.: *Flacourtia* L'Héritier.
Note: if this family is united with *Samydaceae*, the name *Flacourtiaceae* must be used.

Flagellariaceae Dumortier, Anal. Fam. Pl. 59, 60. 1829.
T.: *Flagellaria* Linnaeus.

Fouquieriaceae A.-P. de Candolle, Prodr. 3: 349. 1828 ("*Fouquieraceae*").
T.: *Fouquieria* Kunth.

Francoaceae A. H. L. Jussieu, Ann. Sci. Nat. (Paris) 25: 9. 1832.
T.: *Francoa* Cavanilles.

Frankeniaceae S. F. Gray, Nat. Arr. Brit. Pl. 2: 623, 663. 1821.
T.: *Frankenia* Linnaeus.

Fumariaceae A.-P. de Candolle, Syst. Nat. 2: 105. 1821.
T.: *Fumaria* Linnaeus.

Garryaceae Lindley, Bot. Reg. 20: sub *t. 1686*. 1834.
T.: *Garrya* Douglas ex Lindley.

Geissolomataceae Endlicher, Ench. Bot. 214. 1841 ("*Geissolomeae*").
T.: *Geissoloma* Lindley ex Kunth.

Gentianaceae A. L. Jussieu, Gen. Pl. 141. 1789 ("*Gentianae*").
T.: *Gentiana* Linnaeus.

Geosiridaceae Jonker, Recueil Trav. Bot. Néerl. 36: 477. 1939.
T.: *Geosiris* Baillon.

Geraniaceae A. L. Jussieu, Gen. Pl. 268. 1789 ("*Gerania*").
T.: *Geranium* Linnaeus.

Gesneriaceae Dumortier, Comment. Bot. 57. 1822 ("*Gessneridiae*").
T.: *Gesneria* Linnaeus.

Globulariaceae A.-P. de Candolle in Lamarck et A.-P. de Candolle, Fl. Franç. ed. 3. 3: 427. 1805 ("*Globulariae*").
T.: *Globularia* Linnaeus.

Gomortegaceae Reiche, Ber. Deutsch. Bot. Ges. 14: 232. 1896.
T.: *Gomortega* Ruiz et Pavón.

Gonystylaceae Gilg in Engler et Prantl, Nat. Pflanzenfam. Nachtr. [1]: 231. 1897.
T.: *Gonystylus* Teysmann et Binnendijk.

Goodeniaceae R. Brown, Prodr. 573. 1810 ("*Goodenoviae*").
T.: *Goodenia* J. E. Smith.

Gramineae A. L. Jussieu, Gen. Pl. 28. 1789. – Nom. alt.: *Poaceae*.
T.: *Poa* Linnaeus.

Greyiaceae Hutchinson, Fam. Fl. Pl. 1: 202. 1926.
T.: *Greyia* W. J. Hooker et Harvey.

Grossulariaceae A.-P. de Candolle in Lamarck et A.-P. de Candolle, Fl. Franç. ed. 3. 4(2) [= 5]: 405. 1805 ("*Grossulariae*").
T.: *Grossularia* P. Miller.

Grubbiaceae Endlicher, Gen. Pl. xiv. 1839.
T.: *Grubbia* Bergius.

Gunneraceae Meisner, Pl. Vasc. Gen., Tab. Diagn. 345, 346. Comm. 257. 1841.
T.: *Gunnera* Linnaeus.

Guttiferae A. L. Jussieu, Gen. Pl. 255. 1789. – Nom. alt.: *Clusiaceae*.
T.: *Clusia* Linnaeus.

Gyrostemonaceae Endlicher, Ench. Bot. 509. 1841 ("*Gyrostemoneae*").
T.: *Gyrostemon* Desfontaines.

Haemodoraceae R. Brown, Prodr. 299. 1810.
T.: *Haemodorum* J. E. Smith.

Haloragaceae R. Brown in Flinders, Voy. Terra Austr. 2: 549. 1814 ("*Halorageae*").
T.: *Haloragis* J. R. Forster et G. Forster.

Hamamelidaceae R. Brown in Abel, Narr. Journey China 374. 1818 ("*Hamamelideae*").
T.: *Hamamelis* Linnaeus.

Heliotropiaceae H. A. Schrader, Commentat. Soc. Regiae Sci. Gott. Recent. 4: 192 [Asperifol. Linnei Comm. 22]. 1820 ("*Heliotropiceae*").
T.: *Heliotropium* Linnaeus.

Hernandiaceae Blume, Bijdr. 550. 1826 ("*Hernandieae*").
T.: *Hernandia* Linnaeus.

Heteropyxidaceae Engler et Gilg, Syllabus ed. 8. 281. 1919.
T.: *Heteropyxis* Harvey, nom. cons.

Himantandraceae Diels, Bot. Jahrb. Syst. 55: 126. 1917.
T.: *Himantandra* F. Mueller ex Diels.

Hippocastanaceae A.-P. de Candolle, Prodr. 1: 597. 1824 ("*Hippocastaneae*").
T.: *Hippocastanum* P. Miller, nom. illeg. (*Aesculus* Linnaeus).

Hippocrateaceae A. L. Jussieu, Ann. Mus. Natl. Hist. Nat. 18: 486. 1811 ("*Hippocraticeae*").
T.: *Hippocratea* Linnaeus.
Note: If this family is united with *Celastraceae*, the name *Hippocrateaceae* is rejected in favour of *Celastraceae*.

Hippuridaceae Link, Enum. Hort. Berol. Alt. 1: 5. 1821 ("*Hippurideae*").
T.: *Hippuris* Linnaeus.

Hoplestigmataceae Gilg in Engler et Gilg, Syllabus ed. 9-10. 322. 1924.
T.: *Hoplestigma* Pierre.

Humbertiaceae Pichon, Notul. Syst. (Paris) 13: 23. 1947.
T.: *Humbertia* Lamarck.

Humiriaceae A. H. L. Jussieu in A. Saint-Hilaire, Fl. Bras. Merid. 2: 87. 1829.
T.: *Humiria* Aublet, nom. cons.

Hydnoraceae C. A. Agardh, Aphor. Bot. 88. 1821 ("*Hydnorinae*").
T.: *Hydnora* Thunberg.

Hydrangeaceae Dumortier, Anal. Fam. Pl. 36, 38. 1829.
T.: *Hydrangea* Linnaeus.

Hydrocharitaceae A. L. Jussieu, Gen. Pl. 67. 1789 ("*Hydrocharides*").
T.: *Hydrocharis* Linnaeus.

Hydrocotylaceae Hylander, Uppsala Univ. Årsskr. 1945(7): 20. 1945.
T.: *Hydrocotyle* Linnaeus.

Hydrophyllaceae R. Brown, Bot. Reg. 3: sub t. 242. 1817 ("*Hydrophylleae*").
T.: *Hydrophyllum* Linnaeus.

Hydrostachyaceae Engler, Syllabus ed. 2. 125. 1898 ("*Hydrostachydaceae*").
T.: *Hydrostachys* Du Petit-Thouars.

Hypericaceae A. L. Jussieu, Gen. Pl. 254. 1789 ("*Hyperica*").
T.: *Hypericum* Linnaeus.

Hypoxidaceae R. Brown in Flinders, Voy. Terra Austr. 2: 576. 1814 ("*Hypoxideae*").
T.: *Hypoxis* Linnaeus.

Icacinaceae Miers, Ann. Mag. Nat. Hist. ser. 2. 8: 174. 1851.
T.: *Icacina* A. H. L. Jussieu.

Ilicaceae, see *Aquifoliaceae*.

Illecebraceae R. Brown, Prodr. 413. 1810 ("*Illecebreae*").
T.: *Illecebrum* Linnaeus.

Illiciaceae A. C. Smith, Sargentia 7: 8. 1947.
T.: *Illicium* Linnaeus.

Iridaceae A. L. Jussieu, Gen. Pl. 57. 1789 ("*Irides* ").
T.: *Iris* Linnaeus.

Irvingiaceae Exell et Mendonça, Consp. Fl. Angol. 1: 279, 395. 1951.
T.: *Irvingia* J. D. Hooker.

Iteaceae J. G. Agardh, Theoria Syst. Pl. 151. 1858.
T.: *Itea* Linnaeus.

Ixonanthaceae Exell et Mendonça, Bol. Soc. Brot. ser. 2a. 25: 105. 1951.
T.: *Ixonanthes* Jack.

Juglandaceae A. Richard ex Kunth, Ann. Sci. Nat. (Paris) 2: 343. 1824 ("*Juglandeae*").
T.: *Juglans* Linnaeus.

Julianiaceae Hemsley, J. Bot. 44: 379. 1906.
T.: *Juliania* Schlechtendal non La Llave, nom. illeg. (*Amphipterygium* Standley).

Juncaceae A. L. Jussieu, Gen. Pl. 43. 1789 ("*Junci* ").
T.: *Juncus* Linnaeus.

Juncaginaceae L. C. Richard, Démonstr. Bot. ix. 1808 ('*Juncagines*').
T.: *Juncago* Séguier, nom. illeg. (*Triglochin* Linnaeus).
Note: If this family is united with *Potamogetonaceae*, the name *Juncaginaceae* is rejected in favour of *Potamogetonaceae*.

Koeberliniaceae Engler in Engler et Prantl, Nat. Pflanzenfam. **3**(6): 319. 1895.
T.: *Koeberlinia* Zuccarini.

Krameriaceae Dumortier, Anal. Fam. Pl. 20, 23. 1829.
T.: *Krameria* Linnaeus.

Labiatae A. L. Jussieu, Gen. Pl. 110. 1789. – Nom. alt.: *Lamiaceae*.
T.: *Lamium* Linnaeus.

Lacistemataceae C. F. P. Martius, Nov. Gen. Sp. Pl. **1**: 158. 1826 ("*Lacistemeae*").
T.: *Lacistema* Swartz.

Lactoridaceae Engler in Engler et Prantl, Nat. Pflanzenfam. **3**(2): 19. 1888.
T.: *Lactoris* R. Philippi.

Lamiaceae Lindley, Nat. Syst. Bot. ed. 2. 275. 1836. – Nom. alt.: *Labiatae*.
T.: *Lamium* Linnaeus.

Lardizabalaceae Decaisne, Arch. Mus. Hist. Nat. **1**: 185. 1839 ("*Lardizabaleae*").
T.: *Lardizabala* Ruiz et Pavón.

Lauraceae A. L. Jussieu, Gen. Pl. 80. 1789 ("*Lauri*").
T.: *Laurus* Linnaeus.

Lecythidaceae Poiteau, Mém. Mus. Hist. Nat. **13**: 143. 1825 ("*Lecythidaeae*").
T.: *Lecythis* Loefling.

Leeaceae Dumortier, Anal. Fam. Pl. 21, 27. 1829.
T.: *Leea* Royen ex Linnaeus.

Leguminosae A. L. Jussieu, Gen. Pl. 345. 1789. – Nom. alt.: *Fabaceae*.
T.: *Faba* P. Miller.

Leitneriaceae Bentham in Bentham et J. D. Hooker, Gen. Pl. **3**: vi, 396. 1880 ("*Leitnerieae*").
T.: *Leitneria* Chapman.

Lemnaceae S. F. Gray, Nat. Arr. Brit. Pl. **2**: 729. 1821 ("*Lemnadeae*").
T.: *Lemna* Linnaeus.

Lennoaceae Solms-Laubach, Abh. Naturf. Ges. Halle **11**: 174 [Fam. Lenn. 56]. 1870.
T.: *Lennoa* Lexarza.

Lentibulariaceae L. C. Richard in Poiteau et Turpin, Fl. Paris **1**: 26. 1808 ("*Lentibularieae*"),
T.: *Lentibularia* Séguier, nom. illeg. (*Utricularia* Linnaeus).

Lepidobotryaceae Léonard, Bull. Jard. Bot. Etat **20**: 38. 1950.
T.: *Lepidobotrys* Engler.

Lilaeaceae Dumortier, Anal. Fam. Pl. 62, 65. 1829 ("*Lilaearieae*").
T.: *Lilaea* Humboldt et Bonpland.

Liliaceae A. L. Jussieu, Gen. Pl. 48. 1789 ("*Lilia*").
T.: *Lilium* Linnaeus.

Limnanthaceae R. Brown, London Edinburgh Philos. Mag. & J. Sci. **3**: 70. 1833 ("*Limnantheae*").
T.: *Limnanthes* R. Brown, nom. cons.

Linaceae S. F. Gray, Nat. Arr. Brit. Pl. **2**: 622, 639. 1821 ("*Lineae*").
T.: *Linum* Linnaeus.

Lissocarpaceae Gilg in Engler et Gilg, Syllabus ed. 9.-10. 324. 1924.
T.: *Lissocarpa* Bentham.

Loasaceae Dumortier, Comment. Bot. 58. 1822 ("*Loaseae*").
T.: *Loasa* Adanson.

Lobeliaceae R. Brown, Trans. Linn. Soc. London **12**: 133. 1817.
T.: *Lobelia* Linnaeus.

Loganiaceae C. F. P. Martius, Nov. Gen. Sp. Pl. **2**: 133. 1827 ("*Loganieae*").
T.: *Logania* R. Brown, nom. cons.

Loranthaceae A. L. Jussieu, Ann. Mus. Natl. Hist. Nat. **12**: 292. 1808 ("*Lorantheae*").
T.: *Loranthus* N. J. Jacquin, nom. cons.

Lowiaceae Ridley, Fl. Malay Penins. **4**: 291. 1924.
T.: *Lowia* Scortechini.

Lythraceae Jaume Saint-Hilaire, Expos. Fam. Nat. **2**: 175. 1805 ("*Lythrariae*").
T.: *Lythrum* Linnaeus.

Magnoliaceae A. L. Jussieu, Gen. Pl. 280. 1789 ("*Magnoliae*").
T.: *Magnolia* Linnaeus.

Malaceae Small ex Britton in Small, Fl. South-East. U.S. 529. 1903.
T.: *Malus* P. Miller.

Malesherbiaceae D. Don, Edinburgh New Philos. J. **2**: 321. 1827.
T.: *Malesherbia* Ruiz et Pavón.

Malpighiaceae A. L. Jussieu, Gen. Pl. 252. 1789 ("*Malpighiae*").
T.: *Malpighia* Linnaeus.

Malvaceae A. L. Jussieu, Gen. Pl. 271. 1789.
T.: *Malva* Linnaeus.

Marantaceae Petersen in Engler et Prantl, Nat. Pflanzenfam. 2(6): 33. 1888.
T.: *Maranta* Linnaeus.

Marcgraviaceae Choisy in A.-P. de Candolle, Prodr. 1: 565. 1824.
T.: *Marcgravia* Linnaeus.

Martyniaceae Stapf in Engler et Prantl, Nat. Pflanzenfam. 4(3b): 265. 1895.
T.: *Martynia* Linnaeus.

Mayacaceae Kunth, Abh. Königl. Akad. Wiss. Berlin 1840: Phys. Abh. 93. 1842 ("*Mayaceae*").
T.: *Mayaca* Aublet.

Medusagynaceae Engler et Gilg, Syllabus ed. 9.-10. 280. 1924.
T.: *Medusagyne* J. G. Baker.

Medusandraceae Brenan, Kew Bull. 1952: 228. 1952.
T.: *Medusandra* Brenan.

Melanthiaceae Batsch, Tab. Affin. Regni Veg. 133. 1802 ("*Melanthia*").
T.: *Melanthium* Linnaeus.

Melastomataceae A. L. Jussieu, Gen. Pl. 328. 1789 ("*Melastomae*").
T.: *Melastoma* Linnaeus.

Meliaceae A. L. Jussieu, Gen. Pl. 263. 1789 ("*Meliae*").
T.: *Melia* Linnaeus.

Melianthaceae Link, Handbuch 2: 322. 1831 ("*Meliantheae*").
T.: *Melianthus* Linnaeus.

Menispermaceae A. L. Jussieu, Gen. Pl. 284. 1789 ("*Menisperma*").
T.: *Menispermum* Linnaeus.

Menyanthaceae Dumortier, Anal. Fam. Pl. 20, 25. 1829 ("*Menyanthideae*").
T.: *Menyanthes* Linnaeus.

Mesembryanthemaceae Fenzl, Ann. Wiener Mus. Naturgesch. 1: 349. 1836 ("*Mesembryanthemeae*").
T.: *Mesembryanthemum* Linnaeus.

Mimosaceae R. Brown in Flinders, Voy. Terra Austr. 2: 551. 1814 ("*Mimoseae*").
T.: *Mimosa* Linnaeus.

Misodendraceae J. G. Agardh, Theoria Syst. Pl. 236. 1858 ("*Myzodendreae*").
T.: *Misodendrum* Banks ex A.-P. de Candolle.

Mitrastemonaceae Makino, Bot. Mag. (Tokyo) 25: 252. 1911.
T.: *Mitrastemon* Makino.

Molluginaceae Hutchinson, Fam. Fl. Pl. 1: 128. 1926.
T.: *Mollugo* Linnaeus.

Monimiaceae A. L. Jussieu, Ann. Mus. Natl. Hist. Nat. 14: 133. 1809 ("*Monimieae*").
T.: *Monimia* Du Petit-Thouars.

Monotropaceae Nuttall, Gen. N. Amer. Pl. 1: 272. 1818 ("*Monotropeae*").
T.: *Monotropa* Linnaeus.
Note: If this family is united with *Pyrolaceae*, the name *Montropaceae* is rejected in favour of *Pyrolaceae*.

Montiniaceae Nakai, Ord. Fam. App. 243. 1943.
T.: *Montinia* Thunberg.

Moraceae Link, Handbuch 2: 444. 1831 ("*Moriformes*").
T.: *Morus* Linnaeus.

Moringaceae Dumortier, Anal. Fam. Pl. 43, 48. 1829.
T.: *Moringa* Adanson.

Musaceae A. L. Jussieu, Gen. Pl. 61. 1789 ("*Musae*").
T.: *Musa* Linnaeus.

Myoporaceae R. Brown, Prodr. 514. 1810 ("*Myoporinae*").
T.: *Myoporum* Solander ex G. Forster.

Myricaceae Blume, Fl. Javae 17-18: 3. 1829 ("*Myriceae*"); Dumortier, Anal. Fam. Pl. 11, 12. 1829 ("*Myriceae*").
T.: *Myrica* Linnaeus.

Myristicaceae R. Brown, Prodr. 399. 1810 ("*Myristiceae*").
T.: *Myristica* Gronovius, nom. cons.

Myrothamnaceae Niedenzu in Engler et Prantl, Nat. Pflanzenfam. 3(2a): 103. 1891.
T.: *Myrothamnus* Welwitsch.

Myrsinaceae R. Brown, Prodr. 532. 1810 ("*Myrsineae*").
T.: *Myrsine* Linnaeus.

Myrtaceae A. L. Jussieu, Gen. Pl. 322. 1789 ("*Myrti*").
T.: *Myrtus* Linnaeus.

Myzodendraceae, see *Misodendraceae*.

Najadaceae A. L. Jussieu, Gen. Pl. 18. 1789 ("*Naiades*").
T.: *Najas* Linnaeus.

Nelumbonaceae Dumortier, Anal. Fam. Pl. 53. 1829 ("*Nelumboneae*").
T.: *Nelumbo* Adanson.

Nepenthaceae Dumortier, Anal. Fam. Pl. 14, 16. 1829 ("*Nepenthideae*").
T.: *Nepenthes* Linnaeus.

Nolanaceae Dumortier, Anal. Fam. Pl. 20, 24. 1829.
T.: *Nolana* Linnaeus ex Linnaeus f.

Nyctaginaceae A. L. Jussieu, Gen. Pl. 90. 1789 ("*Nyctagines*").
T.: *Nyctago* A. L. Jussieu, nom. illeg. (*Mirabilis* Linnaeus).

Nymphaeaceae R. A. Salisbury, Ann. Bot. (König & Sims) 2: 70. 1805 ("*Nymphaeeae*").
T.: *Nymphaea* Linnaeus, nom. cons.

Nyssaceae Dumortier, Anal. Fam. Pl. 13. 1829.
T.: *Nyssa* Linnaeus.

Ochnaceae A.-P. de Candolle, Ann. Mus. Natl. Hist. Nat. **17**: 410. 1811.
T.: *Ochna* Linnaeus.

Octoknemaceae Engler, Bot. Jahrb. Syst. **43**: 177. 1909 ("*Octoknemataceae*").
T.: *Octoknema* Pierre.

Oenotheraceae, see *Onagraceae*.

Olacaceae Mirbel ex A.-P. de Candolle, Prodr. 1: 531. 1824 ("*Olacineae*").
T.: *Olax* Linnaeus.

Oleaceae Hoffmannsegg et Link, Fl. Portug. 1: 385. 1813-1820 ("*Oleinae*").
T.: *Olea* Linnaeus.

Oliniaceae Arnott ex Sonder in Harvey et Sonder, Fl. Cap. 2: ix, 519. 1862 ("*Olinieae*").
T.: *Olinia* Thunberg, nom. cons.

Onagraceae A. L. Jussieu, Gen. Pl. 317. 1789 ("*Onagrae*").
T.: *Onagra* P. Miller, nom. illeg (*Oenothera* Linnaeus).

Opiliaceae Valeton, Crit. Overz. Olacin. 136. 1886.
T.: *Opilia* Roxburgh.

Orchidaceae A. L. Jussieu, Gen. Pl. 64. 1789 ("*Orchideae*").
T.: *Orchis* Linnaeus.

Orobanchaceae Ventenat, Tabl. Régne Vég. **2**: 292. 1799 ("*Orobanchoideae*").
T.: *Orobanche* Linnaeus.

Oxalidaceae R. Brown in Tuckey, Narr. Exped. Congo 433. 1818 ("*Oxalideae*").
T.: *Oxalis* Linnaeus.

Paeoniaceae Rudolphi, Syst. Orb. Veg. 61. 1830.
T.: *Paeonia* Linnaeus.

Palmae A. L. Jussieu, Gen. Pl. 37. 1789. – Nom. alt.: *Arecaceae*.
T.: *Areca* Linnaeus.

Pandaceae Engler et Gilg, Syllabus ed. 7. 223. 1912.
T.: *Panda* Pierre.

Pandanaceae R. Brown, Prodr. 340. 1810 ("*Pandaneae*").
T.: *Pandanus* S. Parkinson.

Papaveraceae A. L. Jussieu, Gen. Pl. 235. 1789.
T.: *Papaver* Linnaeus.

Papilionaceae Giseke, Prael. Ord. Nat. Pl. 415. 1792. – Nom. alt.: *Fabaceae*.
T.: *Faba* P. Miller.

Parnassiaceae S. F. Gray, Nat. Arr. Brit. Pl. **2**: 623, 670. 1821 ("*Parnassieae*", "*Parnassiae*" p. 623).
T.: *Parnassia* Linnaeus.

Passifloraceae A. L. Jussieu ex Kunth in Humboldt, Bonpland et Kunth, Nov. Gen. Sp. **2**: ed. fol. 100; ed. qu. 126. 1817 ("*Passifloreae*").
T.: *Passiflora* Linnaeus.

Pedaliaceae R. Brown, Prodr. 519. 1810 ("*Pedalinae*").
T.: *Pedalium* Royen ex Linnaeus.

Penaeaceae Guillemin, Dict. Class. Hist. Nat. **13**: 171. 1828.
T.: *Penaea* Linnaeus.

Pentaphragmataceae J. G. Agardh, Theoria Syst. Pl. 95. 1858 ("*Pentaphragmeae*").
T.: *Pentaphragma* G. Don.

Pentaphylacaceae Engler in Engler et Prantl, Nat. Pflanzenfam. Nachtr. [1]: 214. 1897.
T.: *Pentaphylax* Gardner et Champion.

Penthoraceae Rydberg ex Britton, Man. Fl. N. States 475. 1901.
T.: *Penthorum* Linnaeus.

Peridiscaceae Kuhlmann, Arq. Serv. Florest. 3: 4. 1950.
T.: *Peridiscus* Bentham.

Periplocaceae Schlechter in Schumann et Lauterbach, Nachtr. Fl. Deutsch. Schutzgeb. Südsee 351. 1905.
T.: *Periploca* Linnaeus.

Peripterygiaceae, see *Cardiopteridaceae*.

Petermanniaceae Hutchinson, Fam. Fl. Pl. 2: 113. 1934.
T.: *Petermannia* F. Mueller, nom. cons.

Petrosaviaceae Hutchinson, Fam. Fl. Pl. 2: 36. 1934.
T.: *Petrosavia* Beccari.

Philesiaceae Dumortier, Anal. Fam. Pl. 53, 54. 1829 ("*Phylesiaceae*").
T.: *Philesia* Commerson.

Philydraceae Link, Enum. Hort. Berol. Alt. 1: 5. 1821 ("*Philhydrinae*").
T.: *Philydrum* Banks ex Solander.

Phrymaceae Schauer in A. de Candolle, Prodr. 11: 520. 1847.
T.: *Phryma* Linnaeus.

Phytolaccaceae R. Brown in Tuckey, Narr. Exped. Congo 454. 1818 ("*Phytolaceae*").
T.: *Phytolacca* Linnaeus.

Picrodendraceae Small ex Britton et Millspaugh, Bahama Fl. 102. 1920.
T.: *Picrodendron* J. E. Planchon.

Piperaceae C. A. Agardh, Aphor. Bot. 201. 1824.
T.: *Piper* Linnaeus.

Pittosporaceae R. Brown in Flinders, Voy. Terra Austr. 2: 542. 1814 ("*Pittosporeae*").
T.: *Pittosporum* Banks ex Solander, nom. cons.

Plantaginaceae A. L. Jussieu, Gen. Pl. 89. 1789 ("*Plantagines*").
T.: *Plantago* Linnaeus.

Platanaceae Dumortier, Anal. Fam. Pl. 11, 12, 1829 ("*Plataneae*").
T.: *Platanus* Linnaeus.

Plumbaginaceae A. L. Jussieu, Gen. Pl. 92. 1789 ("*Plumbagines*").
T.: *Plumbago* Linnaeus.

Poaceae Barnhart, Bull. Torrey Bot. Club 22: 7. 1895. – Nom. alt.: *Gramineae*.
T.: *Poa* Linnaeus.

Podophyllaceae A.-P. de Candolle, Syst. Nat. 2: 31. 1821 ("*Podophylleae*").
T.: *Podophyllum* Linnaeus.

Podostemaceae L. C. Richard ex C. A. Agardh, Aphor. Bot. 125. 1822 ("*Podostemeae*").
T.: *Podostemum* A. Michaux.

Polemoniaceae A. L. Jussieu, Gen. Pl. 136. 1789 ("*Polemonia*").
T.: *Polemonium* Linnaeus.

Polygalaceae R. Brown in Flinders, Voy. Terra Austr. 2: 542. 1814 ("*Polygaleae*").
T.: *Polygala* Linnaeus.

Polygonaceae A. L. Jussieu, Gen. Pl. 82. 1789 ("*Polygoneae*").
T.: *Polygonum* Linnaeus.

Pontederiaceae Kunth in Humboldt, Bonpland et Kunth, Nov. Gen. Sp. 1: ed. fol. 211; ed. qu. 265. 1816 ("*Pontedereae*").
T.: *Pontederia* Linnaeus.

Portulacaceae A. L. Jussieu, Gen. Pl. 312. 1789 ("*Portulaceae*").
T.: *Portulaca* Linnaeus.

Posidoniaceae Lotsy, Vortr. Bot. Stammesgesch. 3(1): 658. 1911.
T.: *Posidonia* C. Konig, nom. cons.

Potamogetonaceae Dumortier, Anal. Fam. Pl. 59, 61. 1829 ("*Potamogetoneae*").
T.: *Potamogeton* Linnaeus.
Note: If this family is united with *Juncaginaceae*, the name *Potamogetonaceae* must be used.

Primulaceae Ventenat, Tabl. Règne Vég. 2: 285. 1799.
T.: *Primula* Linnaeus.

Proteaceae A. L. Jussieu, Gen. Pl. 78. 1789 ("*Proteae*").
T.: *Protea* Linnaeus (1771, non 1753), nom. cons.

Pterostemonaceae Small, N. Amer. Fl. 22(2): 183. 1905.
T.: *Pterostemon* Schauer.

Punicaceae Horaninow, Prim. Lin. Syst. Nat. 81. 1834.
T.: *Punica* Linnaeus.

Pyrolaceae Dumortier, Anal. Fam. Pl. 43, 80. 1829.
T.: *Pyrola* Linnaeus.
Note: If this family is united with *Monotropaceae*, the name *Pyrolaceae* must be used.

Quiinaceae Engler in C. F. P. Martius, Fl. Bras. 12(1): 475-476. 1888.
T.: *Quiina* Aublet.

Rafflesiaceae Dumortier, Anal. Fam. Pl. 13, 14. 1829.
T.: *Rafflesia* R. Brown.

Ranunculaceae A. L. Jussieu, Gen. Pl. 231. 1789.
T.: *Ranunculus* Linnaeus.

Rapateaceae Dumortier, Anal. Fam. Pl. 60, 62. 1829.
T.: *Rapatea* Aublet.

Resedaceae S. F. Gray, Nat. Arr. Brit. Pl. 2: 622, 665. 1821.
T.: *Reseda* Linnaeus.

Restionaceae R. Brown, Prodr. 243. 1810 ("*Restiaceae*").
T.: *Restio* Rottböll, nom. cons.

Rhamnaceae A. L. Jussieu, Gen. Pl. 376. 1789 ("*Rhamni*").
T.: *Rhamnus* Linnaeus.

Rhizophoraceae R. Brown in Flinders, Voy. Terra Austr. 2: 549. 1814 ("*Rhizophoreae*").
T.: *Rhizophora* Linnaeus.

Rhoipteleaceae Handel-Mazzetti, Repert. Spec. Nov. Regni Veg. 30: 75. 1932.
T.: *Rhoiptelea* Diels et Handel-Mazzetti.

Roridulaceae Engler et Gilg. Syllabus ed. 9.-10. 226. 1924.
T.: *Roridula* N. L. Burman ex Linnaeus.

Rosaceae A. L. Jussieu, Gen. Pl. 334. 1789.
T.: *Rosa* Linnaeus.

Roxburghiaceae, see *Stemonaceae*.

Rubiaceae A. L. Jussieu, Gen. Pl. 196. 1789.
T.: *Rubia* Linnaeus.

Ruppiaceae Hutchinson, Fam. Fl. Pl. 2: 48. 1934.
T.: *Ruppia* Linnaeus.

Ruscaceae Hutchinson, Fam. Fl. Pl. 2: 109. 1934.
T.: *Ruscus* Linnaeus.

Rutaceae A. L. Jussieu, Gen. Pl. 296. 1789.
T.: *Ruta* Linnaeus.

Sabiaceae Blume, Mus. Bot. 1: 368. 1851.
T.: *Sabia* Colebrooke.

Salicaceae Mirbel, Elém. Phys. Vég. Bot. 2: 905. 1815 ("*Salicineae*").
T.: *Salix* Linnaeus.

Salvadoraceae Lindley, Nat. Syst. Bot. ed. 2. 269. 1836.
T.: *Salvadora* Linnaeus.

Samydaceae Ventenat, Mém. Cl. Sci. Math. Inst. Natl. France 1807(2): 149. 1808 ("*Samydeae*").
T.: *Samyda* N. J. Jacquin, nom. cons.
Note: If this family is united with *Flacourtiaceae*, the name *Samydaceae* is rejected in favour of *Flacourtiaceae*.

Santalaceae R. Brown, Prodr. 350. 1810.
T.: *Santalum* Linnaeus.

Sapindaceae A. L. Jussieu, Gen. Pl. 246. 1789 ("*Sapindi*").
T.: *Sapindus* Linnaeus.

Sapotaceae A. L. Jussieu, Gen. Pl. 151. 1789 ("*Sapotae*").
T.: *Sapota* P. Miller, nom. illeg. (*Achras* Linnaeus).

Sarcolaenaceae Caruel, Atti Reale Accad. Lincei ser. 3. Mem. Cl. Sci. Fis. 10: 226, 248. 1881.
T.: *Sarcolaena* Du Petit-Thouars.

Sarcospermataceae H. J. Lam, Bull. Jard. Bot. Buitenzorg ser. 3. 7: 248. 1925 ("*Sarcospermaceae*").
T.: *Sarcosperma* J. D. Hooker.

Sargentodoxaceae O. Stapf ex Hutchinson, Fam. Fl. Pl. 1: 100. 1926 ("*Sargentadoxaceae*").
T.: *Sargentodoxa* Rehder et Wilson.

Sarraceniaceae Dumortier, Anal. Fam. Pl. 53. 1829.
T.: *Sarracenia* Linnaeus.

Saurauiaceae J. G. Agardh, Theoria Syst. Pl. 110. 1858 ("*Saurajeae*").
T.: *Saurauia* Willdenow, nom. cons.
Note: If this family is united with *Actinidiaceae*, the name *Saurauiaceae* is rejected in favour of *Actinidiaceae*.

Saururaceae E. Meyer, Houttuynia 20. 1827 ("*Saurureae*").
T.: *Saururus* Linnaeus.

Saxifragaceae A. L. Jussieu, Gen. Pl. 308. 1789 ("*Saxifragae*").
T.: *Saxifraga* Linnaeus.

Scheuchzeriaceae Rudolphi, Syst. Orb. Veg. 28. 1830 ("*Scheuchzerieae*").
T.: *Scheuchzeria* Linnaeus.

Schisandraceae Blume, Fl. Javae **32-33**: 3. 1830 ("*Schisandreae*").
T.: *Schisandra* A. Michaux, nom. cons.

Scrophulariaceae A. L. Jussieu, Gen. Pl. 117. 1789 ("*Scrophulariae*").
T.: *Scrophularia* Linnaeus.

Scyphostegiaceae Hutchinson, Fam. Fl. Pl. **1**: 229. 1926.
T.: *Scyphostegia* Stapf.

Scytopetalaceae Engler in Engler et Prantl, Nat. Pflanzenfam. Nachtr. [1]: 242. 1897.
T.: *Scytopetalum* Pierre ex Engler.

Selaginaceae Choisy, Mém. Sélag. 19. 1823 ("*Selagineae*").
T.: *Selago* Linnaeus.

Simaroubaceae A.-P. de Candolle, Ann. Mus. Natl. Hist. Nat. **17**: 422. 1811 ("*Simarubeae*").
T.: *Simarouba* Aublet, nom. cons.

Siphonodontaceae Gagnepain et Tardieu ex Tardieu-Blot, Notul. Syst. (Paris) **14**: 102. 1951.
T.: *Siphonodon* Griffith.

Smilacaceae Ventenat, Tabl. Règne Vég. **2**: 146. 1799 ("*Smilaceae*").
T.: *Smilax* Linnaeus.

Solanaceae A. L. Jussieu, Gen. Pl. 124. 1789 ("*Solaneae*").
T.: *Solanum* Linnaeus.

Sonneratiaceae Engler et Gilg, Syllabus ed. 9.-10. 299. 1924.
T.: *Sonneratia* Linnaeus f., nom. cons.

Sparganiaceae Rudolphi, Syst. Orb. Veg. 27. 1830.
T.: *Sparganium* Linnaeus.

Sphenocleaceae A.-P. de Candolle, Prodr. 7(2): 548. 1839.
T.: *Sphenoclea* J. Gaertner, nom. cons.

Stachyuraceae J. G. Agardh, Theoria Syst. Pl. 152. 1858 ("*Stachyureae*").
T.: *Stachyurus* Siebold et Zuccarini.

Stackhousiaceae R. Brown in Flinders, Voy. Terra Austr. **2**: 555. 1814 ("*Stackhouseae*").
T.: *Stackhousia* J. E. Smith.

Staphyleaceae Lindley, Syn. Brit. Fl. 75. 1829.
T.: *Staphylea* Linnaeus.

Stemonaceae Engler in Engler et Prantl, Nat. Pflanzenfam. 2(5): 8. 1887.
T.: *Stemona* Loureiro.

Stenomeridaceae J. G. Agardh, Theoria Syst. Pl. 66. 1858 ("*Stenomerideae*").
T.: *Stenomeris* J. E. Planchon.

Sterculiaceae Bartling, Ord. Nat. Pl. 255, 340. 1830.
T.: *Sterculia* Linnaeus.
Note: If this family is united with *Byttneriaceae*, the name *Sterculiaceae* must be used.

Stilbaceae Kunth, Handb. Bot. 393. 1831 ("*Stilbineae*").
T.: *Stilbe* P. J. Bergius.

Strasburgeriaceae Engler et Gilg, Syllabus ed. 9.-10. 282. 1924.
T.: *Strasburgeria* Baillon.

Strelitziaceae Hutchinson, Fam. Fl. Pl. **2**: 72. 1934.
T.: *Strelitzia* W. Aiton.

Stylidiaceae R. Brown, Prodr. 565. 1810 ("*Stylideae*").
T.: *Stylidium* Swartz ex Willdenow, nom. cons.

Styracaceae Dumortier, Anal. Fam. Pl. 28, 29. 1829 ("*Styracineae*").
T.: *Styrax* Linnaeus.
Note: If this family is united with *Symplocaceae*, the name *Styracaceae* must be used.

Surianaceae Arnott in Wight et Arnott, Prodr. 360. 1834 ("*Surianeae*").
T.: *Suriana* Linnaeus.

Symplocaceae Desfontaines, Mém. Mus. Hist. Nat. **6**: 9. 1820 ("*Symploceae*").
T.: *Symplocos* N. J. Jacquin.
Note: If this family is united with *Styracaceae*, the name *Symplocaceae* is rejected in favour of *Styracaceae*.

Taccaceae Dumortier, Anal. Fam. Pl. 57, 58. 1829 ("*Tacceae*", p. 58).
T.: *Tacca* J. R. Forster et G. Forster, nom. cons.

Tamaricaceae Link, Enum. Hort. Berol. Alt. **1**: 291. 1821 ("*Tamariscinae*").
T.: *Tamarix* Linnaeus.

Tecophilaeaceae Leybold, Bonplandia **10**: 370. 1862 ("*Tecophileoceae*").
T.: *Tecophilaea* Bertero ex Colla.

Ternstroemiaceae, see *Theaceae.*

Tetracentraceae A. C. Smith, J. Arnold Arbor. **26**: 135. 1945.
T.: *Tetracentron* Oliver.

Tetragoniaceae Nakai, J. Jap. Bot. **18**: 103. 1942.
T.: *Tetragonia* Linnaeus.

Theaceae D. Don, Prodr. Fl. Nepal. 224. 1825.
T.: *Thea* Linnaeus.

Theligonaceae Dumortier, Anal. Fam. Pl. 15, 17. 1829 ("*Theligoneae*").
T.: *Theligonum* Linnaeus.

Theophrastaceae Link, Handbuch **1**: 440. 1829 ("*Theophrasteae*").
T.: *Theophrasta* Linnaeus.

Thismiaceae J. G. Agardh, Theoria Syst. Pl. 99. 1858.
T.: *Thismia* Griffith.

Thurniaceae Engler, Syllabus ed. 5. 94. 1907.
T.: *Thurnia* J. D. Hooker.

Thymelaeaceae A. L. Jussieu, Gen. Pl. 76. 1789 ("*Thymelaeae*").
T.: *Thymelaea* P. Miller, nom. cons.

Tiliaceae A. L. Jussieu, Gen. Pl. 289. 1789.
T.: *Tilia* Linnaeus.

Tovariaceae Pax in Engler et Prantl, Nat. Pflanzenfam. **3**(2): 207. 1891.
T.: *Tovaria* Ruiz et Pavón, nom. cons.

Trapaceae Dumortier, Anal. Fam. Pl. 36, 39. 1829.
T.: *Trapa* Linnaeus.

Tremandraceae R. Brown ex A.-P. de Candolle, Prodr. **1**: 343. 1824 ("*Tremandreae*").
T.: *Tremandra* R. Brown ex A.-P. de Candolle.

Trichopodaceae Hutchinson, Fam. Fl. Pl. **2**: 143. 1934.
T.: *Trichopus* J. Gaertner.

Trigoniaceae Endlicher, Ench. Bot. 570. 1841.
T.: *Trigonia* Aublet.

Trilliaceae Lindley, Veg. Kingd. 218. 1846.
T.: *Trillium* Linnaeus.

Trimeniaceae Gibbs, Fl. Arfak Mts. 135. 1917.
T.: *Trimenia* Seemann.

Triuridaceae G. Gardner, Trans. Linn. Soc. London **19**: 160. 1843 ("*Triuraceae*").
T.: *Triuris* Miers.

Trochodendraceae Prantl in Engler et Prantl, Nat. Pflanzenfam. **3**(2): 21. 1888.
T.: *Trochodendron* Siebold et Zuccarini.

Tropaeolaceae A.-P. de Candolle, Prodr. **1**: 683. 1824 ("*Tropaeoleae*").
T.: *Tropaeolum* Linnaeus.

Turneraceae A.-P. de Candolle, Prodr. **3**: 345. 1828.
T.: *Turnera* Linnaeus.

Typhaceae A. L. Jussieu, Gen. Pl. 25. 1789 ("*Typhae*").
T.: *Typha* Linnaeus.

Ulmaceae Mirbel, Elém. Phys. Vég. Bot. **2**: 905. 1815.
T.: *Ulmus* Linnaeus.

Umbelliferae A. L. Jussieu, Gen. Pl. 218. 1789.
– Nom. alt.: *Apiaceae.*
T.: *Apium* Linnaeus.

Urticaceae A. L. Jussieu, Gen. Pl. 400. 1789 ("*Urticae*").
T.: *Urtica* Linnaeus.

Vacciniaceae S. F. Gray, Nat. Arr. Brit. Pl. **2**: 394, 404. 1821 ("*Vaccinieae*").
T.: *Vaccinium* Linnaeus.

Valerianaceae Batsch, Tab. Affin. Regni Veg. 227. 1802.
T.: *Valeriana* Linnaeus.

Velloziaceae Endlicher, Ench. Bot. 101. 1841 ("*Vellozieae*").
T.: *Vellozia* Vandelli.

Verbenaceae Jaume Saint-Hilaire, Expos. Fam. Nat. **1**: 245. 1805.
T.: *Verbena* Linnaeus.

Violaceae Batsch, Tab. Affin. Regni Veg. 57. 1802 ("*Violariae*").
T.: *Viola* Linnaeus.

Vitaceae A. L. Jussieu, Gen. Pl. 267. 1789 ("*Vites*").
T.: *Vitis* Linnaeus.

Vochysiaceae A. Saint-Hilaire, Mém. Mus. Hist. Nat. **6**: 265. 1820 ("*Vochisieae*").
T.: *Vochysia* Aublet, nom. cons.

Winteraceae Lindley, Intr. Nat. Syst. Bot. 26. 1830 ("*Wintereae*").
T.: *Wintera* Murray, nom. illeg (*Drimys* J. R. Forster et G. Forster).

Xanthorrhoeaceae Dumortier, Anal. Fam. Pl. 60, 62. 1829 ("*Xanthorhaeaceae*", "*Xanthoraea-ceae*").
T.: *Xanthorrhoea* J. E. Smith.

Xyridaceae C. A. Agardh, Aphor. Bot. 158. 1823 ("*Xyrideae*").
T.: *Xyris* Linnaeus.

Zannichelliaceae Dumortier, Anal. Fam. Pl. 59, 61. 1829 ("*Zanichelliaceae*", p. 59).
T.: *Zannichellia* Linnaeus.

Zingiberaceae Lindley, Key Bot. 69. 1835.
T.: *Zingiber* Boehmer, nom. cons.

Zosteraceae Dumortier, Anal. Fam. Pl. 65, 66. 1829.
T.: *Zostera* Linnaeus.

Zygophyllaceae R. Brown in Flinders, Voy. Terra Austr. **2**: 545. 1814 ("*Zygophylleae*").
T.: *Zygophyllum* Linnaeus.

APPENDIX IIIA

In the following lists the **nomina conservanda** have been inserted in the left column; they have been printed in **bold-face** type. Synonyms and earlier homonyms (*nomina rejicienda*) have been listed in the right column.

T.   type (LT. = lectotype).

typ. cons.   typus conservandus, type to be conserved (Arts. 14.3, 14.8 and 10.3).

orth. cons. orthographia conservanda, spelling to be conserved. (In previous editions of the Code "(V)" was used to prefix a rejected original spelling; such cases are now indicated by "(*orth. cons.*)" after the conserved name.) (See Arts. 14.10 and 75.1.)

H   homonym, only the first one being listed (Arts. 14.9 and 64).

≡   nomenclatural synonym, only the earliest legitimate one, if any, being listed. Based on the same nomenclatural type as the conserved name.

=   taxonomic synonym(s) to be rejected only in favour of the conserved name. Based on a type different from that of the conserved name.

*   Conservation approved by the General Committee; use authorized under Art. 15 pending final decision by the next Congress.

Some names listed as conserved have no corresponding nomina rejicienda because they were conserved explicitly to conserve a particular type, because evidence after their conservation may have indicated that conservation was unnecessary, or because they were conserved to eliminate doubt about their legitimacy.

Anabaena Bory de Saint-Vincent ex Bornet et Flahault, Ann. Sci. Nat. Bot. ser. 7. 7: 180, 224. 1886 ("1888").
T.: *A. oscillarioides* Bory de Saint-Vincent ex Bornet et Flahault.

(H) *Anabaena* A. H. L. Jussieu, Euphorb. Gen. 46, *t. 15, f. 48.* 1824 [Euphorb.].
T.: *A. tamnoides* A. H. L. Jussieu.

Aphanothece Nägeli, Neue Denkschr. Allg. Schweiz. Ges. Gesammten Naturwiss. **10**(7): 59. 1849.
T.: *A. microscopica* Nägeli.

(=) *Coccochloris* K. Sprengel, Mant. Prim. Fl. Hal. 14. 1807.
T.: *C. stagnina* K. Sprengel.

Gloeocapsa Kützing, Phycol. General. 173. 1843.
T.: *G. atrata* Kützing, nom. illeg. (*Microcystis atra* Kützing).

(=) *Bichatia* Turpin, Mém. Mus. Hist. Nat. (Paris) **15**: 376. 1827; *ibid.* **16**: 163 (adnot.), *t. 11, f. 10.* 1828; *ibid.* **18**: 177, *t. 5, f. 1-14.* 1829.
T.: *B. vesiculinosa* Turpin.

Homoeothrix (Thuret ex Bornet et Flahault) Kirchner in Engler et Prantl, Nat. Pflanzenfam. 1(1a): 85, 87. 1898.
T.: *H. juliana* (Bornet et Flahault) Kirchner (*Calothrix juliana* Bornet et Flahault).

(=) *Amphithrix* Bornet et Flahault, Ann. Sci. Nat. Bot. ser. 7. **3**: 340, 343. 1886.
LT.: *A. janthina* Bornet et Flahault (vide Geitler in Engler et Prantl, Nat. Pflanzenfam. ed. 2. **1b**: 175. 1942).

(=) *Tapinothrix* Sauvageau, Bull. Soc. Bot. France **39**: cxxiii. 1892.
T.: *T. bornetii* Sauvageau.

Lyngbya C. Agardh ex Gomont, Ann. Sci. Nat. Bot. ser. 7. **16**: 95, 118. 1892 ("1893").
T.: *L. confervoides* C. Agardh ex Gomont.

(H) *Lyngbyea* Sommerfelt, Suppl. Fl. Lapp. 189. 1826 [Bacillarioph.].
T.: non designatus.

Microchaete Thuret ex Bornet et Flahault, Ann. Sci. Nat. Bot. ser. 7. **5**: 82, 83. 1886 ("1887").
T.: *M. grisea* Thuret ex Bornet et Flahault.

(H) *Microchaete* Bentham, Pl. Hartw. 209. 1845 [Comp.].
LT.: *M. pulchella* (Kunth) Bentham (*Cacalia pulchella* Kunth) (vide Pfeiffer, Nomencl. Bot. **2**: 304. 1874).

Microcystis Lemmermann, Kryptogamenfl. Mark Brandenburg **3**: 45, 72. 1907.
T.: *M. aeruginosa* (Kützing) Lemmermann (*Micraloa aeruginosa* Kützing) (*typ. cons.*).

(H) *Microcystis* Kützing, Linnaea **8**: 372. 1833 [Euglenoph.: Euglen.].
≡ *Haematococcus* C. Agardh 1830 (*nom. rej.*).

(≡) *Diplocystis* Trevisan, Sagg. Algh. Coccot. 40. 1848 (vide Drouet et Daily, Butler Univ. Bot. Stud. **12**: 34. 1956).

Nodularia Mertens ex Bornet et Flahault, Ann. Sci. Nat. Bot. ser. 7. **7**: 180, 243. 1886 ("1888").
T.: *N. spumigena* Mertens ex Bornet et Flahault.

(H) *Nodularia* Link ex Lyngbye, Tent. Hydrophytol. Dan. xxx, 99. 1819 [Rhodoph.: Leman.].
≡ *Lemanea* Bory de Saint-Vincent 1808 (*nom. cons.*).

Rivularia C. Agardh ex Bornet et Flahault, Ann. Sci. Nat. Bot. ser. 7. **3**: 341; *ibid.* **4**: 345. 1886.
T.: *R. atra* Roth ex Bornet et Flahault.

(H) *Rivularia* Roth, Catal. Bot. 1: 212. 1797 [Chloroph.: Chaetophor.].
LT.: *R. cornu-damae* Roth (vide Hazen, Mem. Torrey Bot. Club **11**: 210. 1902).

Trichodesmium Ehrenberg ex Gomont, Ann. Sci. Nat. Bot. ser. 7. **16**: 96, 193. 1892 ("1893").
T.: *T. erythraeum* Ehrenberg ex Gomont.

(H) *Trichodesmium* Chevallier, Fl. Gén. Env. Paris **1**: 382. 1826 [Fungi].
≡ *Graphiola* Poiteau 1824.

II. RHODOPHYCEAE

Areschougia W. H. Harvey, Trans. Roy. Irish Acad. **22** (Sci.): 554. 1855.
T.: *A. laurencia* (J. D. Hooker et W. H. Harvey) W. H. Harvey (*Thamnocarpus laurencia* J. D. Hooker et W. H. Harvey).

(H) *Areschougia* Meneghini, Giorn. Bot. Ital. **1**: 293. 1844 [Phaeoph.: Elachist.].
≡ *Centrospora* Trevisan 1845.

Audouinella Bory de Saint-Vincent, Dict. Class. Hist. Nat. **3**: 340. 1823 ("*Auduinella*") (*orth. cons.*).
T.: *A. miniata* Bory de Saint-Vincent [= *A. hermannii* (Roth) Duby (*Conferva hermannii* Roth)].

Bostrychia Montagne in Sagra, Hist. Phys. Cuba, Bot., Pl. Cell. 39. 1842.
T.: *B. scorpioides* (Hudson) Montagne (Dict. Univ. Hist. Nat. **2**: 661. 1842) (*Fucus scorpioides* Hudson).

(H) *Bostrychia* Fries, Kongl. Vetensk. Acad. Handl. **39**: 117. 1818 : Fries, Syst. Mycol. **1**: lii. 1821 [Fungi].
T.: *B. chrysosperma* (Persoon : Fries) Fries (*Sphaeria chrysosperma* Persoon : Fries).
(≡) *Amphibia* Stackhouse, Mém. Soc. Imp. Naturalistes Moscou **2**: 58, 89. 1809.

Botryocladia (J. Agardh) Kylin, Lunds Univ. Årsskr. ser. 2. sect. 2. **27**(11): 17. 1931.
T.: *B. uvaria* Kylin [= *B. botryoides* (Wulfen) J. Feldmann (*Fucus botryoides* Wulfen)].

(=) *Myriophylla* Holmes, Ann. Bot. (London) **8**: 340. 1894.
T.: *M. beckeriana* Holmes.

Calliblepharis Kützing, Phycol. General. 403. 1843.
T.: *C. ciliata* (Hudson) Kützing (*Fucus ciliatus* Hudson).

(≡) *Ciliaria* Stackhouse, Mém. Soc. Imp. Naturalistes Moscou **2**: 54, 70. 1809 (vide Papenfuss, Hydrobiologia **2**: 191. 1950).

Catenella Greville, Alg. Brit. lxiii, 166. 1830.
T.: *C. opuntia* (Goodenough et Woodward) Greville (*Fucus opuntia* Goodenough et Woodward) [= *C. caespitosa* (Withering) L. Irvine (*Ulva caespitosa* Withering)].

(=) *Clavatula* Stackhouse, Mém. Soc. Imp. Naturalistes Moscou **2**: 95, 97. 1809.
T.: *C. caespitosa* Stackhouse (*Fucus caespitosus* Stackhouse, non Forsskål).

Ceramium Roth, Catal. Bot. **1**: 146. 1797.
T.: *C. virgatum* Roth [= *C. rubrum* (Hudson) C. Agardh (*Conferva rubra* Hudson)] (*typ. cons.*).

(H) *Ceramion* Adanson, Fam. Pl. **2**: 13. 1763 [Rhodoph.: Gracilar.].
≡ *Ceramianthemum* Donati ex Léman 1817 (*nom. rej.*).

Chondria C. Agardh, Syn. Alg. Scand. xviii. 1817.
T.: *C. tenuissima* (Withering) C. Agardh (*Fucus tenuissimus* Withering).

(=) *Dasyphylla* Stackhouse, Nereis Brit. ed. 2. ix, xi. 1816.
LT.: *D. woodwardii* Stackhouse (*Fucus dasyphyllus* Woodward) (vide Papenfuss, Hydrobiologia **2**: 192. 1950).

113

Chylocladia Greville in W. J. Hooker, Brit. Fl. 2(1): 256, 297. 1833.
T.: *C. kaliformis* (Withering) Greville (*Fucus kaliformis* Withering) [= *C. verticillata* (Lightfoot) Bliding (*Fucus verticillatus* Lightfoot)].

(≡) *Kaliformis* Stackhouse, Mém. Soc. Imp. Naturalistes Moscou **2**: 56, 78. 1809 (vide Papenfuss, Hydrobiologia **2**: 198. 1950).

Corynomorpha J. Agardh, Lunds Univ. Årsskr. 8 (sect. 3, n. 6): 3. 1872.
T.: *C. prismatica* (J. Agardh) J. Agardh (*Dumontia prismatica* J. Agardh).

(≡) *Prismatoma* (J. Agardh) Harvey, Index Gen. Alg. 11. 1860.

Cryptopleura Kützing, Phycol. General. 444. 1843.
T.: *C. lacerata* (S. G. Gmelin) Kützing (*Fucus laceratus* S. G. Gmelin) [= *C. ramosa* (Hudson) Kylin ex Newton (*Ulva ramosa* Hudson)].

(H) *Cryptopleura* Nuttall, Trans. Amer. Philos. Soc. ser. 2. **7**: 431. 1841 [Comp.].
T.: *C. californica* Nuttall.

(≡) *Papyracea* Stackhouse, Mém. Soc. Imp. Naturalistes Moscou **2**: 56, 76. 1809 (vide Papenfuss, Index Nom. Gen. card 00816. 1955).

Dasya C. Agardh, Syst. Alg. xxxiv, 211. 1824 ("*Dasia*") (*orth. cons.*).
T.: *D. pedicellata* (C. Agardh) C. Agardh (*Sphaerococcus pedicellatus* C. Agardh) [= *D. baillouviana* (S. G. Gmelin) Montagne (*Fucus baillouviana* S. G. Gmelin)].

(=) *Baillouviana* Adanson, Fam. Pl. **2**: 13. 1763.
T.: *Fucus baillouviana* S. G. Gmelin.

Delesseria Lamouroux, Ann. Mus. Hist. Nat. (Paris) **20**: 122. 1813.
T.: *D. sanguinea* (Hudson) Lamouroux (*Fucus sanguineus* Hudson).

(≡) *Hydrolapatha* Stackhouse, Mém. Soc. Imp. Naturalistes Moscou **2**: 54, 67. 1809 (vide Papenfuss, Hydrobiologia **2**: 196. 1950).

Dudresnaya P. Crouan et H. Crouan, Ann. Sci. Nat. Bot. ser. 2. **3**: 98, *t. 2, f. 2-3.* 1835.
T.: *D. coccinea* (C. Agardh) P. Crouan et H. Crouan (*Mesogloia coccinea* C. Agardh) [= *D. verticillata* (Withering) Le Jolis (*Ulva verticillata* Withering)] (*typ. cons.*).

(H) *Dudresnaya* Bonnemaison, J. Phys. Chim. Hist. Nat. Arts **94**: 180. 1822 [Phaeoph.: Chordar.].
T.: *Alcyonidium vermiculatum* (J. E. Smith) Lamouroux (*Rivularia vermiculata* J. E. Smith).

(=) *Borrichius* S. F. Gray, Nat. Arr. Brit. Pl. **1**: 317, 330. 1821.
T.: *B. gelatinosus* S. F. Gray, nom. illeg. (*Ulva verticillata* Withering).

Erythrotrichia J. E. Areschoug, Nova Acta Regiae Soc. Sci. Upsal. ser. 2. **14**: 435. 1850.
T.: *E. ceramicola* (Lyngbye) J. E. Areschoug (*Conferva ceramicola* Lyngbye) [= *E. carnea* (Dillwyn) J. Agardh (*Conferva carnea* Dillwyn)].

(≡) *Goniotrichum* Kützing, Phycol. General. 244. 1843.
(=) *Porphyrostromium* Trevisan, Sagg. Algh. Coccot. 100 (adnot.). 1848.
T.: *P. boryi* Trevisan, nom. illeg. (*Porphyra boryana* Montagne).

Furcellaria Lamouroux, Ann. Mus. Hist. Nat. (Paris) **20**: 45. 1813.
T.: *F. lumbricalis* (Hudson) Lamouroux (*Fucus lumbricalis* Hudson).

(=) *Fastigiaria* Stackhouse, Mém. Soc. Imp. Naturalistes Moscou **2**: 59, 90. 1809.
LT.: *F. linnaei* Stackhouse, nom. illeg. (*Fucus fastigiatus* Linnaeus) (vide Papenfuss, Hydrobiologia **2**: 194. 1950).

Gastroclonium Kützing, Phycol. General. 441. 1843.
T.: *G. ovale* Kützing, nom. illeg. (*Fucus ovalis* Hudson, nom. illeg., *Fucus ovatus* Hudson, *G. ovatum* (Hudson) Papenfuss).

(=) *Sedoidea* Stackhouse, Mém. Soc. Imp. Naturalistes Moscou **2**: 57, 83. 1809.
LT.: *Fucus sedoides* Goodenough et Woodward, nom. illeg. (*Fucus vermicularis* S. G. Gmelin) (vide Papenfuss, Hydrobiologia **2**: 202. 1950).

Gelidium Lamouroux, Ann. Mus. Hist. Nat. (Paris) **20**: 128. 1813.
T.: *G. corneum* (Hudson) Lamouroux (*Fucus corneus* Hudson).

(≡) *Cornea* Stackhouse, Mém. Soc. Imp. Naturalistes Moscou **2**: 57, 83. 1809 (vide Papenfuss, Hydrobiologia **2**: 191-192. 1950).

Gracilaria Greville, Alg. Brit. liv, 121. 1830.
T.: *G. confervoides* (Stackhouse) Greville (*Flagellaria confervoides* Stackhouse, *Fucus confervoides* Linnaeus, non Hudson) [= *G. verrucosa* (Hudson) Papenfuss (*Fucus verrucosus* Hudson)].

(=) *Ceramianthemum* Donati ex Léman, Dict. Sci. Nat. **7**: 421. 1817.
T.: *Gracilaria bursa-pastoris* (S. G. Gmelin) P. C. Silva (*Univ. Calif. Publ. Bot.* **25**: 265. 1952) (*Fucus bursa-pastoris* S. G. Gmelin) (vide Ardissone, Mem. Soc. Crittog. Ital. **1**: 240-241. 1883).
(=) *Plocaria* C. G. Nees, Horae Phys. Berol. 42. 1820.
T.: *P. candida* C. G. Nees.

Helminthocladia J. Agardh, Sp. Alg. **2**: 412. 1851.
T.: *H. purpurea* (W. H. Harvey) J. Agardh (*Mesogloia purpurea* W. H. Harvey) [= *H. calvadosii* (Lamouroux ex Turpin) Setchell (*Dumontia calvadosii* Lamouroux ex Turpin)].

(H) *Helminthocladia* W. H. Harvey, Gen. S. Afr. Pl. 396. 1838 [Phaeoph.: Chordar.].
≡ *Mesogloia* C. Agardh 1817.

Helminthora J. Agardh, Sp. Alg. **2**: 415. 1851.
T.: *H. divaricata* (C. Agardh) J. Agardh (*Mesogloia divaricata* C. Agardh).

(H) *Helminthora* Fries, Syst. Orb. Veg. 341. 1825; Fl. Scan. 311. 1835 [Rhodoph.: Helminthoclad.].
T.: *H. multifida* (Weber et Mohr) Fries (*Rivularia multifida* Weber et Mohr).

Heterosiphonia Montagne, Prodr. Gen. Phyc. 4. 1842.
T.: *H. berkeleyi* Montagne.

(=) *Ellisius* S. F. Gray, Nat. Arr. Brit. Pl. **1**: 317, 333. 1821.
LT.: *E. coccineus* (Hudson) S. F. Gray (*Conferva coccinea* Hudson) (vide Silva, Univ. Calif. Publ. Bot. **25**: 290. 1952).

Hildenbrandia Nardo, Isis (Oken) **1834**: 676. 1834 ("*Hildbrandtia*") (*orth. cons.*).
T.: *H. prototypus* Nardo.

Iridaea Bory de Saint-Vincent, Dict. Class. Hist. Nat. **9**: 15. 1826 ("*Iridaea*", "*Iridea*").
T.: *I. cordata* (D. Turner) Bory de Saint-Vincent (*Fucus cordatus* D. Turner).

(H) *Iridea* Stackhouse, Nereis Brit. ed. 2. ix, xii. 1816 [Phaeoph.: Desmarest.].
≡ *Hyalina* Stackhouse 1809 (*nom. rej.*).

Laurencia Lamouroux, Ann. Mus. Hist. Nat. (Paris) **20**: 130. 1813.
T.: *L. obtusa* (Hudson) Lamouroux (*Fucus obtusus* Hudson).

(=) *Osmundea* Stackhouse, Mém. Soc. Imp. Naturalistes Moscou **2**: 56, 79. 1809.
LT.: *O. expansa* Stackhouse, nom. illeg. (*Fucus osmunda* S. G. Gmelin) (vide Silva, Univ. Calif. Publ. Bot. **25**: 292. 1952).

Lemanea Bory de Saint-Vincent, Ann. Mus. Hist. Nat. (Paris) **12**: 178. 1808.
T.: *L. corallina* Bory de Saint-Vincent, nom. illeg. (*Conferva fluviatilis* Linnaeus, *L. fluviatilis* (Linnaeus) C. Agardh).

(≡) *Apona* Adanson, Fam. Pl. **2**: 2, 519. 1763.

Lenormandia Sonder, Bot. Zeitung (Berlin) **3**: 54. 1845.
T.: *L. spectabilis* Sonder.

(H) *Lenormandia* Delise in Desmazières, Pl. Crypt. N. France no. 1144. 1841 [Fungi: Lich.].
T.: *L. jungermanniae* Delise.

Lithothamnion Heydrich, Ber. Deutsch. Bot. Ges. **15**: 412. 1897 (*orth. cons.*).
T.: *L. muelleri* Lenormand ex Rosanoff (Mém. Soc. Sci. Nat. Cherbourg **12**: 101. 1866) (*typ. cons.*).

(H) *Lithothamnium* Philippi, Arch. Naturgesch. **3**(1): 387. 1837 [Rhodophyc.: Corallin.].
LT.: *L. ramulosum* Philippi (vide Mason, Univ. Calif. Publ. Bot. **26**: 322. 1953).

Martensia Hering, Ann. Mag. Nat. Hist. **8**: 92. 1841.
T.: *M. elegans* Hering.

(H) *Martensia* Giseke, Prael. Ord. Nat. Pl. 207, 227, 249. 1792 [Zingiber].
T.: *M. aquatica* (Retzius) Giseke (*Heritiera aquatica* Retzius).

Nemastoma J. Agardh, Alg. Medit. 89. 1842 ("*Nemostoma*") (*orth. cons.*).
T.: *N. dichotomum* J. Agardh.

Nitophyllum Greville, Alg. Brit. xlvii, 77. 1830.
T.: *N. punctatum* (Stackhouse) Greville (*Ulva punctata* Stackhouse).

(=) *Scutarius* Roussel, Fl. Calvados ed. 2. 91. 1806.
LT.: *Fucus ocellatus* Lamouroux (vide Silva, Univ. Calif. Publ. Bot. **25**: 268. 1952).

Odonthalia Lyngbye, Tent. Hydrophytol. Dan. xxix, 9. 1819.
T.: *O. dentata* (Linnaeus) Lyngbye (*Fucus dentatus* Linnaeus).

(≡) *Fimbriaria* Stackhouse, Mém. Soc. Imp. Naturalistes Moscou **2**: 95, 96. 1809 (vide Silva, Univ. Calif. Publ. Bot. **25**: 269. 1952).

Phacelocarpus Endlicher et Diesing, Bot. Zeitung (Berlin) **3**: 289. 1845.
T.: *P. tortuosus* Endlicher et Diesing.

(=) *Ctenodus* Kützing, Phycol. General. 407. 1843.
T.: *C. labillardierei* (Mertens ex D. Turner) Kützing (*Fucus labillardierei* Mertens ex D. Turner).

Phyllophora Greville, Alg. Brit. lvi, 135. 1830.
T.: *P. crispa* (Hudson) Dixon (Bot. Not. **117**: 63: 1964) (*Fucus crispus* Hudson) (*typ. cons.*).

(≡) *Epiphylla* Stackhouse, Nereis Brit. ed. 2. x, xii. 1816.
(=) *Membranifolia* Stackhouse, Mém. Soc. Imp. Naturalistes Moscou **2**: 55, 75. 1809.
LT.: *M. lobata* Stackhouse, nom. illeg. (*Fucus membranifolius* Goodenough et Woodward, nom. illeg., *Fucus pseudoceranoides* S. G. Gmelin) (vide Papenfuss, Hydrobiologia **2**: 198. 1950).

Pleonosporium Nägeli, Sitzungsber. Bayer. Akad. Wiss. München **1861**(2): 326, 339. 1862.
T.: *P. borreri* (J. E. Smith) Nägeli (*Conferva borreri* J. E. Smith).

Plocamium Lamouroux, Ann. Mus. Hist. Nat. (Paris) **20**: 137. 1813.
T.: *P. vulgare* Lamouroux, nom. illeg. (*Fucus plocamium* S. G. Gmelin, nom. illeg., *Fucus cartilagineus* Linnaeus, *P. cartilagineum* (Linnaeus) Dixon).

(≡) *Nereidea* Stackhouse, Mém. Soc. Imp. Naturalistes Moscou **2**: 58, 86. 1809 (vide Silva, Univ. Calif. Publ. Bot. **25**: 264. 1952).

Plumaria Schmitz, Nuova Notarisia **7**: 5. 1896.
T.: *P. elegans* (Bonnemaison) Schmitz (*Ptilota elegans* Bonnemaison) (*typ. cons.*).

(H) *Plumaria* Stackhouse, Mém. Soc. Imp. Naturalistes Moscou **2**: 58, 86. 1809 [Rhodoph.: Ceram.].
T.: *P. pectinata* (Gunnerus) Stackhouse (*Fucus pectinatus* Gunnerus).

Polyneura (J. Agardh) Kylin, Lunds Univ. Årsskr. ser. 2. sect. 2. **20**(6): 33. 1924.
T.: *P. hilliae* (Greville) Kylin (*Delesseria hilliae* Greville).

(H) *Polyneura* J. Agardh, Lunds Univ. Årsskr. **35** (sect. 2, n. 4): 60. 1899 [Rhodoph.: Kallymen.].
T.: *P. californica* J. Agardh.

Polysiphonia Greville, Scott. Crypt. Fl. *t. 90*. 1823.
T.: *P. urceolata* (Dillwyn) Greville (*Conferva urceolata* Dillwyn) (*typ. cons.*).

(H) *Polysiphonia* Greville, Scott. Crypt. Fl. *t. 90*. 1823 [Rhodoph.: Rhodomel.].
LT.: *P. violacea* (Roth) K. Sprengel (*Ceramium violaceum* Roth) (vide Schmitz, Flora **72**: 448. 1889).
(=) *Grammita* Bonnemaison, J. Phys. Chim. Hist. Nat. Arts **94**: 186. 1822.
T.: *Conferva fucoides* Hudson.
(=) *Vertebrata* S. F. Gray, Nat. Arr. Brit. Pl. **1**: 317, 338. 1821.
T.: *V. fastigiata* S. F. Gray, nom. illeg. (*Conferva polymorpha* Linnaeus).
(=) *Grateloupella* Bory de Saint-Vincent, Dict. Class. Hist. Nat. **3**: 340. 1823 ("*Gratelupella*").
LT.: *Ceramium brachygonium* Lyngbye (vide Bory de Saint-Vincent, Dict. Class. Hist. Nat. **7**: 481. 1825).

Porphyra C. Agardh, Syst. Alg. xxxii, 190. 1824.
T.: *P. purpurea* (Roth) C. Agardh (*Ulva purpurea* Roth).

(H) *Porphyra* Loureiro, Fl. Cochinch. 69. 1790 [Verben.].
T.: *P. dichotoma* Loureiro.
(=) *Phyllona* J. Hill, Hist. Pl. ed. 2. 79. 1773.
T.: non designatus.

Porphyridium Nägeli, Neue Denkschr. Allg. Schweiz. Ges. Gesammten Naturwiss. **10**(7): 71, 138. 1849.
T.: *P. cruentum* (S. F. Gray) Nägeli (*Olivia cruenta* S. F. Gray).

(=) *Chaos* Bory de Saint-Vincent ex Desmazières, Cat. Pl. Omises Botanogr. Belgique 1. 1823.
T.: *C. sanguinarius* Bory de Saint-Vincent ex Desmazières, nom. illeg. (*Phytoconis purpurea* Bory de Saint-Vincent).
(=) *Sarcoderma* Ehrenberg, Ann. Phys. Chem. **94**: 504. 1830.
T.: *S. sanguineum* Ehrenberg.

Prionitis J. Agardh, Sp. Alg. 2: 185. 1851.
T.: *P. ligulata* J. Agardh [= *P. lanceolata* (W. H. Harvey) W. H. Harvey (*Gelidium lanceolatum* W. H. Harvey)].

(H) *Prionitis* Adanson, Fam. Pl. 2: 499, 594. 1763 [Umbell.].
≡ *Falcaria* Fabricius 1759 (*nom. cons.*) (6018).

Ptilota C. Agardh, Syn. Alg. Scand. xix. 1817.
T.: *P. plumosa* (Hudson) C. Agardh (*Fucus plumosus* Hudson).

Rhodomela C. Agardh, Sp. Alg. 1: 368. 1822.
T.: *R. subfusca* (Woodward) C. Agardh (*Fucus subfuscus* Woodward) [= *R. confervoides* (Hudson) P. C. Silva (*Fucus confervoides* Hudson)].

(=) *Fuscaria* Stackhouse, Mém. Soc. Imp. Naturalistes Moscou 2: 59, 93. 1809.
T.: *F. variabilis* Stackhouse, nom. illeg. (*Fucus variabilis* Withering, nom. illeg., *Fucus confervoides* Hudson).

Rhodophyllis Kützing, Bot. Zeitung (Berlin) 5: 23. 1847.
T.: *R. bifida* Kützing, nom. illeg. (*Fucus bifidus* Hudson, non S. G. Gmelin; *Bifida divaricata* Stackhouse, *R. divaricata* (Stackhouse) Papenfuss).

(≡) *Bifida* Stackhouse, Mém. Soc. Imp. Naturalistes Moscou 2: 95, 97. 1809 (vide Silva, Univ. Calif. Publ. Bot. 25: 264. 1952).
(=) *Inochorion* Kützing, Phycol. General. 443. 1843.
T.: *I. dichotomum* Kützing.

Rhodymenia Greville, Alg. Brit. xlviii, 84. 1830 ("*Rhodomenia*") (*orth. cons.*).
T.: *R. palmetta* (Lamouroux) Greville (*Delesseria palmetta* Lamouroux) [= *R. pseudopalmata* (Lamouroux) P. C. Silva (*Fucus pseudopalmatus* Lamouroux)].

Suhria J. Agardh ex Endlicher, Gen. Pl. Suppl. 3: 41. Oct 1843.
T.: *S. vittata* (Linnaeus) J. Agardh ex Endlicher (*Fucus vittatus* Linnaeus).

(=) *Chaetangium* Kützing, Phycol. General. 392. 14-16 Sep 1843.
T.: *C. ornatum* (Linnaeus) Kützing (*Fucus ornatus* Linnaeus).

Vidalia Lamouroux ex J. Agardh, Sp. Alg. 2: 1117. 1863.
T.: *V. spiralis* (Lamouroux) Lamouroux ex J. Agardh (*Delesseria spiralis* Lamouroux).

(=) *Volubilaria* Lamouroux ex Bory de Saint-Vincent, Dict. Class. Hist. Nat. 16: 630. 1830.
T.: *V. mediterranea* Lamouroux ex Bory de Saint-Vincent, nom. illeg. (*Fucus volubilis* Linnaeus).
(=) *Spirhymenia* Decaisne, Arch. Mus. Hist. Nat. (Paris) 2: 177. 1841; Ann. Sci. Nat. Bot. ser. 2. 17: 355. 1842.
T.: *S. serrata* (Suhr) Decaisne (*Carpophyllum serratum* Suhr, "*denticulatum*" lapsu).
(=) *Epineuron* W. H. Harvey, London J. Bot. 4: 532. 1845.
LT.: *E. colensoi* J. D. Hooker et W. H. Harvey (vide Silva, Univ. Calif. Publ. Bot. 25: 293. 1952).

Actinella F. W. Lewis, Proc. Acad. Nat. Sci. Philadelphia 1863: 343. 1864.
T.: *A. punctata* F. W. Lewis.

(H) *Actinella* Persoon, Syn. Pl. 2: 469. 1807 [Comp.].
≡ *Actinea* A. L. Jussieu 1803.

Arachnoidiscus Deane ex Shadbolt, Trans. Microscop. Soc. London 3: 49, *t. 11*. 1852.
T.: *A. japonicus* Shadbolt ex Pritchard (Hist. Infus. ed. 3. 319. 1852).

(H) *Arachnodiscus* J. W. Bailey ex Ehrenberg, Ber. Bekanntm. Verh. Königl. Preuss. Akad. Wiss. Berlin 1849: 63 (adnot.). 1849 [Bacillarioph.].
≡ *Hemiptychus* Ehrenberg 1848 (*nom. rej.*).
(=) *Hemiptychus* Ehrenberg, Ber. Bekanntm. Verh. Königl. Preuss. Akad. Wiss. Berlin 1848: 7. 1848.
T.: *H. ornatus* Ehrenberg.

Aulacodiscus Ehrenberg, Ber. Bekanntm. Verh. Königl. Preuss. Akad. Wiss. Berlin 1844: 73. 1844.
T.: *A. crux* Ehrenberg.

(=) *Tripodiscus* Ehrenberg, Abh. Königl. Akad. Wiss. Berlin, Phys. Kl. 1839: 130. 1841 [praeimpr. 1840. p. 50].
T.: *T. germanicus* Ehrenberg (*T. argus* Ehrenberg, nom. altern.).
(=) *Pentapodiscus* Ehrenberg, Ber. Bekanntm. Verh. Königl. Preuss. Akad. Wiss. Berlin 1843: 165. 1843.
T.: *P. germanicus* Ehrenberg.
(=) *Tetrapodiscus* Ehrenberg, Ber. Bekanntm. Verh. Königl. Preuss. Akad. Wiss. Berlin 1843: 165. 1843.
T.: *T. germanicus* Ehrenberg.

Auricula Castracane, Atti Accad. Pontif. Sci. Nuovi Lincei 26: 407. 1873.
T.: *A. amphitritis* Castracane.

(H) *Auricula* J. Hill, Brit. Herb. 98. 1756 [Primul.].
T.: non designatus.

Brebissonia Grunow, Verh. K. K. Zool.-Bot. Ges. Wien 10: 512. 1860.
T.: *B. boeckii* (Ehrenberg) O'Meara (Proc. Roy. Irish Acad. ser. 2. 2 (Sci.): 338. 1875) (*Cocconema boeckii* Ehrenberg).

(H) *Brebissonia* Spach, Hist. Nat. Vég. 4: 401. 11 Apr 1835 [Onagr.].
LT.: *B. microphylla* (Kunth) Spach (*Fuchsia microphylla* Kunth) (vide Spach, Ann. Sci. Nat. Bot. ser. 2. 4: 175. Sep 1835).

Cerataulina H. Peragallo ex Schütt in Engler et Prantl, Nat. Pflanzenfam. 1(1b): 95. 1896.
T.: *C. bergonii* (H. Peragallo) Schütt (*Cerataulus bergonii* H. Peragallo) [= *C. pelagica* (Cleve) Hendey (*Zygoceras pelagicum* Cleve)].

(=) *Syringidium* Ehrenberg, Ber. Bekanntm. Verh. Königl. Preuss. Akad. Wiss. Berlin 1845: 357. 1845.
T.: *S. bicorne* Ehrenberg.

Cyclotella (Kützing) Brébisson, Consid. Diatom. 19. 1838.
T.: Kützing 139, **BM** (*typ. cons.*).
≡ *Cyclotella tecta* Håkansson et Ross (Taxon 33: 529. 1984).

119

Cymatopleura W. Smith, Ann. Mag. Nat. Hist. ser. 2. **7**: 12. 1851.
T.: *C. solea* (Brébisson) W. Smith (*Cymbella solea* Brébisson) [= *C. librile* (Ehrenberg) Pantocsek (*Navicula librile* Ehrenberg)].

(=) *Sphinctocystis* Hassall, Hist. Brit. Freshwater Alg. 436. 1845.
T.: *S. librile* (Ehrenberg) Hassall (*Navicula librile* Ehrenberg).

Diatoma Bory de Saint-Vincent, Dict. Class. Hist. Nat. **5**: 461. 1824.
T.: *D. vulgaris* Bory de Saint-Vincent (*typ. cons.*).

(H) *Diatoma* Loureiro, Fl. Cochinch. 295. 1790 [Rhizophor.].
T.: *D. brachiata* Loureiro.

Diatomella Greville, Ann. Mag. Nat. Hist. ser. 2. **15**: 259. 1855.
T.: *D. balfouriana* Greville.

(=) *Disiphonia* Ehrenberg, Mikrogeologie 260, *t. 35A, II, f. 7.* 1854.
T.: *D. australis* Ehrenberg.

Didymosphenia M. Schmidt in A. Schmidt, Atlas Diatom.-Kunde *t. 214.* 1899.
T.: *D. geminata* (Lyngbye) M. Schmidt (*Echinella geminata* Lyngbye).

(≡) *Dendrella* Bory de Saint-Vincent, Dict. Class. Hist. Nat. **5**: 393. 1824 (vide Regnum Veg. **3**: 70. 1952).

(=) *Diomphala* Ehrenberg, Ber. Bekanntm. Verh. Königl. Preuss. Akad. Wiss. Berlin **1842**: 336. 1843.
T.: *D. clava-herculis* Ehrenberg.

Eupodiscus J. W. Bailey, Smithson. Contr. Knowl. **2**(8): 39. 1851.
T.: *E. radiatus* J. W. Bailey (*typ. cons.*).

(H) *Eupodiscus* Ehrenberg, Ber. Bekanntm. Verh. Königl. Preuss. Akad. Wiss. Berlin **1844**: 73. 1844 [Bacillarioph.].
≡ *Tripodiscus* Ehrenberg 1841 (*nom. rej.*).

Frustulia Rabenhorst, Süssw.-Diatom. 50. 1853.
T.: *F. saxonica* Rabenhorst (*typ. cons.*).

(H) *Frustulia* C. Agardh, Syst. Alg. xiii, 1. 1824 [Bacillarioph.].
T.: non designatus.

Gomphonema Ehrenberg, Abh. Königl. Akad. Wiss. Berlin, Phys. Kl. **1831**: 87. 1832.
T.: *G. acuminatum* Ehrenberg (*typ. cons.*).

(H) *Gomphonema* C. Agardh, Syst. Alg. xvi, 11. 1824 [Bacillarioph.].
≡ *Didymosphenia* M. Schmidt (*nom. cons.*).

Gyrosigma Hassall, Hist. Brit. Freshwater Alg. 435. 1845.
T.: *G. hippocampus* (Ehrenberg) Hassall (*Navicula hippocampus* Ehrenberg).

(=) *Scalptrum* Corda, Alman. Carlsbad **5**: 193, *t. 5, f. 70.* 1835.
T.: *S. striatum* Corda.

Hantzschia Grunow, Monthly Microscop. J. **18**: 174. 1877.
T.: *H. amphioxys* (Ehrenberg) Grunow (*Eunotia amphioxys* Ehrenberg).

(H) *Hantzschia* Auerswald, Hedwigia **2**: 60. 1862 [Fungi].
T.: *H. phycomyces* Auerswald.

Hemiaulus Heiberg, Krit. Overs. Danske Diatom. 45. 1863.
T.: *H. proteus* Heiberg.

(H) *Hemiaulus* Ehrenberg, Ber. Bekanntm. Verh. Königl. Preuss. Akad. Wiss. Berlin **1844**: 199. 1844 [Bacillarioph.-Biddulph.].
T.: *H. antarcticus* Ehrenberg.

Licmophora C. Agardh, Flora 10: 628. 1827.
T.: *L. argentescens* C. Agardh.

(=) *Styllaria* Draparnaud ex Bory de Saint-Vincent, Dict. Class. Hist. Nat. 2: 129. 1822.
LT.: *S. paradoxa* (Lyngbye) Bory de Saint-Vincent (*Echinella paradoxa* Lyngbye) (vide Bory de Saint-Vincent, Hist. Nat. Zooph. 709. 1827).

(=) *Exilaria* Greville, Scott. Crypt. Fl. *t. 289.* 1827.
T.: *E. flabellata* Greville.

Melosira C. Agardh, Syst. Alg. xiv, 8. 1824 ("*Meloseira*") (*orth. cons.*).
T.: *M. nummuloides* C. Agardh.

(=) *Lysigonium* Link in C. G. Nees, Horae Phys. Berol. 4. 1820.
LT.: *Conferva moniliformis* O. F. Mueller (vide Regnum Veg. 3: 71. 1952).

Nitzschia Hassall, Hist. Brit. Freshwater Alg. 435. 1845.
T.: *N. elongata* Hassall, nom. illeg. (*Bacillaria sigmoidea* Nitzsch, *N. sigmoidea* (Nitzsch) W. Smith).

(≡) *Sigmatella* Kützing, Alg. Aq. Dulc. Germ. no. 2. 1833.

(=) *Homoeocladia* C. Agardh, Flora 10: 629. 1827.
T.: *H. martiana* C. Agardh.

Pantocsekia Grunow ex Pantocsek, Beitr. Kenntn. Foss. Bacill. Ung. 1: 47. 1886.
T.: *P. clivosa* Grunow ex Pantocsek.

(H) *Pantocsekia* Grisebach ex Pantocsek, Österr. Bot. Z. 23: 267. 1873 [Convolvul.].
T.: *P. illyrica* Grisebach ex Pantocsek.

Peronia Brébisson et Arnott ex Kitton, Quart. J. Microscop. Sci. ser. 2. 8: 16. 1868.
T.: *P. erinacea* Brébisson et Arnott ex Kitton, nom. illeg (*Gomphonema fibula* Brébisson ex Kützing, *P. fibula* (Brébisson ex Kützing) Ross).

(H) *Peronia* F. Delaroche in Redouté, Liliac. *t. 342.* 1812 [Marant.].
T.: *P. stricta* F. Delaroche.

Pinnularia Ehrenberg, Ber. Bekanntm. Verh. Königl. Preuss. Akad. Wiss. Berlin 1843: 45. 1843.
T.: *P. viridis* (Nitzsch) Ehrenberg (*Bacillaria viridis* Nitzsch) (*typ. cons.*).

(H) *Pinnularia* Lindley et Hutton, Foss. Fl. Gr. Brit. 2: [81], *t. 111.* 1833 [Foss.].
T.: *P. capillacea* Lindley et Hutton.

(=) *Stauroptera* Ehrenberg, Ber. Bekanntm. Verh. Königl. Preuss. Akad. Wiss. Berlin 1843: 45. 1843.
T.: *S. semicruciata* Ehrenberg.

Pleurosigma W. Smith, Ann. Mag. Nat. Hist. ser. 2. 9: 2. 1852.
T.: *P. angulatum* (Quekett) W. Smith (*Navicula angulata* Quekett).

(=) *Scalptrum* Corda, Alman. Carlsbad 5: 193, *t. 5, f. 70.* 1835.
T.: *S. striatum* Corda.

(=) *Gyrosigma* Hassall, Hist. Brit. Freshwater Alg. 435. 1845 (*nom. cons.*).
T.: *G. hippocampus* (Ehrenberg) Hassall (*Navicula hippocampus* Ehrenberg).

(=) *Endosigma* Brébisson, Dict. Univ. Hist. Nat. 11: 418, 419. 1848.
T.: non designatus.

Podocystis J. W. Bailey, Smithson. Contr. Knowl. 7(3): 11. 1854.
T.: *P. americana* J. W. Bailey.

(H) *Podocystis* Fries, Summa Veg. Scand. 512. 1849 [Fungi].
LT.: *P. capraearum* (A.-P. de Candolle) Fries (*Uredo capraearum* A.-P. de Candolle) (vide Laundon, Mycol. Pap. 99: 14. 1965).

(=) *Euphyllodium* Shadbolt, Trans. Microscop. Soc. London ser. 2. 2: 14. 1854.
T.: *E. spathulatum* Shadbolt.

Rhabdonema Kützing, Kieselschal. Bacill. 126. 1844.
T.: *R. minutum* Kützing.

(=) *Tessella* Ehrenberg, Abh. Königl. Akad. Wiss. Berlin, Phys. Kl. 1835: 173. 1837 [praeimpr. 1836. p. 23].
T.: *T. catena* Ehrenberg.

Rhizosolenia Brightwell, Quart. J. Microscop. Sci. 6: 94. 1858.
T.: *R. styliformis* Brightwell.

(H) *Rhizosolenia* Ehrenberg, Abh. Königl. Akad. Wiss. Berlin, Phys. Kl. 1841: 402. 1843 [Bacillarioph.].
T.: *R. americana* Ehrenberg.

Rhopalodia O. Müller, Bot. Jahrb. Syst. 22: 57. 1897.
T.: *R. gibba* (Ehrenberg) O. Müller (*Navicula gibba* Ehrenberg).

(=) *Pyxidicula* Ehrenberg, Abh. Königl. Akad. Wiss. Berlin, Phys. Kl. 1833: 295. 1834.
T.: *P. operculata* (C. Agardh) Ehrenberg (*Frustulia operculata* C. Agardh).

IV. CHRYSOPHYCEAE

Anthophysa Bory de Saint-Vincent, Dict. Class. Hist. Nat. 1: 427. 1822 ("*Anthophysis*") (*orth. cons.*).
T.: *A. vegetans* (O. F. Müller) F. Stein (*Volvox vegetans* O. F. Müller).

Hydrurus C. Agardh, Syst. Alg. xviii, 24. 1824.
T.: *H. vaucheri* C. Agardh, nom. illeg. (*Conferva foetida* Villars, *H. foetidus* (Villars) Trevisan).

(≡) *Carrodorus* S. F. Gray, Nat. Arr. Brit. Pl. 1: 318, 350. 1821.

(=) *Cluzella* Bory de Saint-Vincent, Dict. Class. Hist. Nat. 3: 14. 1823; *ibid.* 4: 234. 1823.
T.: *C. myosurus* (Ducluzeau) Bory de Saint-Vincent (*Batrachospermum myosurus* Ducluzeau).

V. PHAEOPHYCEAE

Agarum Bory de Saint-Vincent, Dict. Class. Hist. Nat. 9: 193. 1826.
T.: *A. cribrosum* Bory de Saint-Vincent (*Fucus agarum* S. G. Gmelin).

(H) *Agarum* Link, Neues J. Bot. 3(1, 2): 7. 1809 [Rhodoph.: Delesser.].
T.: *A. rubens* (Linnaeus) Link (*Fucus rubens* Linnaeus).

Alaria Greville, Alg. Brit. xxxix, 25. 1830.
T.: *A. esculenta* (Linnaeus) Greville (*Fucus esculentus* Linnaeus).

Ascophyllum Stackhouse, Mém. Soc. Imp. Naturalistes Moscou **2**: 54, 66. 1809 ("*Ascophylla*") (*orth. cons.*).
T.: *A. laevigatum* Stackhouse, nom. illeg. (*Fucus nodosus* Linnaeus, *A. nodosum* (Linnaeus) Le Jolis).

Carpomitra Kützing, Phycol. General. 343. 14-16 Sep 1843.
T.: *C. cabrerae* (Clemente) Kützing (*Fucus cabrerae* Clemente).

Chordaria C. Agardh, Syn. Alg. Scand. xii. 1817.
T.: *C. flagelliformis* (O. F. Müller) C. Agardh (*Fucus flagelliformis* O. F. Müller) (*typ. cons.*).

Cystophora J. Agardh, Linnaea **15**: 3. 1841.
T.: *C. retroflexa* (Labillardière) J. Agardh (*Fucus retroflexus* Labillardière).

Cystoseira C. Agardh, Sp. Alg. **1**: 50. 1820.
T.: *C. concatenata* (Linnaeus) C. Agardh (*Fucus concatenatus* Linnaeus) [= *C. foeniculacea* (Linnaeus) Greville (*Fucus foeniculaceus* Linnaeus)].

Desmarestia Lamouroux, Ann. Mus. Hist. Nat. (Paris) **20**: 43. 1813.
T.: *D. aculeata* (Linnaeus) Lamouroux (*Fucus aculeatus* Linnaeus).

(≡) *Musaefolia* Stackhouse, Mém. Soc. Imp. Naturalistes Moscou **2**: 53, 66. 1809.

(≡) *Nodularius* Roussel, Fl. Calvados ed. 2. 93. 1806 (vide Silva, Univ. Calif. Publ. Bot. **25**: 299. 1952).

(≡) *Dichotomocladia* Trevisan, Atti Riunione Sci. Ital. **4**: 333. 15 Aug 1843.
(=) *Chytraphora* Suhr, Flora **17**: 721. 1834.
T.: *C. filiformis* Suhr.

(H) *Chordaria* Link, Neues J. Bot. **3**(1, 2): 8. 1809 [Phaeoph.: Chord.].
≡ *Chorda* Stackhouse 1797.

(=) *Blossevillea* Decaisne, Bull. Acad. Roy. Sci. Bruxelles **7**(1): 410. 1840 ("*Blosvillea*").
T.: *B. torulosa* (R. Brown ex D. Turner) Decaisne (Ann. Sci. Nat. Bot. ser. 2. **17**: 331. 1842) (*Fucus torulosus* R. Brown ex D. Turner) (vide Silva, Univ. Calif. Publ. Bot. **25**: 279. 1952).

(=) *Gongolaria* Boehmer in Ludwig, Def. Gen. Pl. ed. 3. 503. 1760.
T.: *Fucus abies-marina* S. G. Gmelin.
(=) *Baccifer* Roussel, Fl. Calvados ed. 2. 94. 1806.
T.: *Fucus baccatus* S. G. Gmelin.
(=) *Abrotanifolia* Stackhouse, Mém. Soc. Imp. Naturalistes Moscou **2**: 56, 81. 1809.
LT.: *A. loeflingii* Stackhouse (*Fucus abrotanifolius* Linnaeus) (vide Papenfuss, Hydrobiologia **2**: 184. 1950).
(=) *Ericaria* Stackhouse, Mém. Soc. Imp. Naturalistes Moscou **2**: 56, 80. 1809.
LT.: *Fucus ericoides* Linnaeus (vide Papenfuss, Hydrobiologia **2**: 185. 1950).

(≡) *Hippurina* Stackhouse, Mém. Soc. Imp. Naturalistes Moscou **2**: 59, 89. 1809 (vide Silva, Univ. Calif. Publ. Bot. **25**: 257. 1952).
(=) *Herbacea* Stackhouse, Mém. Soc. Imp. Naturalistes Moscou **2**: 58, 89. 1809.
T.: *H. ligulata* Stackhouse (*Fucus ligulatus* Lightfoot, non S. G. Gmelin).
(=) *Hyalina* Stackhouse, Mém. Soc. Imp. Naturalistes Moscou **2**: 58, 88. 1809.
T.: *H. mutabilis* Stackhouse, nom. illeg. (*Fucus viridis* O. F. Müller).

Desmotrichum Kützing, Phycol. Germ. 244. 1845.

T.: *D. balticum* Kützing.

(H) *Desmotrichum* Blume, Bijdr. 329. 1825 [Orchid.].
≡ *Flickingeria* A. Hawkes 1961.

(=) *Diplostromium* Kützing, Phycol. General. 298. 1843.
LT.: *D. tenuissimum* (C. Agardh) Kützing (*Zonaria tenuissima* C. Agardh) (vide Silva, Univ. Calif. Publ. Bot. 25: 257. 1952).

Dictyopteris Lamouroux, Nouv. Bull. Sci. Soc. Philom. Paris 1: 332. 1809.

T.: *D. polypodioides* (A.-P. de Candolle) Lamouroux (*Ulva polypodioides* A.-P. de Candolle, *Fucus polypodioides* Desfontaines, non S. G. Gmelin).

(≡) *Granularius* Roussel, Fl. Calvados ed. 2. 90. 1806 (vide Silva, Regnum Veg. 101: 745. 1979).

(=) *Neurocarpus* Weber et Mohr, Beitr. Naturk. 1: 300. 1805 (vel 1806).
T.: *N. membranaceus* (Stackhouse) Weber et Mohr (*Polypodoidea membranacea* Stackhouse; *Fucus membranaceus* Stackhouse, non N. Burman).

Dictyosiphon Greville, Alg. Brit. xliii, 55. 1830.

T.: *D. foeniculaceus* (Hudson) Greville (*Conferva foeniculacea* Hudson).

(≡) *Scytosiphon* C. Agardh, Disp. Alg. Suec. 24. 1811 (vide Silva, Regnum Veg. 8: 205. 1956).

Dictyota Lamouroux, J. Bot. (Desvaux) 2: 38. 1809.

T.: *D. dichotoma* (Hudson) Lamouroux (*Ulva dichotoma* Hudson) (*typ. cons.*).

(H) *Dictyota* Lamouroux, J. Bot. (Desvaux) 2: 38. 1809 [Phaeoph.: Dictyot.].
≡ *Padina* Adanson 1763 (*nom. cons.*).

Ectocarpus Lyngbye, Tent. Hydrophytol. Dan. xxxi, 130. 1819.

T.: *E. siliculosus* (Dillwyn) Lyngbye (*Conferva siliculosa* Dillwyn).

(=) *Colophermum* Rafinesque, Précis Découv. Somiol. 49. 1814.
T.: *C. floccosum* Rafinesque.

Elachista Duby, Bot. Gall. 972. 1830 ("*Elachistea*") (*orth. cons.*).

T.: *E. scutellata* Duby, nom. illeg. (*Conferva scutulata* J. E. Smith; *E. scutulata* (J. E. Smith) J. E. Areschoug).

(=) *Opospermum* Rafinesque, Précis Découv. Somiol. 48. 1814.
T.: *O. nigrum* Rafinesque.

Halidrys Lyngbye, Tent. Hydrophytol. Dan. xxix, 37. 1819.

T.: *H. siliquosa* (Linnaeus) Lyngbye (*Fucus siliquosus* Linnaeus) (*typ. cons.*).

(H) *Halidrys* Stackhouse, Mém. Soc. Imp. Naturalistes Moscou 2: 53, 62. 1809 [Phaeoph.: Fuc.].
≡ *Fucus* Linnaeus 1753.

(≡) *Siliquarius* Roussel, Fl. Calvados ed. 2. 94. 1806.

Himanthalia Lyngbye, Tent. Hydrophytol. Dan. xxix, 36. 1819.

T.: *H. lorea* (Linnaeus) Lyngbye (*Fucus loreus* Linnaeus) [= *H. elongata* (Linnaeus) S. F. Gray (*Fucus elongatus* Linnaeus)].

(≡) *Funicularius* Roussel, Fl. Calvados ed. 2. 91. 1806.

(=) *Lorea* Stackhouse, Mém. Soc. Imp. Naturalistes Moscou 2: 60, 94. 1809.
T.: *L. elongata* (Linnaeus) Stackhouse (*Fucus elongatus* Linnaeus).

124

Hormosira (Endlicher) Meneghini, Nuovi Saggi Imp. Regia Accad. Sci. Padova 4: 368. 1838.
T.: *Fucus moniliformis* Labillardière, non Esper [= *H. banksii* (D. Turner) Decaisne (*Fucus banksii* D. Turner)].

(≡) *Moniliformia* Lamouroux, Dict. Class. Hist. Nat. 7: 71. 1825.

Laminaria Lamouroux, Ann. Mus. Hist. Nat. (Paris) 20: 40. 1813.
T.: *L. digitata* (Hudson) Lamouroux (*Fucus digitatus* Hudson).

(=) *Saccharina* Stackhouse, Mém. Soc. Imp. Naturalistes Moscou 2: 53, 65. 1809.
LT.: *S. plana* Stackhouse (*Fucus saccharinus* Linnaeus) (vide Silva, Univ. Calif. Publ. Bot. 25: 259. 1952).

Padina Adanson, Fam. Pl. 2: 13, 586. 1763.
T.: *P. pavonica* (Linnaeus) Lamouroux (Hist. Polyp. Corall. 304. 1816) (*Fucus pavonicus* Linnaeus).

Petalonia Derbès et Solier, Ann. Sci. Nat. Bot. ser. 3. 14: 265. 1850.
T.: *P. debilis* (C. Agardh) Derbès et Solier (*Laminaria debilis* C. Agardh, *P. fascia* var. *debilis* (C. Agardh) Hamel).

(=) *Fasciata* S. F. Gray, Nat. Arr. Brit. Pl. 1: 319 ("*Fascia*"), 383. 1821.
LT.: *F. attenuata* S. F. Gray, nom. illeg. (*Fucus fascia* O. F. Müller) (vide Silva, Univ. Calif. Publ. Bot. 25: 299. 1952).

Saccorhiza Bachelot de la Pylaie, Fl. Terre Neuve 23. 1830 ("1829").
T.: *S. bulbosa* (Hudson) J. Agardh (Sp. Alg. 1: 138. 1848) (*Fucus bulbosus* Hudson) [= *S. polyschides* (Lightfoot) Batters (*Fucus polyschides* Lightfoot)].

(=) *Polyschidea* Stackhouse, Mém. Soc. Imp. Naturalistes Moscou 2: 53, 65. 1809.
LT.: *Fucus polyschides* Lightfoot (vide Papenfuss, Hydrobiologia 2: 189. 1950).

Sargassum C. Agardh, Sp. Alg. 1: 1. 1820.
T.: *S. bacciferum* (D. Turner) C. Agardh (*Fucus bacciferus* D. Turner).

(=) *Acinaria* Donati, Essai Hist. Nat. Mer Adriat. 26, 33, t. 5. 1758.
T.: *Sargassum donatii* (Zanardini) Kützing (*S. vulgare* var. *donatii* Zanardini).

Scytosiphon C. Agardh, Sp. Alg. 1: 160. 1820.
T.: *S. lomentaria* (Lyngbye) Link (Handbuch 3: 232. 1833) (*Chorda lomentaria* Lyngbye, *S. filum* var. *lomentaria* (Lyngbye) C. Agardh) (*typ. cons.*).

(H) *Scytosiphon* C. Agardh, Disp. Alg. Suec. 24. 1811 [Phaeoph.: Dictyosiphon.].
≡ *Dictyosiphon* Greville 1830 (*nom. cons.*).

Spermatochnus Kützing, Phycol. General. 334. 1843.
T.: *S. paradoxus* (Roth) Kützing (*Conferva paradoxa* Roth) (*typ. cons.*).

Stilophora J. Agardh, Linnaea 15: 6. 1841.
T.: *S. rhizodes* (D. Turner) J. Agardh (*Fucus rhizodes* D. Turner) (*typ. cons.*).

(H) *Stilophora* C. Agardh, Flora 10: 642. 1827 [Phaeoph.: Punctar.].
≡ *Hydroclathrus* Bory de Saint-Vincent 1825.

Zonaria C. Agardh, Syn. Alg. Scand. xx. 1817.
T.: *Z. flava* (Clemente) C. Agardh (*Fucus flavus* Clemente) [= *Z. tournefortii* (Lamouroux) Montagne (*Fucus tournefortii* Lamouroux)] (*typ. cons.*).

(H) *Zonaria* Draparnaud ex Weber et Mohr, Beitr. Naturk. 1: 247-253. 1805 (vel 1806) [Phaeoph.: Dictyot.].
≡ *Padina* Adanson 1763 (*nom. cons.*).

125

VI. XANTHOPHYCEAE

Botrydiopsis Borzì, Boll. Soc. Ital. Microscop. 1: 69. 1889.
T.: *B. arhiza* Borzì.

(H) *Botrydiopsis* Trevisan, Nomencl. Alg. 70. 1845 [Plantae inc. sed.].
T.: *B. vulgaris* (Brébisson) Trevisan (*Botrydina vulgaris* Brébisson).

Centritractus Lemmermann, Ber. Deutsch. Bot. Ges. **18**: 274. 1900 ("*Centratractus*") (*orth. cons.*).
T.: *C. belonophorus* (Schmidle) Lemmermann (*Schroederia belonophora* Schmidle).

Ophiocytium Nägeli, Neue Denkschr. Allg. Schweiz. Ges. Gesammten Naturwiss. **10**(7): 87. 1849.
T.: *O. apiculatum* Nägeli [= *O. cochleare* (Eichwald) A. Braun (*Spirogyra cochlearis* Eichwald)].

(=) *Spirodiscus* Ehrenberg, Abh. Königl. Akad. Wiss. Berlin, Phys. Kl. **1831**: 68. 1832.
T.: *S. fulvus* Ehrenberg.

Tetraëdriella Pascher, Arch. Protistenk. **63**: 423. 1930.
T.: *T. acuta* Pascher.

(=) *Polyedrium* Nägeli, Neue Denkschr. Allg. Schweiz. Ges. Gesammten Naturwiss. **10**(7): 83. 1849.
T.: *P. tetraëdricum* Nägeli.

VII. DINOPHYCEAE

Gyrodinium Kofoid et Swezy, Mem. Univ. Calif. **5**: 273. 1921.
T.: *G. spirale* (Bergh) Kofoid et Swezy (*Gymnodinium spirale* Bergh).

(≡) *Spirodinium* Schütt in Engler et Prantl, Nat. Pflanzenfam. 1(1b): 3, 5. 1896.

VIII. EUGLENOPHYCEAE

Anisonema Dujardin, Hist. Nat. Zoophyt. 327, 344. 1841.
T.: *A. acinus* Dujardin.

(H) *Anisonema* A. H. L. Jussieu, Euphorb. Gen. 19. 1824 [Euphorb.].
T.: *A. reticulatum* (Poiret) A. H. L. Jussieu (*Phyllanthus reticulatus* Poiret).

Astasia Dujardin, Hist. Nat. Zoophyt. 353, 356. 1841.
T.: *A. limpida* Dujardin (*typ. cons.*).

(H) *Astasia* Ehrenberg, Ann. Phys. Chem. **94**: 508. 1830 [Euglenoph.: Euglen.].
LT.: *A. haematodes* Ehrenberg (vide Silva, Taxon **9**: 20. 1960).

Lepocinclis Perty, Mitth. Naturf. Ges. Bern **1849**: 28 (adnot.). 1849.
T.: *L. globulus* Perty.

(=) *Crumenula* Dujardin, Ann. Sci. Nat. Zool. ser. 2. **5**: 204, 205. 1836.
T.: *C. texta* Dujardin.

Phacus Dujardin, Hist. Nat. Zoophyt. 327, 334. 1841.

T.: *P. longicauda* (Ehrenberg) Dujardin (*Euglena longicauda* Ehrenberg) (*typ. cons.*).

(H) *Phacus* Nitzsch in Ersch et Gruber, Allg. Encycl. Wiss. Künste sect. 1. **16**: 69. 1827 [Euglenoph.: Euglen.].
≡ *Virgulina* Bory de Saint-Vincent 1823.

IX. CHLOROPHYCEAE

Acetabularia Lamouroux, Nouv. Bull. Sci. Soc. Philom. Paris **3**: 185. 1812.

T.: *A. acetabulum* (Linnaeus) P. C. Silva (Univ. Calif. Publ. Bot. **25**: 255. 1952) (*Madrepora acetabulum* Linnaeus).

(≡) *Acetabulum* Boehmer in Ludwig, Def. Gen. Pl. ed. 3. 504. 1760.

Anadyomene Lamouroux, Nouv. Bull. Sci. Soc. Philom. Paris **3**: 187. 1812 ("*Anadyomena*") (*orth. cons.*).

T.: *A. flabellata* Lamouroux (Hist. Polyp. Corall. 366. 1816) [= *A. stellata* (Wulfen) C. Agardh (*Ulva stellata* Wulfen)].

Aphanochaete A. Braun, Betracht. Erschein. Verjüng. Natur 196 (adnot.). 1850.

T.: *A. repens* A. Braun [= *A. confervicola* (Nägeli) Rabenhorst (*Herposteiron confervicola* Nägeli)].

(=) *Herposteiron* Nägeli in Kützing, Sp. Alg. 424. 1849.

T.: *H. confervicola* Nägeli.

Bambusina Kützing, Sp. Alg. 188. 1849.

T.: *B. brebissonii* Kützing, nom. illeg. (*Didymoprium borreri* Ralfs; *B. borreri* (Ralfs) P. T. Cleve).

(=) *Gymnozyga* Ehrenberg ex Kützing, Sp. Alg. 188. 1849.

T.: *G. moniliformis* Ehrenberg ex Kützing.

Chaetomorpha Kützing, Phycol. Germ. 203. 1845.

T.: *C. melagonium* (Weber et Mohr) Kützing (*Conferva melagonium* Weber et Mohr).

(=) *Chloronitum* Gaillon, Dict. Sci. Nat. **53**: 389. 1828.

LT.: *C. aereum* (Dillwyn) Gaillon (*Conferva aerea* Dillwyn) (vide Silva, Univ. Calif. Publ. Bot. **25**: 270. 1952).

(=) *Spongopsis* Kützing, Phycol. General. 261. 1843.

T.: *S. mediterranea* Kützing.

Chlamydomonas Ehrenberg, Abh. Königl. Akad. Wiss. Berlin, Phys. Kl. **1833**: 288. 1834 ("*Chlamidomonas*") (*orth. cons.*).

T.: *C. pulvisculus* (O. F. Müller) Ehrenberg (*Monas pulvisculus* O. F. Müller).

(=) *Protococcus* C. Agardh, Syst. Alg. xvii, 13. 1824.

LT.: *P. nivalis* (Bauer) C. Agardh (*Uredo nivalis* Bauer) (vide Drouet et Daily, Butler Univ. Bot. Stud. **12**: 167. 1956).

(=) *Sphaerella* Sommerfelt, Mag. Naturvidensk. **4**: 252. 1824.

LT.: *S. nivalis* (Bauer) Sommerfelt (*Uredo nivalis* Bauer) (vide Hazen, Mem. Torrey Bot. Club **6**: 238. 1899).

Chlorococcum Meneghini, Mem. Reale Accad. Sci. Torino ser. 2. **5**: 24. 1842.
T.: *C. infusionum* (Schrank) Meneghini (*Lepra infusionum* Schrank) (*typ. cons.*).

(H) *Chlorococcum* Fries, Syst. Orb. Veg. 356. 1825 [Chloroph.: Chlamydomonad.].
≡ *Protococcus* C. Agardh 1824 (*nom. rej.*).
≡ *Sphaerella* Sommerfelt 1824 (*nom. rej.*).

Chloromonas Gobi, Bot. Zap. **15**: 232, 255. 1899-1900.
T.: *C. reticulata* (Goroschankin) Gobi (*Chlamydomonas reticulata* Goroschankin).

(H) *Chloromonas* Kent, Man. Infus. 369, 401. 1881 [Euglenoph.: Euglen.].
≡ *Cryptoglena* Ehrenberg 1832.
(=) *Tetradonta* Korshikov, Russk. Arh. Protistol. **4**: 183, 195. 1925.
T.: *T. variabilis* Korshikov.
(=) *Platychloris* Pascher, Süssw.-Fl. **4**: 138, 331. 1927.
T.: *P. minima* Pascher (*Chlamydomonas minima* Pascher, non P. A. Dangeard).

Cladophora Kützing, Phycol. General. 262. 1843.
T.: *C. oligoclona* (Kützing) Kützing (*Conferva oligoclona* Kützing).

(=) *Conferva* Linnaeus, Sp. Pl. 1164. 1753.
LT.: *C. rupestris* Linnaeus (vide Bonnemaison, J. Phys. Chim. Hist. Nat. Arts **94**: 198. 1822).
(=) *Annulina* Link in C. G. Nees, Horae Phys. Berol. 4. 1820.
LT.: *A. glomerata* (Linnaeus) C. G. Nees (Horae Phys. Berol. index, 1820) (*Conferva glomerata* Linnaeus) (vide Silva, Univ. Calif. Publ. Bot. **25**: 270. 1952).

Cladophoropsis Børgesen, Overs. Kongel. Danske Vidensk. Selsk. Forh. **1905**: 288. 1905.
T.: *C. membranacea* (C. Agardh) Børgesen (*Conferva membranacea* C. Agardh).

(=) *Spongocladia* J. E. Areschoug, Öfvers. Förh. Kongl. Svenska Vetensk.-Akad. **10**: 202. 1853.
T.: *S. vaucheriiformis* J. E. Areschoug.

Coleochaete Brébisson, Ann. Sci. Nat. Bot. ser. 3. **1**: 29. 1844.
T.: *C. scutata* Brébisson.

(=) *Phyllactidium* Kützing, Phycol. General. 294. 1843.
LT.: *P. pulchellum* Kützing (vide Meneghini, Atti Riunione Sci. Ital. **6**: 457. 1845).

Enteromorpha Link in C. G. Nees, Horae Phys. Berol. 5. 1820.
T.: *E. intestinalis* (Linnaeus) C. G. Nees (Horae Phys. Berol. index, 1820) (*Ulva intestinalis* Linnaeus).

(≡) *Splachnon* Adanson, Fam. Pl. **2**: 13, 607. 1763 ("*Splaknon*") (vide Silva, Univ. Calif. Publ. Bot. **25**: 294. 1952).

Gloeococcus Braun, Betracht. Erschein. Verjüng. Natur 169. 1850.
T.: *G. minor* Braun.

(H) *Gloiococcus* Shuttleworth, Biblioth. Universelle Genève, ser. 2. **25**: 405. 1840 [Algae inc. sed.].
T.: *G. grevillei* (C. Agardh) Shuttleworth (*Haematococcus grevillei* C. Agardh).

Gongrosira Kützing, Phycol. General. 281. 1843.
T.: *G. sclerococcus* Kützing, nom. illeg. (*Stereococcus viridis* Kützing; *G. viridis* (Kützing) De Toni).

(≡) *Stereococcus* Kützing, Linnaea **8**: 379. 1833.

Haematococcus Flotow, Nov. Actorum Acad. Caes. Leop.-Carol. Nat. Cur. **20**: 413 seqq. 1844.
T.: *H. pluvialis* Flotow [= *H. lacustris* (Girod-Chantrans) Rostafinski (*Volvox lacustris* Girod-Chantrans)] (*typ. cons.*).

(H) *Haematococcus* C. Agardh, Icon, Alg. Eur. nos. 22-24. 1830 [Euglenoph.: Euglen.].
LT.: *H. noltei* C. Agardh (vide Trevisan, Sagg. Algh. Coccot. 38. 1848).

(=) *Disceraea* A. Morren et C. Morren, Nouv. Mém. Acad. Roy. Sci. Bruxelles **14**(5): 37. 1841.
T.: *D. purpurea* A. Morren et C. Morren.

Halimeda Lamouroux, Nouv. Bull. Sci. Soc. Philom. Paris **3**: 186. 1812 ("*Halimedea*") (*orth. cons.*).
T.: *H. tuna* (Ellis et Solander) Lamouroux (Hist. Polyp. Corall. 309. 1816) (*Corallina tuna* Ellis et Solander).

(≡) *Sertularia* Boehmer in Ludwig, Def. Gen. Pl. ed. 3. 504. 1760 (vide Silva, Univ. Calif. Publ. Bot. **25**: 294. 1952).

Hydrodictyon Roth, Bemerk. Stud. Crypt. Wassergewächse 48. 1797.
T.: *H. reticulatum* (Linnaeus) Bory de Saint-Vincent (Dict. Class. Hist. Nat. **6**: 506. 1824) (*Conferva reticulata* Linnaeus).

(≡) *Reticula* Adanson, Fam. Pl. **2**: 3. 1763.

Microspora Thuret, Ann. Sci. Nat. Bot. ser. 3. **14**: 221. 1850.
T.: *M. floccosa* (Vaucher) Thuret (*Prolifera floccosa* Vaucher).

(H) *Microspora* Hassall, Ann. Mag. Nat. Hist. **11**: 363. 1843 [Chloroph.: Cladophor.].
T.: non designatus.

Mougeotia C. Agardh, Syst. Alg. xxvi, 83. 1824.
T.: *M. genuflexa* (Roth) C. Agardh (*Conferva genuflexa* Roth).

(H) *Mougeotia* Kunth in Humboldt, Bonpland et Kunth, Nov. Gen. Sp. 5: ed. fol. 253, ed. qu. 326. 1823 ("1821") [Stercul.].
T.: non designatus.

(≡) *Serpentinaria* S. F. Gray, Nat. Arr. Brit. Pl. **1**: 279 ("*Serpentina*"), 299. 1821 (vide Silva, Univ. Calif. Publ. Bot. **25**: 252. 1952).

(=) *Agardhia* S. F. Gray, Nat. Arr. Brit. Pl. **1**: 279 ("*Agardia*"), 299. 1821.
T.: *A. caerulescens* (J. E. Smith) S. F. Gray (*Conferva caerulescens* J. E. Smith).

Prasiola (C. Agardh) Meneghini, Nuovi Saggi Imp. Regia Accad. Sci. Padova **4**: 360. 1838.
T.: *P. crispa* (Lightfoot) Kützing (Phycol. General. 295. 1843) (*Ulva crispa* Lightfoot).

(=) *Humida* S. F. Gray, Nat. Arr. Brit. Pl. **1**: 278, 281. 1821.
LT.: *H. muralis* (Dillwyn) S. F. Gray (*Conferva muralis* Dillwyn) (vide Drouet, Acad. Nat. Sci. Philadelphia Monogr. **15**: 312. 1968).

Sirogonium Kützing, Phycol. General. 278. 1843.
T.: *S. sticticum* (J. E. Smith) Kützing (*Conferva stictica* J. E. Smith).

(≡) *Choaspis* S. F. Gray, Nat. Arr. Brit. Pl. 1: 279 ("*Choaspes*"), 299. 1821.

Sphaerozosma Ralfs, Brit. Desmid. 65. 1848.
T.: *S. vertebratum* Ralfs.

(H) *Sphaerozosma* Corda, Icon. Fung. 5: 27. 1842 [Fungi].
≡ *Sphaerosoma* Klotzsch 1839.

Spirogyra Link in C. G. Nees, Horae Phys. Berol. 5. 1820.
T.: *S. porticalis* (O. F. Müller) Dumortier (Comment. Bot. 99. 1822) (*Conferva porticalis* O. F. Müller).

(=) *Conjugata* Vaucher, Hist. Conferves 3, 37. 1803.
LT.: *C. princeps* Vaucher (vide Bonnemaison, J. Phys. Chim. Hist. Nat. Arts 94: 195. 1822).

Stigeoclonium Kützing, Phycol. General. 253. 1843 ("*Stygeoclonium*") (*orth. cons.*).
T.: *S. tenue* (C. Agardh) Kützing (*Draparnaldia tenuis* C. Agardh).

(=) *Myxonema* Fries, Syst. Orb. Veg. 343. 1825.
LT.: *Conferva lubrica* Dillwyn (vide Hazen, Mem. Torrey Bot. Club 11: 193. 1902).

Struvea Sonder, Bot. Zeitung (Berlin) 3: 49. 1845.
T.: *S. plumosa* Sonder.

(H) *Struvea* H. G. L. Reichenbach, Deutsche Bot. 1(2): 222, 236. 1841 [Tax.].
≡ *Torreya* Arnott 1838 (*nom. cons.*) (17).

Trentepohlia C. F. P. Martius, Fl. Crypt. Erlang. 351. 1817.
T.: *T. aurea* (Linnaeus) C. F. P. Martius (*Byssus aurea* Linnaeus).

(H) *Trentepohlia* Roth, Catal. Bot. 2: 73. 1800 [Cruc.].
T.: non designatus.
(=) *Byssus* Linnaeus, Sp. Pl. 1168. 1753.
LT.: *B. jolithus* Linnaeus (vide Fries, Stirp. Agri Femsion. 42. 1825).

Ulva Linnaeus, Sp. Pl. 1163. 1753.
T.: *U. lactuca* Linnaeus (*typ. cons.*).

Urospora J. E. Areschoug, Nova Acta Regiae Soc. Sci. Upsal. ser. 3. 6(2): 15. 1866.
T.: *U. mirabilis* J. E. Areschoug.

(=) *Hormiscia* Fries, Fl. Scan. 326. 1835.
LT.: *H. penicilliformis* (Roth) Fries (*Conferva penicilliformis* Roth) (vide Silva, Univ. Calif. Publ. Bot. 25: 270. 1952).
(=) *Codiolum* Braun, Alg. Unicell. 19. 1855.
T.: *C. gregarium* Braun.

Zygnema C. Agardh, Syn. Alg. Scand. xxxii. 1817.
T.: *Z. cruciatum* (Vaucher) C. Agardh (*Conjugata cruciata* Vaucher).

(=) *Lucernaria* Roussel, Fl. Calvados ed. 2. 20, 84. 1806.
T.: *L. pellucida* Roussel.

Zygogonium Kützing, Phycol. General. 280. 1843.
T.: *Z. ericetorum* (Roth) Kützing (*Conferva ericetorum* Roth).

(≡) *Leda* Bory de Saint-Vincent, Dict. Class. Hist. Nat. 1: 595. 1822 (vide Silva, Univ. Calif. Publ. Bot. 25: 253. 1952).

X. FUNGI (INCLUDING LICHEN-FORMING FUNGI)

Agaricus Linnaeus, Sp. Pl. 1171. 1753 : Fries,
Syst. Mycol. 1: lvi, 8. 1821.
T.: *A. campestris* Linnaeus : Fries (*typ. cons.*).

Aleurodiscus Rabenhorst ex J. Schröter,
Krypt.-Fl. Schles. 3: 429. 1888.
T.: *A. amorphus* (Persoon : Fries) J. Schröter
(*Peziza amorpha* Persoon : Fries).

(=) *Cyphella* Fries, Syst. Mycol. 2(1): 201.
1822 : Fries, *ibid.*.
T.: *C. digitalis* (Albertini et Schweinitz)
Fries.

Alternaria C. G. Nees, Syst. Pilze 72. 1816 (vel
1817) : Fries, Syst. Mycol. 1: xlvi. 1821.
T.: *A. tenuis* C. G. Nees (*Torula alternata*
Fries : Fries, *A. alternata* (Fries : Fries) Keiss-
ler).

*****Amanita** Persoon, Tent. Disp. Meth. Fung.
65. 1797.
T.: *A. muscaria* (Linnaeus : Fries) Persoon
(*Agaricus muscarius* Linnaeus : Fries).

(H) *Amanita* Boehmer, Defin. Gen. Pl. 490.
1760 [Fungi].
≡ *Agaricus* Linnaeus 1753 (*nom. cons.*)
(vide Earle, Bull. New York Bot. Gard.
5: 382. 1909; Donk, Beih. Nova Hedwigia
5: 20. 1962).

Amanitopsis Roze, Bull. Soc. Bot. France 23:
50, 51, 111. 1876.
T.: *A. vaginata* (Bulliard : Fries) Roze (*Agaricus vaginatus* Bulliard : Fries).

(≡) *****Vaginarius* Roussel, Fl. Calvados ed. 2.
59. 1806 (vide Donk, Beih. Nova Hedwigia 5: 292. 1962).
(=) *Vaginata* S. F. Gray, Nat. Arr. Brit. Pl. 1:
601. 1821.
T.: *V. livida* (Persoon) S. F. Gray (*Amanita livida* Persoon).

Amphisphaeria Cesati et De Notaris, Comment. Soc. Crittog. Ital. 1: 223. 1863.
*T.: *A. umbrina* (Fries) De Notaris (*Sphaeria
umbrina* Fries) (*typ. cons.*).

*****Anisomeridium** (Müller Arg.) M. Choisy,
Icon. Lich. Univ. 3. 1928.
T.: *Arthropyrenia xylogena* Müller Arg.

(=) *Microthelia* Koerber, Syst. Lich. Germ.
327. 1855.
LT.: *M. micula* Koerber, nom. illeg.
Verrucaria biformis Borrer, *Anisomeridium biforme* (Borrer) R. C. Harris)
(vide T. M. Fries, Gen. Heterolich. Eur.
111. 1861).

Anzia Stizenberger, Flora 44: 393. 1861.
T.: *A. colpodes* (Acharius) Sitzenberger (*Lichen colpodes* Acharius).

(=) *Chondrospora* A. Massalongo, Atti Reale Ist. Veneto Sci. ser. 3. 5: 248. 1860.
T.: *C. semiteres* (Montagne et van den
Bosch) A. Massalongo (*Parmelia semiteres* Montagne et van den Bosch).

Aposphaeria Saccardo, Michelia 2: 4. 1880.
T.: *A. pulviscula* Saccardo.

(H) *Aposphaeria* Berkeley, Outl. Brit. Fun-
gol. 315. 1860 [Fungi].
T.: *A. complanata* (Tode : Fries) Berkeley.

131

Arthonia Acharius, Neues J. Bot. **1**(3): 3. 1806.
T.: *A. radiata* (Persoon) Acharius (*Opegrapha radiata* Persoon).

(=) *Coniocarpon* A.-P. de Candolle in Lamarck et A.-P. de Candolle, Fl. Franç. ed. 3. **2**: 323. 1805.
T.: *C. cinnabarinum* A.-P. de Candolle.

Aschersonia Montagne, Ann. Sci. Nat. Bot. ser. 3. **10**: 121. 1848.
T.: *A. taitense* Montagne.

(H) *Aschersonia* Endlicher, Gen. Pl. Suppl. **2**: 103. 1842 [Fungi].
T.: *A. crustacea* (Junghuhn) Endlicher (*Laschia crustacea* Junghuhn).

Boletus Linnaeus, Sp. Pl. 1176. 1753 : Fries, Syst. Mycol. **1**: 385. 1821.
T.: *B. edulis* Bulliard : Fries (*typ. cons.*).

Caloplaca T. M. Fries, Lich. Arct. 218. 1860.
T.: *C. cerina* (Ehrhart ex Hedwig) T. M. Fries (*Lichen cerinus* Ehrhart ex Hedwig).

(=) *Gasparrinia* Tornabene, Lich. Sicula 27. 1849.
T.: *G. murorum* (G. F. Hoffmann) Tornabene (*Lichen murorum* G. F. Hoffmann).

(=) *Pyrenodesmia* A. Massalongo, Atti Reale Ist. Veneto Sci. ser. 2. **3**(App. 3): 119. 1853.
T.: *P. chalybaea* (Fries) A. Massalongo (*Parmelia chalybaea* Fries).

(=) *Xanthocarpia* A. Massalongo et De Notaris in A. Massalongo, Alc. Gen. Lich. 11. 1853.
T.: *X. ochracea* (Schaerer) A. Massalongo et De Notaris (*Lecidea ochracea* Schaerer).

Calvatia Fries, Summa Veg. Scand. **2**: 442. 1849.
T.: *C. craniiformis* (Schweinitz) Fries (*Bovista craniiformis* Schweinitz).

(=) *Langermannia* Rostkovius in Sturm, Deutschl. Fl. Abt. iii. **5**(H. 18): 23. 1839.
T.: *L. gigantea* (Batsch : Persoon) Rostkovius (*Lycoperdon giganteum* Batsch : Persoon).

(=) *Hippoperdon* Montagne, Ann. Sci. Nat. Bot. ser. 2. **17**: 121. 1842.
T.: *H. crucibulum* Montagne.

Candida Berkhout, Schimmelgesl. Monilia 41. 1923.
T.: *C. vulgaris* Berkhout.

(=) *Syringospora* Quinquaud, Arch. Physiol. Norm. Pathol. **1**: 293. 1868.
T.: *S. robinii* Quinquaud, nom. illeg. (*Oidium albicans* Robin).

(=) *Parendomyces* Queyrat et Laroche, Bull. & Mém. Soc. Méd. Hôp. Paris ser. 3. **28**: 136. 1909.
T.: *P. albus* Queyrat et Laroche.

(=) *Parasacharomyces* Beurmann et Gougerot, Tribune Méd. (Paris) **42**: 502. 1909 [nomen provisorium?, vide *ibid.* 518].
T.: non designatus.

(=) *Pseudomonilia* Geiger, Centralbl. Bakteriol. 2. Abth. **27**: 134. 1910.
T.: *P. albomarginata* Geiger.

Cetraria Acharius, Methodus 292. 1803.
T.: *C. islandica* (Linnaeus) Acharius (*Lichen islandicus* Linnaeus).

(≡) *Platyphyllum* Ventenat, Tabl. Règne Vég. 34. 1799.

Ceuthospora Greville, Scott. Crypt. Fl. 5: 253-254. 1826.
T.: *C. lauri* (Greville) Greville (*Cryptosphaeria lauri* Greville).

(H) *Ceuthospora* Fries, Syst. Orb. Veg. 119. 1825 [Fungi].
T.: *C. phaeocomes* (Rebentisch : Fries) Fries (*Sphaeria phaeocomes* Rebentisch : Fries).

Chlo_ociboria Seaver ex Ramamurthi, Korf et Batra, Mycologia 49: 857. 1958 ("1957").
T.: *C. aeruginosa* (Persoon : Fries) Seaver ex Ramamurthi, Korf et Batra (*Peziza aeruginosa* Persoon : Fries) (*typ. cons.*).

*Chondropsis Nylander ex Crombie, J. Linn. Soc., Bot. 17: 397. 1879.
T.: *C. semiviridis* (Nylander) Crombie (*Parmeliopsis semiviridis* Nylander).

(H) *Chondropsis* Rafinesque, Fl. Tellur. 3: 29 ("*Chondropis*"), 97. 1837 [Gentian.].
T.: *C. trinervis* (Linnaeus) Rafinesque (*Chironia trinervis* Linnaeus).

Chrysothrix Montagne, Ann. Sci. Nat. Bot. ser. 3. 18: 312. 1852.
T.: *C. noli-tangere* Montagne, nom. illeg. (*Peribotryon pavonii* ("*pavoni*") Fries : Fries, *C. pavonii* (Fries : Fries) Laundon).

(≡) *Peribotryon* Fries : Fries, Syst. Mycol. 3(2): 287. 1832.

(=) *Pulveraria* Acharius, Methodus 1. 1803.
LT.: *P. chlorina* (Acharius) Acharius (*Lichen chlorinus* Acharius) (vide Laundon, Taxon 30: 663. 1981).

Cistella Quélet, Enchir. Fung. 319. 1886.
T.: *C. dentata* (Persoon : Fries) Quélet (*Peziza dentata* Persoon : Fries).

(H) *Cistella* Blume, Bijdr. 293. 1825 [Orchid.].
T.: *C. cernua* (Willdenow) Blume (*Malaxis cernua* Willdenow).

Cladonia P. Browne, Civ. Nat. Hist. Jamaica 81. 1756.
T.: *C. subulata* (Linnaeus) Wiggers (*Lichen subulatus* Linnaeus).

Clavaria Linnaeus, Sp. Pl. 1182. 1753 : Fries, Syst. Mycol. 1: 465. 1821.
T.: *C. fragilis* Holmskjold : Fries.

Collema Wiggers, Prim. Fl. Holsat. 89. 1780.
T.: *C. lactuca* (Weber) Wiggers (*Lichen lactuca* Weber).

(=) *Gabura* Adanson, Fam. Pl. 2: 6. 1763.
T.: *Lichen fascicularis* Linnaeus.

(=) *Kolman* Adanson, Fam. Pl. 2: 7, 542. 1763.
T.: *Lichen nigrescens* Hudson.

Collybia (Fries) Staude, Schwämme Mitteldeutschl. xxviii, 119. 1857.
T.: *C. tuberosa* (Bulliard : Fries) P. Kummer (*Agaricus tuberosus* Bulliard : Fries).

(=) *Gymnopus* (Persoon) S. F. Gray, Nat. Arr. Brit. Pl. 1: 604. 1821.
T.: *G. fusipes* (Bulliard : Fries) S. F. Gray (*Agaricus fusipes* Bulliard : Fries).

Coniothyrium Corda, Icon. Fung. 4: 38. 1840.
T.: *C. palmarum* Corda.

(=) *Clisosporium* Fries, Novit. Fl. Suec. 80. 1819 : Fries, Syst. Mycol. 1: xlvii. 1821.
T.: *C. lignorum* Fries : Fries.

Conocybe Fayod, Ann. Sci. Nat. Bot. ser. 7. 9: 357. 1889.
T.: *C. tenera* (Schaeffer : Fries) Fayod (*Agaricus tener* Schaeffer : Fries).

(=) *Raddetes* P. Karsten, Hedwigia 26: 112. 1887.
T.: *R. turkestanicus* P. Karsten.

(=) *Pholiotina* Fayod, Ann. Sci. Nat. Bot. ser. 7. 9: 359. 1889.
T.: *P. blattaria* (Fries : Fries) Fayod (*Agaricus blattarius* Fries : Fries).

(=) *Pholiotella* Spegazzini, Bol. Acad. Nac. Ci. 11: 412. 1889.
T.: *P. blattariopsis* Spegazzini.

Cordyceps Fries, Observ. Mycol. 2: 316. 1818 (cancel page).
T.: *C. militaris* (Linnaeus : Fries) Fries (*Clavaria militaris* Linnaeus : Fries).

Cortinarius (Persoon) S. F. Gray, Nat. Arr. Brit. Pl. 1: 627. 1821 ("*Cortinaria*") (*orth. cons.*).
T.: *C. violaceus* (Linnaeus : Fries) S. F. Gray (*Agaricus violaceus* Linnaeus : Fries).

Craterellus Persoon, Mycol. Europ. 2: 4. 1825 ("*Cratarellus*") (*orth. cons.*).
T.: *C. cornucopioides* (Linnaeus : Fries) Persoon (*Peziza cornucopioides* Linnaeus : Fries).

(H) *Craterella* Persoon, Neues Mag. Bot. 1: 112. 1794 [Fungi].
T.: *C. pallida* Persoon.

(≡) *Trombetta* Adanson, Fam. Pl. 2: 6. 1763 (vide Kuntze, Revis. Gen. Pl. 2: 873. 1891).

Crocynia (Acharius) A. Massalongo, Atti Reale Ist. Veneto Sci. ser. 3. 5: 251. 1860.
T.: *C. gossypina* (Swartz) A. Massalongo (*Lichen gossypinus* Swartz).

(≡) *Symplocia* A. Massalongo, Neagen. Lich. 4. 1854.

Cryptothecia Stirton, Proc. Roy. Philos. Soc. Glasgow 10: 164. 1876.
T.: *C. subnidulans* Stirton.

(=) *Myriostigma* Krempelhuber, Lich. Foliicol. 22. 1874; Nuovo Giorn. Bot. Ital. 7: 44. 1875.
T.: *M. candidum* Krempelhuber.

Cylindrocarpon Wollenweber, Phytopathology 3: 225. 1913.
T.: *C. cylindroides* Wollenweber.

(=) *Fusidium* Link, Ges. Naturf. Freunde Berlin Mag. 3: 8. 1809 : Fries, Syst. Mycol. 1: xl. 1821.
T.: *F. candidum* Link : Fries.

Daldinia Cesati et De Notaris, Comment. Soc. Crittog. Ital. 1: 197. 1863.
T.: *D. concentrica* (Bolton : Fries) Cesati et De Notaris (*Sphaeria concentrica* Bolton : Fries).

(≡) *Peripherostoma* S. F. Gray, Nat. Arr. Brit. Pl. 1: 513. 1821 (by lectotypif.).

(≡) *Stromatosphaeria* Greville, Fl. Edin. lxxiii, 355. 1824 (by lectotypif.).

*Debaryomyces Lodder et Kreger-van Rij ex Kreger-van Rij, Yeasts ed. 3. 130, 145. 1984.
T.: *D. hansenii* (Zopf) Lodder et Kreger-van Rij (*Saccharomyces hansenii* Zopf).

(H) *Debaryomyces* Klöcker, Compt.-Rend. Trav. Carlsberg Lab. 7: 273. 1909 [Fungi].
T.: *D. globosus* Klöcker.

(≡) *Debaryozyma* van der Walt et E. Johannsen, Persoonia 10: 147. 1978.

134

Dothiora Fries, Summa Veg. Scand. 2: 418. 1849.
T.: *D. pyrenophora* (Fries : Fries) Fries (*Dothidea pyrenophora* Fries : Fries).

(H) *Dothiora* Fries, Fl. Scan. 347. 1835 [Fungi].
T.: *Variolaria melogramma* Bulliard : Fries.

Drechslera Ito, Proc. Imp. Acad. Japan 6: 355. 1930.
T.: *D. tritici-vulgaris* (Nisikado) Ito ex Hughes (*Helminthosporium tritici-vulgaris* Nisikado).

(=) *Angiopoma* Léveillé, Ann. Sci. Nat. Bot. ser. 2. 16: 235. 1841.
T.: *A. campanulatum* Léveillé.

Encoelia (Fries : Fries) P. Karsten, Bidrag Kännedom Finlands Natur Folk 19: 18. 1871.
T.: *Encoelia furfuracea* (Roth : Fries) P. Karsten (*Peziza furfuracea* Roth : Fries).

(≡) *Phibalis* Wallroth, Fl. Crypt. Germ. 2: 445. 1833.

***Epidermophyton** Sabouraud, Arch. Méd. Exp. Anat. Pathol. 19: 754-762. 1907.
T.: *E. inguinale* Sabouraud.

(H) *Epidermidophyton* E. Lang, Vierteljahrsschr. Dermatol. Syph. 11: 263. 1879 [Fungi].
T.: non designatus.

Gautieria Vittadini, Monogr. Tuber. 25. 1831.
T.: *G. morchelliformis* Vittadini.

(H) *Gautiera* Rafinesque, Med. Fl. 1: 202. 1828 [Eric.].
≡ *Gaultheria* Linnaeus.

***Gloeophyllum** P. Karsten, Bidrag Kännedom Finlands Natur Folk 37: x, 79. 1882 ("*Gleophyllum*") (*orth. cons.*).
T.: *G. sepiarium* (Wulfen : Fries) P. Karsten (*Agaricus sepiarius* Wulfen : Fries).

(≡) *Serda* Adanson, Fam. Pl. 2: 11. 1763 (vide Donk, Persoonia 1: 279. 1960).
(≡) *Sesia* Adanson, Fam. Pl. 2: 10. 1763 (vide Donk, Persoonia 1: 280. 1960).
(=) *Ceratophora* Humboldt, Fl. Friberg. 112. 1793.
T.: *C. fribergensis* Humboldt.

Gymnoderma Nylander, Flora 43: 546. 1860.
T.: *G. coccocarpum* Nylander.

(H) *Gymnoderma* Humboldt, Fl. Friberg. 109. 1793 [Fungi].
T.: *G. sinuatum* Humboldt.

***Gyrodon** Opatowski, Arch. Naturgesch. 2(1): 5. 1836.
T.: *G. sistotremoides* Opatowski, nom. illeg. (*Boletus sistotremoides* Fries, non Albertini et Schweinitz, *Boletus sistotrema* Fries : Fries, *G. sistotrema* (Fries : Fries) Smotlacha) [= *G. lividus* (Bulliard : Fries) Saccardo, *Boletus lividus* Bulliard : Fries].

(=) *Anastomaria* Rafinesque, Ann. Nature 1: 16. 1820.
T.: *A. campanulata* Rafinesque.

Gyromitra Fries, Summa Veg. Scand. 2: 346. 1849.
T.: *G. esculenta* (Persoon : Fries) Fries (*Helvella esculenta* Persoon : Fries).

(=) *Gyrocephalus* Persoon, Mém. Soc. Linn. Paris 3: 77. 1824.
T.: *G. aginnensis* Persoon, nom. illeg. (*Helvella sinuosa* Brondeau).

Helminthosporium Link, Ges. Naturf. Freunde Berlin Mag. 3: 10. 1809 : Fries, Syst. Mycol. 1: xlvi. 1821 ("*Helmisporium*") (*orth. cons.*).
T.: *H. velutinum* Link : Fries.

135

Hirneola Fries, Kongl. Vetensk. Acad. Handl. 1848: 144. 1848.
T.: *H. nigra* Fries, nom. illeg. (*Peziza nigricans* Swartz : Fries, *H. nigricans* (Swartz : Fries) Graff).

Hydnum Linnaeus, Sp. Pl. 1178. 1753 : Fries, Syst. Mycol. 1: 397. 1821.
T.: *H. repandum* Linnaeus : Fries (*typ. cons.*).

Hymenochaete Léveillé, Ann. Sci. Nat. Bot. ser. 3. 5: 150. 1846.
T.: *H. rubiginosa* (Dickson : Fries) Léveillé (*Helvella rubiginosa* Dickson : Fries).

Hypholoma (Fries) P. Kummer, Führer Pilzk. 21, 72. 1871.
T.: *H. fasciculare* (Fries : Fries) P. Kummer (*Agaricus fascicularis* Fries : Fries) (*typ. cons.*).

*Hypoderma De Notaris, Giorn. Bot. Ital. 2(7-8): 13. 1847.
T.: *H. rubi* (Persoon : Fries) A.-P. de Candolle ex Chevallier (*Hysterium rubi* Persoon : Fries).

Hypoxylon Bulliard, Hist. Champ. France 167. 1791.
T.: *H. coccineum* Bulliard [= *H. fragiforme* (Persoon : Fries) Kickx, *Sphaeria fragiformis* Persoon : Fries].

Karstenia Fries, Acta Soc. Fauna Fl. Fenn. 2(6): 166. 1885.
T.: *K. sorbina* (P. Karsten) Fries (*Propolis sorbina* P. Karsten).

Lachnocladium Léveillé, Dict. Univ. Hist. Nat. 8: 487. 1846.
T.: *L. brasiliense* (Léveillé) Patouillard (*Eriocladus brasiliensis* Léveillé).

*Lactarius Persoon, Tent. Disp. Meth. Fung. 63. 1797 ("*Lactaria*") (*orth. cons.*).
T.: *L. piperatus* (Linnaeus : Fries) Persoon ("*Lactaria piperata*") (*Agaricus piperatus* Linnaeus : Fries).

(H) *Hirneola* Fries, Syst. Orb. Veg. 93. 1825 [Fungi].
≡ *Mycobonia* Patouillard 1894 (*nom. cons.*).
(=) *Laschia* Fries, Linnaea 5: 533. 1830 : Fries, Syst. Mycol. 3, index: 107. 1832.
T.: *L. delicata* Fries : Fries.

(H) *Hymenochaeta* Palisot de Beauvois ex T. Lestiboudois, Essai Cypér. 43. 1819 [Cyper.].
T.: non designatus.

(H) *Hypoderma* A.-P. de Candolle in Lamarck et A.-P. de Candolle, Fl. Franç. ed. 3. 2: 304. 1805 [Fungi].
≡ *Lophodermium* Chevallier 1826 (*nom. cons.*) (vide Cannon et Minter, Taxon 32: 580. 1983).

(H) *Hypoxylon* Adanson, Fam. Pl. 2: 9. 1763 [Fungi].
LT.: *Xylaria polymorpha* (Persoon : Fries) Greville (*Sphaeria polymorpha* Persoon : Fries) (vide Donk, Regnum Veg. 34: 16. 1964).
(=) *Sphaeria* Haller, Hist. Stirp. Helv. 3: 120. 1768 : Fries, Syst. Mycol. 1: lii. 1821.
LT.: *Hypoxylon fragiforme* (Persoon : Fries) Kickx (*Sphaeria fragiformis* Persoon : Fries) (vide Donk, Regnum Veg. 34: 16. 1964).

(H) *Karstenia* Goeppert, Nova Acta Phys.-Med. Acad. Caes. Leop.-Carol. Nat. Cur. 17(suppl.): 451. 1836 [Foss.].
T.: non designatus.

(≡) *Eriocladus* Léveillé, Ann. Sci. Nat. Bot. ser. 3. 5: 158. 1846.

136

*Laetinaevia Nannfeldt, Nova Acta Regiae Soc. Sci. Upsal. ser. 4. 8(2): 190. 1932.
T.: *L. lapponica* (Nannfeldt) Nannfeldt (*Naevia lapponica* Nannfeldt).

(=) *Myridium* Clements, Gen. Fungi 67. 1909.
T.: *M. myriospora* (Phillips et Harkness) Clements (*Orbilia myriospora* Phillips et Harkness).

*Lecanactis Koerber, Syst. Lich. Germ. 275. 1855.
T.: *L. abietina* (Acharius) Koerber (*Lichen abietinus* Acharius) (*typ. cons.*).

(H) *Lecanactis* Eschweiler, Syst. Lich. 14, 25. 1824 [Fungi].
T.: *L. lobata* Eschweiler.

Lepraria Acharíus, Methodus 3. 1803.
T.: *L. incana* (Linnaeus) Acharius (*Byssus incana* Linnaeus).

(≡) *Pulina* Adanson, Fam. Pl. 2: 3, 595. 1763 (vide Laundon, Taxon 12: 37. 1963).
(≡) *Conia* Ventenat, Tabl. Règne Vég. 2: 32. 1799 (vide Laundon, Taxon 12: 37. 1963).

*Leptoglossum P. Karsten, Bidrag Kännedom Finlands Natur Folk 32: xvii. 1879.
T.: *L. muscigenum* (Bulliard : Fries) P. Karsten (*Agaricus muscigenus* Bulliard : Fries).

(=) *Boehmia* Raddi, Mem. Mat. Fis. Soc. Ital. Sci. 13(2): 357. 1807.
T.: *B. muscoides* Raddi.

Leptorhaphis Koerber, Syst. Lich. Germ. 371. 1855.
T.: *L. oxyspora* (Nylander) Koerber (*Verrucaria oxyspora* Nylander).

(=) *Endophis* Norman, Nyt Mag. Naturvidensk. 7: 240. 1853.
T.: non designatus.

Leptosphaeria Cesati et De Notaris, Comment. Soc. Crittog. Ital. 1: 234. 1863.
T.: *L. doliolum* (Persoon : Fries) Cesati et De Notaris (*Sphaeria doliolum* Persoon : Fries).

(≡) *Bilimbiospora* Auerswald in Rabenhorst, Fungi Europaei ed. 2. n. 261 (in sched. corr.). 1861(?).
(=) *Nodulosphaeria* Rabenhorst, Herb. Mycol. ed. 2. n. 725 (in sched.). 1858 (*nom. cons.*).

Letharia (T. M. Fries) Zahlbruckner, Hedwigia 31: 34. 1892.
T.: *L. vulpina* (Linnaeus) Hue (*Lichen vulpinus* Linnaeus).

(≡) *Chlorea* Nylander, Mém. Soc. Sci. Nat. Cherbourg 3: 170. 1855.

Lichina C. Agardh, Syn. Alg. Scand. xii. 1817.
T.: *L. pygmaea* (Lightfoot) C. Agardh (*Fucus pygmaeus* Lightfoot).

(=) *Pygmaea* Stackhouse, Mém. Soc. Imp. Naturalistes Moscou 2: 60, 95. 1809.
T.: *Fucus lichenoides* J. F. Gmelin.

Lopadium Koerber, Syst. Lich. Germ. 210. 1855.
T.: *L. pezizoideum* (Acharius) Koerber (*Lecidea pezizoidea* Acharius).

(=) *Brigantiaea* Trevisan, Spighe e Paglie 7. 1853.
T.: *B. tricolor* (Montagne) Trevisan (*Biatora tricolor* Montagne).

Lophiostoma Cesati et De Notaris, Comment. Soc. Crittog. Ital. 1: 219. 1863.
T.: *L. macrostoma* (Tode : Fries) Cesati et De Notaris (*Sphaeria macrostoma* Tode : Fries).

(≡) *Platysphaera* Dumortier, Comment. Bot. 87. 1822.

137

*Lophodermium Chevallier, Fl. Gén. Env.
Paris 1: 435. 1826.
T.: *L. arundinaceum* (Schrader : Fries) Che-
vallier (*Hysterium arundinaceum* Schrader :
Fries) (vide *Hypoderma* De Notaris [Fungi],
nom. cons.).

Marasmius Fries, Fl. Scan. 339. 1835.
T.: *M. rotula* (Scopoli : Fries) Fries (*Agaricus
rotula* Scopoli : Fries).

(=) *Micromphale* S. F. Gray, Nat. Arr. Brit.
Pl. 1:621. 1821.
T.: *M. venosa* (Persoon) S. F. Gray
(*Agaricus venosus* Persoon).

Melanogaster Corda in Sturm, Deutschl. Fl.
Abt. iii. 3(H. 11): 1. 1831.
T.: *M. tuberiformis* Corda.

(=) *Bullardia* Junghuhn, Linnaea 5: 408. 1830
[non *Bulliarda* A.-P. de Candolle 1801].
T.: *B. inquinans* Junghuhn.

*Melanoleuca Patouillard, Cat. Pl. Cell. Tuni-
sie 22. 1897.
T.: *Melanoleuca vulgaris* (Patouillard) Patouillard (*Melaleuca vulgaris* Patouillard, *Agaricus melaleucus* Persoon : Fries, *M. melaleuca*
(Persoon : Fries) Murrill).

(≡) *Psammospora* Fayod, Ann. Reale Accad.
Agric. Torino 35: 91. 1893 ("1892").

*Melanospora Corda, Icon. Fung. 1: 24. 1837.
T.: *M. zamiae* Corda.

(=) *Ceratostoma* Fries, Observ. Mycol. 2:
337. 1818.
LT.: *C. chioneum* (Fries : Fries) Fries
(*Sphaeria chionea* Fries : Fries) (vide
Fries, Summa Veg. Scand., sect. post.
396. 1849).
(=) *Megathecium* Link, Abh. Königl. Akad.
Wiss. Berlin 1824: 176. 1826.
LT.: *Sphaeria chionea* Fries : Fries (vide
Cannon et Hawksworth, Taxon 32: 476.
1983).

Monilia Bonorden, Handb. Mykol. 76. 1851.
T.: *M. cinerea* Bonorden.

(H) *Monilia* Link, Ges. Naturf. Freunde Berlin Mag. 3: 16. 1809 : Fries, Syst. Mycol. 1:
xlvi. 1821 [Fungi].
T.: *M. antennata* (Persoon : Fries) Persoon.

*Mucor Fresenius, Beitr. Mykol. 7. 1850.
T.: Fresenius, Beitr. Mykol. *t. 1. f. 1-12.* 1850,
sub "*Mucor mucedo*" [= *M. murorum* Nau-
mov].

(H) *Mucor* Linnaeus, Sp. Pl. 1185. 1753 :
Fries, Syst. Mycol. 3(2): 317. 1832 [Fun-
gi].
LT.: *M. mucedo* Linnaeus : Fries (vide
Sumstine, Mycologia 2: 127. 1910).
(=) *Hydrophora* Tode, Fungi Mecklenb. Sel.
2: 5. 1791 : Fries, Syst. Mycol. 3(2): 313.
1832.
LT.: *H. stercorea* Tode : Fries (vide
Sumstine, Mycologia 2: 132. 1910).

138

Mutinus Fries, Summa Veg. Scand. 2: 434.
1849.
T.: *M. caninus* (Schaeffer : Persoon) Fries
(*Phallus caninus* Schaeffer : Persoon).

(≡) *Cynophallus* (Fries) Corda, Icon. Fung. 5:
29. 1842.
(=) *Aedycia* Rafinesque, Med. Repos. ser. 2.
5: 358. 1808.
T.: *A. rubra* Rafinesque.
(=) *Ithyphallus* S. F. Gray, Nat. Arr. Brit. Pl.
1: 675. 1821.
T.: *I. inodorus* (Sowerby) S. F. Gray
(*Phallus inodorus* Sowerby).

Mycobonia Patouillard, Bull. Soc. Mycol.
France 10: 76. 1894.
T.: *M. flava* (Swartz : Fries) Patouillard (*Peziza flava* Swartz : Fries) (vide *Hirneola* Fries
[Fungi], *nom. cons.*).

Mycoporum Flotow ex Nylander, Mém. Soc.
Sci. Nat. Cherbourg 3: 186. 1855.
T.: *M. elabens* Flotow ex Nylander.

(H) *Mycoporum* G. Meyer, Nebenst. Beschäft. Pflanzenk. 327. 1825 [Fungi:
Lich.].
T.: *M. melinostigma* G. Meyer.

*****Nectria** Fries, Summa Veg. Scand., sect.
post. 387. 1849.
T.: *N. cinnabarina* (Tode : Fries) Fries
(*Sphaeria cinnabarina* Tode : Fries) (*typ.
cons.*).

(=) *Ephedrosphaera* Dumortier, Comment.
Bot. 90. 1822.
LT.: *E. decolorans* (Persoon) Dumortier
(*Sphaeria decolorans* Persoon) (vide
Cannon et Hawksworth, Taxon 32: 477.
1983).
(=) *Hydropisphaera* Dumortier, Comment.
Bot. 89. 1822.
T.: *H. peziza* (Tode : Fries) Dumortier
(*Sphaeria peziza* Tode : Fries).

Nidularia Fries, Symb. Gasteromyc. 1: 2. 1817.
T.: *N. radicata* Fries.

(H) *Nidularia* Bulliard, Hist. Champ. France
163. 1791 [Fungi].
T.: *N. vernicosa* Bulliard.

*****Nodulosphaeria** Rabenhorst, Herb. Mycol.
ed. 2. n. 725 (in sched.). 1858.
T.: Rabenhorst, Herb. Mycol. ed. 2. n. 725 (S)
[= *N. derasa* (Berkeley et Broome) L. Holm
(*Sphaeria derasa* Berkeley et Broome)] (*typ.
cons.*).

Ocellularia G. Meyer, Nebenst. Beschäft.
Pflanzenk. 327. 1825.
T.: *O. obturata* (Acharius) K. Sprengel (*Thelotrema obturatum* Acharius).

(=) *Ascidium* Fée, Essai Crypt. Ecorc. xlii.
1824.
T.: *A. cinchonarum* Fée.

Oidium Link in Willdenow, Sp. Pl. 6(1): 121.
1824.
T.: *O. monilioides* (Nees : Fries) Link (*Acrosporium monilioides* Nees : Fries).

(H) *Oidium* Link, Ges. Naturf. Freunde Berlin Mag. 3: 18. 1809 : Fries, Syst. Mycol. 1:
xlv. 1821 [Fungi].
T.: *O. aureum* (Persoon : Fries) Link
(*Trichoderma aureum* (Persoon : Fries)
Persoon).
(≡) *Acrosporium* Nees, Syst. Pilze 53. 1816-
1817 : Fries, Syst. Mycol. 1: xlv. 1821.

139

Opegrapha Acharius, Kongl. Vetensk. Acad. Nya Handl. **1808-1811:** 97. 1809.
T.: *O. vulgata* (Acharius) Acharius (*Lichen vulgatus* Acharius).

(H) *Opegrapha* Humboldt, Fl. Friberg. 57. 1793 [Fungi: Lich.].
T.: *O. vulgaris* Humboldt, nom. illeg. (*Lichen scriptus* Linnaeus).

Panaeolus (Fries) Quélet, Mém. Soc. Emul. Montbéliard ser. 2. *5*: 151. 1872.
T.: *P. papilionaceus* (Bulliard : Fries) Quélet (*Agaricus papilionaceus* Bulliard : Fries).

(≡) *Coprinarius* (Fries) P. Kummer, Führer Pilzk. 20, 68. 1871.

Panus Fries, Epicr. Syst. Mycol. 396. 1838.
T.: *P. conchatus* (Bulliard : Fries) Fries (*Agaricus conchatus* Bulliard : Fries).

Parmelia Acharius, Methodus xxxiii, 153. 1803.
T.: *P saxatilis* (Linnaeus) Acharius (*Lichen saxatilis* Linnaeus).

(≡) *Lichen* Linnaeus, Sp. Pl. 1140. 1753.

Peccania A. Massalongo ex Arnold, Flora **41**: 93. 1858.
T.: *P. coralloides* (A. Massalongo) A. Massalongo (*Corinophoros coralloides* A. Massalongo).

(≡) *Corinophoros* A. Massalongo, Flora **39**: 212. 1856.

Peltigera Willdenow, Fl. Berol. Prodr. 347. 1787.
T.: *P. canina* (Linnaeus) Willdenow (*Lichen caninus* Linnaeus).

(=) *Placodion* P. Browne ex Adanson, Fam. Pl. **2**: 7. 1763.
T.: non designatus [*Lichenoides* sp. Dillenius *t. 27, fig. 102*].

Peridermium (Link) J. C. Schmidt et Kunze, Deutschl. Schwämme **6**: 4. 1817.
T.: *P. elatinum* (Albertini et Schweinitz) J. C. Schmidt et Kunze (*Aecidium elatinum* Albertini et Schweinitz) (*typ. cons.*).

Pertusaria A.-P. de Candolle in Lamarck et A.-P. de Candolle, Fl. Franç. ed. 3, **2**: 319. 1805.
T.: *P. communis* A.-P. de Candolle, nom. illeg. (*Lichen pertusus* Linnaeus, nom. illeg., *Lichen verrucosus* Hudson).

(=) *Lepra* Scopoli, Intr. 61. 1777.
T.: non designatus [*Lichen* ordo xxxiv sp. Micheli].
(=) *Variolaria* Persoon, Ann. Bot. (Usteri) 7: 23. 1794.
T.: *V. discoidea* Persoon.
(=) *Leproncus* Ventenat, Tabl. Régne Vég. **2**: 32. 1799.
T.: non designatus [*Lichenoides* sp. Dillenius *t. 18 fig.*].
(=) *Isidium* Acharius, Methodus xxxiii, 136. 1803.
T.: *I. corallinum* (Linnaeus) Acharius (*Lichen corallinus* Linnaeus).

*****Pezicula** L. R. Tulasne et C. Tulasne, Select. Fung. Carpol. **3**: 182. 1865.
T.: *P. carpinea* (Persoon) Saccardo (*Peziza carpinea* Persoon).

(H) *Pezicula* Paulet, Tab. Pl. Fung. 24. 1791 [Fungi].
LT.: *P. cornucopioides* (Linnaeus : Fries) Paulet (*Peziza cornucopioides* Linnaeus : Fries) (vide Cannon et Hawksworth, Taxon **32**: 478. 1983).

140

Phacidium Fries, Observ. Mycol. **1**: 167. 1815 :
Fries, Syst. Mycol. **2**(2): 571. 1823.
T.: *P. lacerum* Fries : Fries (*typ. cons.*).

Phaeotrema Müller Arg., Mém. Soc. Phys. Genève **29**(8): 10. 1887.
T.: *P. subfarinosa* (Fée) Müller Arg. (*Pyrenula subfarinosa* Fée).

(=) *Asteristion* Leighton, Trans. Linn. Soc. London **27**: 163. 1870.
T.: *A. erumpens* Leighton.

***Phellinus** Quélet, Enchir. Fung. 172. 1886.
T.: *P. igniarius* (Linnaeus : Fries) Quélet (*Boletus igniarius* Linnaeus : Fries).

(=) *Mison* Adanson, Fam. Pl. **2**: 10. 1763 [Fungi].
T.: non designatus.

Phillipsia Berkeley, J. Linn. Soc., Bot. **18**: 388. 1881.
T.: *P. domingensis* (Berkeley) Le Gal (*Peziza domingensis* Berkeley).

(H) *Phillipsia* K. B. Presl in Sternberg, Vers. Fl. Vorwelt **2**(7/8): 206. 1838 [Foss.].
T.: *P. harcourtii* (Witham) K. B. Presl (*Lepidodendron harcourtii* Witham).

***Phlyctis** (Wallroth) Flotow, Bot. Zeitung (Berlin) **8**: 571. 1850.
T.: *P. agelaea* (Acharius) Flotow (*Lichen agelaeus* Acharius).

(H) *Phlyctis* Rafinesque, Caratt. Nuovi Gen. 91. 1810 [Algae inc. sed.].
T.: non designatus.

Pholiota (Fries) P. Kummer, Führer Pilzk. 22, 83. 1871.
T.: *P. squarrosa* (Batsch : Fries) P. Kummer (*Agaricus squarrosus* Batsch : Fries).

(≡) *Derminus* (Fries) Staude, Schwämme Mittel-Deutschl. xxvi, 86. 1857.

Phoma Saccardo, Michelia **2**: 4. 1880.
T.: *P. herbarum* Westendorp.

(H) *Phoma* Fries, Novit. Fl. Suec. **5**: 80. 1819 : Fries, Syst. Mycol. **1**: lii. 1821 [Fungi].
T.: *P. pustula* (Persoon : Fries) Fries (*Sphaeria pustula* Persoon : Fries).

***Phomopsis** (Saccardo) Bubák, Österr. Bot. Z. **55**: 78. 1905.
T.: *P. lactucae* (Saccardo) Bubák (*Phoma lactucae* Saccardo).

(H) *Phomopsis* Saccardo et Roumeguère, Rev. Mycol. (Toulouse) **6**: 32. 1884 [Fungi].
T.: *P. brassicae* Saccardo et Roumeguère.

(=) *Myxolibertella* Höhnel, Ann. Mycol. **1**: 526. 1903.
LT.: *M. aceris* Höhnel (vide Clements et Shear, Gen. Fung. 359. 1931).

Phyllachora Nitschke ex Fuckel, Jahrb. Nassauischen Vereins Naturk. **23-24**: 216. 1870.
T.: *P. graminis* (Persoon : Fries) Fuckel (*Sphaeria graminis* Persoon : Fries).

(H) *Phyllachora* Nitschke ex Fuckel, Fungi Rhenani n. 2056 (in sched.). 1867 [Fungi].
T.: *P. agrostis* Fuckel.

Phyllosticta Persoon, Traité Champ. Comest. 55, 147. 1818.
T.: *P. convallariae* Persoon.

***Physconia** Poelt, Nova Hedwigia **9**: 30. 1965.
T.: Germania, Lipsia in *Tilia*, 1767, Schreber sub "*Lichen pulverulentus*" (**M**) (*typ. cons.*).

Pleospora Rabenhorst ex Cesati et De Notaris, Comment. Soc. Crittog. Ital. **1**: 217. 1863.
T.: *P. herbarum* (Fries) Cesati et De Notaris.

(H) *Pleiospora* Harvey, Thes. Cap. **1**: 51. 1859 [Legum.].
T.: *P. cajanifolia* Harvey.

(=) *Clathrospora* Rabenhorst, Hedwigia **1**: 116. 1857.
T.: *C. elynae* Rabenhorst.

Pleurotus (Fries) P. Kummer, Führer Pilzk. 24, 104. 1871.
T.: *P. ostreatus* (Jacquin : Fries) P. Kummer (*Agaricus ostreatus* Jacquin : Fries).

(≡) *Pleuropus* (Persoon) Roussel, Fl. Calvados ed. 2. 67. 1806 (vide Donk, Beih. Nova Hedwigia **5**: 235. 1962).

(≡) *Crepidopus* S. F. Gray, Nat. Arr. Brit. Pl. **1**: 616. 1821.

(=) *Gelona* Adanson, Fam. Pl. **2**: 11. 1763.
T.: non designatus.

(=) *Resupinatus* S. F. Gray, Nat. Arr. Brit. Pl. **1**: 617. 1821.
T.: *R. applicatus* (Batsch : Fries) S. F. Gray (*Agaricus applicatus* Batsch : Fries).

(=) *Pterophyllus* Léveillé, Ann. Sci. Nat. Bot. ser. 3. **2**: 178. 1844.
T.: *P. bovei* Léveillé.

(=) *Hohenbuehelia* Schulzer in Schulzer, Kanitz et Knapp, Verh. Zool.-Bot. Ges. Wien **16** (Abh.): 45. 1866.
T.: *H. petaloides* (Bulliard : Fries) Schulzer (*Agaricus petaloides* Bulliard : Fries).

Podospora Cesati in Rabenhorst, Klotzschii Herb. Mycol. ed. 2. n. 258 (vel 259) (in sched.). 1856; Hedwigia **1**: 103, *pl. 4, fig. A.* 1856.
T.: Rabenhorst, Klotzschii Herb. Mycol. ed. 2. n. 259 (S), type of *P. fimiseda* (Cesati et De Notaris) Niessl (*typ. cons.*).

(=) *Schizothecium* Corda, Icon. Fung. **2**: 29. 1838.
T.: *S. fimicola* Corda.

Polyblastia A. Massalongo, Ric. Auton. Lich. Crost. 147. 1852.
T.: *P. cupularis* A. Massalongo.

(=) *Sporodictyon* A. Massalongo, Flora **35**: 326. 1852.
T.: *S. schaererianum* A. Massalongo.

Porina Müller Arg., Flora **66**: 320. 1883.
T.: *P. nucula* Acharius.

(H) *Porina* Acharius, Kongl. Vetensk. Acad. Nya Handl. **1808-1811**: 158. 1809 [Fungi: Lich.].
T.: *P. pertusa* (Linnaeus) Acharius (*Lichen pertusus* Linnaeus).

(=) *Ophthalmidium* Eschweiler, Syst. Lich. 18. 1824.
T.: *O. hemisphaericum* Eschweiler.

(=) *Segestria* Fries, Syst. Orb. Veg. 263. 1825.
T.: *S. lectissima* Fries.

Pseudographis Nylander, Mém. Soc. Sci. Nat. Cherbourg **3**: 190. 1855.
T.: *P. elatina* (Acharius : Fries) Nylander (*Lichen elatinus* Acharius : Fries).

(=) *Krempelhuberia* A. Massalongo, Geneac. Lich. 34. 1854.
T.: *K. cadubriae* A. Massalongo.

Psora G. F. Hoffmann, Deutschl. Fl. **2**: 161. 1796 (*typ. cons.*).
T.: *P. decipiens* (Hedwig) G. F. Hoffmann (*Lecidea decipiens* Hedwig).

(H) *Psora* J. Hill, Veg. Syst. **4**: 30. 1762 [Comp.].
T.: non designatus.

(H) *Psora* G. F. Hoffmann, Descr. Pl. Cl. Crypt. **1**(2): 37. 1789 [Fungi].
T.: *P. caesia* G. F. Hoffmann.

*Pycnophorus P. Karsten, Rev. Mycol. (Toulouse) **3**(9): 18. 1881.
T.: *P. cinnabarinus* (Jacquin : Fries) P. Karsten (*Boletus cinnabarinus* Jacquin : Fries).

(≡) *Xylometron* Paulet, Prosp. Traité Champ. (Mycétol.) 29. 1808.

*Pyrenula Acharius, Syn. Meth. Lich. 117. 1814.
T.: *P. nitida* (Weigel) Acharius (*Sphaeria nitida* Weigel) (*typ. cons.*).

(H) *Pyrenula* Acharius, Kongl. Vetensk. Akad. Nya Handl. **30**: 160. 1809 [Fungi].
T.: *P. margacea* (Wahlenberg) Acharius (*Thelotrema margaceum* Wahlenberg).

Pythium Pringsheim, Jahrb. Wiss. Bot. **1**: 304. 1858.
T.: *P. monospermum* Pringsheim.

(H) *Pythium* Nees in Carus, Nova Acta Phys.-Med. Acad. Caes. Leop.-Carol. Nat. Cur. **11**: 514. 1823 [Fungi].
T.: non designatus.

(=) **Artotrogus* Montagne, Gard. Chron. **5**: 640. 1845.
T.: *A. hydnosporus* Montagne.

Racodium Fries, Syst. Mycol. **3**(1): 229. 1829.
T.: *R. rupestre* Persoon.

(H) *Racodium* Persoon, Neues Mag. Bot. **1**: 123. 1794 : Fries, Syst. Mycol. **1**. xlvi. 1821 [Fungi].
T.: *R. cellare* Persoon : Fries.

Ramalina Acharius, Lichenogr. Universalis 598. 1810.
T.: *R. fraxinea* (Linnaeus) Acharius (*Lichen fraxineus* Linnaeus) (*typ. cons.*).

Ramaria Fries ex Bonorden, Handb. Mykol. 166. 1851.
T.: *R. botrytis* (Persoon : Fries) Ricken (*Clavaria botrytis* Persoon : Fries).

(H) *Ramaria* Holmskjold, Beata Ruris **1**: xvii. 1790 [Fungi].
T.: non designatus.

(≡) *Cladaria* Ritgen, Schriften Ges. Beförd. Gesammten Naturwiss. Marburg **2**: 94. 1831 [praeimpr. 1828. p. 54].

Rhabdospora (Durieu et Montagne ex Saccardo) Saccardo, Syll. Fung. **3**: 578. 1884.
T.: *R. oleandri* (Durieu et Montagne) Saccardo (*Septoria oleandri* Durieu et Montagne).

(=) *Filaspora* Preuss, Linnaea **26**: 718. 1855.
T.: *F. peritheciiformis* Preuss.

Rhipidium Cornu, Bull. Soc. Bot. France **18**: 58. 1871.
T.: *R. interruptum* Cornu.

(H) *Rhipidium* Wallroth, Fl. Crypt. Germ. **2**: 742. 1833 [Fungi].
T.: *R. stipticum* (Bulliard : Fries) Wallroth (*Agaricus stipticus* Bulliard : Fries).

*Rhizopus Ehrenberg, Nova Acta Phys.-Med. Acad. Caesar. Leop.-Carol. Nat. Cur. **10**: 198. 1821.
T.: *R. nigricans* Ehrenberg, nom. illeg. (*Mucor stolonifer* Ehrenberg, *R. stolonifer* (Ehrenberg) Vuillemin).

(=) *Ascophora* Tode, Fungi Mecklenb. Sel. **1**: 13. 1790 : Fries, Syst. Mycol. **3**(2): 309. 1832.
LT.: *A. mucedo* Tode : Fries (vide Kirk, Taxon **35**: 374. 1986).

Robillarda Saccardo, Michelia 2: 8. 1882.
T.: *R. sessilis* (Saccardo) Saccardo (*Pestalotia sessilis* Saccardo).

*Roccella A.-P. de Candolle in Lamarck et A.-P. de Candolle, Fl. Franç. ed. 3. 2: 334. 1805.
T.: *R. fuciformis* (Linnaeus) A.-P. de Candolle (*Lichen fuciformis* Linnaeus).

*Rutstroemia P. Karsten, Bidrag Kännedom Finlands Natur Folk 19: 12. 1871.
T.: *R. firma* (Persoon : Fries) P. Karsten (*Peziza firma* Persoon : Fries) (*typ. cons.*).

Sclerotinia Fuckel, Jahrb. Nassauischen Vereins Naturk. 23-24: 330. 1870.
T.: *S. libertiana* Fuckel, nom. illeg. (*Peziza sclerotiorum* Libert, *S. sclerotiorum* (Libert) de Bary) (*typ. cons.*).

*Scutellinia (Cooke) Lambotte, Fl. Mycol. Belge, Suppl. 1: 299. 1887.
T.: *S. scutellata* (Linnaeus : Fries) Lambotte (*Peziza scutellata* Linnaeus : Fries).

Septobasidium Patouillard, J. Bot. (Morot) 6: 63. 1892; Bull. Soc. Mycol. France 7: xxxv. 1892 ("1891").
T.: *S. velutinum* Patouillard.

Septoria Saccardo, Syll. Fung. 3: 474. 1884.
T.: *S. cytisi* Desmazières.

Sordaria Cesati et De Notaris, Comment. Soc. Crittog. Ital. 1: 225. 1863.
T.: *S. fimicola* (Roberge ex Desmazières) Cesati et De Notaris (*Sphaeria fimicola* Roberge ex Desmazières) (*typ. cons.*).

Sphaerophorus Persoon, Ann. Bot. (Usteri) 7: 23. 1794.
T.: *S. coralloides* Persoon, nom. illeg. (*Lichen globiferus* Linnaeus) [= *S. globosus* (Hudson) Vainio, *Lichen globosus* Hudson] (*typ. cons.*).

Sphaeropsis Saccardo, Michelia 2: 105. 1880.
T.: *S. visci* (Albertini et Schweinitz : Fries) Saccardo (*Sphaeria atrovirens* var. *visci* Albertini et Schweinitz : Fries).

(H) *Robillarda* Castagne, Cat. Pl. Marseille 205. 1845 [Fungi].
T.: *R. glandicola* Castagne.

(=) *Thamnium* Ventenat, Tabl. Règne Vég. 2: 35. 1799.
LT.: *T. roccella* (Linnaeus) Saint-Hilaire (*Lichen roccella* Linnaeus) (vide Ahti, Taxon 33: 330. 1984).

(=) *Patella* Wiggers, Prim. Fl. Holsat. 106. 1780.
LT.: *P. ciliata* Wiggers (vide Korf et Schumacher, Taxon 35: 378. 1986).

(=) *Gausapia* Fries, Syst. Orb. Veg. 302. 1825.
T.: *Thelephora pedicellata* Schweinitz.
(=) *Glenospora* Berkeley et Desmazières, J. Hort. Soc. London 4: 255. 1849.
T.: *G. curtisii* Berkeley et Desmazières.
(=) *Campylobasidium* Lagerheim ex F. Ludwig, Lehrb. Nied. Krypt. 474. 1892.
T.: non designatus.

(H) *Septaria* Fries, Novit. Fl. Suec. 5: 78. 1819 : Fries, Syst. Mycol. 1. xl. 1821 [Fungi].
T.: *S. ulmi* Fries : Fries.

(H) *Sphaeropsis* Léveillé in A. N. Demidow, Voy. Russie Mér. 2: 112. 1842 [Fungi].
T.: *S. conica* Léveillé.
(=) *Macroplodia* Westendorp, Bull. Acad. Roy. Sci. Belgique ser. 2. 2: 562. 1857.
T.: *M. aquifolia* Westendorp.

Sphaerotheca Léveillé, Ann. Sci. Nat. Bot. ser. 3. **15**: 138. 1851.
T.: *S. pannosa* (Wallroth : Fries) Léveillé (*Alphitomorpha pannosa* Wallroth : Fries).

(H) *Sphaerotheca* Chamisso et Schlechtendal, Linnaea **2**: 605. 1827 [Scrophular.].
T.: *S. scoparioides* Chamisso et Schlechtendal.

*****Spongipellis** Patouillard, Hyménomyc. Eur. 140. 1887.
T.: *S. spumeus* (Sowerby : Fries) Patouillard (*Boletus spumeus* Sowerby : Fries).

(=) *Somion* Adanson, Fam. Pl. **2**: 5. 1763.
LT.: *Hydnum occarium* Batsch : Fries (vide Donk, Verh. Kon. Ned. Akad. Wetensch., Afd. Natuurk., Tweede Sect. **62**: 175. 1974).

Stagonospora (Saccardo) Saccardo, Syll. Fung. **3**: 445. 1884.
T.: *S. paludosa* (Saccardo et Spegazzini) Saccardo (*Hendersonia paludosa* Saccardo et Spegazzini).

(=) *Hendersonia* Berkeley, Ann. Mag. Nat. Hist. **6**: 430. 1841.
T.: *H. elegans* Berkeley.

Staurothele Norman, Nyt Mag. Naturvidensk. **7**: 240. 1853.
T.: *S. clopima* (Wahlenberg) T. M. Fries (*Verrucaria clopima* Wahlenberg).

(=) *Paraphysorma* A. Massalongo, Ric. Auton. Lich. Crost. 116. 1852.
T.: *P. protuberans* (Schaerer) A. Massalongo (*Parmelia cervina* var. *protuberans* Schaerer).

Stereocaulon G. F. Hoffmann, Deutschl. Fl. **2**: 128. 1796.
T.: *S. paschale* (Linnaeus) G. F. Hoffmann (*Lichen paschalis* Linnaeus).

(H) *Stereocaulon* Schrader, Spic. Fl. Germ. 113. 1794 [Fungi: Lich.].
T.: *S. corallinum* (Linnaeus) Schrader (*Lichen corallinus* Linnaeus).

Stilbella Lindau in Engler et Prantl, Nat. Pflanzenfam. 1(1**): 489. 1900.
T.: *S. erythrocephala* (Ditmar : Fries) Lindau (*Stilbum erythrocephalum* Ditmar : Fries).

(=) *Botryonipha* Preuss, Linnaea **25**: 79. 1852.
T.: *B. alba* Preuss.

Telamonia (Fries) Wünsche, Pilze 87, 122. 1877.
T.: *T. torva* (Fries : Fries) Wünsche (*Agaricus torvus* Fries : Fries).

(≡) *Raphanozon* P. Kummer, Führer Pilzk. 22. 1871.

Thamnolia Acharius ex Schaerer, Enum. Crit. Lich. Eur. 243. 1850.
T.: *T. vermicularis* (Swartz) Schaerer (*Lichen vermicularis* Swartz).

(≡) *Cerania* Acharius ex S. F. Gray, Nat. Arr. Brit. Pl. **1**: 413. 1821.

Thelopsis Nylander, Mém. Soc. Sci. Nat. Cherbourg **3**: 194. 1855.
T.: *T. rubella* Nylander.

(=) *Sychnogonia* Koerber, Syst. Lich. Germ. 332. 1855.
T.: *S. bayrhofferi* Zwackh.

Tholurna Norman, Flora **44**: 409. 14 Jul. 1861.
T.: *T. dissimilis* (Norman) Norman (*Podocratera dissimilis* Norman).

(≡) *Podocratera* Norman, Förh. Skand. Naturf. Möte **8**: 726. 6-12 Apr 1861.

145

Tomentella Persoon ex Patouillard, Hyméno-
myc. Eur. 154. 1887.
T.: *T. ferruginea* (Persoon : Fries) Patouillard
(*Thelephora ferruginea* Persoon : Fries).

(=) *Caldesiella* Saccardo, Fungi Ital. *f. 125.*
1877; Michelia **1**: 6. 1877.
T.: *C. italica* Saccardo.

(=) *Odontia* Persoon, Neues Mag. Bot. **1**:
110. 1794.
LT.: *O. ferruginea* Persoon (vide Ban-
ker, Bull. Torrey Bot. Club **29**: 448.
1902).

Tremella Persoon, Neues Mag. Bot. **1**: 111.
1794 : Fries, Syst. Mycol. **2**(1): 210. 1822.
T.: *T. mesenterica* Schaeffer : Fries.

Tricholoma (Fries) Staude, Schwämme Mittel-Deutschl. xxviii, 125. 1857.
T.: *T. flavovirens* (Albertini et Schweinitz :
Fries) Lundell (*Agaricus flavovirens* Albertini
et Schweinitz : Fries).

(H) *Tricholoma* Bentham in A. de Candolle,
Prodr. **10**: 426. 1846 [Scrophular.].
T.: *T. elatinoides* Bentham.

Trypethelium K. Sprengel, Anleit. Kenntn.
Gew. **3**: 350. 1804.
T.: *T. eluteriae* K. Sprengel.

(=) *Bathelium* Acharius, Methodus 111.
1803.
T.: *B. mastoideum* Afzelius ex Acharius.

Tubercularia Tode, Fungi Meckl. **1**: 18. 1790 :
Fries, Syst. Mycol. **1**: xli. 1821.
T.: *T. vulgaris* Tode : Fries.

Urocystis Rabenhorst ex Fuckel, Jahrb. Nassauischen Vereins Naturk. **23-24**: 41. 1870
("1869").
T.: *U. occulta* (Wallroth) Fuckel (*Erysibe occulta* Wallroth).

(=) *Polycystis* Léveillé, Ann. Sci. Nat. Bot.
ser. 3. **5**: 269. 1846.
T.: *P. pompholygodes* (Schlechtendal)
Léveillé (*Caeoma pompholygodes*
Schlechtendal).

(=) *Tuburcinia* Fries, Syst. Mycol. **3** (2): 439.
1832 : Fries, *ibid.*.
T.: *T. orobanches* (Mérat) Fries (*Rhizoctonia orobanches* Mérat).

Uromyces (Link) Unger, Exanth. Pfl. 277.
1833.
T.: *U. appendiculatus* (Persoon : Persoon)
Unger (*Uredo appendiculata* Persoon : Persoon).

(=) *Coeomurus* Link ex S. F. Gray, Nat. Arr.
Brit. Pl. **1**: 541. 1821.
T.: *C. phaseolorum* (R. A. Hedwig ex
A.-P. de Candolle) S. F. Gray (*Puccinia
phaseolorum* R. A. Hedwig ex A.-P. de
Candolle).

(=) *Pucciniola* L. Marchand, Bijdr. Natuurk.
Wetensch. **4**: 47. 1829.
T.: *P. diadelphiae* L. Marchand.

*Valsa Fries, Summa Veg. Scand., sect. post.
410. 1849.
T.: *V. ambiens* (Persoon : Fries) Fries (*Sphaeria ambiens* Persoon : Fries).

(H) *Valsa* Adanson, Fam. Pl. **2**: 9. 1763 [Fun-
gi].
T.: *Sphaeria disciformis* G. F. Hoffmann
(vide Cannon et Hawksworth, Taxon **32**:
478. 1983).

Venturia Saccardo, Syll. Fung. **1**: 586. 1882.
T.: *V. inaequalis* (Cooke) Winter.

(H) *Venturia* De Notaris, Giorn. Bot. Ital. **1**:
332. 1844 [Fungi].
T.: *V. rosae* De Notaris.

Verrucaria Schrader, Spic. Fl. Germ. **1**: 108. 1794.
T.: *V. rupestris* Schrader.

Volutella Fries, Syst. Mycol. **3**: 466. 1832 : Fries, *ibid.*.
T.: *V. ciliata* (Albertini et Schweinitz : Fries) Fries (*Tubercularia ciliata* Albertini et Schweinitz : Fries) (*typ. cons.*).

*****Volvariella** Spegazzini, Anales Mus. Nac. Hist. Nat. Buenos Aires **6**: 119. 1899.
T.: *V. argentina* Spegazzini.

Xanthoria (Fries) T. M. Fries, Lich. Arct. 166. 1860.
T.: *X. parietina* (Linnaeus) T. M. Fries (*Lichen parietinus* Linnaeus).

Xerocomus Quélet in J. A. Mougeot et Ferry, Fl. Vosges, Champ. 477. 1887.
T.: *X. subtomentosus* (Linnaeus : Fries) Quélet (*Boletus subtomentosus* Linnaeus : Fries).

Xylaria Hill ex Schrank, Baier. Fl. **1**: 200. 1789.
T.: *X. hypoxylon* (Linnaeus : Fries) Greville (*Clavaria hypoxylon* Linnaeus : Fries).

(H) *Verrucaria* Scopoli, Intr. 61. 1777 [Fungi: Lich.].
T.: *Baeomyces roseus* Persoon.

(=) *Volvarius* Roussel, Fl. Calvados ed. 2. 59. 1806.
LT.: *Agaricus volvaceus* Bulliard : Fries (vide Earle, Bull. New York Bot. Gard. **5**: 395, 449. 1909).

(≡) *Blasteniospora* Trevisan, Tornab. Blasteniosp. 2. 1853.
(=) *Dufourea* Acharius, Lichenogr. Universalis 103, 524. 1810.
T.: *D. flammea* (Linnaeus f.) Acharius (*Lichen flammeus* Linnaeus f.).

(≡) *Versipellis* Quélet, Ench. Fung. 157. 1886.

XI. MUSCI

Acidodontium Schwägrichen, Sp. Musc. Suppl. 2(2): 152. Mai 1827.
T.: *A. kunthii* Schwägrichen, nom. illeg. (*Bryum megalocarpum* W. J. Hooker, *A. megalocarpum* (W. J. Hooker) Renauld et Cardot).

Aloina Kindberg, Bih. Kongl. Svenska Vetensk.-Akad. Handl. **6**(19): 22. 1882.
T.: *A. aloides* (Schultz) Kindberg (*Trichostomum aloides* Schultz).

Amblyodon Palisot de Beauvois, Mag. Encycl. **5**: 323. 1804 ("*Amblyodum*") (*orth. cons.*).
T.: *A. dealbatus* (Hedwig) Palisot de Beauvois (*Meesia dealbata* Hedwig).

Amphidium Schimper, Coroll. Bryol. Eur. 39. 1856.
T.: *A. lapponicum* (Hedwig) Schimper (*Anictangium lapponicum* Hedwig).

(≡) *Megalangium* Bridel, Bryol. Univ. **2**: 28. 1827.

(=) *Aloidella* Venturi, Comment. Fauna Veneto Trentino 1(3): 124. 1868.
T.: non designatus.

(H) *Amphidium* C. G. Nees in Sturm, Deutschl. Fl. Abt. ii. H. 17: 2. 1819 [Musci].
T.: *A. pulvinatum* C. G. Nees.

147

Anacolia Schimper, Syn. Musc. Eur. ed. 2. 513. 1876.
T.: *A. webbii* (Montagne) Schimper (*Glyphocarpa webbii* Montagne).

(=) *Glyphocarpa* R. Brown, Trans. Linn. Soc. London 12: 575. 1819.
T.: *G. capensis* R. Brown.

Anoectangium Schwägrichen, Sp. Musc. Suppl. 1(1): 33. 1811.
T.: *A. compactum* Schwägrichen [= *A. aestivum* (Hedwig) Mitten, *Gymnostomum aestivum* Hedwig].

(H) *Anoectangium* Röhling, Ann. Wetterauischen Ges. Gesammte Naturk. 1: 199. 1809 [Musci].
T.: non designatus.
(=) *Anictangium* Hedwig, Sp. Musc. 40. 1801.
T.: non designatus.

Atractylocarpus Mitten, J. Linn. Soc., Bot. 12: 13, 71. 1869.
T.: *A. mexicanus* Mitten [= *A. costaricensis* (K. Müller Hal.) W. Wilson, *Leptotrichum costaricense* K. Müller Hal.].

(=) *Metzleria* Schimper ex Milde, Bryol. Siles. 75. 1869.
T.: *M. alpina* Schimper ex Milde.

Atrichum Palisot de Beauvois, Mag. Encycl. 5: 329. 1804.
T.: *A. undulatum* (Hedwig) Palisot de Beauvois (*Polytrichum undulatum* Hedwig).

(≡) *Catharinea* F. Weber et Mohr, Index Mus. Pl. Crypt. 2. 1803.

Aulacomnium Schwägrichen, Sp. Musc. Suppl. 3(1): *t. 215.* 1827 ("*Aulacomnion*") (*orth. cons.*).
T.: *A. androgynum* (Hedwig) Schwägrichen (*Bryum androgynum* Hedwig).

(≡) *Gymnocephalus* Schwägrichen, Sp. Musc. Suppl. 1(2): 87. 1816.
(=) *Arrhenopterum* Hedwig, Sp. Musc. Frond. 198. 1801.
T.: *A. heterostichum* Hedwig.
(=) *Orthopixis* Palisot de Beauvois, Mag. Encycl. 5: 322. 1804.
T.: non designatus.

Barbula Hedwig, Sp. Musc. Frond. 115. 1801.
T.: *B. unguiculata* Hedwig.

(H) *Barbula* Loureiro, Fl. Cochinch. 366. 1790 [Verben.].
T.: *B. sinensis* Loureiro.

Bartramia Hedwig, Sp. Musc. Frond. 164. 1801.
T.: *B. halleriana* Hedwig.

(H) *Bartramia* Linnaeus, Sp. Pl. 389. 1753 [Til.].
T.: *B. indica* Linnaeus.

Bartramidula Bruch et Schimper in Bruch, Schimper et Gümbel, Bryol. Eur. 4: 55. 1846 (fasc. 29-30 Mon. 1).
T.: *B. wilsonii* Bruch et Schimper, nom. illeg. (*Glyphocarpa cernua* W. Wilson, *B. cernua* (W. Wilson) Lindberg).

(=) *Glyphocarpa* R. Brown, Trans. Linn. Soc. London 12: 575. 1819.
T.: *G. capensis* R. Brown.

Bryoxiphium Mitten, J. Linn. Soc., Bot. 12: 580. 1869 ("*Bryoziphium*") (*orth. cons.*).
T.: *B. norvegicum* (Bridel) Mitten (*Phyllogonium norvegicum* Bridel).

(≡) *Eustichium* Bruch et Schimper in Bruch, Schimper et Gümbel, Bryol. Eur. 2: 159. 1849 (fasc. 42).

Crossidium Juratzka, Laubm.-Fl. Oesterr.-Ung. 127. 1882.
T.: *C. squamigerum* (Viviani) Juratzka (*Barbula squamigera* Viviani).

(=) *Chloronotus* Venturi, Comment. Fauna Veneto Trentino 1(3): 124. 1868.
T.: non designatus.

Cynodontium Bruch et Schimper in Schimper, Coroll. Bryol. Eur. 12. 1856.
T.: *C. polycarpum* (Hedwig) Schimper (*Fissidens polycarpus* Hedwig).

(H) *Cynodontium* Bridel, Muscol. Recent. Suppl. 1: 155. 1806 [Musci].
T.: non designatus.

°**Daltonia** W. J. Hooker et T. Taylor, Muscol. Brit. 80. 1818.
T.: *D. splachnoides* W. J. Hooker et T. Taylor (*typ. cons.*).

Distichium Bruch et Schimper in Bruch, Schimper et Gümbel, Bryol. Eur. 2: 153. 1846 (fasc. 29-30 Mon. 1).
T.: *D. capillaceum* (Hedwig) Bruch et Schimper (*Cynontodium capillaceum* Hedwig).

(=) *Cynontodium* Hedwig, Sp. Musc. Frond. 57. 1801.
T.: non designatus.

Ditrichum Hampe, Flora 50: 181. 1867.
T.: *D. homomallum* (Hedwig) Hampe (*Didymodon homomallum* Hedwig) [= *D. heteromallum* (Hedwig) E. G. Britton, *Weissia heteromalla* Hedwig] (*typ. cons.*).

(H) *Ditrichum* Cassini, Bull. Sci. Soc. Philom. Paris 1817: 33. 1817 [Comp.].
T.: *D. macrophyllum* Cassini.

(≡) *Diaphanophyllum* Lindberg, Öfvers. Förh. Kongl. Svenska Vetensk.-Akad. 19: 605. 1863.

(=) *Aschistodon* Montagne, Ann. Sci. Nat. Bot. ser. 3. 4: 109. 1845.
T.: *A. conicus* Montagne.

(=) *Lophiodon* J. D. Hooker et W. Wilson, London J. Bot. 3: 543. 1844.
T.: *L. strictus* J. D. Hooker et W. Wilson.

Drummondia W. J. Hooker in Drummond, Musci Amer. n. 62. 1828.
T.: *D. clavellata* W. J. Hooker [= *D. prorepens* (Hedwig) E. G. Britton, *Gymnostomum prorepens* Hedwig].

(=) *Anodontium* Bridel, Muscol. Recent. Suppl. 1: 41. 1806.
T.: *A. prorepens* (Hedwig) Bridel (*Gymnostomum prorepens* Hedwig).

(=) *Leiotheca* Bridel, Bryol. Univ. 1: 304, 726. 1826.
T.: *L. prorepens* (W. J. Hooker) Bridel (*Orthotrichum prorepens* W. J. Hooker).

Ephemerella K. Müller Hal., Syn. Musc. Frond. 1: 34. 1848.
T.: *E. pachycarpa* (Schwägrichen) K. Müller Hal. (*Phascum pachycarpum* Schwägrichen).

(=) *Physedium* Bridel, Bryol. Univ. 1: 51. 1826.
T.: *P. splachnoides* (Hornschuch) Bridel (*Phascum splachnoides* Hornschuch).

Ephemerum Hampe, Flora 20: 285. 1837.
T.: *E. serratum* (Hedwig) Hampe (*Phascum serratum* Hedwig).

(H) *Ephemerum* P. Miller, Gard. Dict. Abr. ed. 4. 1754 [Commelin.].
≡ *Tradescantia* Linnaeus 1753.

Gymnostomum C. G. Nees et Hornschuch in C. G. Nees, Hornschuch et Sturm, Bryol. Germ. 1: 112, 153. 1823.
T.: *G. calcareum* C. G. Nees et Hornschuch.

(H) *Gymnostomum* Hedwig, Sp. Musc. Frond. 30. 1801 [Musci].
T.: non designatus.

Gyroweisia Schimper, Syn. Musc. Eur. ed. 2. 38. 1876.
T.: *G. tenuis* (Hedwig) Schimper (*Gymnostomum tenue* Hedwig).

Haplohymenium Dozy et Molkenboer, Musc. Frond. Ined. Archip. Ind. 127, *t. 40*. 1846.
T.: *H. sieboldii* (Dozy et Molkenboer) Dozy et Molkenboer (*Leptohymenium sieboldii* Dozy et Molkenboer).

Hedwigia Palisot de Beauvois, Mag. Encycl. 5: 304. 1804.
T.: *H. ciliata* (Hedwig) Palisot de Beauvois (*Anictangium ciliatum* Hedwig).

Helodium Warnstorf, Krypt.-Fl. Brandenburg 2: 675, 692. 1905.
T.: *H. blandowii* (F. Weber et Mohr) Warnstorf (*Hypnum blandowii* F. Weber et Mohr).

Holomitrium Bridel, Bryol. Univ. 1: 226. 1826 ("*Olomitrium*") (*orth. cons.*).
T.: *H. perichaetiale* (W. J. Hooker) Bridel (*Trichostomum perichaetiale* W. J. Hooker) (*typ. cons.*).

Hookeria J. E. Smith, Trans. Linn. Soc. London 9: 275. 23 Nov 1808.
T.: *H. lucens* (Hedwig) J. E. Smith (*Hypnum lucens* Hedwig).

Hygroamblystegium Loeske, Moosfl. Harz. 298. 1903.
T.: *H. irriguum* (W. J. Hooker et W. Wilson) Loeske (*Hypnum irriguum* W. J. Hooker et W. Wilson) [= *H. tenax* (Hedwig) Jennings, *Hypnum tenax* Hedwig].

Hypnum Hedwig, Sp. Musc. Frond. 236. 1801.
T.: *H. cupressiforme* Hedwig.

Lepidopilum (Bridel) Bridel, Bryol. Univ. 2: 267. 1827.
T.: *L. subenerve* Bridel, nom. illeg. (*Neckera scabriseta* Schwägrichen, *L. scabrisetum* (Schwägrichen) Steere).

Leptodon Mohr, Observ. Bot. 27. 1803.
T.: *L. smithii* (Hedwig) F. Weber et Mohr (*Hypnum smithii* Hedwig) (*typ. cons.*).

Leptostomum R. Brown, Trans. Linn. Soc. London 10: 320. 1811.
T.: *L. inclinans* R. Brown.

(=) *Weisiodon* Schimper, Coroll. Bryol. Eur. 9. 1856.
T.: *W. reflexus* (Bridel) Schimper (*Weissia reflexa* Bridel).

(H) *Haplohymenium* Schwägrichen, Sp. Musc. Suppl. 3(2): *t. 271.* 1829 [Musci].
T.: *H. microphyllum* Schwägrichen [= *Thuidium haplohymenium* (W. H. Harvey) Jaeger].

(H) *Hedwigia* Swartz, Prodr. 4, 62. 1788 [Burser.].
T.: *H. balsamifera* Swartz.

(H) *Helodium* Dumortier, Fl. Belg. 77. 1827 [Umbell.].
≡ *Helosciadium* W. D. J. Koch 1824.

(H) *Hookera* R. A. Salisbury, Parad. Lond. *t. 98.* 1 Mar 1808 [Lil.].
≡ *Brodiaea* J. E. Smith 1810 (*nom. cons.*) (1053).

(=) *Drepanophyllaria* K. Müller Hal., Nuov. Giorn. Bot. Ital. 3: 114. 1896.
≡ *Cratoneuron* (Sullivant) Spruce 1867.

(=) *Actinodontium* Schwägrichen, Sp. Musc. Suppl. 2(2): 75. 1826.
T.: *A. ascendens* Schwägrichen.

(=) *Orthopixis* Palisot de Beauvois, Mag. Encycl. 5: 322. 1804.
T.: non designatus.

150

Leucoloma Bridel, Bryol. Univ. 2: 218. 1827.
T.: *L. bifidum* (Bridel) Bridel (*Hypnum bifidum* Bridel).

(≡) *Macrodon* Arnott, Mém. Soc. Linn. Paris 5: 290. 1827 (vel 1826).

(=) *Sclerodontium* Schwägrichen, Sp. Musc. Suppl. 2(1): 124. 1824.
T.: *S. pallidum* (W. J. Hooker) Schwägrichen (*Leucodon pallidus* W. J. Hooker).

Meesia Hedwig, Sp. Musc. Frond. 173. 1801.
T.: *M. longiseta* Hedwig.

(H) *Meesia* J. Gaertner, Fruct. Sem. Pl. 1: 344. 1788 [Ochn.].
T.: *M. serrata* J. Gaertner.

Mittenothamnium Hennings, Hedwigia 41 (Beibl.): 225. 1902.
T.: *M. reptans* (Hedwig) Cardot (*Hypnum reptans* Hedwig) (*typ. cons.*).

Mniobryum Limpricht, Laubm. Deutschl. 2: 272. 1892.
T.: *M. carneum* Limpricht, nom. illeg. (*Bryum delicatulum* Hedwig, *M. delicatulum* (Hedwig) Dixon).

Mnium Hedwig, Sp. Musc. Frond. 188. 1801.
T.: *M. hornum* Hedwig.

(H) *Mnium* Linnaeus, Sp. Pl. 1109. 1753 [Hepat.].
LT.: *M. fissum* Linnaeus (vide Proskauer, Taxon 12: 200. 1963).

Muelleriella Dusén, Bot. Not. 1905: 304. 1905.
T.: *M. crassifolia* (J. D. Hooker et W. Wilson) Dusén (*Orthotrichum crassifolium* J. D. Hooker et W. $Wilson).

(H) *Muelleriella* Van Heurck, Treat. Diatom. 435. 1896 [Foss.].
T.: *M. limbata* (Ehrenberg) Van Heurck (*Pyxidicula limbata* Ehrenberg).

Myrinia Schimper, Syn. Musc. Eur. 482. 1860.
T.: *M. pulvinata* (Wahlenberg) Schimper (*Leskea pulvinata* Wahlenberg).

(H) *Myrinia* Lilja, Fl. Sv. Odl. Vext. Suppl. 25. 1840 [Onagr.].
T.: *M. microphylla* Lilja.

Neckera Hedwig, Sp. Musc. Frond. 200. 1801.
T.: *N. pennata* Hedwig (*typ. cons.*).

(H) *Neckeria* Scopoli, Intr. 313. 1777 [Papaver.].
≡ *Capnoides* P. Miller 1754 (*nom. rej.* sub 2858).

Orthothecium Schimper in Bruch, Schimper et Gümbel, Bryol. Eur. 5: 105. 1851 (fasc. 48 Mon. 1).
T.: *O. rufescens* (J. E. Smith) Schimper (*Hypnum rufescens* J. E. Smith).

(H) *Orthothecium* Schott et Endlicher, Melet. Bot. 31. 1832 [Stercul.].
T.: *O. lhotskyanum* Schott et Endlicher.

Papillaria (K. Müller Hal.) K. Müller Hal. in Ångström, Öfvers. Förh. Kongl. Svenska Vetensk.- Akad. 33(4): 34. 1876.
T.: *P. nigrescens* (Hedwig) Jaeger (*Hypnum nigrescens* Hedwig) (*typ. cons.*).

(H) *Papillaria* Dulac, Fl. Hautes-Pyrénées 45. 1867 [Scheuchzer.].
≡ *Scheuchzeria* Linnaeus 1753.

Platygyrium Schimper in Bruch, Schimper et Gümbel, Bryol. Eur. 5: 95. 1851 (fasc. 46-47 Mon. 1).
T.: *P. repens* (Bridel) Schimper (*Pterigynandrum repens* Bridel).

(=) *Pterigynandrum* Hedwig, Sp. Musc. Frond. 80. 1801.
LT.: *P. filiforme* Hedwig (vide Schimper in Bruch, Schimper et Gümbel, Bryol. Eur. 5: 121. 1851).

(=) *Pterogonium* Swartz, Monthly Rev. 34: 537. 1801.
LT.: *P. gracile* (Hedwig) J. E. Smith (*Pterigynandrum gracile* Hedwig) (vide Schimper in Bruch, Schimper et Gümbel, Bryol. Eur. 5: 125. 1851).

(=) *Leptohymenium* Schwägrichen, Sp. Musc. Suppl. 3(1): *t. 246c*. 1828.
T.: *L. tenue* (W. J. Hooker) Schwägrichen (*Neckera tenuis* W. J. Hooker).

Pleuridium Rabenhorst, Deutschl. Krypt.-Fl. 2(3): 79. 1848.
T.: *P. subulatum* (Hedwig) Rabenhorst (*Phascum subulatum* Hedwig).

(H) *Pleuridium* Bridel, Muscol. Recent. Suppl. 4: 10. 1818-1819 [Musci].
LT.: *P. globiferum* Bridel (vide Snider et Margadant, Taxon 22: 693. 1973).

Pleurozium Mitten, J. Linn. Soc., Bot. 12: 22, 537. 1869.
T.: *P. schreberi* (Bridel) Mitten (*Hypnum schreberi* Bridel).

Pterygoneurum Juratzka, Laubm.-Fl. Oesterr.-Ung. 95. 1882 ("*Pterigoneurum*") (*orth. cons.*).
T.: *P. cavifolium* Juratzka, nom. illeg. (*Pottia cavifolia* Fürnrohr, nom. illeg., *Gymnostomum ovatum* Hedwig, *P. ovatum* (Hedwig) Dixon).

(=) *Pharomitrium* Schimper, Syn. Musc. Eur. 120. 1860.
T.: *P. subsessile* (Bridel) Schimper (*Gymnostomum subsessile* Bridel).

Ptychomitrium Fürnrohr, Flora 12(2), Ergänzungsbl.: 19. 1829 ("*Pthychomitrium*") (*orth. cons.*).
T.: *P. polyphyllum* (Swartz) Bruch et Schimper (*Dicranum polyphyllum* Swartz).

(=) *Brachysteleum* H. G. L. Reichenbach, Consp. 1: 34. 1828.
T.: *B. crispatum* (Hedwig) Hornschuch (*Encalypta crispata* Hedwig).

Timmia Hedwig, Sp. Musc. Frond. 176. 1801.
T.: *T. megapolitana* Hedwig.

(H) *Timmia* J. F. Gmelin, Syst. Nat. 2: 524, 538. 1791 [Amaryllid.].
T.: non designatus.

Tortella (Lindberg) Limpricht, Laubm. Deutschl. 1: 599. 1888.
T.: *T. caespitosa* (Schwägrichen) Limpricht (*Barbula caespitosa* Schwägrichen) [= *T. humilis* (Hedwig) Jennings, *Barbula humilis* Hedwig].

(=) *Pleurochaete* Lindberg, Öfvers. Förh. Kongl. Svenska Vetensk.-Akad. 21: 253. 1864.
T.: *P. squarrosa* (Bridel) Lindberg (*Barbula squarrosa* Bridel).

Tortula Hedwig, Sp. Musc. Frond. 122. 1801.
T.: *T. subulata* Hedwig (*typ. cons.*).

(H) *Tortula* Roxburgh ex Willdenow, Sp. Pl. 3: 359. 1800 [Verben.].
T.: *T. aspera* Roxburgh ex Willdenow.

Trichostomum Bruch, Flora 12: 396. 1829.
T.: *T. brachydontium* Bruch.

(H) *Trichostomum* Hedwig, Sp. Musc. Frond. 107. 1801 [Musci].
T.: non designatus.

(=) *Plaubelia* Bridel, Bryol. Univ. 1: 522. 1826.
T.: *P. tortuosa* Bridel.

XII. HEPATICAE

Acrolejeunea (Spruce) Schiffner in Engler et Prantl, Nat. Pflanzenfam. 1(3): 119, 128. 1893.
T.: *A. torulosa* (Lehmann et Lindenberg) Schiffner (*Jungermannia torulosa* Lehmann et Lindenberg).

(H) *Acro-lejeunea* Stephani, Bot. Gaz. 15: 286. 1890 [Hepat.].
T.: *A. parviloba* Stephani [= *Schiffneriolejeunea parviloba* (Stephani) S. R. Gradstein].

Adelanthus Mitten, J. Proc. Linn. Soc., Bot. 7: 243. 1864.
T.: *A. falcatus* (W. J. Hooker) Mitten (*Jungermannia falcata* W. J. Hooker).

(H) *Adelanthus* Endlicher, Gen. Pl. 1327. 1840 [Icacin.].
T.: *A. scandens* (Thunberg) Endlicher ex Baillon (*Cavanilla scandens* Thunberg).

Bazzania S. F. Gray, Nat. Arr. Brit. Pl. 1: 704, 775. 1821 ("*Bazzanius*") (*orth. cons.*).
T.: *B. trilobata* (Linnaeus) S. F. Gray (*Jungermannia trilobata* Linnaeus).

Calypogeia Raddi, Jungermanniogr. Etrusca 31. 1818 ("*Calypogeja*") (*orth. cons.*).
T.: *C. fissa* (Linnaeus) Raddi (*Mnium fissum* Linnaeus).

Cephaloziella (Spruce) Schiffner in Engler et Prantl, Nat. Pflanzenfam. 1(3): 98. 1893.
T.: *C. divaricata* (J. E. Smith) Schiffner (*Jungermannia divaricata* J. E. Smith).

(=) *Dichiton* Montagne, Syll. Gen. Sp. Crypt. 52. 1856.
T.: *D. perpusillus* Montagne, nom. illeg. (*Jungermannia calyculata* Durieu et Montagne, *D. calyculatus* (Durieu et Montagne) Trevisan).

Chiloscyphus Corda in Opiz, Beitr. Naturgesch. [Naturalientausch 12:] 651. 1829 ("*Cheilocyphos*") (*orth. cons.*).
T.: *C. polyanthos* (Linnaeus) Corda (*Jungermannia polyanthos* Linnaeus).

Conocephalum Wiggers, Prim. Fl. Holsat. 82. 1780.
T.: *C. trioicum* Wiggers, nom. illeg. (*Marchantia conica* Linnaeus, *C. conicum* (Linnaeus) Dumortier ex Cogniaux).

(≡) *Conicephala* J. Hill, Gener. Nat. Hist. ed. 2. 2: 118. 1773.

Diplophyllum (Dumortier) Dumortier, Recueil Observ. Jungerm. 15. 1835.
T.: *D. albicans* (Linnaeus) Dumortier (*Jungermannia albicans* Linnaeus).

(H) *Diplophyllum* Lehmann, Ges. Naturf. Freunde Berlin Mag. 8: 310. 1818 [Scrophular.].
T.: *D. veroniciforme* Lehmann.

153

Gymnomitrion Corda in Opiz, Beitr. Natur-gesch. [Naturalientausch **12**:] 651. 1829.
T.: *G. concinnatum* (Lightfoot) Corda (*Jungermannia concinnata* Lightfoot).

(≡) *Cesius* S. F. Gray, Nat. Arr. Brit. Pl. **1**: 705. 1821.

Haplomitrium C. G. Nees, Naturgesch. Eur. Leberm. **1**: 109. 1833.
T.: *H. hookeri* (J. E. Smith) C. G. Nees (*Jungermannia hookeri* J. E. Smith).

(≡) *Scalius* S. F. Gray, Nat. Arr. Brit. Pl. **1**: 704. 1821.

Heteroscyphus Schiffner, Österr. Bot. Z. **60**: 171. 1910.
T.: *H. aselliformis* (Reinwardt, Blume et C. G. Nees) Schiffner (*Jungermannia aselliformis* Reinwardt, Blume et C. G. Nees).

(≡) *Gamoscyphus* Trevisan, Mem. Reale Ist. Lombardo Cl. Sci. ser. 3. **4**: 422. 1877.

Lejeunea Libert, Ann. Gén. Sci. Phys. **6**: 372. 1820 ("*Lejeunia*") (*orth. cons.*).
T.: *L. serpillifolia* Libert (non *Jungermannia serpillifolia* Scopoli 1772; non *Jungermannia serpyllifolia* Dickson 1801) [= *L. cavifolia* (Ehrhart) Lindberg, *Jungermannia cavifolia* Ehrhart].

Lembidium Mitten in J. D. Hooker, Handb. N. Zeal. Fl. 754. 1867.
T.: *L. nutans* (J. D. Hooker et T. Taylor) A. W. Evans (*Jungermannia nutans* J. D. Hooker et T. Taylor).

(H) *Lembidium* Koerber, Syst. Lich. Germ. **5**: 358. 1855 [Fungi: Lich.].
T.: *L. polycarpum* Körber.

Lepidozia (Dumortier) Dumortier, Recueil Observ. Jungerm. 19. 1835.
T.: *L. reptans* (Linnaeus) Dumortier (*Jungermannia reptans* Linnaeus) (etiam vide *Mastigophora* C. G. Nees [Hepat.], *nom. cons.*).

Lopholejeunea (Spruce) Schiffner in Engler et Prantl, Nat. Pflanzenfam. **1**(3): 119, 129. 1893.
T.: *L. sagraeana* (Montagne) Schiffner (*Phragmicoma sagraeana* Montagne).

(H) *Lopho-lejeunea* Stephani, Bot. Gaz. **15**: 285. 1890 [Hepat.].
T.: *L. multilacera* Stephani.

Marchesinia S. F. Gray, Nat. Arr. Brit. Pl. **1**: 679 ("*Marchesinius*"), 689, 817 ("*Marchesinus*"). 1821 (*orth. cons.*).
T.: *M. mackaii* (W. J. Hooker) S. F. Gray (*Jungermannia mackaii* W. J. Hooker).

Mastigophora C. G. Nees, Naturgesch. Eur. Leberm. **3**: 89. 1838.
T.: *M. woodsii* (W. J. Hooker) C. G. Nees (*Jungermannia woodsii* W. J. Hooker).

(H) *Mastigophora* C. G. Nees, Naturgesch. Eur. Leberm. **1**: 95, 101. 1833 [Hepat.].
≡ *Lepidozia* (Dumortier) Dumortier 1835 (*nom. cons.*).

Mylia S. F. Gray, Nat. Arr. Brit. Pl. **1**: 693. 1821 ("*Mylius*") (*orth. cons.*).
T.: *M. taylorii* (W. J. Hooker) S. F. Gray (*Jungermannia taylorii* W. J. Hooker).

Nardia S. F. Gray, Nat. Arr. Brit. Pl. **1**: 694. 1821 ("*Nardius*") (*orth. cons.*).
T.: *N. compressa* (W. J. Hooker) S. F. Gray (*Jungermannia compressa* W. J. Hooker).

Pallavicinia S. F. Gray, Nat. Arr. Brit. Pl. **1**: 775. 1821 ("*Pallavicinius*") (*orth. cons.*).
T.: *P. lyellii* (W. J. Hooker) Carruthers (*Jungermannia lyellii* W. J. Hooker).

Pellia Raddi, Jungermanniogr. Etrusca 38. 1818.
T.: *P. fabroniana* Raddi, nom. illeg. (*Jungermannia epiphylla* Linnaeus, *P. epiphylla* (Linnaeus) Corda).

(≡) *Merkia* Borkhausen, Tent. Disp. Pl. German. 156. 1792 (vide Grolle, Taxon **24**: 693. 1975).

Plagiochasma Lehmann et Lindenberg in Lehmann, Nov. Stirp. Pug. **4**: 13. 1832.
T.: *P. cordatum* Lehmann et Lindenberg.

(=) *Aytonia* J. R. Forster et G. Forster, Char. Gen. Pl. 74. 1775.
T.: *A. rupestris* J. R. Forster et G. Forster.

Plagiochila (Dumortier) Dumortier, Recueil Observ. Jungerm. 14. 1835.
T.: *P. asplenioides* (Linnaeus) Dumortier (*Jungermannia asplenioides* Linnaeus).

(=) *Carpolepidum* Palisot de Beauvois, Fl. Oware **1**: 21. 1805.
LT.: *C. dichotomum* Palisot de Beauvois (vide Bonner, Index Hepat. **3**: 526. 1963).

Radula Dumortier, Comment. Bot. 112. 1822.
T.: *R. complanata* (Linnaeus) Dumortier (*Jungermannia complanata* Linnaeus).

(≡) *Martinellius* S. F. Gray, Nat. Arr. Brit. Pl. **1**: 690. 1821.

Reboulia Raddi, Opusc. Sci. **2**: 357. 1818 ("*Rebouillia*") (*orth. cons.*).
T.: *R. hemisphaerica* (Linnaeus) Raddi (*Marchantia hemisphaerica* Linnaeus).

Riccardia S. F. Gray, Nat. Arr. Brit. Pl. **1**: 683. 1821 ("*Riccardius*") (*orth. cons.*).
T.: *R. multifida* (Linnaeus) S. F. Gray (*Jungermannia multifida* Linnaeus).

Saccogyna Dumortier, Comment. Bot. 113. 1822.
T.: *S. viticulosa* (Linnaeus) Dumortier (*Jungermannia viticulosa* Linnaeus).

(≡) *Lippius* S. F. Gray, Nat. Arr. Brit. Pl. **1**: 679, 706. 1821.

Scapania (Dumortier) Dumortier, Recueil Observ. Jungerm. 14. 1835.
T.: *S. undulata* (Linnaeus) Dumortier (*Jungermannia undulata* Linnaeus) (*typ. cons.*).

Solenostoma Mitten, J. Linn. Soc., Bot. **8**: 51. 1864.
T.: *S. tersum* (C. G. Nees) Mitten (*Jungermannia tersa* C. G. Nees).

(=) *Gymnoscyphus* Corda in Sturm, Deutschl. Fl. Abt. ii. H. **26-27**: 158. 1835.
T.: *G. repens* Corda.

Trachylejeunea (Spruce) Schiffner in Engler et Prantl, Nat. Pflanzenfam. **1**(3): 119, 126. 1893.
T.: *T. acanthina* (Spruce) Schiffner (*Lejeunea acanthina* Spruce).

(H) *Trachylejeunea* Stephani, Hedwigia **28**: 262. 1889 [Hepat.].
T.: *T. elegantissima* Stephani.

Treubia Goebel, Ann. Jard. Bot. Buitenzorg **9**: 1. 1890.
T.: *T. insignis* Goebel.

Trichocolea Dumortier, Comment. Bot. 113. 1822 ("*Thricholea*") (*orth. cons.*).
T.: *T. tomentella* (Ehrhart) Dumortier (*Jungermannia tomentella* Ehrhart).

XIII. PTERIDOPHYTA

Anemia Swartz, Syn. Fil. 6, 155. 1806.
T.: *A. phyllitidis* (Linnaeus) Swartz (*Osmunda phyllitidis* Linnaeus).

(=) *Ornithopteris* Bernhardi, Neues J. Bot. **1**(2): 40. 1805.
LT.: *O. adiantifolia* (Linnaeus) Bernhardi (*Osmunda adiantifolia* Linnaeus) (vide Reed, Bol. Soc. Brot. ser. 2, **21**: 153. 1947).

Angiopteris G. F. Hoffmann, Commentat. Soc. Regiae Sci. Gott. **12** (Cl. Phys.): 29. 1796.
T.: *A. evecta* (G. Forster) G. F. Hoffmann (*Polypodium evectum* G. Forster).

(H) *Angiopteris* Adanson, Fam. Pl. **2**: 21, 518. 1763 [Pteridoph.: Onocl.].
≡ *Onoclea* Linnaeus 1753.

Araiostegia Copeland, Philipp. J. Sci. **34**: 240. 1927.
T.: *A. hymenophylloides* (Blume) Copeland (*Aspidium hymenophylloides* Blume).

(=) *Gymnogrammitis* Griffith, Ic. Pl. Asiat. **2**: t. *129, f. 1.* 1849; Notul. Pl. Asiat. **2**: 608. 1849.
T.: *G. dareiformis* (W. J. Hooker) Ching ex Tardieu et C. Christensen (*Polypodium dareiforme* W. J. Hooker).

Ceterach Willdenow, Anleit. Selbststud. Bot. 578. 1804.
T.: *C. officinarum* Willdenow.

(H) *Ceterac* Adanson, Fam. Pl. **2**: 20, 536. 1763 [Pteridoph.: Asplen.].
T.: non designatus.

Cheilanthes Swartz, Syn. Fil. 5, 126. 1806.
T.: *C. micropteris* Swartz.

(=) *Allosorus* Bernhardi, Neues J. Bot. **1**(2): 36. 1805.
LT.: *A. pusillus* (Willdenow ex Bernhardi) Bernhardi (*Adiantum pusillum* Willdenow ex Bernhardi) (vide Pichi Sermolli, Webbia **9**: 394. 1953).

Coniogramme Fée, Mém. Foug. **5**: 167. 1852.
T.: *C. javanica* (Blume) Fée (*Gymnogramma javanica* Blume).

(=) *Dictyogramme* Fée, Mém. Soc. Mus. Hist. Nat. Strasbourg **4**(1): 202. 1850.
T.: *D. japonica* (Thunberg) Fée (*Hemionitis japonica* Thunberg).

Cystodium John Smith in W. J. Hooker, Gen. Fil. *t. 96*. 1841.
T.: *C. sorbifolium* (J. E. Smith) John Smith (*Dicksonia sorbifolia* J. E. Smith).

(H) *Cystodium* Fée, Essai Crypt. Ecorc. **2**: 13. 1837 [Fungi].
≡ *Gassicurtia* Fée 1824.

Cystopteris Bernhardi, Neues J. Bot. **1**(2): 5, 26. 1805.
T.: *C. fragilis* (Linnaeus) Bernhardi (*Polypodium fragile* Linnaeus).

Danaea J. E. Smith, Mém. Acad. Sci. (Turin) **5**: 420, *t. 9, f. 11*. 1793.
T.: *D. nodosa* (Linnaeus) J. E. Smith (*Acrostichum nodosum* Linnaeus).

(H) *Danaa* Allioni, Fl. Pedem. **2**: 34, *t. 63*. 1785 [Umbell.].
T.: *D. aquilegiifolia* (Allioni) Allioni (*Coriandrum aquilegiifolium* Allioni) [= *Physospermum cornubiense* (Linnaeus) A.-P. de Candolle, *Ligusticum cornubiense* Linnaeus].

Doryopteris John Smith, J. Bot. (Hooker) **3**: 404. 1841.
T.: *D. palmata* (Willdenow) John Smith (*Pteris palmata* Willdenow).

(=) *Cassebeera* Kaulfuss, Enum. Filic. 216. 1824.
LT.: *C. triphylla* (Lamarck) Kaulfuss (*Adiantum triphyllum* Lamarck) (vide Fée, Mém. Foug. **5**: 119. 1852).

Drymoglossum K. B. Presl, Tent. Pterid. 227, *t. 10, fig. 5, 6*. 1836.
T.: *D. piloselloides* (Linnaeus) K. B. Presl (*Pteris piloselloides* Linnaeus).

(=) *Pteropsis* Desvaux, Mém. Soc. Linn. Paris **6**(3): 218. 1827.
LT.: *Acrostichum heterophyllum* Linnaeus (vide Pichi Sermolli, Webbia **9**: 403. 1953).

Drynaria (Bory de Saint-Vincent) John Smith, J. Bot. (Hooker) **4**: 60. 1841.
T.: *D. quercifolia* (Linnaeus) John Smith (*Polypodium quercifolium* Linnaeus).

Dryopteris Adanson, Fam. Pl. **2**: 20, 551. 1763.
T.: *D. filix-mas* (Linnaeus) Schott (*Polypodium filix-mas* Linnaeus).

(≡) *Filix* Séguier, Pl. Veron. **3**: 53. 1754.

Elaphoglossum Schott ex John Smith, J. Bot. (Hooker) **4**: 148. 1841.
T.: *E. conforme* (Swartz) John Smith (*Acrostichum conforme* Swartz) (*typ. cons.*).

(=) *Aconiopteris* K. B. Presl, Tent. Pterid. 236, *t. 10, fig. 17*. 1836.
T.: *A. subdiaphana* (Hooker et Greville) K. B. Presl (*Acrostichum subdiaphanum* Hooker et Greville).

Gleichenia J. E. Smith, Mém. Acad. Sci. (Turin) **5**: 419. 1793.
T.: *G. polypodioides* (Linnaeus) J. E. Smith (*Onoclea polypodioides* Linnaeus).

Lygodium Swartz, J. Bot. (Schrader) **1800**(2): 7, 106. 1801.
T.: *L. scandens* (Linnaeus) Swartz (*Ophioglossum scandens* Linnaeus).

(=) *Ugena* Cavanilles, Icon. **6**: 73. 1801.
LT.: *U. semihastata* Cavanilles, nom. illeg. (*Ophioglossum flexuosum* Linnaeus) (vide Pichi Sermolli, Webbia **9**: 418. 1953).

Matteuccia Todaro, Giorn. Sci. Nat. Econ. Palermo **1**: 235. 1866.
T.: *M. struthiopteris* (Linnaeus) Todaro (*Osmunda struthiopteris* Linnaeus).

(≡) *Pteretis* Rafinesque, Amer. Monthly Mag. & Crit. Rev. **2**: 268. 1818.

Pellaea Link, Fil. Spec. 59. 1841.
T.: *P. atropurpurea* (Linnaeus) Link (*Pteris atropurpurea* Linnaeus).

Polystichum Roth, Tent. Fl. Germ. **3**: 31, 69. 1799.
T.: *P. lonchitis* (Linnaeus) Roth (*Polypodium lonchitis* Linnaeus).

(=) *Hypopeltis* A. Michaux, Fl. Bor.-Amer. **2**: 266. 1803.
T.: *H. lobulata* Bory de Saint-Vincent.

Pteridium Gleditsch ex Scopoli, Fl. Carn. 169. 1760.
T.: *P. aquilinum* (Linnaeus) Kuhn (*Pteris aquilina* Linnaeus).

Schizaea J. E. Smith, Mém. Acad. Sci. (Turin) **5**: 419, *t. 9, f. 9.* 1793.
T.: *S. dichotoma* (Linnaeus) J. E. Smith (*Acrostichum dichotomum* Linnaeus).

(=) *Lophidium* L. C. Richard, Actes Soc. Hist. Nat. Paris **1**: 114. 1792.
T.: *L. latifolium* L. C. Richard [= *L. elegans* (Vahl) K. B. Presl].

Selaginella Palisot de Beauvois, Prodr. 101. 1805.
T.: *S. spinosa* Palisot de Beauvois, nom. illeg. (*Lycopodium selaginoides* Linnaeus, *S. selaginoides* (Linnaeus) Link).

(≡) *Selaginoides* Séguier, Pl. Veron. **3**: 51. 1754.

(=) *Lycopodioides* Boehmer in Ludwig, Defin. Gen. Pl. ed. 3. 485. 1760.
LT.: *L. denticulata* (Linnaeus) O. Kuntze (*Lycopodium denticulatum* Linnaeus) (vide Rothmaler, Feddes Repert. Spec. Nov. Regni Veg. **54**: 69. 1944).

(=) *Stachygynandrum* Palisot de Beauvois ex Mirbel in Lamarck et Mirbel, Hist. Nat. Vég. **3**: 477. 1802; **4**: 312. 1802.
LT.: *S. flabellatum* (Linnaeus) Palisot de Beauvois (Prodr. 113. 1805) (*Lycopodium flabellatum* Linnaeus) (vide Pichi Sermolli, Webbia **26**: 164. 1971).

Sphenomeris Maxon, J. Wash. Acad. Sci. **3**: 144. 1913.
T.: *S. clavata* (Linnaeus) Maxon (*Adiantum clavatum* Linnaeus).

(≡) *Stenoloma* Fée, Mém. Foug. **5**: 330. 1852 (vide Morton, Taxon **8**: 29. 1959).

Thelypteris Schmidel, Icon. Pl. ed. J. C. Keller 3, 45, *t. 11, 13.* Oct 1763.
T.: *Acrostichum thelypteris* Linnaeus (*T. palustris* Schott).

(H) *Thelypteris* Adanson, Fam. Pl. **2**: 20, 610, Jul-Aug 1763 [Pteridoph.: Pterid.].
≡ *Pteris* Linnaeus 1753.

158

XIV. SPERMATOPHYTA

The number assinged to each genus is that of Dalla Torre et Harms, Gen. Siphonogam.

The present list is based on "Nomina generica conservanda et rejicienda spermatophytorum" by H. W. Rickett and F. A. Stafleu, 1959-1961, published as follows:

I: nos. 7-1490 in Taxon **8**(7): 213-243. 12 Aug 1959.
II: nos. 1494-2858 in Taxon **8**(8): 256-274. 20 Oct 1959.
III: nos. 2884-5311 in Taxon **8**(9): 282-314. 21 Dec 1959.
IV: nos. 5320-7414 in Taxon **9**(3): 67-86. 29 Mar 1960.
V: nos. 7485-8918 in Taxon **9**(4): 111-124. 10 Mai 1960.
VI: nos. 8919-9604 in Taxon **9**(5): 153-161. 30 Jun 1960.
VII: Bibliography in Taxon **10**(3): 70-91. 25 Apr 1961.
VIII: Bibliography in Taxon **10**(4): 111-121. 2 Jun 1961.
IX: Bibliography in Taxon **10**(5): 132-149. 30 Jun 1961.
X: Serials and index in Taxon **10**(6): 170-194. 30 Aug 1961.

The list as published here contains the text of the above mentioned publication, parts I-VI, with the exception of the notes, and as amended, corrected and approved by the Committee for Spermatophyta. The new cases of conservation adopted by the IXth, Xth, XIth, XIIth, XIIIth, and XIVth International Botanical Congresses and those approved by the General Committee subsequent to the XIVth Congress have been added.

CYCADACEAE

4 **Dioon** Lindley, Edward's Bot. Reg. 29(Misc.): 59. 1843 (*"Dion"*) (*orth. cons.*).
T.: *D. edule* Lindley.

7 **Zamia** Linnaeus, Sp. Pl. ed. 2. 1659. Jul-Aug 1763.
T.: *Z. pumila* Linnaeus.

(≡) *Palma-filix* Adanson, Fam. Pl. **2**: 21, 587. Jul-Aug 1763 (vide Florin, Taxon **5**: 189. 1956).

TAXACEAE

13 **Podocarpus** L'Héritier ex Persoon, Syn. Pl. **2**: 580. 1807.
T.: *P. elongatus* (W. Aiton) L'Héritier ex Persoon (*Taxus elongata* W. Aiton) (*typ. cons.*).

(H) *Podocarpus* Labillardière, Nov. Holl. Pl. 2: 71. *t. 221.* 1806 [Podocarp.].
≡ *Phyllocladus* L. C. Richard 1826 (*nom. cons.*) (15).

(=) *Nageia* J. Gaertner, Fruct. Sem. Pl. 1: 191. 1788.
T.: *N. japonica* J. Gaertner, nom. illeg. (*Myrica nagi* Thunberg).

159

15 **Phyllocladus** L. C. Richard et Mirbel,
Mém. Mus. Hist. Nat. **13**: 48. 1825.
T.: *P. billardierei* Mirbel, nom. illeg.
(*Podocarpus aspleniifolia* Labillardière, *Phyllocladus aspleniifolius* (Labillardière) J. D. Hooker) (etiam vide
13).

17 **Torreya** Arnott, Ann. Nat. Hist. **1**:
130. 1838.
T.: *T. taxifolia* Arnott.

(H) *Torreya* Rafinesque, Amer. Monthly
Mag. & Crit. Rev. **3**: 356. 1818 [Lab.].
T.: *T. grandiflora* Rafinesque.

PINACEAE

20 **Agathis** R. A. Salisbury, Trans. Linn.
Soc. London **8**: 311. 1807.
T.: *A. loranthifolia* R. A. Salisbury,
nom. illeg. (*Pinus dammara* A. B.
Lambert, *Agathis dammara* (A. B.
Lambert) L. C. Richard).

23 **Cedrus** Trew, Cedr. Lib. Hist. **1**: 6. 1757.
T.: *C. libani* A. Richard (Dict. Class.
Hist. Nat. **3**: 299. 1823) (*Pinus cedrus*
Linnaeus).

(H) *Cedrus* Duhamel, Traité Arbr. Arbust. **1**:
xxviii, 139. *t. 52.* 1755 [Cupress.].
T.: non designatus.

25 **Pseudolarix** Gordon, Pinetum 292. 1858.
T.: [Cult. in England], "Herb. George
Gordon" (**K** 0003455) [= *P. amabilis*
Rehder] (*typ. cons.*).

31 **Cunninghamia** R. Brown in L. C. Richard, Comm. Bot. Conif. Cycad. 149.
1826.
T.: *C. sinensis* R. Brown, nom. illeg.
(*Pinus lanceolata* A. B. Lambert, *C.
lanceolata* (A. B. Lambert) W. J.
Hooker).

(H) *Cunninghamia* Schreber, Gen. 789. 1791
[Rub.].
≡ *Malanea* Aublet 1775.
(≡) *Belis* R. A. Salisbury, Trans. Linn. Soc.
London **8**: 315. 1807.

32 **Sequoia** Endlicher, Syn. Conif. 197.
1847.
T.: *S. sempervirens* (D. Don) Endlicher (*Taxodium sempervirens* D.
Don).

32a **Metasequoia** Hu et Cheng, Bull. Fan
Mem. Inst. Biol. ser. 2. **1**: 154. 1948.
T.: *M. glyptostroboides* Hu et Cheng
(*typ. cons.*).

(H) *Metasequoia* Miki, Jap. J. Bot. **11**: 261.
1941 [Foss.].
T.: *M. disticha* (Heer) Miki (*Sequoia
disticha* Heer).

160

GNETACEAE

48 **Welwitschia** J. D. Hooker, Gard. Chron. **1862:** 71. 1862.
T.: *W. mirabilis* J. D. Hooker.

(H) *Welwitschia* H. G. L. Reichenbach, Handb. 194. 1837 [Polemon.].
T.: *Hugelia densifolia* Bentham (*Eriastrum densifolium* (Bentham) H. L. Mason).

(≡) *Tumboa* Welwitsch, Gard. Chron. **1861:** 75. 1861.

POTAMOGETONACEAE

57 **Posidonia** C. König, Ann. Bot. (König & Sims) **2:** 95. 1805 ("1806").
T.: *P. caulinii* C. König, nom. illeg. (*Zostera oceanica* Linnaeus, *P. oceanica* (Linnaeus) Delile).

(=) *Alga* Boehmer in Ludwig, Defin. Gen. Pl. ed. 3. 503. 1760.
T.: non designatus.

60 **Cymodocea** C. König, Ann. Bot. (König & Sims) **2:** 96. 1805 ("1806").
T.: *C. aequorea* C. König.

(=) *Phucagrostis* Cavolini, Phucagr. Theophr. Anth. xiii. 1792.
T.: *P. major* Cavolini.

APONOGETONACEAE

65 **Aponogeton** Linnaeus f., Suppl. Pl. 32. 1782.
T.: *A. monostachyon* Linnaeus f., nom. illeg. (*A. natans* (Linnaeus) Engler et Krause, *Saururus natans* Linnaeus).

(H) *Aponogeton* J. Hill, Brit. Herb. 480. 1756 [Potamogeton.].
≡ *Zannichellia* Linnaeus 1753.

GRAMINEAE (POACEAE)

124 **Vossia** Wallich et Griffith, J. Asiat. Soc. Bengal **5:** 572. 1836.
T.: *V. procera* Wallich et Griffith, nom. illeg. (*Ischaemum cuspidatum* Roxburgh, *V. cuspidata* (Roxburgh) Griffith).

(H) *Vossia* Adanson, Fam. Pl. **2:** 243. 1763 [Aiz.].

127 **Rottboellia** Linnaeus f., Suppl. Pl. 13, 114. 1782.
T.: *R. exaltata* Linnaeus f. 1782, non (Linnaeus) Linnaeus f. 1779 [= *R. cochinchinensis* (Loureiro) Clayton] (*typ. cons.*).

(H) *Rottboelia* Scopoli, Intr. 233. 1777 [Olac.].
≡ *Heymassoli* Aublet 1775.

(=) *Manisuris* Linnaeus, Mant. Pl. **2:** 164, 300. 1771.
T.: *M. myurus* Linnaeus.

134a **Diectomis** Kunth, Mém. Mus. Hist. Nat. **2**: 69. 1815.
T.: *D. fastigiata* (Swartz) Palisot de Beauvois (Essai Agrost. 132, 160. 1812). (*Andropogon fastigiatus* Swartz) (*typ. cons.*).

(H) *Diectomis* Palisot de Beauvois, Essai Agrost. 132, 160. 1812 [Gram.].
T.: *D. fasciculata* Palisot de Beauvois.

134b **Sorghum** Moench, Methodus 207. 1794.
T.: *S. bicolor* (Linnaeus) Moench (*Holcus bicolor* Linnaeus).

(H) *Sorgum* Adanson, Fam. Pl. **2**: 38, 606. 1763 [Gram.].
≡ *Holcus* Linnaeus 1753 (*nom. cons.*) (257).

134c **Chrysopogon** Trinius, Fund. Agrost. 187. 1820.
T.: *C. gryllus* (Linnaeus) Trinius (*Andropogon gryllus* Linnaeus) (*typ. cons.*).

(≡) *Pollinia* K. Sprengel, Pugill. **2**: 10. 1815.
(=) *Rhaphis* Loureiro, Fl. Cochinch. 552. 1790.
T.: *R. trivialis* Loureiro.
(=) *Centrophorum* Trinius, Fund. Agrost. 106. 1820.
T.: *C. chinense* Trinius.

143 **Tragus** Haller, Hist. Stirp. Helv. **2**: 203. 1768.
T.: *T. racemosus* (Linnaeus) Allioni (Fl. Pedem. **2**: 241. 1785) (*Cenchrus racemosus* Linnaeus).

(≡) *Nazia* Adanson, Fam. Pl. **2**: 31, 581. 1763.

150 **Zoysia** Willdenow, Ges. Naturf. Freunde Berlin Neue Schr. **3**: 440. 1801.
T.: *Z. pungens* Willdenow.

166 **Echinochloa** Palisot de Beauvois, Essai Agrost. 53. 1812.
T.: *E. crus-galli* (Linnaeus) Palisot de Beauvois (*Panicum crus-galli* Linnaeus).

(≡) *Tema* Adanson, Fam. Pl. **2**: 496. 1763.

166a **Digitaria** Haller, Stirp. Helv. **2**: 244. 1768.
T.: *D. sanguinalis* (Linnaeus) Scopoli (Fl. Carn. ed. 2. **1**: 52. 1771) (*Panicum sanguinale* Linnaeus) (*typ. cons.*).

(H) *Digitaria* Fabricius, Enum. 207. 1759 [Gram.].
T.: non designatus [*Paspalum* sp.].

169 **Oplismenus** Palisot de Beauvois, Fl. Oware **2**: 14. Aug 1810 ("1807").
T.: *O. africanus* Palisot de Beauvois.

(=) *Orthopogon* R. Brown, Prodr. 194. Apr 1810.
LT.: *O. compositus* (Linnaeus) R. Brown (*Panicum compositum* Linnaeus) (vide Hitchcock, U.S.D.A. Bull. **772**: 238. 1920).

171 **Setaria** Palisot de Beauvois, Essai Agrost. 51, 178. 1812.
T.: *S. viridis* (Linnaeus) Palisot de Beauvois (*Panicum viride* Linnaeus) (*typ. cons.*).

(H) *Setaria* Acharius ex A. Michaux, Fl. Bor.-Amer. **2**: 331. 1803 [Fungi: Lich.].
T.: *S. trichodes* A. Michaux.

194 **Leersia** Swartz, Prodr. **1**: 21. 1788.
T.: *L. oryzoides* (Linnaeus) Swartz
(*Phalaris oryzoides* Linnaeus) (*typ. cons.*).

(≡) *Homalocenchrus* Mieg, Acta Helv. Phys.-Math. **4**: 307. 1760.

201 **Ehrharta** Thunberg, Kongl. Vetensk. Akad. Handl. Stockholm **40**: 217. *t. 8.* 1779 sem. 2.
T.: *E. capensis* Thunberg.

(=) *Trochera* L. C. Richard, Obs. Phys. Chim. Hist. Nat. Arts **13**: 225. Mar 1779.
T.: *T. striata* L. C. Richard.

206 **Hierochloë** R. Brown, Prodr. 208. 1810.
T.: *H. odorata* (Linnaeus) Palisot de Beauvois (*Holcus odoratus* Linnaeus) (*typ. cons.*).

(=) *Savastana* Schrank, Baiersche Fl. **1**: 100, 337. 1789.
T.: *S. hirta* Schrank.
(=) *Torresia* Ruiz et Pavón, Prodr. 125. 1794.
T.: *T. utriculata* Ruiz et Pavón (Syst. 251. 1798).
(=) *Disarrenum* Labillardière, Nov. Holl. Pl. **2**: 82, *t. 232.* 1806.
T.: *D. antarcticum* (G. Forster) Labillardière (*Aira antarctica* G. Forster).

212 **Piptochaetium** J. S. Presl in K. B. Presl, Reliq. Haenk. **1**: 222. 1830.
T.: *P. setifolium* J.S. Presl.

(=) *Podopogon* Rafinesque, Neogenyton 4. 1825.
LT.: *Stipa avenacea* Linnaeus (vide Clayton, Taxon **32**: 649. 1983).

221 **Crypsis** W. Aiton, Hortus Kew. **1**: 48. 1789.
T.: *C. aculeata* (Linnaeus) W. Aiton (*Schoenus aculeatus* Linnaeus).

228 **Coleanthus** Seidel in Roemer et Schultes, Syst. Veg. **2**: 11, 276. 1817.
T.: *C. subtilis* (Trattinick) Seidel (*Schmidtia subtilis* Trattinick).

257 **Holcus** Linnaeus, Sp. Pl. 1047. 1753.
T.: *H. lanatus* Linnaeus (*typ. cons.*).

269 **Corynephorus** Palisot de Beauvois, Essai Agrost. 90, 159. 1812.
T.: *C. canescens* (Linnaeus) Palisot de Beauvois (*Aira canescens* Linnaeus).

(≡) *Weingaertneria* Bernhardi, Syst. Verz. 23, 51. 1800.

272 **Ventenata** Koeler, Descr. Gram. 272. 1802 (*orth. cons.*).
T.: *V. avenacea* Koeler, nom. illeg. (*Avena dubia* Leers, *V. dubia* (Leers) Cosson) (*typ. cons.*).

(H) *Ventenatia* Cavanilles, Icon. **4**: 28. 1797 ("*Vintenatia* ") [Epacrid.].
T.: non designatus.
(=) *Heteranthus* Borkhausen, Botaniker **16-18**: 71. 1796(?).
T.: non designatus.

278a **Loudetia** Hochstetter ex Steudel, Syn. Pl. Glum. **1**: 238. 1854.
T.: *L. elegans* Hochstetter ex A. Braun.

(H) *Loudetia* Hochstetter ex A. Braun, Flora **24**: 713. 1841 [Gram.].
≡ *Tristachya* C. G. Nees 1829.

280 **Danthonia** A.-P. de Candolle in La-
marck et A.-P. de Candolle, Fl.
Fran = c. ed. 3. 3: 32. 1805.
T.: *D. spicata* (Linnaeus) Roemer et
Schultes (Syst. Veg. 2: 690. 1817)
(*Avena spicata* Linnaeus) (*typ. cons.*).

(=) *Sieglingia* Bernhardi, Syst. Verz. 44.
1800.
T.: *S. decumbens* (Linnaeus) Bernhardi
(*Danthonia decumbens* (Linnaeus) A.-P.
de Candolle).

282 **Cynodon** L. C. Richard in Persoon,
Syn. Pl. 1: 85. 1805.
T.: *C. dactylon* (Linnaeus) Persoon
(*Panicum dactylon* Linnaeus).

(≡) *Dactilon* Villars, Hist. Pl. Dauphiné 2:
69. 1787.

(=) *Capriola* Adanson, Fam. Pl. 2: 31, 532.
1763.
T.: non designatus ["*Gramen dactylon*
Offic."].

286 **Ctenium** Panzer, Ideen Rev. Gräser
36, 61. 1813.
T.: *C. carolinianum* Panzer, nom. illeg.
(*Chloris monostachya* A. Michaux)
[= *Ctenium aromaticum* (Walter)
Wood, *Aegilops aromatica* Walter].

(≡) *Campulosus* Desvaux, Nouv. Bull. Sci.
Soc. Philom. Paris 2: 189. 1810.
T.: *C. gracilior* Desvaux, nom. illeg.
(*Chloris monostachya* A. Michaux, *Cam-
pulosus monostachyus* (A. Michaux) Pa-
lisot de Beauvois).

295 **Bouteloua** Lagasca, Varied. Ci. 2(4):
134. 1805 ("*Botelua*") (*orth. cons.*).
T.: *B. racemosa* Lagasca.

308 **Buchloe** Engelmann, Trans. Acad.
Sci. St. Louis 1: 432. 1859.
T.: *B. dactyloides* (Nuttall) Engel-
mann (*Sesleria dactyloides* Nuttall).

312 **Schmidtia** Steudel ex J. A. Schmidt,
Beitr. Fl. Cap Verd. Ins. 144. 1852.
T.: *S. pappophoroides* Steudel ex J. A.
Schmidt.

(H) *Schmidtia* Moench, Suppl. Meth. 217.
1802 [Comp.].
T.: *S. fruticosa* Moench.

320 **Echinaria** Desfontaines, Fl. Atlant. 2:
385. 1799.
T.: *E. capitata* (Linnaeus) Desfontaines (*Cenchrus capitatus* Linnaeus).

(H) *Echinaria* Heister ex Fabricius, Enum.
206. 1759 [Gram.].
≡ *Cenchrus* Linnaeus 1753.

(≡) *Panicastrella* Moench, Methodus 205.
1794.

329 **Cortaderia** O. Stapf, Gard. Chron. ser.
3. 22: 378, 397. 1897.
T.: *C. selloana* (J. A. Schultes et J. H.
Schultes) Ascherson et Graebner (Syn.
Mitteleur. Fl. 2(1): 325. 1900) (*Arundo
selloana* J. A. Schultes et J. H. Schultes).

(=) *Moorea* Lemaire, Ill. Hort. 2 (misc.): 15.
1854.
T.: *M. argentea* (C. G. Nees) Lemaire
(*Gynerium argenteum* C. G. Nees).

356 **Diarrhena** Palisot de Beauvois, Essai
Agrost. 142, 160 162. 1812.
T.: *D. americana* Palisot de Beauvois
(*Festuca diandra* A. Michaux 1803, non
Moench 1794).

358 **Zeugites** P. Browne, Civ. Nat. Hist. Jamaica 341. 1756.
T: *Z. americanus* Willdenow (*Apluda zeugites* Linnaeus).

374 **Lamarckia** Moench, Methodus 201. 1794. ("*Lamarkia*") (*orth. cons.*).
T.: *L. aurea* (Linnaeus) Moench (*Cynosurus aureus* Linnaeus).

(H) *Lamarckia* Olivi, Zool. Adriat. 258. 1792 [Chloroph.].
T.: non designatus.
(≡) *Achyrodes* Boehmer in Ludwig, Defin. Gen. Pl. ed. 3. 420. 1760.

381 **Scolochloa** Link, Hort. Berol. 1: 1136. 1827.
T.: *S. festucacea* (Willdenow) Link (*Arundo festucacea* Willdenow).

(H) *Scolochloa* Mertens et Koch in Röhling, Deutschl. Fl. ed. 3. 1: 374, 528 [Gram.].
T.: *S. arundinacea* (Palisot de Beauvois) Mertens et Koch (*Donax arundinaceus* Palisot de Beauvois).

383 **Glyceria** R. Brown, Prodr. 179. 1810.
T.: *G. fluitans* (Linnaeus) R. Brown (*Festuca fluitans* Linnaeus.

384 **Puccinellia** Parlatore, Fl. Ital. 1: 366. 1848.
T.: *P. distans* (N. J. Jacquin) Parlatore (*Poa distans* N. J. Jacquin) (*typ. cons.*).

(≡) *Atropis* Ruprecht, Beitr. Pflanzenk. Russ. Reich. 2: 64. 1845.

417 **Phyllostachys** Siebold et Zuccarini, Abh. Math.-Phys. Cl. Königl. Bayer. Akad. Wiss. 3: 745. 1843 (1844?).
T.: *P. bambusoides* Siebold et Zuccarini.

424 **Bambusa** Schreber, Gen. 1: 236. 1789.
T.: *B. arundinacea* (Retzius) Willdenow (Sp. Pl. 2: 245. 1799) (*Bambos arundinacea* Retzius).

(≡) *Bambos* Retzius, Obs. Bot. 5: 24. 1788.

CYPERACEAE

452 **Lipocarpha** R. Brown in Tuckey, Narr. Exp. Congo 459. 1818.
T.: *L. argentea* R. Brown, nom. illeg. (*Hypaelyptum argenteum* Vahl, nom. illeg., *Scirpus senegalensis* Lamarck, *Lipocarpha senegalensis* (Lamarck) T. Durand et H. Durand).

(=) *Hypaelyptum* Vahl, Enum. 2: 283. 1806.
T.: non designatus.

454 **Ascolepis** C. G. Nees ex Steudel, Syn. Pl. Cyp. 105. 1855.
T.: *A. eriocauloides* (Steudel) C. G. Nees ex Steudel (*Kyllinga eriocauloides* Steudel) (*typ. cons.*).

165

459 **Mariscus** Vahl, Enum. 2: 372. 1806.
T.: *M. capillaris* (Swartz) Vahl
(*Schoenus capillaris* Swartz) (*typ. cons.*).

(H) *Mariscus* Scopoli, Meth. Pl. 22. 1754 [Cyper.].
T.: *Schoenus mariscus* Linnaeus (*Cladium mariscus* (Linnaeus) Pohl).

462 **Kyllinga** Rottböll, Descr. Ic. Nov. Pl. 12. 1773.
T.: *K. nemoralis* (J. R. Forster et G. Forster) Dandy ex Hutchinson et Dalziel (*Thryocephalon nemorale* J. R. Forster et G. Forster) [= *K. monocephala* sensu Rottböll, non Rottböll 1773, nom illeg.] (*typ. cons.*).

(H) *Kyllinga* Adanson, Fam. Pl. 2: 498, 539. 1763 [Umbell.].
≡ *Athamantha* Linnaeus 1753.

465 **Ficinia** H. A. Schrader, Commentat. Soc. Regiae Sci. Gott. Recent. 7: 143. 1832.
T.: *F. filiformis* (Lamarck) H. A. Schrader (*Schoenus filiformis* Lamarck) (*typ. cons.*).

(≡) *Melancranis* Vahl, Enum. 2: 239. 1806.
(=) *Hemichlaena* Schrader, Gött. Gel. Anz. 1821: 2066. 1821.
LT.: *H. capillifolia* Schrader (vide Levyns, J. S. African Bot. 13: 65. 1947).

468a **Blysmus** Panzer ex J. A. Schultes, Mant. 2: 41. 1824.
T.: *B. compressus* (Linnaeus) Panzer ex Link (Hort. Berol. 1: 278. 1827) (*Schoenus compressus* Linnaeus).

(≡) *Nomochloa* Palisot de Beauvois ex Lestiboudois, Essai Cypér. 37. 1819.

468b **Schoenoplectus** (Reichenbach) Palla, Verh. K. K. Zool.-Bot. Ges. Wien 38, Sitzungsber.: 49. 1888.
T.: *S. lacustris* (Linnaeus) Palla (*Scirpus lacustris* Linnaeus).

(=) *Heleophylax* Palisot de Beauvois ex Lestiboudois, Essai Cypér. 41. 1819.
T.: non designatus.
(=) *Elytrospermum* C. A. Meyer, Mém. Acad. Imp. Sci. St.-Pétersbourg Divers Savans 1: 200. 1831.
T: *E. californicum* C. A. Meyer.

471 **Fimbristylis** Vahl, Enum. 2: 285. 1806.
T.: *F. dichotoma* (Linnaeus) Vahl (*Scirpus dichotomus* Linnaeus) (*typ. cons.*).

(=) *Iria* (L. C. Richard) R. A. Hedwig, Gen. Pl. 360. Jul 1806.
T.: *Cyperus monostachyos* Linnaeus.

471a **Bulbostylis** Kunth, Enum. Pl. 2: 205. 1837.
T.: *B. capillaris* (Linnaeus) C. B. Clarke (*Scirpus capillaris* Linnaeus) (*typ. cons.*).

(H) *Bulbostylis* Steven, Mém. Soc. Imp. Naturalistes Moscou 5: 355. 1817 [Cyper.].
T.: non designatus.
(=) *Stenophyllus* Rafinesque, Neogenyton 4. 1825.
T.: *S. cespitosus* Rafinesque (*Scirpus stenophyllus* Elliott).

492 **Rhynchospora** Vahl, Enum. 2: 229. 1806 ("*Rynchospora*") (*orth. cons.*).
T.: *R. alba* (Linnaeus) Vahl (*Schoenus albus* Linnaeus) (*typ. cons.*).

(=) *Dichromena* A. Michaux, Fl. Bor.-Amer. 1: 37. 1803.
T.: *D. leucocephala* A. Michaux.

166

542 **Pritchardia** Seemann et Wendland, Bonplandia **10**: 197. 1862.
T.: *P. pacifica* Seemann et Wendland.

(H) *Pritchardia* Unger ex Endlicher, Gen. Pl. Suppl. **2**: 102. 1842 [Foss.].
T.: *P. insignis* Unger ex Endlicher.

543 **Washingtonia** H. Wendland, Bot. Zeitung (Berlin) **37**: lxi, 68, 148. 1879.
T.: *W. filifera* (Linden ex André) H. Wendland (*Pritchardia filifera* Linden ex André).

565 **Metroxylon** Rottböll, Nye Saml. Kongel. Danske Vidensk. Selsk. Skr. **2**: 527. 1783.
T.: *M. sagu* Rottböll.

(=) *Sagus* Steck, Diss. de Sagu 21. 1757.
T.: *S. genuina* Giseke.

567 **Pigafetta** (Blume) Beccari, Malesia **1**: 89. 1877 ("*Pigafettia*") (*orth. cons.*).
T.: *P. filaria* (Giseke) Beccari (*Sagus filaria* Giseke) (*typ. cons.*).

(H) *Pigafetta* Adanson, Fam. Pl. **2**: 223, 590. 1763 [Acanth.].
≡ *Eranthemum* Linnaeus 1753.

575 **Arenga** Labillardière in A.-P. de Candolle, Bull. Sci. Soc. Philom. Paris **2**: 161. 1800.
T.: *Arenga saccharifera* Labillardière [= *A. pinnata* (Wurmb) Merrill, *Saguerus pinnatus* Wurmb].

(=) *Saguerus* Steck, Diss. de Sagu 15. 1757.
T.: *S. pinnatus* Wurmb (Verh. Batav. Genootsch. **1**: 351. 1781).

594 **Chamaedorea** Willdenow, Sp. Pl. **4**: 638, 800. 1806.
T.: *C. gracilis* Willdenow, nom. illeg. (*Borassus pinnatifrons* N. J. Jacquin, *C. pinnatifrons* (N. J. Jacquin) Oersted).

(=) *Morenia* Ruiz et Pavón, Prodr. 150. 1794.
T.: *M. fragrans* Ruiz et Pavón.
(=) *Nunnezharia* Ruiz et Pavón, Prodr. 147. 1794.
T.: *N. fragrans* Ruiz et Pavón.

612 **Prestoea** J. D. Hooker in Bentham et J. D. Hooker, Gen. Pl. **3**: 899. 1883.
T.: *Hyospathe pubigera* Grisebach et H. Wendland.

(=) *Martinezia* Ruiz et Pavón, Prodr. 148. 1794.
T.: *M. ensiformis* Ruiz et Pavón.
(=) *Oreodoxa* Willdenow, Mém. Acad. Roy. Sci. Hist. (Berlin) **1804**: 34. 1807.
T.: *O. acuminata* Willdenow.

631 **Euterpe** C. F. P. Martius, Hist. Nat. Palm. **2**: 28. 1823; emend. **3**: 165. 1837.
T.: *E. oleracea* C. F. P. Martius.

(H) *Euterpe* J. Gaertner, Fruct. Sem. Pl. **1**: 24. 1788 [Palmae].
T.: *E. pisifera* J. Gaertner.
(=) *Martinezia* Ruiz et Pavón, Prodr. 148. 1794.
T.: *M. ensiformis* Ruiz et Pavón.
(=) *Oreodoxa* Willdenow, Mém. Acad. Roy. Sci. Hist. (Berlin) **1804**: 34. 1807.
T.: *O. acuminata* Willdenow.

639 **Veitchia** H. Wendland in Seemann, Fl. Vitiens. 270. 1868.
T.: *V. joannis* H. Wendland (*typ. cons.*).

(H) *Veitchia* Lindley, Gard. Chron. **1861**: 265. 1861 [Pin.].
T.: *V. japonica* Lindley [= *Picea jezoënsis* (Siebold et Zuccarini) Carrière].

657 **Orbignya** C. F. P. Martius ex Endlicher, Gen. Pl. 257. 1837.
T.: *O. phalerata* C. F. P. Martius (Hist. Nat. Palm. **3**: 302. 1845).

(H) *Orbignya* Bertero, Mercurio Chileno **16**: 737. 1829 [Euphorb.].
T.: *O. trifolia* Bertero.

660 **Maximiliana** C. F. P. Martius, Palm. Fam. 20. 1824.
T.: *M. regia* C. F. P. Martius (Hist. Nat. Palm. **2**: 131. 1826) non *Maximilianea regia* C. F. P. Martius 1819 (*M. martiana* H. Karsten) (*typ. cons.*).

(H) *Maximilianea* C. F. P. Martius, Flora **2**: 452. 1819 [Cochlosperm.].
T.: *M. regia* C. F. P. Martius.

668 **Astrocaryum** G. F. W. Meyer, Prim. Fl. Esseq. 265. 1818.
T.: *A. aculeatum* G. F. W. Meyer.

(=) *Avoira* Giseke, Prael. Ord. Nat. Pl. 38, 53. 1792.
T.: *A. vulgaris* Giseke.

670 **Desmoncus** C. F. P. Martius, Palm. Fam. 20. 1824.
T.: *D. polyacanthos* C. F. P. Martius (Hist. Nat. Palm. **2**: 84. 1824) (*typ. cons.*).

CYCLANTHACEAE

682 **Ludovia** A. T. Brongniart, Ann. Sci. Nat. Bot. ser. 4. **15**: 361. 1861.
T.: *L. lancifolia* A. T. Brongniart.

(H) *Ludovia* Persoon, Syn. Pl. **2**: 576. 1807 [Cyclanth.].
T.: non designatus.

ARACEAE

690 **Culcasia** Palisot de Beauvois, Fl. Oware 3. 1805.
T.: *C. scandens* Palisot de Beauvois (*typ. cons.*).

700 **Monstera** Adanson, Fam. Pl. **2**: 470. 1763.
T.: *M. adansonii* Schott (*Dracontium pertusum* Linnaeus) (*typ. cons.*).

708 **Symplocarpus** R. A. Salisbury ex Nuttall, Gen. N. Amer. Pl. **1**: 105. 1818.
T.: *S. foetidus* (Linnaeus) Nuttall (*Dracontium foetidum* Linnaeus.

723 **Amorphophallus** Blume ex Decaisne, Nouv. Ann. Mus. Hist. Nat. **3**: 366. 1834.
T.: *A. campanulatus* Decaisne.

(=) *Thomsonia* Wallich, Pl. Asiat. Rar. **1**: 83. 1830.
T.: *T. napalensis* Wallich.
(=) *Pythion* Martius, Flora **14**: 458. 1831.
T.: *Arum campanulatum* Roxburgh, nom. illeg. (*Dracontium paeoniifolium* Dennstedt, *Amorphophallus paeoniifolius* (Dennstedt) Nicolson).

730 **Montrichardia** Crüger, Bot. Zeitung (Berlin) **12**: 25. 1854.
T.: *M. aculeata* (G. F. W. Meyer) Schott (*Caladium aculeatum* G. F. W. Meyer).

(=) *Pleurospa* Rafinesque, Fl. Tell. **4**: 8. 1838.
T.: *P. reticulata* Rafinesque, nom. illeg. (*Arum arborescens* Linnaeus).

739 **Philodendron** Schott, Wiener Z. Kunst **1829** (3): 780. 1829 ("*Philodendrum*") (*orth. cons.*).
T.: *P. grandifolium* (N. J. Jacquin) Schott (*Arum grandifolium* N. J. Jacquin).

747 **Peltandra** Rafinesque, J. Phys. Chim. Hist. Nat. Arts **89**: 103. 1819.
T.: *P. undulata* Rafinesque.

748 **Zantedeschia** K. Sprengel, Syst. Veg. **3**: 756, 765. 1826.
T.: *Z. aethiopica* (Linnaeus) K. Sprengel (*Calla aethiopica* Linnaeus) (etiam vide 755).

752 **Alocasia** (Schott) G. Don in Sweet, Hort. Brit. ed. **3**. 631. 1839.
T.: *A. cucullata* (Loureiro) G. Don (*Arum cucullatum* Loureiro) (*typ. cons.*).

(H) *Alocasia* Rafinesque, Fl. Tell. **3**: 64. 1837 ("1836") [Ar.].
T.: non designatus.

755 **Colocasia** Schott in Schott et Endlicher, Melet. Bot. 18. 1832.
T.: *C. antiquorum* Schott (*Arum colocasia* Linnaeus) (*typ. cons.*).

(H) *Colocasia* Link, Diss. Bot. 77. 1795. [Ar.].
≡ *Zantedeschia* K. Sprengel 1826 (*nom. cons.*) (748).

756 **Hapaline** Schott, Gen. Aroid. 44. 1858.
T.: *H. benthamiana* (Schott) Schott (*Hapale benthamiana* Schott).

(≡) *Hapale* Schott, Österr. Bot. Wochenbl. **7**: 85. 1857.

779 **Helicodiceros** Schott in Klotzsch, App. Gen. Sp. Nov. **1855**: 2. 1855 vel 1856.
T.: *H. muscivorus* (Linnaeus f.) Engler (in A. de Candolle et C. de Candolle, Monogr. Phan. **2**: 605. 1879) (*Arum muscivorum* Linnaeus f.).

(≡) *Megotigea* Rafinesque, Fl. Tell. **3**: 64. 1837 ("1836").

784 **Biarum** Schott in Schott et Endlicher, Melet. Bot. 17. 1832.
T.: *B. tenuifolium* (Linnaeus) Schott (*Arum tenuifolium* Linnaeus).

(≡) *Homaid* Adanson, Fam. Pl. **2**: 470. 1763.

787 **Pinellia** Tenore, Atti Reale Accad. Sci. Sez. Soc. Reale Borbon. **4**: 69. 1839.
T.: *P. tuberifera* Tenore, nom. illeg. (*Arum subulatum* Desfontaines) [= *P. ternata* (Thunberg) Makino, *Arum ternatum* Thunberg].

(=) *Atherurus* Blume, Rumphia **1**: 135. 1837 ("1835").
LT.: *A. tripartitus* Blume (vide Nicolson, Taxon **16**: 515. 1967).

LEMNACEAE

796 **Wolffia** Horkel ex Schleiden, Beitr. **1**: 233. 1844.
T.: *W. michelii* Schleiden.

(H) *Wolfia* Schreber, Gen. Pl. ed. 8. **2**: 801. 1791 [Flacourt.].
T.: non designatus.

RESTIONACEAE

800 **Lyginia** R. Brown, Prodr. 248. 1810.
T.: *L. barbata* R. Brown (*typ. cons.*).

804 **Restio** Rottböll, Descr. Pl. Rar. 9. 1772.
T.: *R. triticeus* Rottböll.

(H) *Restio* Linnaeus, Syst. Nat. ed. 12. **2**: 735. 1767 [Restion.].
T.: *R. dichotomus* Linnaeus.

808 **Leptocarpus** R. Brown, Prodr. 250. 1810.
T.: *L. aristatus* R. Brown (*typ. cons.*).

(=) *Schoenodum* Labillardière. Nov. Holl. Pl. **2**: 79. 1806.
T.: *S. tenax* Labillardière (vide Kunth, Enum. Pl. **3**: 445. 1841).

815 **Hypolaena** R. Brown, Prodr. 251. 1810.
T.: *H. fastigiata* R. Brown (*typ. cons.*).

(=) *Calorophus* Labillardière, Nov. Holl. Pl. **2**: 78, *t. 228*. 1806.
T.: *C. elongata* Labillardière.

816 **Hypodiscus** C. G. Nees in Lindley, Intr. Nat. Syst. Bot. 450. 1836.
T.: *H. aristatus* (Thunberg) Masters (J. Linn. Soc., Bot. **10**: 252. 1868) (*Restio aristatus* Thunberg).

(=) *Lepidanthus* C. G. Nees, Linnaea **5**: 665. 1830.
T.: *L. willdenowia* C. G. Nees, nom. illeg. (*Willdenowia striata* Thunberg).

ERIOCAULACEAE

830 **Paepalanthus** Kunth, Enum. Pl. **3**: 498. 1841.
T.: *P. lamarckii* Kunth (*typ. cons.*).

(H) *Paepalanthus* C. F. P. Martius, Nova Acta Phys.-Med. Acad. Caes. Leop.-Carol. Nat. Cur. **17**(1): 13. 1835 [Eriocaul.].
LT.: *P. corymbosus* (Bongard) Kunth (*Eriocaulon corymbosum* Bongard) (vide Moldenke, N. Amer. Fl. **19**: 37. 1937).
(=) *Dupatya* Vellozo, Fl. Flum. 35. 1825.
T.: non designatus.

846 **Cryptanthus** Otto et Dietrich, Allg. Gartenzeitung **4**: 297. 1836.
T.: *C. bromelioides* Otto et Dietrich.

(H) *Cryptanthus* Osbeck, Dagb. Ostind. Resa 215. 1757 [Spermatoph.].
T.: *C. chinensis* Osbeck.

861 **Aechmea** Ruiz et Pavón, Prodr. 47. 1794.
T.: *A. paniculata* Ruiz et Pavón (Fl. Peruv. Chil. **3**: 37. 1802).

(=) *Hoiriri* Adanson, Fam. Pl. **2**: 67, 584. 1763.
T.: *Bromelia nudicaulis* Linnaeus.

878 **Pitcairnia** L'Héritier, Sert. Angl. 7, *t. 11.* 1789 ("1788").
T.: *P. bromeliifolia* L'Héritier.

(=) *Hepetis* Swartz, Prodr. 4, 56. 1788.
T.: *H. angustifolia* Swartz.

891 **Vriesea** Lindley, Bot. Reg. **29**: *t. 10.* 1843 ("*Vriesia*") (*orth. cons.*).
T.: *V. psittacina* (W. J. Hooker) Lindley (*Tillandsia psittacina* W. J. Hooker).

(≡) *Hexalepis* Rafinesque, Fl. Tell. **4**: 24. 1838.

COMMELINACEAE

894 **Palisota** H. G. L. Reichenbach ex Endlicher, Gen. Pl. 125. 1836.
T.: *P. ambigua* (Palisot de Beauvois) C. B. Clarke (in A. de Candolle et C. de Candolle, Monogr. Phan. **3**: 131. 1881) (*Commelina ambigua* Palisot de Beauvois).

(=) *Duchekia* Kosteletzky, Allg. Med.-Pharm. Fl. **1**: 213. 1831.
T.: *D. hirsuta* (Thunberg) Kosteletzky (*Dracaena hirsuta* Thunberg).

899a **Murdannia** Royle, Ill. Bot. Himal. *t. 95, fig. 3.* 1839; 403. 1840.
T.: *M. scapiflora* (Roxburgh) Royle (*Commelina scapiflora* Roxburgh).

(=) *Dilasia* Rafinesque, Fl. Tell. **4**: 122. 1838.
T.: *D. vaginata* (Linnaeus) Rafinesque (*Commelina vaginata* Linnaeus).

(=) *Streptylis* Rafinesque, Fl. Tell. **4**: 122. 1838.
T.: *S. bracteolata* Rafinesque.

904 **Cyanotis** D. Don, Prodr. Fl. Nepal. 45. 1825.
T.: *C. barbata* D. Don.

909 **Dichorisandra** Mikan, Del. Fl. Faun. Bras. *t. 3.* 1820.
T.: *D. thyrsiflora* Mikan.

910 **Tinantia** Scheidweiler, Allg. Garten-zeitung 7: 365. 1839.
T.: *T. fugax* Scheidweiler.

(H) *Tinantia* Dumortier, Anal. Fam. Pl. 58. 1829 [Irid.].
T.: non designatus.

(=) *Pogomesia* Rafinesque, Fl. Tell. 3: 67. 1837 ("1836").
T.: *P. undata* (Humboldt et Bonpland) Rafinesque (*Tradescantia undata* Humboldt et Bonpland).

PONTEDERIACEAE

921 **Eichhornia** Kunth, Eichhornia. 1842.
T.: *E. azurea* (Swartz) Kunth (*Pontederia azurea* Swartz) (*typ. cons.*).

(=) *Piaropus* Rafinesque, Fl. Tell. 2: 81. 1837 ("1836").
T.: non designatus.

923 **Reussia** Endlicher, Gen. Pl. 139. 1836.
T.: *R. triflora* Seubert (in C. F. P. Martius, Fl. Bras. 3(1):96. 1847).

924 **Heteranthera** Ruiz et Pavón, Prodr. 9. 1794.
T.: *H. reniformis* Ruiz et Pavón (Fl. Peruv. Chil. 1: 43. 1798).

JUNCACEAE

937 **Luzula** A.-P. de Candolle in Lamarck et A.-P. de Candolle, Fl. Franç. ed. 3. 3: 158. 1805.
T.: *L. campestris* (Linnaeus) A.-P. de Candolle (*Juncus campestris* Linnaeus) (*typ. cons.*).

(≡) *Juncoides* Séguier, Pl. Veron. 3: 88. 1754.

LILIACEAE

944 **Narthecium** Hudson, Fl. Angl. 127. 1762.
T.: *N. ossifragum* (Linnaeus) Hudson (*Anthericum ossifragum* Linnaeus).

(H) *Narthecium* L. Gérard, Fl. Gallo-Prov. 142. 1761 [Lil.].
T.: *Anthericum calyculatum* Linnaeus.

951 **Chionographis** Maximowicz, Bull. Acad. Imp. Sci. St.-Pétersbourg 11: 435. 1867.
T.: *C. japonica* (Willdenow) Maximowicz (*Melanthium japonicum* Willdenow).

(=) *Siraitos* Rafinesque, Fl. Tell. 4: 26. 1838.
T.: *S. aquaticus* Rafinesque.

172

952 **Heloniopsis** A. Gray, Mem. Amer.
Acad. Arts ser. 2. 6: 416. 1859.
T.: *H. pauciflora* A. Gray.

(=) *Hexonix* Rafinesque, Fl. Tell. 2: 13. 1837.
T.: *H. japonica* (Thunberg) Rafinesque
(*Scilla japonica* Thunberg).

(=) *Kozola* Rafinesque, Fl. Tell. 2: 25. 1837.
T.: *K. japonica* (Thunberg) Rafinesque
(*Scilla japonica* Thunberg).

955 **Amianthium** A. Gray, Ann. Lyceum
Nat. Hist. New York 4: 121 1837
("1848").
T.: *A. muscaetoxicum* (Walter) A.
Gray (*Melanthium muscaetoxicum*
Walter) (*typ. cons.*).

(=) *Chrosperma* Rafinesque, Neogenyton 3.
1825.
T.: *Melanthium laetum* W. Aiton.

957 **Stenanthium** (A. Gray) Kunth, Enum.
Pl. 4: 189. 1843.
T.: *S. angustifolium* (Pursh) Kunth
(*Veratrum angustifolium* Pursh).

(≡) *Anepsa* Rafinesque, Fl. Tell. 2: 31. 1837
("1836"); 4: 27. 1838.

962 **Schelhammera** R. Brown, Prodr. 273.
1810.
T.: *S. undulata* R. Brown (*typ. cons.*).

(H) *Schelhameria* Heister ex Fabricius,
Enum. 161. 1759 [Cruc.].
T.: non designatus.

967 **Tricyrtis** Wallich, Tent. Fl. Napal. 61,
t. 46. 1826.
T.: *T. pilosa* Wallich.

(=) *Composoa* D. Don, Prodr. Fl. Nepal. 50.
1825.
T.: *C. maculata* D. Don.

968 **Burchardia** R. Brown, Prodr. 272.
1810.
T.: *B. umbellata* R. Brown.

(H) *Burcardia* Duhamel, Traité Arbr. Arbust. 1: xxx, 111. 1755 [Verben.].
≡ *Callicarpa* Linnaeus 1753.

974 **Anguillaria** R. Brown. Prodr. 273.
1810.
T.: *A. dioica* R. Brown (*typ. cons.*).

(H) *Anguillaria* J. Gaertner, Fruct. Sem. Pl.
1: 372. 1788 [Myrsin.].
≡ *Heberdenia* Banks ex A. de Candolle
1841 (*nom. cons.*) (6288) (vide Rickett et
Stafleu, Taxon 8: 234. 1959).

975 **Iphigenia** Kunth, Enum. Pl. 4: 212.
1843.
T.: *I. indica* (Linnaeus) Kunth (*Melanthium indicum* Linnaeus).

(=) *Aphoma* Rafinesque, Fl. Tell. 2: 31. 1837
("1836").
T.: *A. angustiflora* Rafinesque.

982 **Paradisea** Mazzucato, Viaggio Bot.
Alp. Giulie 27. 1811.
T.: *P. hemeroanthericoides* Mazzuca-
to, nom. illeg. (*Hemerocallis liliastrum*
Linnaeus, *P. liliastrum* (Linnaeus)
Bertoloni).

(≡) *Liliastrum* Fabricius, Enum. 4. 1759.

985 **Bulbine** Wolf, Gen. Pl. Vocab. Char.
Def. 84. 1776.
T.: *B. frutescens* (Linnaeus) Willdenow (*Anthericum frutescens* (Linnaeus).

985a **Trachyandra** Kunth, Enum. **4**: 573. 1843.
T.: *T. hispida* (Linnaeus) Obermeyer (*Anthericum hispidum* Linnaeus).

(≡) *Obsitila* Rafinesque, Fl. Tell. **2**: 27. 1837 (vide Merrill, Ind. Rafinesq. 92. 1949).

(=) *Lepicaulon* Rafinesque, Fl. Tell. **2**: 27. 1837.
T.: *L. squameum* (Linnaeus f.) Rafinesque (*Anthericum squameum* Linnaeus f.).

987 **Simethis** Kunth, Enum. Pl. **4**: 618. 1843.
T.: *S. bicolor* Kunth, nom. illeg. (*Anthericum planifolium* Vandelli ex Linnaeus, *S. planifolia* (Vandelli ex Linnaeus) Grenier et Godron).

992 **Thysanotus** R. Brown, Prodr. 282. 1810.
T.: *T. junceus* R. Brown, nom. illeg. (*Chlamysporum juncifolium* R. A. Salisbury, *T. juncifolius* (R. A. Salisbury) Willis et Court).

(≡) *Chlamysporum* R. A. Salisbury, Parad. Lond. *t. 103.* 1808.

1006 **Schoenolirion** Torrey, J. Acad. Nat. Sci. Philadelphia ser. 2. **3**: 103. 1855.
T.: *S. croceum* (A. Michaux) A. Gray (*Phalangium croceum* A. Michaux).

(=) *Amblostima* Rafinesque, Fl. Tell. **2**: 26. 1837 ("1836").
T.: non designatus.

(=) *Oxytria* Rafinesque, Fl. Tell. **2**: 26. 1837 ("1836").
T.: *O. crocea* Rafinesque (*Phalangium croceum* Nuttall 1818, non A. Michaux 1803).

1007 **Chlorogalum** (Lindley) Kunth, Enum. Pl. **4**: 681. 1843.
T.: *C. pomeridianum* (A.-P. de Candolle) Kunth (*Scilla pomeridiana* A.-P. de Candolle) (*typ. cons.*).

(≡) *Laothoë* Rafinesque, Fl. Tell. **3**: 53. 1837 ("1836").

1011 **Bowiea** W. H. Harvey ex J. D. Hooker, Bot. Mag. **93**: *t. 5619.* 1867.
T.: *B. volubilis* W. H. Harvey ex J. D. Hooker.

(H) *Bowiea* Haworth, Philos. Mag. J. **64**: 299. 1824 [Lil.].
T.: *B. africana* Haworth.

1018 **Hosta** Trattinick, Arch. Gewächsk. **1**: 55, *t. 89.* 1812 (1814?).
T.: *H. japonica* Trattinick.

(H) *Hosta* N. J. Jacquin, Pl. Hort. Schoenbr. **1**: 60. 1797 [Verben.].
T.: non designatus.

1021 **Blandfordia** J. E. Smith, Exot. Bot. **1**: 5. *t. 4.* 1 Dec 1804.
T.: *B. nobilis* J. E. Smith.

(H) *Blandfordia* H. Andrews, Bot. Repos. **5**: *t. 343.* 9 Feb 1804 [Diapens.].
T.: *B. cordata* Andrews.

1024 **Kniphofia** Moench, Methodus 631. 1794.
T.: *K. alooides* Moench, nom. illeg. (*Aloë uvaria* Linnaeus, *K. uvaria* (Linnaeus) W. J. Hooker).

(H) *Kniphofia* Scopoli, Intr. 327. 1777 [Combret.].
T.: non designatus.

1029 **Haworthia** H. A. Duval, Pl. Succ. Horto Alencon. 7. 1809.
T.: *H. arachnoidea* (Linnaeus) H. A. Duval (*Aloë pumila* var. *arachnoidea* Linnaeus).

(=) *Catevala* Medikus, Theodora 67. 1786.
T.: non designatus.

1032 **Laxmannia** R. Brown, Prodr. 285. 1810.
T.: *L. gracilis* R. Brown (*typ. cons.*).

(H) *Laxmannia* J. R. Forster et G. Forster, Char. Gen. Pl. 47. 1775 [Comp.].
T.: *L. arborea* J. R. Forster et G. Forster.

1037 **Johnsonia** R. Brown, Prodr. 287. 1810.
T.: *J. lupulina* R. Brown.

(H) *Johnsonia* P. Miller, Gard. Dict. Abr. ed. 4. 1754 [Verben.].
T.: non designatus.

1044 **Baxteria** R. Brown ex W. J. Hooker, London J. Bot. 2: 492. 1843.
T.: *B. australis* R. Brown ex W. J. Hooker.

(H) *Baxtera* H. G. L. Reichenbach, Consp. 131. 1828 [Asclepiad.].
T.: *B. loniceroides* (W. J. Hooker) Steudel (*Harrisonia loniceroides* W. J. Hooker).

1046 **Agapanthus** L'Héritier, Sert. Angl. 17. 1789 ("1788").
T.: *A. umbellatus* L'Héritier, nom. illeg. (*Crinum africanum* Linnaeus, *A. africanus* (Linnaeus) Hoffmannsegg) (etiam vide 1047).

(≡) *Abumon* Adanson, Fam. Pl. 2: 54, 511-512. 1763.

1047 **Tulbaghia** Linnaeus, Mant. Pl. 2: 148, 223. 1771 ("*Tulbagia*") (*orth. cons.*).
T.: *T. capensis* Linnaeus.

(H) *Tulbaghia* Heister, Beschr. Neu. Geschl. 15. 1755 [Lil.].
≡ *Agapanthus* L'Héritier 1789 (*nom. cons.*) (1046).

1050 **Nothoscordum** Kunth, Enum. Pl. 4: 457. 1843.
T.: *N. striatum* Kunth, nom. illeg. (*Ornithogalum bivalve* Linnaeus, *N. bivalve* (Linnaeus) Britton) (*typ. cons.*).

1053 **Brodiaea** J. E. Smith, Trans. Linn. Soc. London 10: 2. 1810.
T.: *B. grandiflora* J. E. Smith, nom. illeg. (*Hookera coronaria* R. A. Salisbury, *B. coronaria* (R. A. Salisbury) Jepson) (*typ. cons.*) (etiam vide *Hookeria* J. E. Smith [Musci], *nom. cons.*).

1055 **Bessera** J. H. Schultes, Linnaea 4: 121. 1829.
T.: *B. elegans* J. H. Schultes.

(H) *Bessera* J. A. Schultes, Obs. Bot. 27. 1809 [Boragin.].
T.: *B. azurea* J. A. Schultes.

1077 **Lloydia** H. G. L. Reichenbach, Fl. G-erm. Excurs. 102. 1830.
T.: *L. serotina* (Linnaeus) H. G. L. Reichenbach (*Anthericum serotinum* Linnaeus).

1087 **Camassia** Lindley, Edward's Bot. Reg. 18: *t. 1486.* 1832.
T.: *C. esculenta* Lindley, nom. illeg. (*Phalangium quamash* Pursh, *C. quamash* (Pursh) Greene).

(=) *Cyanotris* Rafinesque, Amer. Monthly Mag. & Crit. Rev. 3: 356. 1818.
T.: *C. scilloides* Rafinesque.

1088 **Eucomis** L'Héritier, Sert. Angl. 17. 1789 ("1788").
T.: *E. regia* (Linnaeus) L'Héritier (*Fritillaria regia* Linnaeus) (*typ. cons.*).

(≡) *Basilaea* A. L. Jussieu ex Lamarck, Encycl. 1: 382. 1785 ("1783").

1093 **Bellevalia** Lapeyrouse, J. Phys. Chim. Hist. Nat. Arts 67: 425. 1808.
T.: *B. opercula* Lapeyrouse.

(H) *Bellevalia* Scopoli, Introd. 198. 1777. [Verben.].
T.: *Clerodendron speciosissimum* Van Geert.

1095a **Leopoldia** Parlatore, Fl. Palerm. 435. 1845.
T.: *L. comosa* (Linnaeus) Parlatore (*Hyacinthus comosus* Linnaeus).

(H) *Leopoldia* Herbert, Trans. Hort. Soc. London 4: 181. 1821 [Amaryllid.].
T.: non designatus.

1108 **Cordyline** Commerson ex R. Brown, Prodr. 280. 1810.
T.: *C. cannifolia* R. Brown (*typ. cons.*).

(H) *Cordyline* Adanson, Fam. Pl. 2: 54, 543. 1763 [Lil.].
≡ *Sansevieria* Thunberg 1794 (*nom. cons.*) (1110).
(=) *Taetsia* Medikus, Theodora 82. 1786.
T.: *T. ferrea* Medikus, nom. illeg. (*Dracaena ferrea* Linnaeus, nom. illeg., *Convallaria fruticosa* Linnaeus).

1110 **Sansevieria** Thunberg, Prodr. Pl. Cap. [xii], 65. 1794.
T.: *S. thyrsiflora* Thunberg, nom. illeg. (*Aloë hyacinthoides* Linnaeus, *S. hyacinthoides* (Linnaeus) Druce) (etiam vide 1108).

(≡) *Acyntha* Medikus, Theodora 76. 1786.
(=) *Sanseverinia* Petagna, Inst. Bot. 3: 643. 1787.
T.: *S. thyrsiflora* Petagna.

1111 **Astelia** Banks et Solander ex R. Brown, Prodr. 291. 1810.
T.: *A. alpina* R. Brown.

(=) *Funckia* Willdenow, Ges. Naturf. Freunde Berlin Mag. 2: 19. 1808.
T.: *F. magellanica* Willdenow, nom. illeg. (*Melanthium pumilum* G. Forster).

1112 **Milligania** J. D. Hooker, Hooker's J. Bot. Kew Gard. Misc. 5: 296. 1853.
T.: *M. longifolia* J. D. Hooker (*typ. cons.*).

(H) *Milligania* J. D. Hooker, Icon. Pl. 3: *t.299.* 1840 [Gunner.].
T.: *M. cordifolia* J. D. Hooker.

1118 **Smilacina** Desfontaines, Ann. Mus. Natl. Hist. Nat. **9**: 51. 1807.
T.: *S. stellata* (Linnaeus) Desfontaines (*Convallaria stellata* Linnaeus) (*typ. cons.*).

(=) *Vagnera* Adanson, Fam. Pl. **2**: 496. 1763.
T.: non designatus ["*Polygonatum*. Corn. t. 33. 37. Mor. s. 13. *t. 4*. f. 7. 9"].

(=) *Polygonastrum* Moench, Methodus 637. 1794.
T.: *P. racemosum* (Linnaeus) Moench (*Convallaria racemosa* Linnaeus).

1119 **Maianthemum** Wiggers, Prim. Fl. Holsat. 14. 1780.
T.: *M. convallaria* Wiggers, nom. illeg. (*Convallaria bifolia* Linnaeus, *M. bifolium* (Linnaeus) F. W. Schmidt).

(≡) *Unifolium* Ludwig, Inst. Regn. Veg. ed. 2. 124. 1757.

1129 **Reineckea** Kunth, Abh. Königl. Akad. Wiss. Berlin **1842**: 29. 1844.
T.: *R. carnea* (Andrews) Kunth (*Sansevieria carnea* Andrews).

1140 **Ophiopogon** Ker-Gawler, Bot. Mag. **27**: t. *1063*. 1807.
T.: *O. japonicus* (Linnaeus f.) Ker-Gawler (*Convallaria japonica* Linnaeus f.).

(=) *Mondo* Adanson, Fam. Pl. **2**: 496, 578. 1763.
T.: non designatus ["Kaempf. Amoen. t. 824"].

1146 **Luzuriaga** Ruiz et Pavón, Fl. Peruv. Chil. **3**: 65. 1802.
T.: *L. radicans* Ruiz et Pavón.

(=) *Enargea* Banks et Solander ex J. Gaertner, Fruct. Sem. Pl. **1**: 283. 1788.
T.: *E. marginata* J. Gaertner.

(=) *Callixene* Commerson ex A. L. Jussieu, Gen. Pl. 41. 1789.
T.: non designatus.

HAEMODORACEAE

1161 **Lachnanthes** S. Elliott, Sketch Bot. S. Carolina **1**: 47. 1816 ("1821").
T.: *L. tinctoria* Elliott.

AMARYLLIDACEAE

1166 **Hessea** Herbert, Amaryllidaceae 289. 1837.
T.: *H. stellaris* (N. J. Jacquin) Herbert (*Amaryllis stellaris* N. J. Jacquin).

(H) *Hessea* P. J. Bergius ex Schlechtendal, Linnaea **1**: 252. 1826 [Amaryllid.].
≡ *Carpolyza* R. A. Salisbury 1806.

1175 **Nerine** Herbert, Bot. Mag. **47**: t. *2124*. 1820.
T.: *N. sarniensis* (Linnaeus) Herbert (*Amaryllis sarniensis* Linnaeus) (*typ. cons.*).

(≡) *Imhofia* Heister, Beschr. Neu. Geschl. 29. 1755.

*1176 **Amaryllis** Linnaeus, Sp. Pl. 292. 1753.
T.: *A. belladonna* Linnaeus (typus in
Herb. Cliffort., **BM**) (*typ. cons.*).

1178 **Vallota** R. A. Salisbury ex Herbert,
Appendix [to Bot. Reg. 7] 29. 1821.
T.: *V. purpurea* Herbert, nom. illeg.
(*Crinum speciosum* Linnaeus f., *V.
speciosa* (Linnaeus f.) Voss).

(H) *Valota* Adanson, Fam. Pl. **2**: 495. 1763
[Gram.].
T.: *V. insularis* (Linnaeus) Chase (*An-
dropogon insularis* Linnaeus).

1181 **Zephyranthes** Herbert, Appendix [to
Bot. Reg. 7] 36. 1821.
T.: *Z. atamasca* (Linnaeus) Herbert
(*Amaryllis atamasca* Linnaeus) (*typ.
cons.*).

(≡) *Atamosco* Adanson, Fam. Pl. **2**: 57, 522.
1763.

*1196 **Eucharis** Planchon et Linden in Linden, Cat. Pl. Exot. **8**: 3. 1852-1853.
T.: *E. candida* Planchon et Linden.

(=) *Caliphruria* Herbert, Edward's Bot. Reg.
30(Misc.): 87. 1844.
T.: *C. hartwegiana* Herbert.

1208 **Hippeastrum** Herbert, Appendix [to
Bot. Reg. 7] 31. 1821.
T.: *H. reginae* (Linnaeus) Herbert
(*Amaryllis reginae* Linnaeus) (*typ.
cons.*).

(=) *Leopoldia* Herbert, Trans, Hort. Soc.
London **4**: 181. 1821 [Amaryllid.].
T.: non designatus.

1211 **Urceolina** H. G. L. Reichenbach,
Consp. 61. 1828.
T.: *Urceolaria pendula* Herbert, nom.
illeg. (*Crinum urceolatum* Ruiz et Pa-
vón, *U. urceolata* (Ruiz et Pavón) M.
L. Green).

(=) *Leperiza* Herbert, Appendix [to Bot.
Reg. 7] 41. 1821.
T.: *L. latifolia* (Ruiz et Pavón) Herbert
(*Pancratium latifolium* Ruiz et Pavón).

1236 **Lanaria** W. Aiton, Hortus Kew. **1**: 462.
1789, post 7 Aug.
T.: *L. plumosa* W. Aiton, nom. illeg.
(*Hyacinthus lanatus* Linnaeus, *L. lanata* (Linnaeus) Druce).

(H) *Lanaria* Adanson, Fam. Pl. **2**: 255, 568.
1763 [Caryophyll.].
≡ *Gypsophila* Linnaeus 1753.
(=) *Argolasia* A. L. Jussieu, Gen. 60. 4 Aug
1789.
T.: non designatus.

TACCACEAE

1248 **Tacca** J. R. Forster et G. Forster, Char.
Gen. Pl. 35. 1775.
T.: *T. pinnatifida* J. R. Forster et G.
Forster.

(=) *Leontopetaloides* Boehmer in Ludwig,
Defin. Gen. Pl. ed. 3. 512. 1760.
T.: *Leontice leontopetaloides* Linnaeus.

178

DIOSCOREACEAE

1258 **Petermannia** F. Mueller, Fragm. 2: 92. 1860.
T.: *P. cirrosa* F. Mueller.

(H) *Petermannia* H. G. L. Reichenbach, Deut. Bot. Herb.-Buch [1]: 236. 1841 [Chenopod.].
≡ *Cycloloma* Moquin-Tandon 1840.

IRIDACEAE

1260 **Syringodea** J. D. Hooker, Bot. Mag. 99: t. 6072. 1873.
T.: *S. pulchella* J. D. Hooker.

(H) *Syringodea* D. Don, Edinburgh New Philos. J. 17: 155. 1834 [Eric.].
T.: *S. vestita* (Thunberg) D. Don (*Erica vestita* Thunberg).

1261 **Romulea** Maratti, Pl. Romul. Saturn. 13. 1772.
T.: *R. bulbocodium* (Linnaeus) Sebastiani et Mauri (*Crocus bulbocodium* Linnaeus) (*typ. cons.*).

(≡) *Ilmu* Adanson, Fam. Pl. 2: 497. 1763.

1265 **Moraea** P. Miller, Fig. Pl. Gard. Dict. 159, t. 238. 1758 ("*Morea*") (*orth. cons.*).
T.: *M. vegeta* Linnaeus (*typ. cons.*).

1265a **Dietes** R. A. Salisbury ex Klatt, Linnaea 34: 583. 1866.
T.: *D. compressa* (Linnaeus f.) Klatt (*Iris compressa* Linnaeus f.) [= *D. iridioides* (Linnaeus) Klatt, *Moraea iridioides* Linnaeus].

(=) *Naron* Medikus, Hist. & Commentat. Acad. Elect. Sci. Theod.-Palat. 6 (Phys.): 419. 1790.
T.: *N. orientale* Medikus, nom. illeg. (*Moraea iridioides* Linnaeus).

1283 **Libertia** K. Sprengel, Syst. Veg. 1: 127. 1824.
T.: *L. ixioides* (G. Forster) K. Sprengel (*Sisyrinchium ixioides* G. Forster).

(H) *Libertia* Dumortier, Comment. Bot. 9. 1822 [Lil.].
T.: *L. recta* Dumortier, nom illeg. (*Hemerocallis caerulea* Andrews).

(=) *Tekel* Adanson, Fam. Pl. 2: 497. 1763.
T.: non designatus ["Feuillé t. 4"].

1284 **Bobartia** Linnaeus, Sp. Pl. 54. 1753.
T.: *B. indica* Linnaeus (typus in Herb. Hermann vol. 4, fol. 80 [top left-hand specimen]) (*typ. cons.*).

1285 **Belamcanda** Adanson, Fam. Pl. 2: 60 ("*Belam-Canda*"), 524 ("*Belamkanda*"). 1763 (*orth. cons.*).
T.: *B. chinensis* (Linnaeus) A.-P. de Candolle (in Redouté, Lil. 3: t. 121. 1805) (*Ixia chinensis* Linnaeus) (*typ. cons.*).

1289 **Patersonia** R. Brown, Bot. Mag. 26: t. 1041. 1807.
T.: *P. sericea* R. Brown.

(=) *Genosiris* Labillardière, Nov. Holl. Pl. 1: 13. 1804.
T.: *G. fragilis* Labillardière.

1292 **Eleutherine** Herbert, Edward's Bot.
Reg. **29**: *t. 57*. 1843.
T.: *Marica plicata* Ker-Gawler, nom.
illeg. (*Moraea plicata* Swartz, nom. illeg., *Sisyrinchium latifolium* Swartz)
[= *E. bulbosa* (P. Miller) Urban (*Sisyrinchium bulbosum* P. Miller)].

1302 **Ixia** Linnaeus, Sp. Pl. ed. 2. 51. 1762.
T.: *I. polystachya* Linnaeus ("*polysta-
chia*") (*typ. cons.*).

(H) *Ixia* Linnaeus, Sp. Pl. 36. 1753 [Irid.].
LT.: *I. africana* Linnaeus.

1310 **Babiana** Ker-Gawler ex Sims, Bot.
Mag. **15**: sub *t. 539*. 1801.
T.: *B. plicata* Ker-Gawler (Bot. Mag.
16: *t. 576*. 1802), nom. illeg. (*Gladiolus
fragrans* N. J. Jacquin, *B. fragrans* (N.
J. Jacquin) Ecklon).

(=) *Beverna* Adanson, Fam. Pl. **2**: (20). 1763.
T.: non designatus ["Iris du Cap. B. Esp.
a fleurs bleues"].

1313 **Micranthus** (Persoon) Ecklon, Topogr.
Verz. Pflanzensamml. Ecklon 43. 1827.
T.: *M. alopecuroides* (Linnaeus) Ecklon (*Gladiolus alopecuroides* Linnaeus) (*typ. cons.*).

(H) *Micranthus* J. C. Wendland, Bot. Beob.
38. 1798 [Acanth.].
≡ *Phaulopsis* Willdenow 1800 (*nom.
cons.*) (7932).

1315 **Watsonia** P. Miller, Fig. Pl. Gard. Dict.
184, *t. 276*. 1758.
T.: *W. meriana* (Linnaeus) P. Miller
(Gard. Dict. ed. 8. (1768) (*Antholyza
meriana* Linnaeus) (etiam vide 5692).

1316 **Freesia** Klatt, Linnaea **34**: 672. 1866.
T.: *F. refracta* (Jacquin) Klatt (*Gladiolus refractus* Jacquin) (*typ. cons.*).

MUSACEAE

1321 **Heliconia** Linnaeus, Mant. Pl. **2**: 147,
211. 1771.
T.: *H. bihai* (Linnaeus) Linnaeus
(*Musa bihai* Linnaeus).

(≡) *Bihai* P. Miller, Gard. Dict. Abr. ed. 4.
1754.

ZINGIBERACEAE

1324 **Zingiber** Boehmer in Ludwig, Defin.
Gen. Pl. ed. 3. 89. 1760.
T.: *Z. officinale* Roscoe (Trans. Linn.
Soc. London **8**: 358. 1807) (*Amomum
zingiber* Linnaeus) (etiam vide 1344).

(≡) *Zinziber* P. Miller, Gard. Dict. Abr. ed.
4. 1754.

1328 **Alpinia** Roxburgh, Asiat. Res. **11**: 350. 1810.
T.: *A. galanga* (Linnaeus) Willdenow (Sp. Pl. 1: 12. 1797) (*Maranta galanga* Linnaeus) (*typ. cons.*).

(H) *Alpinia* Linnaeus, Sp. Pl. 2. 1753 [Zingiber.].
T.: *A. racemosa* Linnaeus.

(=) *Albina* Giseke, Prael. Ord. Nat. Pl. 207, 227, 248. 1792.
T.: non designatus.

(=) *Buekia* Giseke, Prael. Ord. Nat. Pl. 204, 216, 239. 1792.
T.: *B. malaccensis* (Koenig) Raeuschel (Nom. ed. 3. 1. 1797) (*Costus malaccensis* Koenig).

(=) *Zerumbet* Wendland, Sert. Hann. **4**: 3. 1798.
T.: *Z. speciosum* Wendland.

1331 **Renealmia** Linnaeus f., Suppl. Pl. 7, 79. 1782.
T.: *R. exaltata* Linnaeus f.

(H) *Renealmia* Linnaeus, Sp. Pl. 286. 1753 [Bromel.].
LT.: *R. paniculata* Linnaeus.

1332 **Riedelia** Oliver, Hooker's Icon. Pl. **15**: 1883.
T.: *R. curviflora* Oliver.

(H) *Riedelia* Chamisso, Linnaea **7**: 240 ("224"). 1832 [Verben.].
T.: *R. lippioides* Chamisso.

(=) *Nyctophylax* Zippelius, Alg. Konst- Lett.-Bode **1829** (1): 298. 1829.
T.: *N. alba* Zippelius.

1337a **Nicolaia** Horaninow, Prodr. Monogr. Scitam. 32. 1862.
T.: *N. imperialis* Horaninow [= *N. elatior* (Jack) Horaninow] (*typ. cons.*).

(=) *Diracodes* Blume, Enum. Pl. Javae **1**: 55. 1827.
T.: *D. javanica* Blume.

1344 **Amomum** Roxburgh, Fl. Ind. **1**: 317. 1820.
T.: *A. subulatum* Roxburgh (Pl. Corom. **3**: 75. 1819) (*typ. cons.*).

(H) *Amomum* Linnaeus, Sp. Pl. 1. 1753 [Zingiber.].
≡ *Zingiber* Boehmer 1760 (*nom. cons.*) (1324) (vide Taxon **17**: 730. 1968).

(=) *Etlingera* Giseke, Prael. Ord. Nat. Pl. 209. 1792.
T.: *E. littoralis* (Koenig) Raeuschel (Nomencl. Bot. ed. 3. 1. 1797) (*Amomum littorale* Koenig).

(=) *Meistera* Giseke, Prael. Ord. Nat. Pl. 205. 1792.
T.: *Amomum koenigii* J. F. Gmelin.

(=) *Paludana* Giseke, Prael. Ord. Nat. Pl. 207. 1792.
T.: *Amomum globba* J. F. Gmelin.

(=) *Wurfbainia* Giseke, Prael. Ord. Nat. Pl. 206. 1792.
T.: *W. uliginosa* (Koenig) Giseke (*Amomum uliginosum* Koenig).

1351 **Curcuma** Linnaeus, Sp. Pl. 2. 1753.
T.: *C. longa* Linnaeus (*typ. cons.*).

1360 **Tapeinochilos** Miquel, Ann. Mus.
Lugd.-Bat. **4**: 101. 1869 ("1868") ("*Ta-
peinocheilos*") (*orth. cons.*).
T.: *T. pungens* (Teysmann et Binnendijk) Miquel (*Costus pungens* Teysmann et Binnendijk).

MARANTACEAE

1368 **Phrynium** Willdenow, Sp. Pl. **1**: 17.
1797.
T.: *P. capitatum* Willdenow, nom. illeg. (*Pontederia ovata* Linnaeus, *P.
rheedei* Suresh et Nicolson).

(=) *Phyllodes* Loureiro, Fl. Cochinch. 13.
1790.
T.: *P. placentaria* Loureiro.

BURMANNIACEAE

1386 **Arachnitis** R. Philippi, Bot. Zeitung
(Berlin) **22**: 217. 1864.
T.: *A. uniflora* R. Philippi.

(H) *Arachnites* F. W. Schmidt, Fl. Boëm. **1**:
74. 1793 [Orchid.].
T.: non designatus.

ORCHIDACEAE

1393a **Paphiopedilum** Pfitzer, Morph. Stud.
Orch. 11. 1886.
T.: *P. insigne* (Wallich ex Lindley)
Pfitzer (*Cypripedium insigne* Wallich
ex Lindley) (*typ. cons.*).

(≡) *Cordula* Rafinesque, Fl. Tell. **4**: 46. 1838
("1836").

(=) *Stimegas* Rafinesque, Fl. Tell. **4**: 45. 1838
("1836").
T.: *S. venustum* (Wallich ex Sims) Rafinesque (*Cypripedium venustum* Wallich
ex Sims).

1393b **Phragmipedium** Rolfe, Orchid Rev. **4**:
330. 1896.
T.: *P. caudatum* (Lindley) Rolfe (*Cypripedium caudatum* Lindley).

(=) *Uropedium* Lindley, Orchid. Linden. 28.
1846.
T.: *U. lindenii* Lindley.

1397 **Serapias** Linnaeus, Sp. Pl. 949. 1753.
T.: *S. lingua* Linnaeus (*typ. cons.*)
(etiam vide Swartz, Kongl. Vetensk.
Acad. Nya Handl. **21**: 224. 1800).

1399 **Himantoglossum** Koch, Syn. Fl. Germ.
Helv. 689. 1837 ("*Himanthoglossum*")
(*orth. cons.*).
T.: *H. hircinum* (Linnaeus) Koch (*Satyrium hircinum* Linnaeus).

(H) *Himantoglossum* K. Sprengel, Syst. Veg.
3: 675, 694. 1826 [Orchid.].
≡ *Aceras* R. Brown 1813.

1403a **Peristylus** Blume, Bijdr. 404. Jun-Dec 1825.
T.: *P. grandis* Blume.

(=) *Glossula* Lindley, Bot. Reg. **10**: *t. 862.* Feb 1825.
T.: *G. tentaculata* Lindley.

1408 **Holothrix** L. C. Richard ex Lindley, Gen. Sp. Orchid. Pl. 257. Aug 1835; 283. Sep 1835.
T.: *H. parvifolia* Lindley, nom. illeg. (*Orchis hispidula* Linnaeus f., *H. hispidula* (Linnaeus f.) Durand et Schinz) (*typ. cons.*).

(=) *Monotris* Lindley, Edward's Bot. Reg. **20**: sub *t. 1701.* 1834 ("1835").
T.: *M. secunda* Lindley.

(=) *Scopularia* Lindley, Edward's Bot. Reg. **20**: sub *t. 1701.* 1834 ("1835").
T.: *S. burchellii* Lindley.

(=) *Saccidium* Lindley, Gen. Sp. Orchid. Pl. 258. Aug 1835; 301. Oct 1835.
T.: *S. pilosum* Lindley.

(=) *Tryphia* Lindley, Gen. Sp. Orchid. Pl. 258. Aug 1835; 333. Oct. 1835.
T.: *T. secunda* (Thunberg) Lindley (*Orchis secunda* Thunberg).

1410 **Platanthera** L. C. Richard, Orch. Eur. Annot. 20. 26, 35. 1817; Mém. Mus. Hist. Nat. **4**: 42, 48, 57. 1818.
T.: *P. bifolia* (Linnaeus) L. C. Richard (*Orchis bifolia* Linnaeus).

1430 **Satyrium** Swartz, Kongl. Vetensk. Acad. Nya Handl. **21**: 214. 1800.
T.: *S. bicorne* (Linnaeus) Thunberg (Prodr. Pl. Cap. 6. 1794) (*Orchis bicornis* Linnaeus) (*typ. cons.*).

(H) *Satyrium* Linnaeus, Sp. Pl. 944. 1753 [Orchid.].
LT.: *S. viride* Linnaeus (vide M. L. Green, Prop. Brit. Bot. 185. 1929).

1449 **Pterostylis** R. Brown, Prodr. 326. Apr 1810.
T.: *P. curta* R. Brown (*typ. cons.*).

(=) *Diplodium* Swartz, Ges. Naturf. Freunde Berlin Mag. **4**: 84. 1810.
T.: *Disperis alata* Labillardière (Nov. Holl. Pl. **2**: 59. 1806).

1468 **Nervilia** Commerson ex Gaudichaud, Voy. Uranie 421. 1829.
T.: *N. aragoana* Gaudichaud (*typ. cons.*).

(=) *Stellorkis* Du Petit-Thouars, Nouv. Bull. Sci. Soc. Philom. Paris **1**: 317. 1809.
T.: *Arethusa simplex* Linnaeus (vide Du Petit-Thouars, Hist. Pl. Orch. *tabl. t. 24*).

1482 **Epipactis** Zinn, Catal. 85. 1757.
T.: *E. helleborine* (Linnaeus) Crantz (Stirp. Austr. ed. 2. 467. 1769) (*Serapias helleborine* Linnaeus) (*typ. cons.*).

(H) *Epipactis* Séguier, Pl. Veron. **3**: 253. 1754 [Orchid.].
T.: *Satyrium repens* Linnaeus.

(≡) *Helleborine* P. Miller, Gard. Dict. Abr. ed. 4. 1754.

1483 **Limodorum** Boehmer in Ludwig, Defin. Gen. Pl. ed. 3. 358. 1760.
T.: *L. abortivum* (Linnaeus) Swartz (Nova Acta Soc. Sci. Upsal. **6**: 80. 1799) (*Orchis abortiva* Linnaeus).

(H) *Limodorum* Linnaeus, Sp. Pl. 950. 1753 [Orchid.].
≡ *Calopogon* R. Brown 1813 (*nom. cons.*) (1534).

1488 **Pelexia** Poiteau ex Lindley, Bot. Reg. **12**: *t. 985.* 1826.
T.: *P. spiranthoides* Lindley, nom. illeg. (*Satyrium adnatum* Swartz, *P. adnata* (Swartz) K. Sprengel).

(≡) *Collea* Lindley, Bot. Reg. **9**: sub *t. 760.* 1823.

1490 **Spiranthes** L. C. Richard, Orch. Eur.
Annot. 20, 28, 36. 1817; Mém. Mus.
Hist. Nat. 4: 42, 50, 58. 1818.
T.: *S. autumnalis* L. C. Richard, nom.
illeg. (*Ophrys spiralis* Linnaeus, *S. spiralis* (Linnaeus) Chevalier) (*typ. cons.*).

(≡) *Orchiastrum* Séguier, Pl. Veron. **3**: 252. 1754.

1494 **Listera** R. Brown in W. T. Aiton, Hortus Kew. ed. 2. **5**: 201. 1813.
T.: *L. ovata* (Linnaeus) R. Brown (*Ophrys ovata* Linnaeus) (*typ. cons.*).

(H) *Listera* Adanson, Fam. Pl. **2**: 321, 571. 1763 [Legum.].
T.: non designatus.

(=) *Diphryllum* Rafinesque, Med. Repos. ser. 2. **5**: 357. 1808.
T.: *D. bifolium* Rafinesque.

1495 **Neottia** Guettard, Hist. Acad. Roy. Sci. Mém. Math. Phys. (Paris, 4°) **1750**: 374. 1754.
T.: *N. nidus-avis* (Linnaeus) L. C. Richard (*Ophrys nidus-avis* Linnaeus).

1500 **Anoectochilus** Blume, Bijdr. 411. 1825 ("*Anecochilus*") (*orth. cons.*).
T.: *A. setaceus* Blume.

1502 **Zeuxine** Lindley, Collect. Bot. Append. n. 18. 1826 (?1825); Orchid. Scelet. 9. 1826 ("*Zeuxina*") (*orth. cons.*).
T.: *Z. sulcata* (Roxburgh) Lindley (Gen. Sp. Orchid. Pl. 485. 1840) (*Pterygodium sulcatum* Roxburgh).

1516 **Platylepis** A. Richard, Mém. Soc. Hist. Nat. Paris **4**: 34. 1828.
T.: *P. goodyeroides* A. Richard, nom. illeg. (*Goodiera occulta* Du Petit-Thouars, *Platylepis occulta* (Du Petit-Thouars) H. G. Reichenbach).

(=) *Erporkis* Du Petit-Thouars, Nouv. Bull. Sci. Soc. Philom. Paris **1**: 317. 1809.
T.: non designatus.

1533 **Bletilla** H. G. Reichenbach, Fl. Serres **8**: 246. 1852-53.
T.: *B. gebinae* (Lindley) H. G. Reichenbach (*Bletia gebinae* Lindley) (*typ. cons.*).

(=) *Jimensia* Rafinesque, Fl. Tell. **4**: 38. 1838.
T.: *J. nervosa* Rafinesque, nom. illeg. (*Limodorum striatum* Thunberg, *J. striata* (Thunberg) Garay et R. E. Schultes).

1534 **Calopogon** R. Brown in W. T. Aiton, Hortus Kew. ed. 2. **5**: 204. 1813.
T.: *C. pulchellus* R. Brown, nom. illeg. (*Limodorum tuberosum* Linnaeus, *C. tuberosus* (Linnaeus) Britton, Sterns et Poggenburg) (etiam vide 1483).

184

1553 **Microstylis** (Nuttall) A. Eaton, Man. Bot. ed. 3. 115 ("*Microstylus*"), 347, 353. 1822.
T.: *M. ophioglossoides* A. Eaton, nom. illeg. (*Malaxis unifolia* A. Michaux, *Microstylis unifolia* (A. Michaux) Britton, Sterns et Poggenburg).

(≡) *Achroanthes* Rafinesque, Amer. Monthly Mag. & Crit. Rev. **4**: 195. 1819.

1556 **Liparis** L. C. Richard, Orch. Eur. Annot. 21, 30, 38. 1817; Mém. Mus. Hist. Nat. **4**: 43, 52, 60. 1818.
T.: *L. loeselii* (Linnaeus) L. C. Richard (*Ophrys loeselii* Linnaeus).

(=) *Leptorkis* Du Petit-Thouars, Nouv. Bull. Sci. Soc. Philom. Paris **1**: 317. 1809.
T.: non designatus.

1558 **Oberonia** Lindley, Gen. Sp. Orchid. Pl. 15. 1830.
T.: *O. iridifolia* Lindley, nom. illeg. (*Malaxis ensiformis* J. E. Smith, *O. ensiformis* (J. E. Smith) Lindley) (*typ. cons.*).

(=) *Iridorkis* Du Petit-Thouars, Nouv. Bull. Sci. Soc. Philom. Paris **1**: 319. 1809.
T.: non designatus.

1559 **Calypso** R. A. Salisbury, Parad. Lond. *t. 89.* 1807.
T.: *C. borealis* R. A. Salisbury, nom. illeg. (*Cypripedium bulbosum* Linnaeus, *Calypso bulbosa* (Linnaeus) Oakes).

(H) *Calypso* Du Petit-Thouars, Hist. Vég. Iles France 29, *t. 6.* 1804 [Hippocrat.].
T.: *Epidendrum distichum* Lamarck.

1565 **Polystachya** W. J. Hooker, Exot. Fl. *t. 103.* 1824.
T.: *Epidendrum minutum* Aublet [= *Polystachya extinctoria* H. G. Reichenbach].

(=) *Dendrorkis* Du Petit-Thouars, Nouv. Bull. Sci. Soc. Philom. Paris **1**: 318, 1809.
T.: non designatus.

1569 **Claderia** J. D. Hooker, Fl. Brit. India **5**: 810. 1890.
T.: *C. viridiflora* J. D. Hooker.

(H) *Claderia* Rafinesque, Sylva Tell. 12. 1838 [Rut.].
T.: *C. parviflora* Rafinesque.

1587 **Stelis** Swartz, J. Bot. (Schrader) **1799** (4): 239. 1800.
T.: *S. ophioglossoides* (N. J. Jacquin) Swartz (*Epidendrum ophioglossoides* N. J. Jacquin) (*typ. cons.*).

1614 **Epidendrum** Linnaeus, Sp. Pl. ed. 2. 1347. 1763.
T.: *E. nocturnum* N. J. Jacquin (Enum. Pl. Carib. 29. 1760) (*typ. cons.*).

(H) *Epidendrum* Linnaeus, Sp. Pl. 952. 1753 [Orchid.].
LT.: *E. nodosum* Linnaeus (vide M. L. Green, Prop. Brit. Bot. 186. 1929).

1617 **Laelia** Lindley, Gen. Sp. Orchid. Pl. 96, 115. 1831.
T.: *L. grandiflora* (La Llave et Lexarza) Lindley (*Bletia grandiflora* La Llave et Lexarza) (*typ. cons.*).

(H) *Laelia* Adanson, Fam. Pl. **2**: 423. 1763 [Cruc.].
T.: *Bunias orientalis* Linnaeus.

1619 **Brassavola** R. Brown in W. T. Aiton, Hortus Kew. ed. 2. 5: 216. 1813.
T.: *B. cucullata* (Linnaeus) R. Brown (*Epidendrum cucullatum* Linnaeus).

(H) *Brassavola* Adanson, Fam. Pl. 2: 127, 527. 1763 [Comp.].
≡ *Helenium* Linnaeus 1753.

1631 **Calanthe** Ker-Gawler, Bot. Reg. 7: sub t. 573. 1821.
T.: *C. veratrifolia* R. Brown, nom. illeg. (*Orchis triplicata* Willemet, *C. triplicata* (Willemet) Ames).

(=) *Alismorkis* Du Petit-Thouars, Nouv. Bull. Sci. Soc. Philom. Paris 1: 318. 1809.
T.: non designatus.

1648 **Eulophia** R. Brown ex Lindley, Bot. Reg. 8: t. 686. 1823.
T.: *E. guineensis* Lindley.

(H) *Eulophia* C. A. Agardh, Aphor. Bot. 109. 1822 [Musci].
T.: *Leskea cristata* Hedwig.
(=) *Graphorkis* Du Petit-Thouars. Nouv. Bull. Sci. Soc. Philom. Paris 1: 318. 1809 (*nom. cons.*) (1648a).
(=) *Eulophus* R. Brown, Bot. Reg. 7: sub t. 573. 1821.
T.: *Limodorum virens* Roxburgh.
(=) *Lissochilus* R. Brown, Bot. Reg. 7: sub t. 573. 1821.
T.: *L. speciosus* R. Brown.

1648a **Graphorkis** Du Petit-Thouars, Nouv. Bull. Sci. Soc. Philom. Paris 1: 318. 1809.
T.: [Réunion or Madagascar] Thouars s.n. (**P**) [= *Graphorkis concolor* (Thouars) O. Kuntze (*Limodorum concolor* Thouars)] (*typ. cons.*) (etiam vide 1648).

1694 **Dendrobium** Swartz, Nova Acta Regiae Soc. Sci. Upsal. ser. 2. 6: 82. 1799.
T.: *D. moniliforme* (Linnaeus) Swartz (*Epidendrum moniliforme* Linnaeus) (*typ. cons.*).

(=) *Callista* Loureiro, Fl. Cochinch. 519. 1790.
T.: *C. amabilis* Loureiro.
(=) *Ceraia* Loureiro, Fl. Cochinch. 518. 1790.
T.: *C. simplicissima* Loureiro.

1697 **Eria** Lindley, Bot. Reg. 11: t. 904. 1825.
T.: *E. stellata* Lindley.

1704 **Cirrhopetalum** Lindley, Gen. Sp. Orchid. Pl. 45, 58. 1830.
T.: *C. thouarsii* Lindley, nom. illeg. (*Epidendrum umbellatum* G. Forster, *C. umbellatum* (G. Forster) Frappier ex Cordemoy).

(=) *Ephippium* Blume, Bijdr. 308. 1825.
T.: non designatus.
(=) *Zygoglossum* Reinwardt, Syll. Pl. 2: 4. 1825-1826 ("1828").
T.: *Z. umbellatum* Reinwardt.

1705 **Bulbophyllum** Du Petit-Thouars, Hist. Orchid. t. esp. 3. sub u. 1822.
T.: *B. nutans* Du Petit-Thouars (*typ. cons.*).

(≡) *Phyllorkis* Du Petit-Thouars, Nouv. Bull. Sci. Soc. Philom. 1: 319. 1809.

1714 **Panisea** (Lindley) Lindley, Fol. Orchid. 5. 1854.
T.: *P. parviflora* (Lindley) Lindley (*Coelogyne parviflora* Lindley).

(≡) *Androgyne* Griffith, Not. Pl. Asiat. 3: 279. 1851.

1739 **Warmingia** H. G. Reichenbach, Otia Bot. Hamb. 87. 1881.
T.: *W. eugenii* H. G. Reichenbach.

(H) *Warmingia* Engler in C. F. P. Martius, Fl. Bras. **12**(2): 281. 1 Sep 1874 [Anacard.].
T.: *W. pauciflora* Engler.

(H) *Warmingia* Engler in C. F. P. Martius, Fl. Bras. **12**(2): 86, 92. 1 Sep 1874 [Rut.].
≡ *Ticorea* Aublet 1775.

1751 **Brachtia** H. G. Reichenbach, Linnaea **22**: 853. 1849.
T.: *B. glumacea* H. G. Reichenbach.

(H) *Brachtia* Trevisan, Saggio Monogr. Alghe Cocc. 57. 1848 [Chloroph.].
T.: *B. crassa* (Naccari) Trevisan (*Palmella crassa* Naccari).

1778 **Miltonia** Lindley, Edward's Bot. Reg. **23**: sub *t. 1976.* 1837.
T.: *M. spectabilis* Lindley.

1779 **Oncidium** Swartz, Kongl. Vetensk. Acad. Nya Handl. **21**: 239. 1800.
T.: *O. altissimum* (Jacquin) Swartz (*Epidendrum altissimum* Jacquin) (*typ. cons.*).

1822 **Saccolabium** Blume, Bijdr. 292. 1825 sem. 2.
T.: *S. pusillum* Blume.

(=) *Gastrochilus* D. Don, Prodr. Fl. Nepal. 32. Feb 1825.
T.: *G. calceolaris* D. Don.

1824 **Acampe** Lindley, Fol. Orchid. 4. 1853.
T.: *A. multiflora* (Lindley) Lindley (*Vanda multiflora* Lindley).

(=) *Sarcanthus* Lindley, Bot. Reg. **10**: sub *t. 817.* 1824.
T.: *Epidendrum praemorsum* Roxburgh.

1834 **Oeonia** Lindley, Bot. Reg. **10**: sub *t. 817.* 1824 ("*Aeonia*") (*orth. cons.*).
T.: *O. aubertii* Lindley, nom. illeg. (*Epidendrum volucre* Du Petit-Thouars, *O. volucris* (Du Petit-Thouars) Durand et Schinz).

1834a **Symphyglossum** Schlechter, Orchis **13**: 8. 1919.
T.: *S. sanguineum* (H. G. Reichenbach) Schlechter (*Mesospinidium sanguineum* H. G. Reichenbach).

(H) *Symphyoglossum* Turczaninow, Bull. Soc. Imp. Naturalistes Moscou **21**(1): 255. 1848 [Asclepiad.].
T.: *S. hastatum* (Bunge) Turczaninow (*Asclepias hastata* Bunge).

SAURURACEAE

1857 **Houttuynia** Thunberg, Kongl. Vetensk. Acad. Nya Handl. **4**: 149, 151. 1783 ("*Houtuynia*") (*orth. cons.*).
T.: *H. cordata* Thunberg.

(H) *Houttuynia* Houttuyn, Nat. Hist. 2(12): 448. 1780 [Irid.].
T.: *H. capensis* Houttuyn.

JUGLANDACEAE

1882 **Carya** Nuttall, Gen. N. Amer. Pl. 2: 220. 1818.
T.: *C. tomentosa* (Poiret) Nuttall (*Juglans tomentosa* Poiret) (*typ. cons.*).

(=) *Hicorius* Rafinesque, Fl. Ludov. 109. 1817.
T.: non designatus.

BETULACEAE

1885 **Ostrya** Scopoli, Fl. Carn. 414. 1760.
T.: *O. carpinifolia* Scopoli (*Carpinus ostrya* Linnaeus).

(H) *Ostrya* J. Hill, Brit. Herb. 513. 1757 [Betul.].
≡ *Carpinus* Linnaeus 1753.

FAGACEAE

1889 **Nothofagus** Blume, Mus. Bot. 1: 307. 1850.
T.: *N. antarctica* (G. Forster) Oersted (Kongel. Norske Vidensk. Selsk. Skr. ser. 5. 9: 354. 1871) (*Fagus antarctica* G. Forster) (*typ. cons.*).

(=) *Fagaster* Spach, Hist. Nat. Vég. Phan. 11: 142. 1841.
T.: *Fagus dombeyi* Mirbel.
(=) *Calucechinus* Hombron et Jacquinot in Urville, Voy. Pôle Sud, Bot. Atlas (Dicot.) 6: *t. 6.* 1844.
T.: *C. antarctica* Hombron et Jacquinot.
(=) *Calusparassus* Hombron et Jacquinot in Urville, Voy. Pôle Sud, Bot. Atlas (Dicot.) 6: *t. 6.* 1844.
T.: *C. forsteri* Hombron et Jacquinot.

1891a **Castanopsis** (D. Don) Spach, Hist. Nat. Vég. Phan. 11: 142, 185. 1842.
T.: *C. armata* (Roxburgh) Spach (*Quercus armata* Roxburgh).

(=) *Balanoplis* Rafinesque, Alsogr. 29. 1838.
LT.: *B. tribuloides* (J. E. Smith) Rafinesque (*Quercus tribuloides* J. E. Smith).

1893 **Cyclobalanopsis** Oersted, Vidensk. Meddel. Dansk Naturhist. Foren. Kjøbenhavn 1866: 69. 1867.
T.: *C. velutina* Oersted (*Quercus velutina* Lindley ex Wallich, non Lamarck) (*typ. cons.*).

(=) *Perytis* Rafinesque, Alsogr. 29. 1838.
LT.: *P. lamellosa* (J. E. Smith) Rafinesque (*Quercus lamellosa* J. E. Smith).

ULMACEAE

1901 **Zelkova** Spach, Ann. Sci. Nat. Bot. ser. 2. 15: 356. 1841.
T.: *Z. crenata* Spach, nom. illeg. (*Rhamnus carpinifolius* Pallas, *Z. carpinifolia* (Pallas) K. Koch).

1904 **Aphananthe** J. E. Planchon, Ann. Sci. Nat. Bot. ser. 3. 10: 265. 1848.
T.: *A. philippinensis* J. E. Planchon.

(H) *Aphananthe* Link, Enum. Hort. Berol. Alt. 1: 383. 1821 [Phytolacc.].
T.: *A. celosioides* (K. Sprengel) Link (*Galenia celosioides* K. Sprengel).

188

1917 **Trophis** P. Browne, Civ. Nat. Hist. Jamaica 357. 1756.
T.: *T. americana* Linnaeus (Syst. Nat. ed. 10. 1289. 1759).

(=) *Bucephalon* Linnaeus, Sp. Pl. 1190. 1753.
T.: *B. racemosum* Linnaeus.

1918 **Maclura** Nuttall, Gen. N. Amer. Pl. 2: 233. 1818.
T.: *M. aurantiaca* Nuttall.

(=) *Ioxylon* Rafinesque, Amer. Monthly Mag. & Crit. Rev. 2: 118. 1817.
T.: *I. pomiferum* Rafinesque.

1923 **Broussonetia** L'Héritier ex Ventenat, Tabl. Règne Vég. 3: 547. 1799.
T.: *B. papyrifera* (Linnaeus) Ventenat (*Morus papyrifera* Linnaeus).

(H) *Broussonetia* Ortega, Nov. Pl. Descr. Dec. 61. 1798 [Legum.].
T.: *B. secundiflora* Ortega.

(=) *Papyrius* Lamarck, Tabl. Encycl. t. 762. 1797.
T.: non designatus.

1937 **Clarisia** Ruiz et Pavón, Prodr. 116. t. 28. 1794.
T.: *C. racemosa* Ruiz et Pavón (Syst. 255. 1798) (*typ. cons.*).

(H) *Clarisia* Abat, Mem. Acad. Soc. Med. Sevilla 10: 418. 1792 [Basell.].
≡ *Anredera* A. L. Jussieu 1789.

1942 **Cudrania** Trécul, Ann. Sci. Nat. Bot. ser. 3. 8: 122. 1847.
T.: *C. javanensis* Trécul (*typ. cons.*).

(=) *Vanieria* Loureiro, Fl. Cochinch. 564. 1790.
LT.: *V. cochinchinensis* Loureiro (vide Merrill, Trans. Amer. Philos. Soc. ser. 2. 24: 134. 1935).

1946 **Artocarpus** J. R. Forster et G. Forster, Char. Gen. Pl. 51. 1775.
T.: *A. communis* J. R. Forster et G. Forster.

(=) *Sitodium* S. Parkinson, J. Voy. South Seas 45. 1773.
T.: *S. altile* S. Parkinson.

1956 **Antiaris** Leschenault, Ann. Mus. Natl. Hist. Nat. 16: 478. 1810.
T.: *A. toxicaria* Leschenault.

(=) *Ipo* Persoon, Syn. Pl. 2: 566. 1807.
T.: *I. toxicaria* Persoon.

1957 **Brosimum** Swartz, Prodr. 1. 1788.
T.: *B. alicastrum* Swartz (*typ. cons.*).

(≡) *Alicastrum* P. Browne, Civ. Nat. Hist. Jamaica 372. 1756.
(=) *Piratinera* Aublet, Hist. Pl. Guiane 888. 1775.
T.: *P. guianensis* Aublet.

(=) *Ferolia* Aublet, Hist. Pl. Guiane Suppl. 7. 1775.
T.: *F. guianensis* Aublet.

1971 **Cecropia** Loefling, Iter Hispan. 272. 1758.
T.: *C. peltata* Linnaeus (Syst. Nat. ed. 10. 1286. 1759).

(≡) *Coilotapalus* P. Browne, Civ. Nat. Hist. Jamaica 111. 1756.

1980 **Laportea** Gaudichaud, Voy. Uranie 498. 1830 ("1826").
T.: *L. canadensis* (Linnaeus) Weddell (Ann. Sci. Nat. Bot. ser. 4. 1: 181. 1854) (*Urtica canadensis* Linnaeus).

(≡) *Urticastrum* Heister ex Fabricius, Enum. 204. 1759.

1984 **Pilea** Lindley, Collect. Bot. sub *t. 4.* 1821.
T.: *P. muscosa* Lindley, nom. illeg. (*Parietaria microphylla* Linnaeus, *Pilea microphylla* (Linnaeus) Liebmann).

1987 **Pellionia** Gaudichaud, Voy. Uranie 494. 1830 ("1826").
T.: *P. elatostemoides* Gaudichaud (*typ. cons.*).

(=) *Polychroa* Loureiro, Fl. Cochinch. 538, 559. 1790.
T.: *P. repens* Loureiro.

1988 **Elatostema** J. R. Forster et G. Forster, Char. Gen. Pl. 53. 1775.
T.: *E. sessile* J. R. Forster et G. Forster (*typ. cons.*).

PROTEACEAE

2023 **Persoonia** J. E. Smith, Trans. Linn. Soc. London 4: 215. 24 Mai 1798.
T.: *P. lanceolata* Andrews (Bot. Repos. *t. 74.* 1799) (*typ. cons.*).

(=) *Linkia* Cavanilles, Icon. **4:** 61. 14 Mai 1798.
T.: *L. levis* Cavanilles.

2026 **Isopogon** R. Brown ex Knight, Cult. Prot. 93. 1809.
T.: *I. anemonifolius* (R. A. Salisbury) Knight (*Protea anemonifolia* R. A. Salisbury) (*typ. cons.*).

(=) *Atylus* R. A. Salisbury, Parad. Lond. sub *t. 67.* 1807.
T.: non designatus.

2028 **Sorocephalus** R. Brown, Trans. Linn. Soc. London **10:** 139. 1810.
T.: *S. imbricatus* (Thunberg) R. Brown (*Protea imbricata* Thunberg) (*typ. cons.*).

(=) *Soranthe* R. A. Salisbury ex Knight, Cult. Prot. 71. 1809.
T.: non designatus.

2035 **Protea** Linnaeus, Mant. Pl. **2:** 187, 328. 1771.
T.: *P. cynaroides* (Linnaeus) Linnaeus (*Leucadendron cynaroides* Linnaeus, "*cinaroides*") (*typ. cons.*).

(H) *Protea* Linnaeus, Sp. Pl. 94. 1753 [Prot.].
≡ *Leucadendron* R. Brown 1810 (*nom. cons.*) (2037) (vide Hitchcock, Prop. Brit. Bot. 113. 1929).
(=) *Lepidocarpus* Adanson, Fam. Pl. **2:** 284, 569. 1763.
≡ *Leucadendron* Linnaeus 1753 (*nom. rej.* sub 2037).

2036 **Leucospermum** R. Brown, Trans. Linn. Soc. London **10**: 95. 1810.
T.: *L. hypophyllum* R. Brown, nom. illeg. (*Leucadendron hypophyllocarpodendron* Linnaeus, *Leucospermum hypophyllocarpodendron* (Linnaeus) Druce) (*typ. cons.*).

2037 **Leucadendron** R. Brown, Trans. Linn. Soc. London **10**: 50. 1810.
T.: *L. argenteum* (Linnaeus) R. Brown (*Protea argentea* Linnaeus) (*typ. cons.*) (etiam vide 2035).

(H) *Leucadendron* Linnaeus, Sp. Pl. 91. 1753 [Prot.].
LT.: *L. lepidocarpodendron* Linnaeus (vide Hitchcock, Prop. Brit. Bot. 122. 1929).

2045 **Grevillea** R. Brown ex Knight, Cult. Prot. 120. 1809 ("*Grevillia*") (*orth. cons.*).
T.: *G. aspleniifolia* R. Brown ex Knight.

(=) *Lysanthe* R. A. Salisbury ex Knight, Cult. Prot. 116. 1809.
T.: non designatus.
(=) *Stylurus* R. A. Salisbury ex Knight, Cult. Prot. 115. 1809.
T.: non designatus.

2062 **Telopea** R. Brown, Trans. Linn. Soc. London **10**: 197. 1810.
T.: *T. speciosissima* (J. E. Smith) R. Brown (*Embothrium speciosissimum* J. E. Smith) (*typ. cons.*).

(≡) *Hylogyne* R. A. Salisbury ex Knight, Cult, Prot. 126. 1809.

2063 **Lomatia** R. Brown, Trans. Linn. Soc. London **10**: 199. 1810.
T.: *L. silaifolia* (J. E. Smith) R. Brown (*Embothrium silaifolium* J. E. Smith) (*typ. cons.*).

(=) *Tricondylus* R. A. Salisbury ex Knight, Cult. Prot. 121. 1809.
T.: non designatus.

2064 **Knightia** R. Brown, Trans. Linn. Soc. London **10**: 193. 1810.
T.: *K. excelsa* R. Brown.

(=) *Rymandra* R. A. Salisbury ex Knight, Cult. Prot. 124. 1809.
T.: *R. excelsa* Knight.

2066 **Stenocarpus** R. Brown, Trans. Linn. Soc. London **10**: 201. 1810.
T.: *S. forsteri* R. Brown, nom. illeg. (*S. umbellifer* (J. R. Forster et G. Forster) Druce, *Embothrium umbelliferum* J. R. Forster et G. Forster) (*typ. cons.*).

(≡) *Cybele* R. A. Salisbury ex Knight, Cult. Prot. 123. 1809.

2068 **Banksia** Linnaeus f., Suppl. Pl. 15. 126. 1782.
T.: *B. serrata* Linnaeus f.

(H) *Banksia* J. R. Forster et G. Forster, Char. Gen. Pl. 4. 1775 [Thymel.].
T.: non designatus.

2069 **Dryandra** R. Brown, Trans. Linn. Soc. London **10**: 211. 1810.
T.: *D. formosa* R. Brown (*typ. cons.*).

(H) *Dryandra* Thunberg, Nova Gen. Pl. 60. 1783 [Euphorb.].
T.: *D. cordata* Thunberg.
(=) *Josephia* R. Brown ex Knight, Cult. Prot. 110. 1809.
T.: non designatus.

191

2074 **Loranthus** N. J. Jacquin, Enum. Stirp. Vindob. 55, 230, *t. 3*. 1762.
T.: *L. europaeus* N. J. Jacquin.

(H) *Loranthus* Linnaeus, Sp. Pl. 331. 1753 [Loranth.].
T.: *L. americanus* Linnaeus.

(=) *Scurrula* Linnaeus, Sp. Pl. 110. 1753.
T.: *S. parasitica* Linnaeus.

2074a **Tapinanthus** (Blume) H. G. L. Reichenbach, Deut. Bot. Herb.-Buch [1]: 73. 1841.
T.: *T. sessilifolius* (Palisot de Beauvois) Van Tieghem (Bull. Soc. Bot. France 42: 267. 1895) (*Loranthus sessilifolius* Palisot de Beauvois).

(H) *Tapeinanthus* Herbert, Amaryll. 59, 73, 190, 414. 1837 [Amaryllid.].
T.: *T. humilus* (Cavanilles) Herbert (*Pancratium humile* Cavanilles).

2078 **Struthanthus** C. F. P. Martius, Flora 13: 102. 1830.
T.: *S. syringifolius* (C. F. P. Martius) C. F. P. Martius (*Loranthus syringifolius* C. F. P. Martius) (*typ. cons.*).

(=) *Spirostylis* K. B. Presl in J. A. Schultes et J. H. Schultes, Syst. Veg. 7 (1): xvii, 163. 1829.
T.: *S. haenkei* K. B. Presl.

2091 **Arceuthobium** Marschall von Bieberstein, Fl. Taur.-Caucas. Suppl. 629. 1819.
T.: *A. oxycedri* (A.-P. de Candolle) Marschall von Bieberstein (*Viscum oxycedri* A.-P. de Candolle).

(=) *Razoumofskya* G. F. Hoffmann, Hort. Mosq. 1. 1808.
T.: *R. caucasica* G. F. Hoffmann.

SANTALACEAE

2097 **Exocarpos** Labillardière, Voy. Rech. Pérouse 1: 155, *t. 14*. 1799.
T.: *E. cupressiformis* Labillardière.

(=) "Xylophyllos" Rumphius, Auct. 19, *t. 12*. 1755 [*nom. inval.*].

2103 **Scleropyrum** Arnott, Mag. Zool. Bot. 2: 549. 1838.
T.: *S. wallichianum* (Wight et Arnott) Arnott (*Sphaerocarya wallichiana* Wight et Arnott).

(=) *Heydia* Dennstedt ex Kosteletzky, Allg. Med.-Pharm. Fl. 5: 2005. 1836.
T.: *H. horrida* Dennstedt ex Kosteletzky.

2109 **Buckleya** Torrey, Amer. J. Sci. Arts 45: 170. 1843.
T.: *B. distichophylla* (Nuttall) Torrey (*Borya distichophylla* Nuttall).

(=) *Nestronia* Rafinesque, New Fl. 3: 12. 1838 ("1836").
T.: *N. umbellula* Rafinesque.

2120 **Quinchamalium** Molina, Sag. Stor. Nat. Chili 151, 350. 1782.
T.: *Q. chilense* Molina.

OPILIACEAE

2124 **Cansjera** A. L. Jussieu, Gen. Pl. 448. 1789.
T.: *C. rheedei* J. F. Gmelin (Syst. Nat. 280. 1791, "*rheedii* ").

(≡) *Tsjeru-caniram* Adanson, Fam. Pl. **2**: 80, 614. 1763.

OLACACEAE

2147 **Heisteria** N. J. Jacquin, Enum. Syst. Pl. 4, 20. 1760.
T.: *H. coccinea* N. J. Jacquin.

(H) *Heisteria* Linnaeus, Opera Var. 242. 1758 [Polygal.].
≡ *Muraltia* A.-P. de Candolle 1824 (*nom. cons.*) (4278).

BALANOPHORACEAE

2163 **Helosis** L. C. Richard, Mém. Mus. Hist. Nat. **8**: 416, 432. 1822.
T.: *H. guyannensis* L. C. Richard, nom. illeg. (*Cynomorium cayanense* Swartz, *H. cayanensis* (Swartz) K. Sprengel).

RAFFLESIACEAE

2180 **Cytinus** Linnaeus, Gen. Pl. ed. 6. 576 (err. 566). 1764.
T.: *C. hypocistis* (Linnaeus) Linnaeus (*Asarum hypocistis* Linnaeus).

(≡) *Hypocistis* P. Miller, Gard. Dict. Abr. ed. 4. 1754.

POLYGONACEAE

2194 **Emex** Campderá, Monogr. Rumex 56. 1819.
T.: *E. spinosa* (Linnaeus) Campderá. (*Rumex spinosus* Linnaeus) (*typ. cons.*).

(≡) *Vibo* Medikus, Philos. Bot. **1**: 178. 1789.

2201 **Polygonum** Linnaeus, Sp. Pl. 359. 1753.
T.: *P. aviculare* Linnaeus (*typ. cons.*).

2202 **Fagopyrum** P. Miller, Gard. Dict. Abr. ed. 4. 1754.
T.: *Fagopyrum esculentum* Moench (Methodus 290. 1794) (*Polygonum fagopyrum* Linnaeus) (*typ. cons.*).

2208 **Muehlenbeckia** Meisner, Pl. Vasc. Gen.
1: 316. 1841; 2: 227. 1841.
T.: *M. australis* (G. Forster) Meisner
(*Coccoloba australis* G. Forster).

(≡) *Calacinum* Rafinesque, Fl. Tell. 2: 33.
1837 ("1836").

(=) *Karkinetron* Rafinesque, Fl. Tell. 3: 11.
1837 ("1836").
T.: non designatus.

2209 **Coccoloba** P. Browne, Civ. Nat. Hist.
Jamaica 209. 1756 ("*Coccolobis*") (*orth. cons.*).
T.: *C. uvifera* (Linnaeus) Linnaeus
(*Polygonum uvifera* Linnaeus) (*typ. cons.*).

(=) *Guaiabara* P. Miller, Gard. Dict. Abr. ed.
4. 1754.
T.: non designatus.

CHENOPODIACEAE

2261 **Suaeda** Forsskål ex Scopoli, Intr. 333.
1777.
T.: *S. vera* Forsskål ex J. F. Gmelin
(Syst. 2: 503. 1791) (*typ. cons.*).

AMARANTHACEAE

2297 **Chamissoa** Kunth in Humboldt, Bonpland et Kunth, Nov. Gen. Sp. 2: ed. fol.
158, ed. qu. 196, t. *125*. 1818.
T.: *C. altissima* (N. J. Jacquin) Kunth
(*Achyranthes altissima* N. J. Jacquin)
(*typ. cons.*).

(=) *Kokera* Adanson, Fam. Pl. 2: 269, 541.
1763.
T.: non designatus.

2312 **Cyathula** Blume, Bijdr. 548. 1825 (or
1826).
T.: *C. prostrata* (Linnaeus) Blume
(*Achyranthes prostrata* Linnaeus).

(H) *Cyathula* Loureiro, Fl. Cochinch. 93, 101.
1790 [Amaranth.].
T.: *C. geniculata* Loureiro.

2314 **Pupalia** A. L. Jussieu, Ann. Mus. Natl.
Hist. Nat. 2: 132. 1803.
T.: *P. lappacea* (Linnaeus) A. L. Jussieu (*Achyranthes lappacea* Linnaeus).

(≡) *Pupal* Adanson, Fam. Pl. 2: 268, 596.
1763.

2317 **Aerva** Forsskål, Fl. Aegypt.-Arab. cxxii,
170. 1775.
T.: *A. tomentosa* Forsskål (*typ. cons.*).

(=) *Ouret* Adanson, Fam. Pl. 2: 268, 586.
1763.
T.: *Achyranthes lanata* Linnaeus.

2339 **Iresine** P. Browne, Civ. Nat. Hist. Jamaica 358. 1756.
T.: *I. celosioides* Nuttall (Gen. N.
Amer. Pl. 2: 237. 1818) (*Celosia paniculata* Linnaeus, *I. paniculata* (Linnaeus) O. Kuntze, non Poiret) [= *I. diffusa* Humboldt et Bonpland ex Willdenow, Sp. Pl. 4: 765. 1805].

NYCTAGINACEAE

2348 **Allionia** Linnaeus, Syst. Nat. ed. 10. 890, 1361. 1759.
T.: *A. incarnata* Linnaeus (*typ. cons.*) (etiam vide 9192).

(H) *Allionia* Loefling, Iter Hispan. 181. 1758 [Nyctagin.].
T.: *A. violacea* Linnaeus (Syst. Nat. ed. 10. 890. 1759).

2350 **Bougainvillea** Commerson ex A. L. Jussieu, Gen. Pl. 91 1789 ("*Buginvillaea*") (*orth. cons.*).
T.: *B. spectabilis* Willdenow (Sp. Pl. 2: 348. 1799) (*typ. cons.*).

AIZOACEAE

2405 **Mesembryanthemum** Linnaeus, Sp. Pl. 480. 1753.
T.: *M. nodiflorum* Linnaeus (*typ. cons.*).

2405a **Lampranthus** N. E. Brown, Gard. Chron. ser. 3. **87**: 71. 1930.
T.: *L. multiradiatus* (Jacquin) N. E. Brown (*Mesembryanthemum multiradiatum* Jacquin.

(=) *Oscularia* Schwantes, Möller's Deutsche Gärtn.-Zeitung **42**: 187. 1927.
T.: *O. deltoides* (Linnaeus) Schwantes (*Mesembryanthemum deltoides* Linnaeus).

PORTULACACEAE

2406 **Talinum** Adanson, Fam. Pl. **2**: 245, 609. 1763.
T.: *T. triangulare* (N. J. Jacquin) Willdenow (*Portulaca triangularis* N. J. Jacquin) (*typ. cons.*).

2407 **Calandrinia** Kunth in Humboldt, Bonpland et Kunth, Nov. Gen. Sp. **6**: ed. fol. 62, ed. qu. 77. 1823.
T.: *C. caulescens* Kunth (*typ. cons.*).

(=) *Baitaria* Ruiz et Pavón, Prodr. 63. 1794.
T.: *B. acaulis* Ruiz et Pavón (Syst. 111. 1798).

2412 **Anacampseros** Linnaeus, Opera Var. 232. 1758.
T.: *A. telephiastrum* A.-P. de Candolle (Cat. Pl. Horti Monsp. 77. 1813) (*Portulaca anacampseros* Linnaeus).

(H) *Anacampseros* P. Miller, Gard. Dict. Abr. ed. 4. 1754 [Crassul.].
T.: non designatus.

CARYOPHYLLACEAE

2432 **Moenchia** Ehrhart, Neues Mag. Aertzte **5**: 203. 1783.
T.: *M. quaternella* Ehrhart, nom. illeg. (*Sagina erecta* Linnaeus, *M. erecta* (Linnaeus) P. Gaertner, B. Meyer et Scherbius).

2450 **Spergularia** (Persoon) J. S. Presl et K. B. Presl, Fl. Cech. 94. 1819.
T.: *S. rubra* (Linnaeus) J. S. Presl et K. B. Presl (*Arenaria rubra* Linnaeus) (*typ. cons.*).

(≡) *Tissa* Adanson, Fam. Pl. 2: 507, 611. 1763 (vide Swart in Regnum Veg. 102: 1764. 1979).

(=) *Buda* Adanson, Fam. Pl. 2: 507, 528. 1763.
T.: non designatus.

2455 **Polycarpaea** Lamarck, J. Hist. Nat. 2: 3 ("*Polycarpea*"), 5. 1792.
T.: *P. teneriffae* Lamarck (*typ. cons.*).

(=) *Polia* Loureiro, Fl. Cochinch. 97, 164. 1790.
T.: *P. arenaria* Loureiro.

2467 **Pollichia** W. Aiton, Hortus Kew. 1: 5. 1789.
T.: *P. campestris* W. Aiton.

(H) *Polichia* Schrank, Acta Acad. Elect. Mogunt. Sci. Util. Erfurti 1781: 35. 1781 [Lab.].
T.: *P. galeobdolon* (Linnaeus) Willdenow (Fl. Berol. Prodr. 198. 1787) (*Galeopsis galeobdolon* Linnaeus).

2477 **Siphonychia** Torrey et A. Gray, Fl. N. Amer. 1: 173. 1838.
T.: *S. americana* (Nuttall) Torrey et A. Gray (*Herniaria americana* Nuttall).

NYMPHAEACEAE

2513 **Nymphaea** Linnaeus, Sp. Pl. 510. 1753.
T.: *N. alba* Linnaeus (*typ. cons.*).

2514 **Nuphar** J. E. Smith in Sibthorp et J. E. Smith, Fl. Graecae Prodr. 1: 361. 1809 ("1806").
T.: *N. lutea* (Linnaeus) J. E. Smith (*Nymphaea lutea* Linnaeus).

(≡) *Nymphozanthus* L. C. Richard, Demonstr. Bot. Anal. Fruits 63, 68 ("*Nymphosanthus*"), 103. 1808.

*2515 **Barclaya** Wallich, Trans. Linn. Soc. London 15: 442. Dec 1827.
T.: *B. longifolia* Wallich.

(≡) *Hydrostemma* Wallich, Philos. Mag. Ann. Chem. 1: 454. Jun 1827.

RANUNCULACEAE

2528 **Eranthis** R. A. Salisbury, Trans. Linn. Soc. London 8: 303, 1807.
T.: *E. hyemalis* (Linnaeus) R. A. Salisbury (*Helleborus hyemalis* Linnaeus).

(=) *Cammarum* J. Hill, Brit. Herb. 47. 1756.

2542 **Naravclia** Adanson, Fam. Pl. 2: 460.
1763 (*"Naravel"*) (*orth. cons.*).
T.: *N. zeylanica* (Linnaeus) A.-P. de
Candolle (*Atragene zeylanica* Linnaeus) (*typ. cons.*).

LARDIZABALACEAE

2551 **Decaisnea** J. D. Hooker et Thomson,
Proc. Linn. Soc. London 2: 350. 1855.
T.: *D. insignis* (Griffith) J. D. Hooker
et Thomson (*Slackia insignis* Griffith).

(H) *Decaisnea* A. T. Brongniart in Duperrey,
Voyage Monde, Bot. (Phan.) 192. 1834
("1829") [Orchid.].
T.: *D. densiflora* A. T. Brongniart.

BERBERIDACEAE

2566 **Mahonia** Nuttall, Gen. N. Amer. Pl. 1:
211. 1818.
T.: *M. aquifolium* (Pursh) Nuttall
(*Berberis aquifolium* Pursh) (*typ. cons.*).

MENISPERMACEAE

2568 **Pericampylus** Miers, Ann. Mag. Nat.
Hist. ser. 2. 7: 36, 40. 1851.
T.: *P. incanus* (Colebrooke) J. D.
Hooker et T. Thomson (Fl. Ind. 194.
1855) (*Cocculus incanus* Colebrooke).

(=) *Pselium* Loureiro, Fl. Cochinch. 600, 621.
1790.
T.: *P. heterophyllum* Loureiro.

2570 **Cocculus** A.-P. de Candolle, Syst. Nat.
1: 515. 1817 ("1818").
T.: *C. hirsutus* (Linnaeus) Diels (in
Engler, Pflanzenr. iv.94 (Heft **46**): 236.
1910) (*Menispermum hirsutum* Linnaeus) (*typ. cons.*).

(=) *Cebatha* Forsskål, Fl. Aegypt.-Arab. 171.
1775.
T.: *Cocculus cebatha* A.-P. de Candolle
(Syst. Nat. 1: 527. 1817).

(=) *Leaeba* Forsskål, Fl. Aegypt.-Arab. 172.
1775.
T.: *L. dubia* J. F. Gmelin (Syst. Veg. 2:
567. 1791).

(=) *Epibaterium* J. R. Forster et G. Forster,
Char. Gen. Pl. 54. 1775.
T.: *E. pendulum* J. R. Forster et G. Forster.

(=) *Nephroia* Loureiro, Fl. Cochinch. 539,
565. 1790.
T.: *N. sarmentosa* Loureiro.

(=) *Baumgartia* Moench, Methodus 650.
1794.
T.: *B. scandens* Moench.

(=) *Androphylax* J. C. Wendland, Bot. Beob.
37, 38. 1798.
T.: *A. scandens* J. C. Wendland.

*2577 **Tiliacora** Colebrook, Trans. Linn. Soc. London **13**: 53, 67. 1821.
T.: *T. racemosa* Colebrook.

(=) *Braunea* Willdenow, Sp. Pl. 4(2): 797. 1806.
T.: *B. menispermoides* Willdenow.

*2583 **Tinospora** Miers, Ann. Mag. Nat. Hist. ser. 2. **7**: 35, 38. 1851.
T.: *T. cordifolia* (Willdenow) Miers ex J. D. Hooker et Thomson (*Menispermum cordifolium* Willdenow).

(=) *Campylus* Loureiro, Fl. Cochinch. 113. 1790.
T.: *C. sinensis* Loureiro.

2611 **Hyperbaena** Miers ex Bentham, J. Linn. Soc., Bot. **5**, Suppl. **2**: 47, 50. 1861.
T.: *H. domingensis* (A.-P. de Candolle) Bentham (*Cocculus domingensis* A.-P. de Candolle) (*typ. cons.*).

(=) *Alina* Adanson, Fam. Pl. **2**: 84, 515. 1763.
T.: non designatus ["Vimen Brown. t. 22. f. 5"].

MAGNOLIACEAE

2656 **Schisandra** A. Michaux, Fl. Bor.-Amer. **2**: 218. Mar 1803.
T.: *S. coccinea* A. Michaux.

(=) *Stellandria* Brickell, Med. Repos. **6**: 327. Feb-Mar 1803.
T.: *S. glabra* Brickell.

2658 **Drimys** J. R. Forster et G. Forster, Char. Gen. Pl. 42. 1775.
T.: *D. winteri* J. R. Forster et G. Forster (*typ. cons.*).

CALYCANTHACEAE

2663 **Calycanthus** Linnaeus, Syst. Nat. ed.10. 1053, 1066, 1371. 1759.
T.: *C. floridus* Linnaeus.

(=) *Basteria* P. Miller, Fig. Pl. Gard. Dict. 40, t. 60. 1755.
T.: non designatus.

2663a **Chimonanthus** Lindley, Bot. Reg. **5**: sub t. 404. 1819.
T.: *C. fragrans* Lindley (Bot. Reg. **6**: 451. 1820) nom. illeg. (*Calycanthus praecox* Linnaeus, *Chimonanthus praecox* (Linnaeus) Link).

(≡) *Meratia* Loiseleur, Herb. Gén. Amat. **3**: t. 173. 1818 ("1819").

ANNONACEAE

2679 **Guatteria** Ruiz et Pavón, Prodr. 85, t. 17. 1794.
T.: *G. eriopoda* A.-P. de Candolle (Syst. Nat. **1**: 505. 1817, "1818").

198

2680 **Duguetia** A. Saint-Hilaire, Fl. Bras. Mer. 1: ed. qu. 35, ed. fol. 28. 1824 ("1825"). T.: *D. lanceolata* A. Saint-Hilaire.

(=) *Aberemoa* Aublet, Hist. Pl. Guiane 610. 1775. T.: *A. guianensis* Aublet.

2684 **Cananga** (A.-P. de Candolle) J. D. Hooker et Thomson, Fl. Ind. 1: 129. 1855. T.: *C. odorata* (Lamarck) J. D. Hooker et Thomson (*Uvaria odorata* Lamarck).

(H) *Cananga* Aublet, Hist. Pl. Guiane 607. 1775 [Annon.]. T.: *C. ouregou* Aublet.

2691a **Enneastemon** Exell, J. Bot. 70 (suppl. 1): 209. 1932. T.: *E. angolensis* Exell.

(=) *Clathrospermum* J. E. Planchon ex Bentham in Bentham et J. D. Hooker, Gen. Pl. 1: 29. 1862. T.: *C. vogelii* (J. D. Hooker) Bentham (*Uvaria vogelii* J. D. Hooker).

2717 **Xylopia** Linnaeus, Syst. Nat. ed. 10. 1250, 1378. 1759. T.: *X. muricata* Linnaeus (*typ. cons.*).

(≡) *Xylopicrum* P. Browne, Civ. Nat. Hist. Jamaica 250, 254. 1756.

MYRISTICACEAE

2750 **Myristica** Gronovius, Fl. Orient. 141. 1755. T.: *M. fragrans* Houttuyn (Nat. Hist. 2(3): 333. 1774).

MONIMIACEAE

2759 **Peumus** Molina, Sag. Stor. Nat. Chili 185, 350. 1782. T.: *P. boldus* Molina (*typ. cons.*).

(=) *Boldu* Adanson, Fam. Pl. 2: 446, 526. 1763. T.: non designatus ["Feuill. t. 6"].

2775 **Laurelia** A. L. Jussieu, Ann. Mus. Natl. Hist. Nat. 14: 134. 1809. T.: *L. sempervirens* (Ruiz et Pavón) Tulasne (*Pavonia sempervirens* Ruiz et Pavón) (*typ. cons.*).

LAURACEAE

2782 **Cinnamomum** Schaeffer, Bot. Exped. 74. 1760. T.: *C. zeylanicum* Blume (*Laurus cinnamomum* Linnaeus).

(=) *Camphora* Fabricius, Enum. 218. 1759. T.: *Laurus camphora* Linnaeus.

2783 **Persea** P. Miller, Gard. Dict. Abr. ed. 4. 1754.

T.: *P. americana* P. Miller (Gard. Dict. ed. 8. 1768) (*Laurus persea* Linnaeus).

2790 **Nectandra** Rolander ex Rottböll, Acta Lit. Univ. Hafn. 1: 279. 1778.

T.: *N. sanguinea* Rolander ex Rottböll (*typ. cons.*).

(H) *Nectandra* P. J. Bergius, Descr. Pl. Cap. 131. 1767 [Thymel.].

LT.: *N. sericea* P. J. Bergius (vide Mansfeld, Bull. Misc. Inform. 1935: 439. 1935).

2793 **Eusideroxylon** Teysmann et Binnendijk, Natuurk. Tijdschr. Ned. Indie 25: 292. 1863.

T.: *E. zwageri* Teysmann et Binnendijk.

(=) *Salgada* Blanco, Fl. Filip. ed. 2. 221. 1845.

T.: *S. lauriflora* Blanco.

2797 **Neolitsea** (Bentham) Merrill, Philipp. J. Sci. 1 (suppl.): 56. 1906.

T.: *N. zeylanica* (C. G. Nees) Merrill (*Tetradenia zeylanica* C. G. Nees) (*typ. cons.*).

(=) *Bryantea* Rafinesque, Sylva Tell. 165. 1838.

T.: *B. dealbata* (R. Brown) Rafinesque (*Tetranthera dealbata* R. Brown).

2798 **Litsea** Lamarck, Encycl. 3: 574. 1791. ("1789").

T.: *L. chinensis* Lamarck.

(=) *Malapoenna* Adanson, Fam. Pl. 2: 447, 573. 1763.

T.: *Darwinia quinqueflora* Dennstedt.

2804 **Bernieria** Baillon, Bull. Mens. Soc. Linn. Paris 1: 434. 1884.

T.: *B. madagascariensis* Baillon.

(H) *Berniera* A.-P. de Candolle, Prodr. 7: 18. 1838 [Comp.].

T.: *B. nepalensis* A.-P. de Candolle, nom. illeg. (*Chaptalia maxima* D. Don).

2811a **Endlicheria** C. G. Nees, Linnaea 8: 37. 1833.

T.: *E. hirsuta* (Schott) C. G. Nees (*Cryptocarpa hirsuta* Schott) (*typ. cons.*).

(H) *Endlichera* K. B. Presl, Symb. Bot. 1: 73. 1832 [Rub.].

T.: *E. brasiliensis* K. B. Presl.

2821 **Lindera** Thunberg, Nova Gen. Pl. 64. 1783.

T.: *L. umbellata* Thunberg.

(H) *Lindera* Adanson, Fam. Pl. 2: 499, 571. 1763 [Umbell.].

T.: *Chaerophyllum coloratum* Linnaeus (Mant. Pl. 1: 57. 1767).

(=) *Benzoin* Schaeffer, Bot. Exped. 60. 1760.

T.: *B. odoriferum* C. G. Nees (in Wallich, Pl. As. Rar. 2: 63. 1831) (*Laurus benzoin* Linnaeus) (vide Fabricius, Enum. ed. 2. 401. 1763).

2856 **Dicentra** Bernhardi, Linnaea 8: 457, 468. 1833.
T.: *D. cucullaria* (Linnaeus) Bernhardi (*Fumaria cucullaria* Linnaeus).

(≡) *Diclytra* Borkhausen, Arch. Bot. (Leipzig) 1(2): 46. 1797.

(=) *Capnorchis* P. Miller, Gard. Dict. Abr. ed. 4. 1754.
T.: non designatus ["*Capnorchis Americana* Boerh. Ind."].

(=) *Bikukulla* Adanson, Fam. Pl. 2: (23). 1763.
T.: non designatus ["*Bicucullata* March. Mém. Acad. 1733, t. 20"].

(=) *Dactylicapnos* Wallich, Tent. Fl. Napal. 51. 1826.
T.: *D. thalictrifolia* Wallich.

2857 **Adlumia** Rafinesque ex A.-P. de Candolle, Syst. Nat. 2: 111. 1821.
T.: *A. cirrhosa* Rafinesque ex A.-P. de Candolle, nom. illeg. (*Fumaria fungosa* W. Aiton, *A. fungosa* (W. Aiton) Greene ex Britton, Sterns et Poggenburg).

2858 **Corydalis** A.-P. de Candolle in Lamarck et A.-P. de Candolle, Fl. Franç. ed. 3. 4: 637. 1805.
T.: *C. bulbosa* (Linnaeus) A.-P. de Candolle (*Fumaria solida* Linnaeus) (typus: herb. Linnaeus 881.5, LINN) (etiam vide *Neckera* Hedwig [Musci], *nom. cons.*).

(H) *Corydalis* Medikus, Philos. Bot. 1: 96. 1789 [Papaver.].
≡ *Cysticapnos* P. Miller 1754 (*nom. rej.*).

(≡) *Pistolochia* Bernhardi, Syst. Verz. 57, 74. 1800.

(=) *Capnoides* P. Miller, Gard. Dict. Abr. ed. 4. 1754.
T.: *C. sempervirens* (Linnaeus) Borckhausen (*Fumaria sempervirens* Linnaeus).

(=) *Cysticapnos* P. Miller, Gard. Dict. Abr. ed. 4. 1754.
T.: *C. vesicarius* (Linnaeus) Fedde (*Fumaria vesicaria* Linnaeus).

(=) *Pseudo-fumaria* Medikus, Philos. Bot. 1: 110. 1789.
T.: *P. lutea* (Linnaeus) Borckhausen (*Fumaria lutea* Linnaeus).

CRUCIFERAE (BRASSICACEAE)

2884 **Coronopus** Zinn, Cat. Pl. Gotting. 325. 1757.
T.: *C. ruellii* Allioni (*Cochlearia coronopus* Linnaeus).

(H) *Coronopus* P. Miller, Gard. Dict. Abr. ed. 4. 1754 [Plantagin.].
T.: non designatus.

2902 **Bivonaea** A.-P. de Candolle, Mém. Mus. Hist. Nat. **7**: 241. Apr 1821.
T.: *B. lutea* (Bivona-Bernardi) A.-P. de Candolle (Syst. Nat. **2**: 255. Mai 1821) (*Thlaspi luteum* Bivona-Bernardi).

(H) *Bivonea* Rafinesque, Specchio **1**: 156. 1814 [Euphorb.].
T.: *B. stimulosa* (A. Michaux) Rafinesque (*Jatropha stimulosa* A. Michaux).

2908 **Kernera** Medikus, Pfl.-Gatt. 77, 95. 1792.
T.: *K. myagrodes* Medikus, nom. illeg. (*Cochlearia saxatilis* Linnaeus, *K. saxatilis* (Linnaeus) H. G. L. Reichenbach).

(H) *Kernera* F. P. Schrank, Baier. Reise 50. 1786 [Scrophular.].
T.: *K. bavarica* F. P. Schrank.

2923 **Goldbachia** A.-P. de Candolle, Mém. Mus. Hist. Nat. **7**: 242. Apr 1821.
T.: *G. laevigata* (Marschall von Bieberstein) A.-P. de Candolle (Syst. Nat. **2**: 577. Mai 1821) (*Raphanus laevigatus* Marschall von Bieberstein).

(H) *Goldbachia* Trinius in K. Sprengel, Neue Entdeck. **2**: 42. 1820 [Gram.].
T.: *G. mikanii* Trinius.

2936 **Carrichtera** A.-P. de Candolle, Mém. Mus. Hist. Nat. **7**: 244. Apr 1821.
T.: *C. annua* (Linnaeus) A.-P. de Candolle (*Vella annua* Linnaeus) (*typ. cons.*).

(H) *Carrichtera* Adanson, Fam. Pl. **2**: 421, 533. 1763 [Cruc.].
T.: *Vella pseudocytisus* Linnaeus.

2940 **Schouwia** A.-P. de Candolle, Mém. Mus. Hist. Nat. **7**: 244. Apr 1821.
T.: *S. arabica* A.-P. de Candolle (Syst. Nat. **2**: 644. Mai 1821), nom. illeg. (*Subularia purpurea* Forsskål, *Schouwia purpurea* (Forsskål) Schweinfurth).

2956 **Rapistrum** Crantz, Cl. Crucif. Emend. 105. 1769.
T.: *R. hispanicum* (Linnaeus) Crantz (*Myagrum hispanicum* Linnaeus) (*typ. cons.*).

(H) *Rapistrum* Scopoli, Meth. Pl. 13. 1754 [Cruc.].
≡ *Neslia* Desvaux 1814 (*nom. cons.*) (2988).

2961 **Barbarea** R. Brown in W. T. Aiton, Hortus Kew. ed. 2. **4**: 109. 1812.
T.: *B. vulgaris* R. Brown (*Erysimum barbarea* Linnaeus).

(H) *Barbarea* Scopoli, Fl. Carniol. 522. 1760 [Cruc.].
T.: *Dentaria bulbifera* Linnaeus.

2965 **Nasturtium** R. Brown in W. T. Aiton, Hortus Kew. ed. 2. **4**: 109. 1812.
T.: *N. officinale* R. Brown (*Sisymbrium nasturtium-aquaticum* Linnaeus) (*typ. cons.*).

(H) *Nasturtium* P. Miller, Gard. Dict. Abr. ed. 4. 1754 [Cruc.].
T.: non designatus.
(≡) *Cardaminum* Moench, Methodus 262. 1794.

2965a **Armoracia** P. Gaertner, B. Meyer et Scherbius, Oekon. Fl. Wetterau **2**: 426. 1800.
T.: *A. rusticana* P. Gaertner, B. Meyer et Scherbius (*Cochlearia armoracia* Linnaeus).

(≡) *Raphanis* Moench, Methodus 267. 1794.

2968 **Ricotia** Linnaeus, Sp. Pl. ed. 2. 912. 1763.
T.: *R. aegyptiaca* Linnaeus, nom. illeg. (*Cardamine lunaria* Linnaeus, *R. lunaria* (Linnaeus) A.-P. de Candolle) (etiam vide 7393).

2973 **Mancoa** Weddell, Chloris And. **2**: *t. 86d*. 1859 ("1857").
T.: *M. hispida* Weddell.

(H) *Mancoa* Rafinesque, Fl. Tell. **3**: 56. 1837 ("1836") [Phytolacc.].
T.: *M. secunda* (Ruiz et Pavón) Rafinesque (*Rivina secunda* Ruiz et Pavón).

2986 **Capsella** Medikus, Pfl.-Gatt. 85, 99. 1792.
T.: *C. bursa-pastoris* (Linnaeus) Medikus (*Thlaspi bursa-pastoris* Linnaeus) (*typ. cons.*).

(≡) *Bursa-pastoris* Séguier, Pl. Veron. **3**: 166. 1754.

2988 **Neslia** Desvaux, J. Bot. Agric. **3**: 162. 1814.
T.: *N. paniculata* (Linnaeus) Desvaux (*Myagrum paniculatum* Linnaeus) (etiam vide 2956).

2989a **Erophila** A.-P. de Candolle, Mém. Mus. Hist. Nat. **7**: 244. 1821.
T.: *E. verna* (Linnaeus) A.-P. de Candolle (*Draba verna* Linnaeus) (*typ. cons.*).

(≡) *Gansblum* Adanson, Fam. Pl. **2**: 420, 561. 1763.

2997 **Descurainia** Webb et Berthelot, Hist. Nat. Iles Canaries 3(2, 1): 72. 1836.
T.: *D. sophia* (Linnaeus) Webb ex Prantl (*Sisymbrium sophia* Linnaeus) (*typ. cons.*).

(≡) *Sophia* Adanson, Fam. Pl. **2**: 417. 1763.
(=) *Hugueninia* H. G. L. Reichenbach, Fl. Germ. Excurs. 691. 1832.
T.: *H. tanacetifolia* (Linnaeus) H. G. L. Reichenbach (*Sisymbrium tanacetifolium* Linnaeus).

2999 **Arabidopsis** Heynhold in Holl et Heynhold, Fl. Sachsen **2**: 538. 1842.
T.: *A. thaliana* (Linnaeus) Heynhold (*Arabis thaliana* Linnaeus) (*typ. cons.*).

3013 **Lobularia** Desvaux, J. Bot. Agric. **3**: 162. 1814.
T.: *L. maritima* (Linnaeus) Desvaux (*Clypeola maritima* Linnaeus).

(≡) *Aduseton* Adanson, Fam. Pl. **2**: (23) [sic!], 420. 1763.

3022 **Lepidostemon** J. D. Hooker et Thomson, J. Linn. Soc., Bot. **5**: 131. 1861.
T.: *L. pedunculosus* J. D. Hooker et Thomson (*typ. cons.*).

(H) *Lepidostemon* Hasskarl, Cat. Hort. Bot. Bogor. 140. 1844 [Convolvul.].
≡ *Lepistemon* Blume 1825.

3032 **Malcolmia** R. Brown in W. T. Aiton, Hortus Kew. ed. 2. **4**: 121. 1812 ("*Malcomia*") (*orth. cons.*).
T.: *M. maritima* (Linnaeus) R. Brown (*Cheiranthus maritimus* Linnaeus) (*typ. cons.*).

(≡) *Wilckia* Scopoli, Intr. 317. 1777.

3038 **Euclidium** R. Brown in W. T. Aiton, Hortus Kew. ed. 2. **4**: 74. 1812.
T.: *E. syriacum* (Linnaeus) R. Brown (*Anastatica syriaca* Linnaeus).

(≡) *Hierochontis* Medikus, Pfl. Gatt. 51. 1792.
(=) *Soria* Adanson, Fam. Pl. **2**: 421. 1763.
T.: non designatus.

3042 **Matthiola** R. Brown in W. T. Aiton, Hortus Kew. ed. 2. **4**: 119. 1812 ("*Mathiola*") (*orth. cons.*).
T.: *M. incana* (Linnaeus) R. Brown (*Cheiranthus incanus* Linnaeus) (*typ. cons.*).

(H) *Matthiola* Linnaeus, Sp. Pl. 1192. 1753 [Rub.].
T.: *M. scabra* Linnaeus.

3050 **Dontostemon** Andrzejowski ex C. A. Meyer in Ledebour, Fl. Altaic. **3**: 4, 118. 1831.
T.: *D. integrifolius* (Linnaeus) C. A. Meyer (*Sisymbrium integrifolium* Linnaeus).

3051 **Chorispora** R. Brown ex A.-P. de Candolle, Mém. Mus. Hist. Nat. **7**: 237. Apr 1821.
T.: *C. tenella* (Pallas) A.-P. de Candolle (Syst. Nat. **2**: 435. Mai 1821) (*Raphanus tenellus* Pallas).

(≡) *Chorispermum* R. Brown in W. T. Áiton, Hortus Kew. ed. 2. **4**: 129. 1812.

TOVARIACEAE

3081 **Tovaria** Ruiz et Pavón, Prodr. 49. 1794.
T.: *T. pendula* Ruiz et Pavón (Fl. Peruv. Chil. **1**: 73. 1802).

(H) *Tovara* Adanson, Fam. Pl. **2**: 276, 612. 1763 [Polygon.].
T.: *Polygonum virginianum* Linnaeus.

CAPPARACEAE

3087 **Gynandropsis** A.-P. de Candolle, Prodr. **1**: 237. 1824.
T.: *G. pentaphylla* A.-P. de Candolle, nom. illeg. (*Cleome gynandra* Linnaeus, *G. gynandra* (Linnaeus) Briquet).

(≡) *Pedicellaria* Schrank, Bot. Mag. (Römer & Usteri) **3** (8): 10. 1790.

3103 **Steriphoma** K. Sprengel, Syst. Veg. 4 (2): 130, 139. 1827.
T.: *S. cleomoides* K. Sprengel, nom. illeg. (*Capparis paradoxa* N. J. Jacquin, *S. paradoxum* (N. J. Jacquin) Endlicher).

(=) *Hermupoa* Loefling, Iter Hispan. 307. 1758.
T.: *H. loeflingiana* A.-P. de Candolle (Prodr. **1**: 254. 1824, "*laeflingiana*").

204

3106 **Boscia** Lamarck, Tabl. Encycl. *t. 395.*
1793.
T.: *B. senegalensis* Lamarck (Tabl. Encycl. 2: 517. 1819.

RESEDACEAE

3122 **Caylusea** A. Saint-Hilaire, 2me Mém.
Réséd. 29. 1837 vel 1838.
T.: *C. canescens* Webb (in W. J. Hooker, Niger Flora 101. 1849), non (Linnaeus) Walpers 1843 [= *C. hexagyna* (Forsskål) M. L. Green].

3126 **Oligomeris** Cambessèdes in Jacquemont, Voy. Inde 4, Bot. 23. 1838 vel 1839 ("1844").
T.: *O. glaucescens* Cambessèdes.

(=) *Dipetalia* Rafinesque, Fl. Tell. 3: 73. 1837 ("1836").
T.: *D. capensis* (N. L. Burman) Rafinesque (*Reseda capensis* N. L. Burman).

(=) *Ellimia* Nuttall ex Torrey et A. Gray, Fl. N. Amer. 1: 125. Jul 1838.
T.: *E. ruderalis* Nuttall ex Torrey et A. Gray.

SARRACENIACEAE

3131 **Darlingtonia** Torrey, Smithson. Contr. Knowl. 6(4): 4. 1853.
T.: *D. californica* Torrey.

(H) *Darlingtonia* A.-P. de Candolle, Ann. Sci. Nat. (Paris) 4: 97. 1825 [Legum.].
LT.: *D. brachyloba* (Willdenow) A.-P. de Candolle (*Acacia brachyloba* Willdenow) (vide Regnum Veg. 8: 241. 1956).

CRASSULACEAE

3171 **Rochea** A.-P. de Candolle, Pl. Hist. Succ. sub. *t. 103* in adnot. 1802.
T.: *R. coccinea* (Linnaeus) A.-P. de Candolle (Pl. Hist. Succ. in indice) (*Crassula coccinea* Linnaeus) (*typ. cons.*).

(H) *Rochea* Scopoli, Intr. 296. 1777 [Legum.].
T.: non designatus.

3172 **Diamorpha** Nuttall, Gen. 1: 293. 1818.
T.: "north of Camden, South Carolina, winter 1816", Nuttall s.n. (**PH**) [= *D. smallii* Britton] (*typ. cons.*).

3176 **Cephalotus** Labillardière, Nov. Holl. Pl. 2: 6. 1807.
T.: *C. follicularis* Labillardière.

(H) *Cephalotos* Adanson, Fam. Pl. 2: 189, 534. 1763 [Lab.].
T.: *Thymus cephalotos* Linnaeus.

SAXIFRAGACEAE

3182 **Bergenia** Moench, Methodus 664. 1794.
T.: *B. bifolia* Moench, nom. illeg. (*B. crassifolia* (Linnaeus) Fritsch, *Saxifraga crassifolia* Linnaeus).

(H) *Bergena* Adanson, Fam. Pl. 2: 345. 1763 [Lecythid.].
≡ *Lecythis* Loefling 1758.

3185 **Boykinia** Nuttall, J. Acad. Nat. Sci. Philadelphia 7(1): 113. 1834.
T.: *B. aconitifolia* Nuttall.

(H) *Boykiana* Rafinesque, Neogenyton 2. 1825 [Lythr.].
T.: *B. humilis* (A. Michaux) Rafinesque (*Ammannia humilis* A. Michaux).

3187 **Suksdorfia** A. Gray, Proc. Amer. Acad. Arts 15: 41. 1880.
T.: *S. violacea* A. Gray.

(=) *Hemieva* Rafinesque, Fl. Tell. 2: 70. 1837 ("1836").
LT.: *H. ranunculifolia* (W. J. Hooker) Rafinesque (*Saxifraga ranunculifolia* W. J. Hooker) (vide Rydberg, N. Amer. Fl. 22: 121. 1905).

3196 **Tolmiea** Torrey et A. Gray, Fl. N. Amer. 1: 582. 1840.
T.: *T. menziesii* (W. J. Hooker) Torrey et A. Gray (*Heuchera menziesii* W. J. Hooker).

(H) *Tolmiea* W. J. Hooker, Fl. Bor.-Amer. 2: 44. 1834 [Eric.].
T.: *T. occidentalis* W. J. Hooker.
(≡) *Leptaxis* Rafinesque, Fl. Tell. 2: 75. 1837 ("1836").

3197 **Lithophragma** (Nuttall) Torrey et A. Gray, Fl. N. Amer. 1: 583. 1840.
T.: *L. parviflorum* (W. J. Hooker) Torrey et A. Gray (*Tellima parviflora* W. J. Hooker) (*typ. cons.*).

(≡) *Pleurendotria* Rafinesque, Fl. Tell. 2: 73. 1837 ("1836").

3201 **Vahlia** Thunberg, Nov. Gen. Pl. 2: 36. 1782.
T.: *V. capensis* (Linnaeus f.) Thunberg (*Russelia capensis* Linnaeus f.).

(=) *Bistella* Adanson, Fam. Pl. 2: 226. 1763.
T.: *B. geminiflora* Delile.

3204 **Donatia** J. R. Forster et G. Forster, Char. Gen. Pl. 5. 1775.
T.: *D. fascicularis* J. R. Forster et G. Forster.

3209 **Jamesia** Torrey et A. Gray, Fl. N. Amer. 1: 593. 1840.
T.: *J. americana* Torrey et A. Gray.

(H) *Jamesia* Rafinesque, Atlantic J. 1: 145. 1832 [Legum.].
T.: *J. obovata* Rafinesque, nom. illeg. (*Psoralea jamesii* Torrey).

3225 **Brexia** Noronha ex Du Petit-Thouars, Gen. Nov. Madagasc. 20. 1808.
T.: *B. madagascariensis* (Lamarck) Ker-Gawler (Bot. Reg. 730. 1823) (*Venana madagascariensis* Lamarck).

(≡) *Venana* Lamarck, Tabl. Encycl. 2: 99. 1797.

PITTOSPORACEAE

3252 **Pittosporum** Banks ex Solander in J. Gaertner, Fruct. Sem. Pl. 1: 286. 1788.
T.: *P. tenuifolium* J. Gaertner (*typ. cons.*).

(=) *Tobira* Adanson, Fam. Pl. 2: 449. 1763.
T.: non designatus ["Tobira, frutex arborescens, sagapeni foetoris . . ." Kaempfer].

CUNONIACEAE

3269 **Platylophus** D. Don, Edinburgh New Philos. J. 9: 92. 1830.
T.: *P. trifoliatus* (Linnaeus f.) D. Don (*Weinmannia trifoliata* Linnaeus f.).

(H) *Platylophus* Cassini, Dict. Sci. Nat. 44: 36. 1826 [Comp.].
T.: *Centaurea nigra* Linnaeus.

3275 **Cunonia** Linnaeus, Syst. Nat. ed. 10. 1025, 1368. 1759.
T.: *C. capensis* Linnaeus.

(H) *Cunonia* P. Miller, Fig. Pl. Gard. Dict. 1: 75. 1756 [Irid.].
T.: *Antholyza cunonia* Linnaeus.

3276 **Weinmannia** Linnaeus, Syst. Nat. ed. 10. 1005, 1367. 1759.
T.: *W. pinnata* Linnaeus.

(≡) *Windmannia* P. Browne, Civ. Nat. Hist. Jamaica 212. 1756.

BRUNIACEAE

3284 **Thamnea** Solander ex A. T. Brongniart, Ann. Sci. Nat. (Paris) 8: 386. 1826.
T.: *T. uniflora* Solander ex A. T. Brongniart (etiam vide 6157).

(H) *Thamnia* P. Browne, Civ. Nat. Hist. Jamaica 245. 1756 [Flacourt.].
T.: *Laetia thamnia* Linnaeus (Pl. Jamaic. Pug. 31. 1759).

3285 **Tittmannia** A. T. Brongniart, Ann. Sci. Nat. (Paris) 8: 385. 1826.
T.: *T. lateriflora* A. T. Brongniart.

(H) *Tittmannia* H. G. L. Reichenbach, Ic. Bot. Exot. 1: 26, *t. 38.* 1824. [Scrophular.].
T.: *T. viscosa* (Hornemann) H. G. L. Reichenbach (*Gratiola viscosa* Hornemann).

3286 **Lonchostoma** Wikström, Kongl. Vetensk. Acad. Handl. **1818**: 350. 1818.
T.: *L. obtusiflorum* Wikström, nom. illeg. (*Passerina pentandra* Thunberg, *L.
pentandrum* (Thunberg) Druce).

(=) *Ptyxostoma* Vahl, Skr. Naturhist.-Selsk.
6: 95. 1810.
T.: non designatus.

3292 **Brunia** Lamarck, Encycl. 1(2): 474.
1785.
T.: *B. paleacea* Bergius.

(H) *Brunia* Linnaeus, Sp. Pl. 199. 1753
[Brun.].
T.: *B. lanuginosa* Linnaeus.

ROSACEAE

3316 **Physocarpus** (Cambessèdes) Maximowicz, Trudy Imp. S.-Peterburgsk. Bot.
Sada **6**: 219. 1879.
T.: *P. opulifolius* (Linnaeus) Maximowicz (*Spiraea opulifolia* Linnaeus).

(≡) *Physocarpa* Rafinesque, Sylva Tell. 151.
1838.
(=) *Epicostorus* Rafinesque, Atlantic J. 1:
144. 1832.
T.: *E. montanus* Rafinesque.

3323 **Sorbaria** (Seringe ex A.-P. de Candolle)
A. Braun in Ascherson, Fl. Branden-
burg **1**: 177. 1860.
T.: *S. sorbifolia* (Linnaeus) A. Braun
(*Spiraea sorbifolia* Linnaeus).

(≡) *Schizonotus* Lindley, Intr. Nat. Syst. Bot.
81. 1830.

3328 **Lindleya** Kunth in Humboldt, Bonpland et Kunth, Nova Gen. Sp. **6**: ed.
fol. 188, ed. qu. 239. 1824.
T.: *L. mespiloides* Kunth.

(H) *Lindleya* C. G. Nees, Flora **4**: 299. 21 Mai
1821 [The.].
≡ *Wikstroemia* Schrader, 5 Mai 1821
(*nom. rej.* sub 5446).

3332 **Holodiscus** (C. Koch) Maximowicz,
Trudy Imp. S.-Peterburgsk. Bot. Sada **6**:
253. 1879.
T.: *H. discolor* (Pursh) Maximowicz
(*Spiraea discolor* Pursh) (*typ. cons.*).

3338a **Aronia** Medikus, Philos. Bot. **1**: 140,
155. 1789.
T.: *A. arbutifolia* (Linnaeus) Persoon
(Syn. **2**: 39. 1806) (*Mespilus arbutifolia*
Linnaeus).

(H) *Aronia* Mitchell, Diss. Brev. Bot. Zool.
28. 1769 [Ar.].
≡ *Orontium* Linnaeus 1753.

3339 **Rhaphiolepis** Lindley, Bot. Reg. **6**: *t.
468.* 1820 ("*Raphiolepis*") (*orth. cons.*).
T.: *R. indica* (Linnaeus) Lindley (*Crataegus indica* Linnaeus).

(=) *Opa* Loureiro, Fl. Cochinch. 308. 1790.
LT.: *O. metrosideros* Loureiro (vide
Seemann, J. Bot. **1**: 280. 1863; McVaugh,
Taxon **5**: 144. 1956).

3377 **Aremonia** Necker ex Nestler, Monogr.
Potentilla iv, 17. 1816.
T.: *A. agrimonoides* (Linnaeus) A.-P.
de Candolle (Prodr. **2**: 588. 1825)
(*Agrimonia agrimonoides* Linnaeus).

(≡) *Agrimonoides* P. Miller, Gard. Dict. Abr.
ed. 4. 1754.

208

CONNARACEAE

3424 **Rourea** Aublet, Hist. Pl. Guiane 467. 1775.
T.: *R. frutescens* Aublet.

3424a **Santaloides** Schellenberg, Beitr. Conn. 38. 1910.
T.: *S. minus* (J. Gaertner) Schellenberg (*Aegiceras minus* J. Gaertner) (*typ. cons.*).

(H) *Santalodes* O. Kuntze, Revis. Gen. Pl. 1: 155. 1891 [Connar.].
T.: *S. hermannianum* O. Kuntze, nom. illeg. (*Connarus santaloides* Vahl).
(≡) *Kalawael* Adanson, Fam. Pl. 2: 344, 530. 1763.

LEGUMINOSAE (FABACEAE)

3441 **Pithecellobium** C. F. P. Martius, Flora 20(2) (Beibl. 8): 114. 1837 ("*Pithecollobium*") (*orth. cons.*).
T.: *P. unguis-cati* (Linnaeus) Bentham (London J. Bot. 3: 200. 1844, "*Pithecolobium*") (*Mimosa unguis-cati* Linnaeus) (*typ. cons.*).

(=) *Zygia* P. Browne, Civ. Nat. Hist. Jamaica 279. 1756.
LT.: *Z. latifolia* (Linnaeus) Fawcett et Rendle (Fl. Jamaica 4: 150. 1920, q.v.) (*Mimosa latifolia* Linnaeus).

3444 **Calliandra** Bentham, J. Bot. (Hooker) 2: 138. 1840.
T.: *C. houstonii* Bentham, nom. illeg. (*Gleditsia inermis* Linnaeus, *C. inermis* (Linnaeus) Druce).

3448 **Schrankia** Willdenow, Sp. Pl. 4: 1041. 1806.
T.: *S. aculeata* Willdenow, nom. illeg. (*Mimosa quadrivalvis* Linnaeus, *S. quadrivalvis* (Linnaeus) Merrill).

(H) *Schranckia* J. F. Gmelin, Syst. Nat. 2: 312, 515. 1791 [Celastr.].
T.: *S. quinquefaria* J. F. Gmelin.

3450 **Desmanthus** Willdenow, Sp. Pl. 4: 1044. 1806.
T.: *D. virgatus* (Linnaeus) Willdenow (*Mimosa virgata* Linnaeus).

(≡) *Acuan* Medikus, Theodora 62. 1786.

3452 **Dichrostachys** (A.-P. de Candolle) Wight et Arnott, Prodr. 271. 1834.
T.: *D. cinerea* (Linnaeus) Wight et Arnott (*Mimosa cinerea* Linnaeus).

(=) *Cailliea* Guillemin et Perrottet in Guillemin, Perrottet et A. Richard, Fl. Seneg. Tent. 239. 1832.
T.: *C. dicrostachys* Guillemin et Perrottet, nom. illeg. (*Mimosa nutans* Persoon).

3466 **Plathymenia** Bentham, J. Bot. (Hooker) 2: 134. 1840.
T.: *P. foliolosa* Bentham.

(=) *Echyrospermum* Schott in Schreibers, Nachr. Österr. Naturf. Bras. 2(Anhang): 55. 1822.
T.: non designatus.

3468 **Entada** Adanson, Fam. Pl. 2: 318, 554. 1763.
T.: *E. rheedei* K. Sprengel (Syst. Veg. 2: 325. 1825, "*rheedii*") (*Mimosa entada* Linnaeus).

(=) *Gigalobium* P. Browne, Civ. Nat. Hist. Jamaica 362. 1756.
LT.: *Entada gigas* (Linnaeus) Fawcett et Rendle (*Mimosa gigas* Linnaeus) (vide Panigrahi, Taxon **34**: 714. 1985).

3490 **Copaifera** Linnaeus, Sp. Pl. ed. 2. 557. 1762.
T.: *C. officinalis* (N. J. Jacquin) Linnaeus (*Copaiva officinalis* N. J. Jacquin).

(≡) *Copaiva* N. J. Jacquin, Enum. Syst. Pl. 4, 21. 1760.
(=) *Copaiba* P. Miller, Gard. Dict. Abr. ed. 4. 1754.
T.: non designatus.

3495 **Crudia** Schreber, Gen. 1: 282. 1789.
T.: *C. spicata* (Aublet) Willdenow (Sp. Pl. 2: 539. 1799) (*Apalatoa spicata* Aublet) (*typ. cons.*).

(≡) *Apalatoa* Aublet, Hist. Pl. Guiane 382. 1775.
(=) *Touchiroa* Aublet, Hist. Pl. Guiane 384. 1775.
T.: *T. aromatica* Aublet.

3500 **Peltogyne** Vogel, Linnaea **11**: 410. 1837.
T.: *P. discolor* Vogel.

(=) *Orectospermum* Schott in Schreibers, Nachr. Österr. Naturf. Bras. 2(Anhang): 54. 1822.
T.: non designatus.

3506 **Schotia** N. J. Jacquin, Collectanea **1**: 93. 1787 ("1786").
T.: *S. speciosa* N. J. Jacquin, nom. illeg. (*Guaiacum afrum* Linnaeus, *S. afra* (Linnaeus) Thunberg).

(≡) *Theodora* Medikus, Theodora 16. 1786.

3509 **Afzelia** J. E. Smith, Trans. Linn. Soc. London **4**: 221. 1798.
T.: *A. africana* J. E. Smith ex Persoon (Syn. Pl. 1: 455. 1805).

(H) *Afzelia* J. F. Gmelin, Syst. Nat. **2**: 927. 1791 [Scrophular.].
≡ *Seymeria* Pursh 1814 (*nom. cons.*) (7602).

3516 **Berlinia** Solander ex J. D. Hooker in W. J. Hooker, Niger Fl. 326. 1849.
T.: *B. acuminata* Solander ex J. D. Hooker.

(=) *Westia* Vahl, Skr. Naturhist.-Selsk. **6**: 117. 1810.
T.: non designatus.

3517 **Macrolobium** Schreber, Gen. **1**: 30. 1789.
T.: *M. vuapa* J. F. Gmelin (Syst. Nat. **2**: 93. 1791), nom. illeg. (*Vouapa bifolia* Aublet, *M. bifolium* (Aublet) Persoon) (*typ. cons.*).

(≡) *Vouapa* Aublet, Hist. Pl. Guiane 25. 1775.
(=) *Outea* Aublet, Hist. Pl. Guiane 28. 1775.
T.: *O. guianensis* Aublet.

3518 **Humboldtia** Vahl, Symb. **3**: 106. 1794.
T.: *H. laurifolia* Vahl.

(H) *Humboltia* Ruiz et Pavón, Prodr. 121. Oct 1794 [Orch.].
LT.: *H. purpurea* Ruiz et Pavón (vide Garay et Sweet, J. Arnold Arbor. **53**: 522. 1972).

3524 **Brownea** N. J. Jacquin, Enum. Syst. Pl. 6, 26. 1760 ("*Brownaea*") (*orth. cons.*).
T.: *B. coccinea* N. J. Jacquin.

(=) *Hermesias* Loefling, Iter Hispan. 278. 1758.
T.: non designatus.

3528 **Piliostigma** Hochstetter, Flora **29**: 598. 1846.
T.: *P. reticulatum* (A.-P. de Candolle) Hochstetter (*Bauhinia reticulata* A.-P. de Candolle) (*typ. cons.*).

3532 **Apuleia** C. F. P. Martius, Flora **20**(2) (Beibl. 8): 123. 1837 ("*Apuleja*") (*orth. cons.*).
T.: *A. praecox* C. F. P. Martius.

3553 **Pterolobium** R. Brown ex Wight et Arnott, Prodr. 283. 1834.
T.: *P. lacerans* (Roxburgh) Wight et Arnott (*Caesalpinia lacerans* Roxburgh).

3557 **Hoffmannseggia** Cavanilles, Icon. **4**: 63, tt. *392, 393*. 1798 ("*Hoffmanseggia*") (*orth. cons.*).
T.: *H. falcaria* Cavanilles, nom. illeg. (*Larrea glauca* Ortega, *H. glauca* (Ortega) Eifert) (etiam vide 3973).

3558 **Zuccagnia** Cavanilles, Icon. **5**: 2. 1799.
T.: *Z. punctata* Cavanilles.

3561 **Peltophorum** (T. Vogel) Bentham, J. Bot. (Hooker) **2**: 75. 1840.
T.: *P. dubium* (K. Sprengel) Taubert (*Caesalpinia dubia* K. Sprengel).

3574 **Swartzia** Schreber, Gen. **2**: 518. 1791.
T.: *S. alata* Willdenow (Sp. Pl. **2**: 1220. 1800).

3575 **Aldina** Endlicher, Gen. Pl. 1322. 1840.
T.: *A. insignis* (Bentham) Endlicher (*Allania insignis* Bentham).

3582 **Sweetia** K. Sprengel, Syst. Veg. **2**: 171, 23. 1825.
T.: *S. fruticosa* K. Sprengel.

(=) *Elayuna* Rafinesque, Sylva Tell. 145. 1838.
T.: *E. biloba* Rafinesque, nom. illeg. (*Bauhinia tamarindacea* Delile).

(H) *Apuleja* J. Gaertner, Fruct. Sem. Pl. **2**: 439. 1791 [Comp.].
≡ *Berkheya* Ehrhart 1788 (*nom. cons.*). (9438).

(H) *Pterolobium* Andrzejowski ex C. A. Meyer, Verz. Pfl. Casp. Meer. 185. 1831 [Cruc.].
T.: *P. biebersteinii* Andrzejowski ex C. A. Meyer, nom. illeg. (*Thlaspi latifolium* Marschall von Bieberstein).
(=) *Cantuffa* J. F. Gmelin, Syst. Nat. **2**: 677. 1791.
T.: *C. excelsa* J. F. Gmelin.

(H) *Zuccangnia* Thunberg, Nov. Gen. Pl. 127. 1798 [Lil.].
T.: *Z. viridis* (Linnaeus) Thunberg (*Hyacinthus viridis* Linnaeus).

(=) *Baryxylum* Loureiro, Fl. Cochinch, 266. 1790.
T.: *B. rufum* Loureiro.

(=) *Possira* Aublet, Hist. Pl. Guiane 934. 1775.
T.: *P. arborescens* Aublet.
(=) *Tounatea* Aublet, Hist. Pl. Guiane 549. 1775.
T.: *T. guianensis* Aublet.

(H) *Aldina* Adanson, Fam. Pl. **2**: 328. 1763 [Legum.].
≡ *Brya* P. Browne 1756.

3584 **Myroxylon** Linnaeus f., Suppl. Pl. 34, 233. 1782.
T.: *M. peruiferum* Linnaeus f.

(H) *Myroxylon* J. R. Forster et G. Forster, Char. Gen. Pl. 63. 1775 [Flacourt.].
≡ *Xylosma* G. Forster 1786 (*nom. cons.*) (5320).

(=) *Toluifera* Linnaeus, Sp. Pl. 384. 1753.
T.: *T. balsamum* Linnaeus.

3589 **Camoensia** Welwitsch ex Bentham in Bentham et J. D. Hooker, Gen. Pl. 1: 557. 1865.
T.: *C. maxima* Welwitsch ex Bentham (Trans. Linn. Soc. London 25: 301. 1865) (*typ. cons.*).

(=) *Giganthemum* Welwitsch, Anais Cons. Ultramarino Parte Não Offic. 1858: 585. 1859.
T.: *G scandens* Welwitsch.

3597 **Ormosia** G. Jackson, Trans. Linn. Soc. London 10: 360. 1811.
T.: *O. coccinea* (Aublet) G. Jackson (*Robinia coccinea* Aublet) (*typ. cons.*).

(=) *Toulichiba* Adanson, Fam. Pl. 2: 326. 1763.
T.: non designatus ["Plum. M.S. Vol. 7. t. 145."].

3608 **Virgilia** Poiret in Lamarck, Encycl. 8: 677. 1808.
T.: *V. capensis* (Linnaeus) Poiret (*Sophora capensis* Linnaeus) (*typ. cons.*).

(H) *Virgilia* L'Héritier, Virgilia. 1788 [Comp.].
T.: *V. helioides* L'Héritier.

3619 **Pickeringia** Nuttall ex Torrey et A. Gray, Fl. N. Amer. 1: 388. 1840.
T.: *P. montana* Nuttall ex Torrey et A. Gray.

(H) *Pickeringia* Nuttall, J. Acad. Nat. Sci. Philadelphia 7: 95. 1834 [Myrsin.].
T.: *P. paniculata* (Nuttall) Nuttall (*Cyrilla paniculata* Nuttall).

3621 **Podalyria** Willdenow, Sp. Pl. 2: 501. 1799.
T.: *P. retzii* (J. F. Gmelin) Rickett et Stafleu (*Sophora retzii* J. F. Gmelin, *Sophora biflora* Retzius 1779, non Linnaeus 1759) (*typ. cons.*).

3624 **Oxylobium** H. Andrews, Bot. Repos. 7: t. 492. 1807.
T.: *O. cordifolium* H. Andrews.

(=) *Callistachys* Ventenat, Jard. Malm. t. 115. 1805.
T.: *C. lanceolata* Ventenat.

3629 **Burtonia** R. Brown in W. T. Aiton, Hortus Kew. ed. 2. 3: 12. 1811.
T.: *B. scabra* (J. E. Smith) R. Brown (*Gompholobium scabrum* J. E. Smith).

(H) *Burtonia* R. A. Salisbury, Parad. Lond. 73. 1807 [Dillen.].
T.: *B. grossulariifolia* R. A. Salisbury ("*grossulariaefolia*").

3647 **Walpersia** W. H. Harvey in W. H. Harvey et Sonder, Fl. Cap. 2: 26. 1862.
T.: *W. burtonioides* W. H. Harvey.

(H) *Walpersia* Reissek ex Endlicher, Gen. Pl. 1100. 1840 [Rhamn.].
≡ *Trichocephalus* A. T. Brongniart 1827.

3657 **Lotononis** (A.-P. de Candolle) Ecklon et Zeyher, Enum. Pl. Afric. Austral. 176. 1836.
T.: *L. vexillata* (E. Meyer) Ecklon et Zeyher (*Crotalaria vexillata* E. Meyer) [= *Ononis prostrata* Linnaeus, *L. prostrata* (Linnaeus) Bentham] (*typ. cons.*).

(=) *Amphinomia* A.-P. de Candolle, Prodr. 2: 522. 1825.
T.: *A. decumbens* (Thunberg) A.-P. de Candolle (*Connarus decumbens* Thunberg).

(=) *Leobordea* Delile in Laborde, Voy. Arabie Pétrée 82, 86. 1830.
T.: *L. lotoidea* Delile.

3659 **Rothia** Persoon, Syn. Pl. 2: 638, et corrig. 1807.
T.: *R. trifoliata* (Roth) Persoon (*Dillwynia trifoliata* Roth) (*typ. cons.*).

(H) *Rothia* Schreber, Gen. 2: 531. 1791 [Comp.].
T.: non designatus.

3661 **Wiborgia** Thunberg, Nov. Gen. Pl. 137. 1800.
T.: *W. obcordata* Thunberg (*typ. cons.*).

(H) *Viborgia* Moench, Methodus 132. 1794 [Legum.].
T.: non designatus.

3673 **Argyrolobium** Ecklon et Zeyher, Enum. Pl. Afric. Austral. 184. 1836.
T.: *A. argenteum* (N. J. Jacquin) Ecklon et Zeyher (*Crotalaria argentea* N. J. Jacquin) (*typ. cons.*).

(=) *Lotophyllus* Link, Handbuch 2: 156. 1831.
T.: *L. argenteus* Link.

3675a **Retama** Rafinesque, Sylva Tell. 22. 1838.
T.: *R. monosperma* (Linnaeus) Boissier (*Spartium monospermum* Linnaeus [var. *b*]).

(≡) *Lygos* Adanson, Fam. Pl. 2: 321, 573. 1763.

3676 **Petteria** K. B. Presl, Abh. Königl. Böhm. Ges. Wiss. ser. 5. 3: 569. 1845.
T.: *P. ramentacea* (Sieber) K. B. Presl (*Cytisus ramentaceus* Sieber).

(H) *Pettera* H. G. L. Reichenbach, Icon. Fl. German. 5: 33, *t. 220.* 1841 [Caryophyll.].
T.: *P. graminifolia* (Arduino) H. G. L. Reichenbach (*Arenaria graminifolia* Arduino).

3682 **Cytisus** Desfontaines, Fl. Atlant. 2: 139. 1798.
T.: *C. triflorus* L'Héritier 1791, non Lamarck 1786 [= *C. villosus* Pourret] (*typ. cons.*).

(H) *Cytisus* Linnaeus, Sp. Pl. 739. 1753.
LT.: *C. sessiliflorus* Linnaeus (vide Hitchcock et Green, Nomencl. Prop. Brit. Botanists 175. 1929; Polhill et al., Taxon 27: 556. 1978).

3682a **Sarothamnus** C. F. H. Wimmer, Fl. Schles. 278. 1832.
T.: *S. vulgaris* C. F. H. Wimmer, nom. illeg. (*Spartium scoparium* Linnaeus, *S. scoparius* (Linnaeus) W. D. J. Koch).

(≡) *Cytisogenista* Duhamel, Traité Arbr. Arbust. 1: 203. 1755.

*3688 **Medicago** Linnaeus, Sp. Pl. 778. 1753.
T.: *M. sativa* Linnaeus (*typ. cons.*).

3693 **Hymenocarpos** Savi, Fl. Pisana 2: 205. 1798.
T.: *H. circinnatus* (Linnaeus) Savi (*Medicago circinnata* Linnaeus).

(≡) *Circinnus* Medikus, Vorles. Churpfälz. Phys.-Öcon. Ges. 2: 384. 1787.

3694 **Securigera** A.-P. de Candolle in Lamarck et A.-P. de Candolle, Fl. Franç. ed. 3. 4: 609. 1805.
T.: *S. coronilla* A.-P. de Candolle, nom. illeg. (*Coronilla securidaca* Linnaeus, *S. securidaca* (Linnaeus) Dalla Torre et Sarntheim).

(≡) *Bonaveria* Scopoli, Intr. 310. 1777.

3699 **Tetragonolobus** Scopoli, Fl. Carn. ed. 2. 2: 87 ("*Tetragonobolus*"), 507. 1772.
T.: *T. scandalida* Scopoli, nom. illeg. (*Lotus siliquosus* Linnaeus, *T. siliquosus* (Linnaeus) Roth).

(≡) *Scandalida* Adanson, Fam. Pl. 2: 326, 602. 1763.

3708 **Eysenhardtia** Kunth in Humboldt, Bonpland et Kunth, Nova Gen. Sp. 6: ed. fol. 382, ed. qu. 489. 1824.
T.: *E. amorphoides* Kunth.

(=) *Viborquia* Ortega, Nov. Pl. Descr. Dec. 66. 1798.
T.: *V. polystachya* Ortega.

3709 **Dalea** Linnaeus, Opera Var. 244. 1758.
T.: *D. alopecuroides* Willdenow (*Psoralea dalea* Linnaeus).

(H) *Dalea* P. Miller, Gard. Dict. Abr. ed. 4. 1754 [Solan.].
≡ *Browallia* Linnaeus 1753.

3710 **Petalostemon** A. Michaux, Fl. Bor.-Amer. 2: 48. 1803 ("*Petalostemum*") (*orth. cons.*).
T.: *P. candidum* (Willdenow) A. Michaux (*Dalea candida* Willdenow).

(=) *Kuhnistera* Lamarck, Encycl. 3: 370. 1792 ("1789").
T.: *K. caroliniensis* Lamarck.

3718 **Tephrosia** Persoon, Syn. Pl. 2: 328. 1807.
T.: *T. villosa* (Linnaeus) Persoon (*Cracca villosa* Linnaeus) (etiam vide 3745).

(=) *Erebinthus Mitchell, Diss. Brev. Bot. Zool. 32. 1769.
T.: *Tephrosia spicata* (Walter) Torrey et A. Gray (*Galega spicata* Walter) (vide Wood, Rhodora 51: 292. 1948).

(=) *Needhamia* Scopoli, Intr. 310. 1777.
T.: *Vicia littoralis* N. J. Jacquin (Enum. Syst. Pl. 27. 1760).

(=) *Reineria* Moench, Suppl. Meth. Pl. 44. 1802.
T.: *R. reflexa* Moench.

3720 **Millettia** Wight et Arnott, Prodr. 263. 1834.
T.: *M. rubiginosa* Wight et Arnott.

(=) *Pongam* Adanson, Fam. Pl. 2: 322, 593. 1763 (*nom. rej.* sub 3836).

(=) *Pongamia* Ventenat, Jard. Malm. t. 28. 1803 (*nom. cons.*) (3836).

3722 **Wisteria** Nuttall, Gen. N. Amer. Pl. 2: 115. 1818.
T.: *W. speciosa* Nuttall, nom. illeg. (*Glycine frutescens* Linnaeus, *W. frutescens* (Linnaeus) Poiret).

(≡) *Phaseoloides* Duhamel, Traité Arbr. Arbust. 2: 115. 1755.

(=) *Diplonyx* Rafinesque, Fl. Ludov. 101. 1817.
T.: *D. elegans* Rafinesque.

3745 **Cracca** Bentham in Bentham et Oersted, Vidensk. Meddel. Dansk Naturhist. Foren. Kjøbenhavn **1853**: 8. 1853.
T.: *C. glandulifera* Bentham.

(H) *Cracca* Linnaeus, Sp. Pl. 752. 1753 [Legum.].
≡ *Tephrosia* Persoon 1807 (*nom. cons.*) (3718).

3747 **Sesbania** Scopoli, Intr. 308. 1777.
T.: *S. sesban* (Linnaeus) Merrill (*Aeschynomene sesban* Linnaeus).

(≡) *Sesban* Adanson, Fam. Pl. 2: 327, 604. 1763.
(=) *Agati* Adanson, Fam. Pl. 2: 326, 513. 1763.
T.: *Robinia grandiflora* Linnaeus.

3753 **Clianthus** Solander ex Lindley, Edward's Bot. Reg. **21**: t. *1775*. 1835.
T.: *C. puniceus* (G. Don) Solander ex Lindley (*Donia punicea* G. Don).

(=) *Sarcodum* Loureiro, Fl. Cochinch. 461. 1790.
T.: *S. scandens* Loureiro.

3754 **Sutherlandia** R. Brown in W. T. Aiton, Hortus Kew. ed. 2. 4: 327. 1812.
T.: *S. frutescens* (Linnaeus) R. Brown (*Colutea frutescens* Linnaeus).

(H) *Sutherlandia* J. F. Gmelin, Syst. Nat. 2: 1027. 1791 [Stercul.].
≡ *Heritiera* W. Aiton 1789.

3756 **Lessertia** A.-P. de Candolle, Astragalogia ed. fol. 4, 15, 37; ed. qu. 5, 19, 47. 1802.
T.: *L. perennans* (N. J. Jacquin) A.-P. de Candolle (*Colutea perennans* N. J. Jacquin).

(≡) *Sulitra* Medikus, Vorles. Churpfälz. Phys.-Öcon. Ges. 2: 366. 1787 (vide Brummitt, Regnum Veg. 40: 24. 1965).
(=) *Coluteastrum* Fabricius, Enum. ed. 2. 317. 1763.
LT.: *C. herbaceum* (Linnaeus) O. Kuntze (*Colutea herbacea* Linnaeus).

3767 **Oxytropis** A.-P. de Candolle, Astragalogia ed. fol. 3, 19, 53, ed. qu. 4, 24, 66. 1802.
T.: *O. montana* (Linnaeus) A.-P. de Candolle (*Astragalus montanus* Linnaeus) (*typ. cons.*).

3784 **Nissolia** N. J. Jacquin, Enum. Syst. Pl. 7, 27. 1760.
T.: *N. fruticosa* N. J. Jacquin (*typ. cons.*).

(H) *Nissolia* P. Miller, Gard. Dict. Abr. ed. 4. 1754 [Legum.].
T.: non designatus.

3789 **Poiretia** Ventenat, Mém. Cl. Sci. Math. Inst. Natl. France **1807**(1): 4. 1807.
T.: *P. scandens* Ventenat.

(H) *Poiretia* J. F. Gmelin, Syst. Nat. 2: 263. 1791 [Rub.].
T.: non designatus.

3792 **Ormocarpum** Palisot de Beauvois, Fl. Oware 1: 95. 1810(?) ("1806").
T.: *O. verrucosum* Palisot de Beauvois.

(=) *Diphaca* Loureiro, Fl. Cochinch. 453. 1790.
T.: *D. cochinchinensis* Loureiro.

3796 **Smithia** W. Aiton, Hortus Kew. 3: 496. 1789.
T.: *S. sensitiva* W. Aiton.

(H) *Smithia* Scopoli, Intr. 322. 1777 [Clus.].
≡ *Quapoya* Aublet 1775.
(≡) *Damapana* Adanson, Fam. Pl. 2: 323. 1763.

3800 **Adesmia** A.-P. de Candolle, Ann. Sci. Nat. (Paris) 4: 94. 1825 ("1824").
T.: *A. muricata* (N. J. Jacquin) A.-P. de Candolle (*Hedysarum muricatum* N. J. Jacquin) (*typ. cons.*).

(≡) *Patagonium* Schrank, Denkschr. Königl. Akad. Wiss. München **1808**: 93. 1809.

3807 **Desmodium** Desvaux, J. Bot. Agric. 1: 112. 1813.
T.: *D. scorpiurus* (Swartz) Desvaux (*Hedysarum scorpiurus* Swartz) (*typ. cons.*).

(=) *Meibomia* Heister ex Fabricius, Enum. 168. 1759.
LT.: *Hedysarum canadense* Linnaeus (vide Adanson, Fam. Pl. 2: 509, 575. 1763).

(=) *Grona* Loureiro, Fl. Cochinch. 424, 459. 1790.
T.: *G. repens* Loureiro.

(=) *Pleurolobus* Jaume Saint-Hilaire, Nouv. Bull. Sci. Soc. Philom. Paris 3: 192. 1812.
T.: non designatus.

3810 **Alysicarpus** Desvaux, J. Bot. Agric. 1: 120. 1813.
T.: *A. bupleurifolius* (Linnaeus) A.-P. de Candolle (*Hedysarum bupleurifolium* Linnaeus) (*typ. cons.*).

3821 **Dalbergia** Linnaeus f., Suppl. Pl. 52, 316. 1782.
T.: *D. lanceolaria* Linnaeus f.

(=) *Amerimnon* P. Browne, Civ. Nat. Hist. Jamaica 288. 1756.
T.: *A. brownei* N. J. Jacquin (Enum. Syst. Pl. 27. 1760).

(=) *Ecastaphyllum* P. Browne, Civ. Nat. Hist. Jamaica 299. 1756.
T.: *E. brownei* Persoon (Syn. Pl. 2: 277. 1807) (*Hedysarum ecastaphyllum* Linnaeus).

(=) *Acouroa* Aublet, Hist. Pl. Guiane 753. 1775.
T.: *A. violacea* Aublet.

3823 **Machaerium** Persoon, Syn. Pl. 2: 276. 1807.
T.: *M. ferrugineum* (Willdenow) Persoon (*Nissolia ferruginea* Willdenow).

(=) *Nissolius* Medikus, Vorles. Churpfälz. Phys.-Öcon. Ges. 2: 389. 1787.
T.: *N. arboreus* (N. J. Jacquin) Medikus (*Nissolia arborea* N. J. Jacquin).

(=) *Quinata* Medikus, Vorles. Churpfälz. Phys.-Öcon. Ges. 2: 389. 1787.
T.: *Q. violacea* Medikus (*Nissolia quinata* Aublet).

3828 **Pterocarpus** N. J. Jacquin, Select. Stirp. Amer. Hist. 283. 1763.
T.: *P. officinalis* N. J. Jacquin (*typ. cons.*).

(H) *Pterocarpus* Linnaeus, Herb. Amb. 10. 1754 [Legum.].
T.: non designatus ["Fl. Zeyl. 417"].

216

3834 **Lonchocarpus** Kunth in Humboldt, Bonpland et Kunth, Nova Gen. Sp. 6: ed. fol. 300, ed. qu. 383. 1824.
T.: *L. sericeus* (Poiret) A.-P. de Candolle (*Robinia sericea* Poiret) (*typ. cons.*).

(=) *Clompanus* Aublet, Hist. Pl. Guiane 773. 1775.
T.: *C. paniculata* Aublet.
(=) *Coublandia* Aublet, Hist. Pl. Guiane 937. 1775 (*nom. rej.* sub 3837).
(=) *Muellera* Linnaeus f., Suppl. Pl. 53, 329. 1782 (*nom. cons.*) (3837).

3836 **Pongamia** Ventenat, Jard. Malm. *t. 28.* 1803.
T.: *P. glabra* Ventenat, nom. illeg. (*Cytisus pinnatus* Linnaeus, *P. pinnata* (Linnaeus) Pierre) (etiam vide 3720).

(=) *Pongam* Adanson, Fam. Pl. 2: 322, 593. 1763.
T.: *Dalbergia arborea* Willdenow.
(=) *Galedupa* Lamarck, Encycl. 2: 594. 1788 ("1786").
T.: *G. indica* Lamarck.
(=) *Pungamia* Lamarck, Tabl. Encycl. *t. 603.* 1796.
T.: non designatus.

3837 **Muellera** Linnaeus f., Suppl. Pl. 53, 329. 1782.
T.: *M. moniliformis* Linnaeus f. (etiam vide 3834).

(=) *Coublandia* Aublet, Hist. Pl. Guiane 937. 1775.
T.: *C. frutescens* Aublet.

3838 **Derris** Loureiro, Fl. Cochinch. 432. 1790.
T.: *D. trifoliata* Loureiro (*typ. cons.*).

(=) *Salken* Adanson, Fam. Pl. 2: 322, 600. 1763.
T.: non designatus ["H.M. 8. t. 46"].
(=) *Solori* Adanson, Fam. Pl. 2: 327, 606. 1763.
T.: non designatus ["H.M. 6. t. 22"].
(=) *Deguelia* Aublet, Hist. Pl. Guiane 750. 1775.
T.: *D. scandens* Aublet.

3839 **Piscidia** Linnaeus, Syst. Nat. ed. 10. 1155, 1376. 1759.
T.: *P. erythrina* Linnaeus, nom. illeg. (*Erythrina piscipula* Linnaeus, *P. piscipula* (Linnaeus) Sargent).

(≡) *Ichthyomethia* P. Browne, Civ. Nat. Hist. Jamaica 296. 1756.

3841 **Andira** A. L. Jussieu, Gen. Pl. 363. 1789.
T.: *A. racemosa* Lamarck ex Jaume Saint-Hilaire (Dict. Sci. Nat. 2: 137. 1804).

(=) *Vouacapoua* Aublet, Hist. Pl. Guiane, Suppl. 9. 1775.
T.: *V. americana* Aublet.

3845 **Dipteryx** Schreber, Gen. 2: 485. 1791.
T.: *D. odorata* (Aublet) Willdenow (*Coumarouna odorata* Aublet) (*typ. cons.*).

(≡) *Coumarouna* Aublet, Hist. Pl. Guiane 740. 1775.
(=) *Taralea* Aublet, Hist. Pl. Guiane 745. 1775.
T.: *T. oppositifolia* Aublet.

3848 **Inocarpus** J. R. Forster et G. Forster, Char. Gen. Pl. 33. 1775.
T.: *I. edulis* J. R. Forster et G. Forster.

(=) *Aniotum* S. Parkinson, J. Voy. South Seas 39. 1773.
T.: *A. fagiferum* S. Parkinson.

3853 **Lens** P. Miller, Gard. Dict. Abr. ed. 4. 1754.
T.: *L. culinaris* Medikus (Vorles. Chur-pfälz. Phys.-Öcon. Ges. **2**: 361. 1787) (*Ervum lens* Linnaeus).

3858 **Centrosema** (A.-P. de Candolle) Bentham, Comm. Legum. Gen. 53. 1837.
T.: *C. brasilianum* (Linnaeus) Bentham (*Clitoria brasiliana* Linnaeus) (*typ. cons.*).

(=) *Steganotropis* J. G. C. Lehmann, Sem. Hort. Bot. Hamburg. **1826**: 18. 1826; Linnaea **3** (Litt.): 11. 1828.
T.: *S. conjugata* J. G. C. Lehmann.

3860 **Amphicarpaea** S. Elliott ex Nuttall, Gen. N. Amer. Pl. **2**: 113. 1818 ("*Amphicarpa*") (*orth. cons.*).
T.: *A. monoica* S. Elliott ex Nuttall, nom. illeg. (*Glycine monoica* Linnaeus, nom. illeg., *Glycine bracteata* Linnaeus, *A. bracteata* (Linnaeus) Fernald).

(=) *Falcata* J. F. Gmelin, Syst. Nat. **2**: 1131. 1791.
T.: *F. caroliniana* J. F. Gmelin.

3863 **Shuteria** Wight et Arnott, Prodr. 207. Oct 1834.
T.: *S. vestita* Wight et Arnott (*typ. cons.*).

(H) *Shutereia* Choisy, Convolv. Orient. 103. Aug 1834; Mém. Soc. Phys. Genève **6**: 485. 1834 [Convolvul.].
T.: *S. bicolor* Choisy (*Convolvulus bicolor* Vahl 1794, non Lamarck 1788).

3864 **Glycine** Willdenow, Sp. Pl. ed. 4. **3**(2): 1053. 1802.
T.: *G. clandestina* Wendland (Bot. Beobacht. 54. 1798).

(H) *Glycine* Linnaeus, Sp. Pl. 753, 754. 1753 [Legum.].
LT.: *G. javanica* Linnaeus (vide Green, Nomencl. Prop. Brit. Botanists 176. 1929).
(=) *Soja* Moench, Meth. Pl. 153, index. 1794.
T.: *S. hispida* Moench./.

3868 **Kennedia** Ventenat, Jard. Malm. *t. 104*. 1805 ("1804").
T.: *K. rubicunda* (Schneevoigt) Ventenat (*Glycine rubicunda* Schneevoigt).

3871 **Rhodopis** Urban, Symb. Antill. **2**: 304. 1900.
T.: *R. planisiliqua* (Linnaeus) Urban (*Erythrina planisiliqua* Linnaeus).

(H) *Rhodopsis* Lilja, Fl. Sv. Odl. Vext. Suppl. **1**: 42. 1840 [Portulac.].
≡ *Tegneria* Lilja 1839.

3874 **Apios** Fabricius, Enum. 176. 1759.
T.: *A. americana* Medikus (*Glycine apios* Linnaeus).

3876 **Butea** Roxburgh ex Willdenow, Sp. Pl. **3**: 917. 1802.
T.: *B. frondosa* Roxburgh ex Willdenow, nom. illeg. (*Erythrina monosperma* Lamarck, *B. monosperma* (Lamarck) Taubert) (*typ. cons.*).

(≡) *Plaso* Adanson, Fam. Pl. **2**: 325, 592. 1763 (vide Panigrahi et Mishra, Taxon **33**: 119. 1984).

3877 **Mucuna** Adanson, Fam. Pl. 2: 325, 579. 1763.
T.: *M. urens* (Linnaeus) A.-P. de Candolle (*Dolichos urens* Linnaeus) (*typ. cons.*).

(≡) *Zoophthalmum* P. Browne, Civ. Nat. Hist. Jamaica 295. 1756.
(=) *Stizolobium* P. Browne, Civ. Nat. Hist. Jamaica 290. 1756.
LT.: *S. pruriens* (Linnaeus) Medikus (*Dolichos pruriens* Linnaeus) (vide O. Kuntze, Revis. Gen. Pl. 207. 1891; Piper, U.S.D.A. Bur. Pl. Industr. Bull. 179: 9. 1910).

3891 **Canavalia** A.-P. de Candolle, Prodr. 2: 403. 1825.
T.: *C. ensiformis* (Linnaeus) A.-P. de Candolle (*Dolichos ensiformis* Linnaeus).

(≡) *Canavali* Adanson, Fam. Pl. 2: 325, 531. 1763.

3892 **Cajanus** A.-P. de Candolle, Cat. Pl. Horti Monsp. 85. 1813.
T.: *C. cajan* (Linnaeus) Millspaugh (*Cytisus cajan* Linnaeus).

(≡) *Cajan* Adanson, Fam. Pl. 2: 326, 529. 1763.

3897 **Rhynchosia** Loureiro, Fl. Cochinch. 460. 1790.
T.: *R. volubilis* Loureiro.

(=) *Dolicholus* Medikus, Vorles. Churpfälz. Phys.-Öcon. Ges. 2: 354. 1787.
T.: *D. flavus* Medikus, nom. illeg. (*Dolichos minimus* Linnaeus).
(=) *Cylista* W. Aiton, Hortus Kew. 3: 36, 512. 1789.
T.: *C. villosa* W. Aiton.

3898 **Eriosema** (A.-P. de Candolle) H. G. L. Reichenbach, Consp. 150. 1828.
T.: *E. rufum* (Kunth) G. Don (Gen. Hist. 2: 347. 1832) (*Glycine rufa* Kunth).

(=) *Euriosma* Desvaux, Ann. Sci. Nat. (Paris) 9: 421. 1826.
T.: *E. sessiliflora* Desvaux.

3899 **Flemingia** Roxburgh ex W. T. Aiton, Hortus Kew. ed. 2. 4: 349. Dec 1812.
T.: *F. strobilifera* (Linnaeus) W. T. Aiton (*Hedysarum strobiliferum* Linnaeus).

(H) *Flemingia* Roxburgh ex Rottler, Ges. Naturf. Freunde Berlin Neue Schriften 4: 202. 1803 [Acanth.].
T.: *F. grandiflora* Roxburgh ex Rottler, nom. illeg. (*Thunbergia fragrans* Roxburgh).
(≡) *Lourea* ("*Luorea*") Necker ex Jaume Saint-Hilaire, Bull. Sci. Soc. Philom. Paris ser. 2. 3: 193. Dec 1812.

3905 **Vigna** Savi, Nuovo Giorn. Lett. 8: 113. 1824.
T.: *V. luteola* (N. J. Jacquin) Bentham (*Dolichos luteolus* N. J. Jacquin).

(=) *Voandezia* Du Petit-Thouars, Gen. Nov. Madagasc. 23. 1806.
T.: *V. subterranea* (Linnaeus f.) A.-P. de Candolle (*Glycine subterranea* Linnaeus f.).

3908 **Pachyrhizus** L. C. Richard ex A.-P. de Candolle, Prodr. 2: 402. 1825.
T.: *P. angulatus* L. C. Richard ex A.-P. de Candolle, nom. illeg (*Dolichos erosus* Linnaeus, *P. erosus* (Linnaeus) Urban) (*typ. cons.*).

(≡) *Cacara* Du Petit-Thouars, Dict. Sci. Nat. 6: 35. 1805 (1806?).

3910 **Dolichos** Linnaeus, Sp. Pl. 725. 1753.
T.: *D. trilobus* Linnaeus (*typ. cons.*).

3910a **Macrotyloma** (Wight et Arnott) Verd-
court, Kew Bull. 24: 322. 1970.
T.: *M. uniflorum* (Lamarck) Verdcourt
(*Dolichos uniflorus* Lamarck).

(=) *Kerstingiella* Harms, Ber. Deutsch. Bot.
Ges. 26a: 230. 1908.
T.: *K. geocarpa* Harms.

3914 **Psophocarpus** A.-P. de Candolle,
Prodr. 2: 403. 1825.
T.: *P. tetragonolobus* (Linnaeus) A.-P.
de Candolle (*Dolichos tetragonolobus*
Linnaeus).

(≡) *Botor* Adanson, Fam. Pl. 2: 326. 1763.

GERANIACEAE

3931 **Wendtia** Meyen, Reise 1: 307. 1834.
T.: *W. gracilis* Meyen.

(H) *Wendia* G. F. Hoffmann, Gen. Pl. Umbell. 136. 1814 [Umbell.].
T.: *W. chorodanum* G. F. Hoffmann.

3932 **Balbisia** Cavanilles, Anales Ci. Nat. 7:
61. 1804.
T.: *B. verticillata* Cavanilles.

(H) *Balbisia* Willdenow, Sp. Pl. 3: 2214. 1803
[Comp.].
T.: *B. elongata* Willdenow.

LINACEAE

3947 **Durandea** J. E. Planchon, London J.
Bot. 6: 594. 1847.
T.: *D. serrata* J. E. Planchon (London
J. Bot. 7: 528. 1848).

(H) *Durandea* Delarbre, Fl. Auvergne ed. 2.
365. 1800 [Cruc.].
T.: *D. unilocularis* Delarbre, nom. illeg.
(*Raphanus raphanistrum* Linnaeus).

HUMIRIACEAE

3953 **Humiria** Aublet, Hist. Pl. Guiane 564.
1775 ("*Houmiri*") (*orth. cons.*).
T.: *H. balsamifera* Aublet.

ZYGOPHYLLACEAE

3967 **Augea** Thunberg, Prodr. Pl. Cap. 1:
[viii], 80. 1794.
T.: *A. capensis* Thunberg.

(H) *Augia* Loureiro, Fl. Cochinch. 337. 1790
[Anacard.].
T.: *A. sinensis* Loureiro.

3973 **Larrea** Cavanilles, Anales Hist. Nat. 2:
119. 1800.
T.: *L. nitida* Cavanilles (*typ. cons.*).

(H) *Larrea* Ortega, Nov. Pl. Descr. Dec. 15, t.
2. 1797 [Legum.].
≡ *Hoffmannseggia* Cavanilles 1798 (*nom.
cons.*) (3557).

3980 **Balanites** Delile, Descr. Egypte, Hist. Nat. 77. 1813.
T.: *B. aegyptiaca* (Linnacus) Delile (*Ximenia aegyptiaca* Linnaeus).

(≡) *Agialid* Adanson, Fam. Pl. 2: 508, 514. 1763.

RUTACEAE

3991 **Fagara** Linnaeus, Syst. Nat. ed. 10. 2: 897, 1362. 1759.
T.: *F. pterota* Linnaeus (*typ. cons.*).

(H) *Fagara* Duhamel, Traité Arbr. Arbust. 1: 229. 1755 [Rut.].
≡ *Zanthoxylum* Linnaeus 1753.
(≡) *Pterota* P. Browne, Civ. Nat. Hist. Jamaica 146. 1756.

3998 **Pentaceras** J. D. Hooker in Bentham et J. D. Hooker, Gen. Pl. 1: 298. 1862.
T.: *P. australis* (F. Mueller) Bentham (Fl. Austral. 1: 365. 1863) (*Cookia australis* F. Mueller).

(H) *Pentaceros* G. F. W. Meyer, Prim. Fl. Esseq. 136. 1818 [Byttner.].
T.: *P. aculeatus* G. F. W. Meyer.

4011 **Boenninghausenia** H. G. L. Reichenbach ex Meisner, Pl. Vasc. Gen. 1: 60. 1837; 2: 44. 1837.
T.: *B. albiflora* (W. J. Hooker) Meisner (*Ruta albiflora* W. J. Hooker).

(H) *Boenninghausia* K. Sprengel, Syst. Veg. 3: 153, 245. 1826 [Legum.].
T.: *B. vincentina* (Ker-Gawler) K. Sprengel (*Glycine vincentina* Ker-Gawler).

4012a **Haplophyllum** A. H. L. Jussieu, Mém. Mus. Hist. Nat. 12: 464. 1825 ("*Aplophyllum*") (*orth. cons.*).
T.: *H. tuberculatum* (Forsskål) A. H. L. Jussieu (*Ruta tuberculata* Forsskål).

(H) *Aplophyllum* Cassini, Dict. Sci. Nat. 33: 463. 1824 [Comp.].
T.: non designatus.

4020 **Myrtopsis** Engler in Engler et Prantl, Nat. Pflanzenfam. 3(4): 137. 1896.
T.: *M. novae-caledoniae* Engler.

(H) *Myrtopsis* O. Hoffmann, Linnaea 43: 133. 1881 [Barrington.].
T.: *M. malangensis* O. Hoffmann.

4031 **Correa** H. Andrews, Bot. Repos. 1: t. 18. 1798.
T.: *C. alba* H. Andrews.

(H) *Correia* Vellozo in Vandelli, Fl. Lusit. Brasil. 28. 1788 [Ochn.].
T.: non designatus.

4035 **Calodendrum** Thunberg, Nov. Gen. Pl. 41. 1782.
T.: *C. capense* Thunberg.

(=) *Pallassia* Houttuyn, Nat. Hist. 2(4): 382. 1775.
T.: *P. capensis* Christmann (Vollst. Pflanzensyst. 3: 318. 1778).

4036 **Barosma** Willdenow, Enum. Pl. Hort. Berol. 257. 1809.
T.: *B. serratifolia* (W. Curtis) Willdenow (*Diosma serratifolia* W. Curtis).

(=) *Parapetalifera* J. C. Wendland, Coll. Pl. 1: 49. 1806 ("1808").
T.: *P. odorata* J. C. Wendland.

4037 **Agathosma** Willdenow, Enum. Pl. Hort. Berol. 259. 1809.
T.: *A. villosa* (Willdenow) Willdenow (*Diosma villosa* Willdenow).

(≡) *Bucco* J. C. Wendland, Coll. Pl. 1: 13. 1805 ("1808").
(=) *Hartogia* Linnaeus, Syst. Nat. ed. 10. 2: 939, 1365. 1759.
T.: *H. capensis* Linnaeus.

4038 **Adenandra** Willdenow, Enum. Pl. Hort. Berol. 256. 1809.
T.: *A. uniflora* (Linnaeus) Willdenow (*Diosma uniflora* Linnaeus).

(=) *Haenkea* F. W. Schmidt, Neue Selt. Pfl. 19. 1793.
T.: non designatus.

(=) *Glandulifolia* J. C. Wendland, Coll. Pl. 1: 35. 1805 ("1808").
T.: *G. umbellata* J. C. Wendland.

4060 **Naudinia** Planchon et Linden, Ann. Sci. Nat. Bot. ser. 3. 19: 79. 1853.
T.: *N. amabilis* Planchon et Linden.

(H) *Naudinia* A. Richard in Sagra, Hist. Cuba 4(1): 561. 1845 [Melastomat.].
LT.: *N. argyrophylla* A. Richard (vide Mansfeld, Bull. Misc. Inform. 1935: 438. 1935).

4063 **Dictyoloma** A. H. L. Jussieu, Mém. Mus. Hist. Nat. 12: 499. 1825.
T.: *D. vandellianum* A. H. L. Jussieu.

(=) *Benjamina* Vellozo, Fl. Flum. 92. Mai-Dec 1825.
T.: *B. alata* Vellozo.

4065 **Chloroxylon** A.-P. de Candolle, Prodr. 1: 625. 1824.
T.: *C. swietenia* A.-P. de Candolle (*Swietenia chloroxylon* Roxburgh).

(H) *Chloroxylum* P. Browne, Civ. Nat. Hist. Jamaica 187. 1756 [Rhamn.].
T.: *Ziziphus chloroxylon* (Linnaeus) Oliver (*Laurus chloroxylon* Linnaeus).

4066 **Spathelia** Linnaeus, Sp. Pl. ed. 2. 386. 1762.
T.: *S. simplex* Linnaeus.

4073 **Araliopsis** Engler in Engler et Prantl, Nat. Pflanzenfam. 3(4): 175. 1896.
T.: *A. soyauxii* Engler.

4074 **Sargentia** S. Watson, Proc. Amer. Acad. Arts 25: 144. 1890.
T.: *S. greggii* S. Watson.

(H) *Sargentia* Wendland et Drude ex Salomon, Palmen 160. 1887 [Palmae].
≡ *Pseudophoenix* Wendland ex Sargent 1886.

4077 **Toddalia** A. L. Jussieu, Gen. Pl. 371. 1789.
T.: *T. asiatica* (Linnaeus) Lamarck (Tabl. Encycl. 2: 116. 1797) (*Paullinia asiatica* Linnaeus).

4079 **Acronychia** J. R. Forster et G. Forster, Char. Gen. Pl. 27. 1775.
T.: *A. laevis* J. R. Forster et G. Forster.

(=) *Jambolifera* Linnaeus, Sp. Pl. 349. 1753.
T.: *J. pedunculata* Linnaeus.

(=) *Cunto* Adanson, Fam. Pl. 2: 446, 547. 1763.
T.: non designatus ["H.M. 5. t. 15"].

4083 **Skimmia** Thunberg, Nov. Gen. Pl. 58. 1783.
T.: *S. japonica* Thunberg.

*4085 **Teclea** Delile, Ann. Sci. Nat. sér. 2. 20: 90. 1843.
T.: *T. nobilis* Delile.

(=) *Aspidostigma* Hochstetter in Schimper, Iter Abyss. sect. 2. n. 1293. 1842-1843 (in sched.).
T.: *A. acuminatum* Hochstetter.

*4087 **Glycosmis** Correa, Ann. Mus. Natl. Hist. Nat. **6**: 384. 1805.
T.: *G. arborea* (Roxburgh) A.-P. de Candolle 1824 (*Limonia arborea* Roxburgh).

(=) *Panel* Adanson, Fam. Pl. **2**: 447, 587. 1763.
T.: *Limonia winterlia* Steudel [= *Glycosmis pentaphylla* (Retzius) A.-P. de Candolle (*Limonia pentaphylla* Retzius)].

4089 **Micromelum** Blume, Bijdr. 137. 1825.
T.: *M. pubescens* Blume.

(=) *Aulacia* Loureiro, Fl. Cochinch. 273. 1790.
T.: *A. falcata* Loureiro.

4090 **Murraya** J. G. Koenig ex Linnaeus, Mant. Pl. **2**: 554, 563. 1771 ("*Murraea*") (*orth. cons.*).
T.: *M. exotica* Linnaeus.

(=) *Bergera* J. G. Koenig ex Linnaeus, Mant. Pl. **2**: 555, 563. 1771.
T.: *B. koenigii* Linnaeus.

4096 **Atalantia** Correa, Ann. Mus. Natl. Hist. Nat. **6**: 383. 1805.
T.: *A. monophylla* A.-P. de Candolle (Prodr. **1**: 535. 1824).

(=) *Malnaregam* Adanson, Fam. Pl. **2**: 345, 574. 1763.
T.: *M. malabarica* Rafinesque (Sylva Tell. 143. 1838).

4099 **Aegle** Correa, Trans. Linn. Soc. London **5**: 222. 1800.
T.: *A. marmelos* (Linnaeus) Correa (*Crateva marmelos* Linnaeus).

(≡) *Belou* Adanson, Fam. Pl. **2**: 408, 525. 1763.

SIMAROUBACEAE

4109 **Samadera** J. Gaertner, Fruct. Sem. Pl. **2**: 352. 1791.
T.: *S. indica* J. Gaertner.

(=) *Locandi* Adanson, Fam. Pl. **2**: 449. 1763.
T.: *Niota pentapetala* Poiret (in Lamarck, Encycl. **4**: 490. 1798).

4111 **Simarouba** Aublet, Hist. Pl. Guiane 859. 1775.
T.: *S. amara* Aublet.

(H) *Simaruba* Boehmer in Ludwig, Defin. Gen. Pl. ed. 3. 513. 1760 [Burser.].
T.: *Pistacia simaruba* Linnaeus (Sp. Pl. 1026. 1753) (*Bursera simaruba* (Linnaeus) Sargent).

4117 **Harrisonia** R. Brown ex A. H. L. Jussieu, Mém. Mus. Hist. Nat. **12**: 517, 540, *t. 28, no. 47*. 1825.
T.: *H. brownii* A. H. L. Jussieu.

(H) *Harrissona* Adanson ex Leman, Dict. Sci. Nat. **20**: 290. 1821 [Musci].
T.: non designatus.

4118 **Castela** Turpin, Ann. Mus. Natl. Hist. Nat. **7**: 78. 1806.
T.: *C. depressa* Turpin (*typ. cons.*).

(H) *Castelia* Cavanilles, Anales Ci. Nat. **3**: 134. 1801 [Verben.].
T.: *C. cuneato-ovata* Cavanilles.

4120 **Brucea** J. F. Miller, [Icon. Anim. Pl.] *t. 25*. 1779.
T.: *B. antidysenterica* J. F. Miller.

4124 **Ailanthus** Desfontaines, Mém. Acad. Sci. (Paris) **1786**: 265. 1788.
T.: *A. glandulosa* Desfontaines.

(=) *Pongelion* Adanson, Fam. Pl. **2**: 319, 593. 1763.
T.: *Ailanthus triphysa* (Dennstedt) Alston (*Adenanthera triphysa* Dennstedt).

4131 **Picramnia** Swartz, Prodr. 2, 27. 1788.
T.: *P. antidesma* Swartz.

(=) *Pseudo-brasilium* Adanson, Fam. Pl. **2**: 341, 595. 1763.
T.: *Brasiliastrum americanum* Lamarck (Encycl. **1**: 462. 1785).
(=) *Tariri* Aublet, Hist. Pl. Guiane 37. 1775.
T.: *T. guianensis* Aublet.

4134 **Picrodendron** Grisebach, Fl. Brit. W. I. **2**: 176. 1860.
T.: Jamaica, Macfadyen s.n. (**K**) [= *P. baccatum* (Linnaeus) Krug (*Juglans baccata* Linnaeus)] (*typ. cons.*).

(H) *Picrodendron* Planchon, London J. Bot. **5**: 579. 1846 [Sapind.].
T.: *P. arboreum* (P. Miller) Planchon (*Toxicodendron arboreum* P. Miller).

BURSERACEAE

4137 **Protium** N. L. Burman, Fl. Indica 88. 1768.
T.: *P. javanicum* N. L. Burman.

4150 **Bursera** N. J. Jacquin ex Linnaeus, Sp. Pl. ed. 2. 471. 1762.
T.: *B. gummifera* Linnaeus, nom. illeg. (*Pistacia simaruba* Linnaeus, *Bursera simaruba* (Linnaeus) Sargent).

(≡) *Simaruba* Boehmer in Ludwig, Defin. Gen. Pl. ed. 3. 513. 1760.
(=) *Elaphrium* N. J. Jacquin, Enum. Syst. Pl. 3, 19. 1760.
LT.: *E. tomentosum* N. J. Jacquin (vide J. N. Rose, N. Amer. Fl. **25**: 241. 1911).

4151 **Commiphora** N. J. Jacquin, Pl. Hort. Schoenbr. **2**: 66. 1797.
T.: *C. madagascarensis* N. J. Jacquin.

(=) *Balsamea* Gleditsch, Schriften Berlin. Ges. Naturf. Freunde **3**: 127. 1782.
T.: *B. meccanensis* Gleditsch.

MELIACEAE

4172 **Naregamia** Wight et Arnott, Prodr. 116. 1834.
T.: *N. alata* Wight et Arnott.

(≡) *Nelanaregam* Adanson, Fam. Pl. **2**: 343. 1763.

4189 **Aglaia** Loureiro, Fl. Cochinch. 173. 1790.
T.: *A. odorata* Loureiro.

(H) *Aglaia* Allamand, Nova Acta Phys.-Med. Acad. Caes. Leop.-Carol. Nat. Cur. **4**: 93. 1770 [Cyper.].
T.: non designatus.
(=) *Nialel* Adanson, Fam. Pl. **2**: 446, 582. 1763.
T.: *Nyalelia racemosa* Dennstedt ex Kosteletzky (Allg. Med.-Pharm. Fl. **5**: 2005. 1836) (vide Nicolson et Suresh, Taxon **35**: 388. 1986).

4190 **Guarea** Allamand ex Linnaeus, Mant. Pl. **2**: 150, 228 ("*Guara*"). 1771.
T.: *G. trichilioides* Linnaeus, nom. illeg. (*Melia guara* N. J. Jacquin, *G. guara* (N. J. Jacquin) P. Wilson).

(=) *Elutheria* P. Browne, Civ. Nat. Hist. Jamaica 369. 1756.
T.: *E. microphylla* (W. J. Hooker) M. J. Roemer (Fam. Nat. Syn. Monogr. **1**: 122. 1846) (*Guarea microphylla* W. J. Hooker).

4195 **Trichilia** P. Browne, Civ. Nat. Hist. Jamaica 278. 1756.
T.: *T. hirta* Linnaeus (Syst. Nat. ed. 10. 1020. 1759) (*typ. cons.*).

MALPIGHIACEAE

*4208 **Hiptage** J. Gaertner, Fruct. Sem. Pl. **2**: 169. 1790.
T.: *H. madablota* J. Gaertner, nom. illeg. (*Banisteria tetraptera* Sonnerat) (etiam vide 8428).

4222 **Rhyssopteris** Blume ex A. H. L. Jussieu in Delessert, Icon. Sel. Pl. **3**: 21. 1837 ("*Ryssopterys*") (*orth. cons.*).
T.: *R. timorensis* A. H. L. Jussieu.

4226 **Heteropteris** Kunth in Humboldt, Bonpland et Kunth, Nova Gen. Sp. **5**: ed. fol. 126, ed. qu. 163. 1822.
T.: *H. purpurea* (Linnaeus) Kunth (*Banisteria purpurea* Linnaeus) (*typ. cons.*).

(=) *Banisteria* Linnaeus, Sp. Pl. 427. 1753.
LT.: *B. brachiata* Linnaeus (vide T. A. Sprague, Gard. Chron. ser. 3. **75**: 104. 1924; M. L. Green, Prop. Brit. Bot. 156. 1929).

4234 **Ptilochaeta** Turczaninow, Bull. Soc. Imp. Naturalistes Moscou **16**: 52. 1843.
T.: *P. bahiensis* Turczaninow.

(H) *Ptilochaeta* C. G. Nees in C. F. P. Martius, Fl. Bras. **2**(1): 147. 1842 [Cyper.].
T.: *P. diodon* C. G. Nees.

4244 **Thryallis** C. F. P. Martius, Nov. Gen. Sp. Pl. **3**: 77. 1829.
T.: *T. longifolia* C. F. P. Martius.

(H) *Thryallis* Linnaeus, Sp. Pl. ed. 2. 554. 1762 [Malpigh.].
T.: *T. brasiliensis* Linnaeus.

4247 **Lophanthera** A. H. L. Jussieu, Ann. Sci. Nat. Bot. ser. 2. **13**: 328. 1840.
T.: *L. kunthiana* A. H. L. Jussieu, nom. illeg. (*Galphimia longifolia* Kunth, *L. longifolia* (Kunth) Grisebach).

(H) *Lophanthera* Rafinesque, New Fl. **2**: 58. 1837 ("1836") [Scrophular.].
T.: *Gerardia delphiniifolia* Linnaeus.

TRIGONIACEAE

4264 **Trigoniastrum** Miquel, Fl. Ned. Ind., Eerste Bijv. 394. 1861.
T.: *T. hypoleucum* Miquel.

4266 **Vochysia** Aublet, Hist. Pl. Guiane 18.
1775 ("*Vochy*") (*orth. cons.*).
T.: *V. guianensis* Aublet.

POLYGALACEAE

4275 **Securidaca** Linnaeus, Syst. Nat. ed. 10. 1155. 1759.
T.: *S. volubilis* Linnaeus 1759 non Linnaeus 1753 [= *S. diversifolia* (Linnaeus) Blake, *Polygala diversifolia* Linnaeus].

(H) *Securidaca* Linnaeus, Sp. Pl. 707. 1753 [Legum.].
T.: *S. volubilis* Linnaeus 1753.

4277 **Salomonia** Loureiro, Fl. Cochinch. 1, 14. 1790.
T.: *S. cantoniensis* Loureiro.

(H) *Salomonia* Heister ex Fabricius, Enum. 20. 1759 [Lil.].
≡ *Polygonatum* P. Miller 1754.

4278 **Muraltia** A.-P. de Candolle, Prodr. 1: 335. 1824.
T.: *M. heisteria* (Linnaeus) A.-P. de Candolle (*Polygala heisteria* Linnaeus) (*typ. cons.*) (etiam vide 2147).

(H) *Muralta* Adanson, Fam. Pl. 2: 460, 580. 1763 [Ranuncul.].
T.: *Clematis cirrhosa* Linnaeus.

4281 **Xanthophyllum** Roxburgh, Pl. Corom. 3: 81. 1820 ("1819").
T.: *X. flavescens* Roxburgh (*typ. cons.*).

(=) *Pelae* Adanson, Fam. Pl. 2: 448. 1763.
T.: non designatus ["Banisterioïdes. Lin."].
(=) *Eystathes* Loureiro, Fl. Cochinch. 234. 1790.
T.: *E. sylvestris* Loureiro.

EUPHORBIACEAE

4297 **Securinega** Commerson ex A. L. Jussieu, Gen. Pl. 388. 1789.
T.: *S. durissima* J. F. Gmelin (Syst. Nat. 2: 1008. 1791) (*typ. cons.*).

4299a **Androstachys** Prain, Bull. Misc. Inform. **1908**: 438. 1908.
T.: *A. johnsonii* Prain.

(H) *Androstachys* Grand'Eury, Mém. Divers Savants Acad. Roy. Sci. Inst. Roy. France, Sci. Math. **24**(1): 190, *t. 17*. 1877 [Foss.].
T.: *A. frondosus* Grand'Eury.

4302 **Glochidion** J. R. Forster et G. Forster, Char. Gen. Pl. 57. 1775.
T.: *G. ramiflorum* J. R. Forster et G. Forster.

(=) *Agyneia* Linnaeus, Mant. Pl. 2: 161, 296, 576. 1771.
T.: *A. pubera* Linnaeus.

4303 **Breynia** J. R. Forster et G. Forster, Char. Gen. Pl. 73. 1775.
T.: *B. disticha* J. R. Forster et G. Forster.

(H) *Breynia* Linnaeus, Sp. Pl. 503. 1753 [Cappar.].
T.: *B. indica* Linnaeus.

4331 **Buraeavia** Baillon, Adansonia 11: 83. 1873.
T.: *B. carunculata* (Baillon) Baillon (*Baloghia carunculata* Baillon).

(H) *Bureava* Baillon, Adansonia 1: 71. 1861 [Combret.].
T.: *B. crotonoides* Baillon.

4349 **Julocroton** C. F. P. Martius, Flora 20(2) (Beibl. 8): 119. 1837.
T.: *J. phagedaenicus* C. F. P. Martius.

(=) *Cieca* Adanson, Fam. Pl. 2: 355. 1763.
T.: *Croton argenteus* Linnaeus.

4355 **Chrozophora** A. H. L. Jussieu, Euphorb. Gen. 27. 1824 ("*Crozophora*") (*orth. cons.*).
T.: *C. tinctoria* (Linnaeus) A. H. L. Jussieu (*Croton tinctorius* Linnaeus).

(≡) *Tournesol* Adanson, Fam. Pl. 2: 356, 612. 1763.

4397 **Adelia** Linnaeus, Syst. Nat. ed. 10. 1285, 1298. 1759.
T.: *A. ricinella* Linnaeus (*typ. cons.*).

(H) *Adelia* P. Browne, Civ. Nat. Hist. Jamaica 361. 1756 [Ol.].
T.: non designatus.

(=) *Bernardia* P. Miller, Gard. Dict. Abr. ed. 4. 1754.
T.: *B. carpinifolia* Grisebach (*Adelia bernardia* Linnaeus).

4415 **Acidoton** Swartz, Prodr. 6, 83. 1788.
T.: *A. urens* Swartz.

(H) *Acidoton* P. Browne, Civ. Nat. Hist. Jamaica 355. 1756 [Euphorb.].
T.: *Adelia acidoton* Linnaeus (Syst. Nat. ed. 10. 1298. 1759).

4421 **Pterococcus** Hasskarl, Flora 25(2) (Beibl.): 41. 1842.
T.: *P. glaberrimus* Hasskarl, nom. illeg. (*Plukenetia corniculata* J. E. Smith, *Pterococcus corniculatus* (J. E. Smith) Pax et K. Hoffmann).

(H) *Pterococcus* Pallas, Reise Russ. Reich. 2: 738. 1773 [Polygon.].
T.: *P. aphyllus* Pallas.

4435 **Micrandra** Bentham, Hooker's J. Bot. Kew Gard. Misc. 6: 371. 1854.
T.: *M. siphonioides* Bentham (*typ. cons.*).

(H) *Micrandra* Bennett et R. Brown, Pl. Jav. Rar. 237. 1844 [Euphorb.].
T.: *M. ternata* R. Brown et Bennett.

4449 **Trigonostemon** Blume, Bijdr. 600. 1825 ("*Trigostemon*") (*orth. cons.*).
T.: *T. serratus* Blume.

(=) *Enchidium* Jack, Malayan Misc. 2: 89. 1822.
T.: *E. verticillatum* Jack.

4452 **Sagotia** Baillon, Adansonia 1: 53. 1860.
T.: *S. racemosa* Baillon.

(H) *Sagotia* Duchassaing et Walpers, Linnaea 23: 737. 1850 [Legum.].
T.: *S. triflora* (Linnaeus) Duchassaing et Walpers (*Hedysarum triflorum* Linnaeus).

4454 **Codiaeum** A. H. L. Jussieu, Euphorb. Gen. 33. 1824.
T.: *C. variegatum* (Linnaeus) A. H. L. Jussieu (*Croton variegatus* Linnaeus) (*typ. cons.*).

(≡) *Phyllaurea* Loureiro, Fl. Cochinch. 375. 1790.

4455 **Galearia** Zollinger et Moritzi in Moritzi, Syst. Verz. 19. 1846.
T.: *G. pedicellata* Zollinger et Moritzi.

(H) *Galearia* K. B. Presl, Symb. Bot. 1: 49. 1831 [Legum.].
LT.: *G. fragifera* (Linnaeus) K. B. Presl (*Trifolium fragiferum* Linnaeus) (vide Hossain, Notes Roy. Bot. Gard. Edinburgh 23: 446. 1961).

4467 **Chaetocarpus** Thwaites, Hooker's J. Bot. Kew Gard. Misc. 6: 300, *t. 10a* 1854.
T.: *C. castanicarpus* (Roxburgh) Thwaites (Enum. Pl. Zeyl. 275. 1861) (*Adelia castanicarpa* Roxburgh).

(H) *Chaetocarpus* Schreber, Gen. 1: 75. 1789 [Sapot.].
≡ *Pouteria* Aublet 1775.

4470 **Endospermum** Bentham, Fl. Hongk. 304. 1861.
T.: *E. chinense* Bentham.

(H) *Endespermum* Blume, Catalogus 24. 1823 [Legum.].
T.: *E. scandens* Blume.

4472 **Omphalea** Linnaeus, Syst. Nat. ed. 10. 1264, 1378. 1759.
T.: *O. triandra* Linnaeus (*typ. cons.*).

(≡) *Omphalandria* P. Browne, Civ. Nat. Hist. Jamaica 334. 1756.

4498a **Tithymalus** J. Gaertner, Fruct. Sem. Pl. 2: 115. 1791.
T.: *T. peplus* (Linnaeus) J. Gaertner (*Euphorbia peplus* Linnaeus).

(H) *Tithymalus* P. Miller, Gard. Dict. Abr. ed. 4. 1754 [Euphorb.].
≡ *Pedilanthus* Poiteau 1812 (*nom. cons.*) (4501).

4501 **Pedilanthus** Poiteau, Ann. Mus. Natl. Hist. Nat. 19: 388. 1812.
T.: *P. tithymaloides* (Linnaeus) Poiteau (*Euphorbia tithymaloides* Linnaeus) (etiam vide 4498a).

(≡) *Tithymaloides* Ortega, Tabl. Bot. 9. 1773.

4516 **Botryophora** J. D. Hooker, Fl. Brit. India 5: 476. 1888.
T.: *B. kingii* J. D. Hooker.

(H) *Botryophora* Bompard, Hedwigia 6: 129. 1867 [Chloroph.].
T.: *B. dichotoma* Bompard.

LIMNANTHACEAE

4542 **Limnanthes** R. Brown, London Edinburgh Philos. Mag. & J. Sci. 3: 71. 1833.
T.: *L. douglassii* R. Brown.

(H) *Limnanthes* Stokes, Bot. Mater. Med. 1: 300. 1812 [Gentian.].
≡ *Limnanthemum* S. G. Gmelin 1770.

4563 **Lannea** A. Richard in Guillemin, Perrottet et A. Richard, Fl. Seneg. Tent. **1:** 153. 1831.
T.: *L. velutina* A. Richard (*typ. cons.*).

(=) *Calesiam* Adanson, Fam. Pl. **2:** 446, 530. 1763.
T.: *Calsiama malabarica* Rafinesque (Sylva Tell. 12. 1838).

4578 **Campnosperma** Thwaites, Hooker's J. Bot. Kew Gard. Misc. **6:** 65. 1854.
T.: *C. zeylanicum* Thwaites.

(=) *Coelopyrum* Jack, Malayan Misc. **2(7):** 65. 1822.
T.: *C. coriaceum* Jack.

4600 **Nothopegia** Blume, Mus. Bot. 203. 1850.
T.: *N. colebrookiana* (R. Wight) Blume (*Pegia colebrookiana* R. Wight).

(=) *Glycycarpus* Dalzell, J. Roy. Asiat. Soc. Bombay **3(1):** 69. 1849.
T.: *G. edulis* Dalzell.

4604 **Holigarna** Buchanan-Hamilton ex Roxburgh, Pl. Corom. **3:** 79. 1820 ("1819").
T.: *H. longifolia* Buchanan-Hamilton ex Roxburgh.

(=) *Katou-tsjeroe* Adanson, Fam. Pl. **2:** 84, 534. 1763.
T.: non designatus ["H.M. 4. t. 9"].

AQUIFOLIACEAE

4615 **Nemopanthus** Rafinesque, Amer. Monthly Mag. & Crit. Rev. **4:** 357. 1819.
T.: *N. fascicularis* Rafinesque, nom. illeg. (*Ilex canadensis* A. Michaux, *N. canadensis* (A. Michaux) A.-P. de Candolle).

(=) *Ilicioides* Dumont de Courset, Bot. Cult. **4:** 27. 1802.
T.: non designatus.

CELASTRACEAE

4621 **Microtropis** Wallich ex Meisner, Pl. Vasc. Gen. 68. 1837.
T.: *M. discolor* (Wallich) Meisner.

(H) *Microtropis* E. Meyer, Comm. Pl. Afr. Austr. 65. 1836 [Legum.].
≡ *Euchlora* Ecklon et Zeyher 1836 (prius).

4623 **Denhamia** Meisner, Pl. Vasc. Gen. **1:** 18. 1837; **2:** 16. 1837.
T.: *D. obscura* (A. Richard) Meisner ex Walpers (Repert. **1:** 203. 1842) (*Leucocarpum obscurum* A. Richard).

(H) *Denhamia* Schott in Schott et Endlicher, Melet. Bot. 19. 1832 [Ar.].
T.: *D. scandens* (Palisot de Beauvois) Schott (*Culcasia scandens* Palisot de Beauvois).

(≡) *Leucocarpum* A. Richard, Sert. Astrolab. 46. 1834.

4627 **Gymnosporia** (Wight et Arnott) J. D. Hooker in Bentham et J. D. Hooker, Gen. Pl. 1: 365. 1862.
T.: *G. montana* (Roth ex Roemer et J. A. Schultes) Bentham (Fl. Austral. 1: 400. 1863) (*Celastrus montanus* Roth ex Roemer et J. A. Schultes) (*typ. cons.*).

(=) *Catha* Forsskål ex Schreber, Gen. 147. 1789.
T.: non designatus.

(=) *Scytophyllum* Ecklon et Zeyher, Enum. Pl. Afric. Austral. 124. 1834 vel 1835.
T.: non designatus.

(=) *Encentrus* K. B. Presl, Abh. Königl. Böhm. Ges. Wiss. ser 5. 3: 463. 1845.
T.: *E. linearis* (Linnaeus f.) K. B. Presl (*Celastrus linearis* Linnaeus f.).

(=) *Polyacanthus* K. B. Presl, Abh. Königl. Böhm. Ges. Wiss. ser. 5. 3: 463. 1845.
T.: *P. stenophyllus* (Ecklon et Zeyher) K. B. Presl (*Celastrus stenophyllus* Ecklon et Zeyher).

4637 **Plenckia** Reisseck in C. F. P. Martius, Fl. Bras. 11(1): 29. 1861.
T.: *P. populnea* Reisseck.

(H) *Plenckia* Rafinesque, Specchio 1: 194. 1814 [Aiz.].
T.: *P. setiflora* (Forsskål) Rafinesque (*Glinus setiflorus* Forsskål).

HIPPOCRATEACEAE

*4662 **Salacia** Linnaeus, Mant. 159, 293. 1771.
T.: *S. chinensis* Linnaeus.

(=) *Courondi* Adanson, Fam. Pl. 2: 446, 545. 1763.
T.: *Christmannia corondi* Dennstedt ex Kosteletzky, Allg. Med.-Pharm. Fl. 5: 2005. 1836 (vide Nicolson et Suresh, Taxon 35: 181. 1986).

STAPHYLEACEAE

4666 **Turpinia** Ventenat, Mém. Cl. Sci. Math. Phys. Inst. Natl. France 1807(1): 3. Jul 1807.
T.: *T. paniculata* Ventenat.

(H) *Turpinia* Humboldt et Bonpland, Pl. Aequinoct. 1: 113. Apr 1807 ("1808") [Comp.].
T.: *T. laurifolia* Bonpland.

(=) *Triceros* Loureiro, Fl. Cochinch. 184. 1790.
T.: *T. cochinchinensis* Loureiro.

4667 **Euscaphis** Siebold et Zuccarini, Fl. Jap. 1: 122. 1840.
T.: *E. staphyleoides* Siebold et Zuccarini, nom. illeg. (*Sambucus japonica* Thunberg, *E. japonica* (Thunberg) Kanitz).

(≡) *Hebokia* Rafinesque, Alsogr. 47. 1838.

230

ICACINACEAE

4693 **Mappia** N. J. Jacquin, Pl. Hort. Schoenbr. 1: 22. 1797.
T.: *M. racemosa* N. J. Jacquin.

(H) *Mappia* Heister ex Fabricius, Enum. 58. 1759 [Lab.].
≡ *Cunila* P. Miller 1754.

4709 **Pyrenacantha** W. J. Hooker, Bot. Misc. 2: 109. 1830.
T.: *P. volubilis* W. J. Hooker.

4712 **Phytocrene** Wallich, Pl. As. Rar. 3: 11. 1831.
T.: *P. gigantea* Wallich [= *P. macrophylla* (Blume) Blume].

(=) *Gynocephalum* Blume, Bijdr. 483. 1825.
T.: *G. macrophyllum* Blume.

4713 **Miquelia** Meisner, Pl. Vasc. Gen. 1: 152. Sep 1838.
T.: *M. kleinii* Meisner.

(H) *Miquelia* Blume, Bull. Sci. Phys. Nat. Néerl. 1: 94. Jun 1838 [Gesner.].
T.: *M. caerulea* Blume.

4715 **Stachyanthus** Engler in Engler et Prantl, Nat. Pflanzenfam. Nachtr. [1]: 227. 1897.
T.: *S. zenkeri* Engler.

(H) *Stachyanthus* A.-P. de Candolle, Prodr. 5: 84. 1836 [Comp.].
T.: *S. martii* A.-P. de Candolle.

HIPPOCASTANACEAE

4722 **Billia** Peyritsch, Bot. Zeitung (Berlin) 16: 153. 1858.
T.: *B. hippocastanum* Peyritsch.

(H) *Billya* Cassini, Dict. Sci. Nat. 34: 38. 1825 [Comp.].
T.: *B. bergii* Cassini, nom. illeg. (*Gnaphalium crispum* Linnaeus).

SAPINDACEAE

4730 **Bridgesia** Bertero ex Cambessèdes, Nouv. Ann. Mus. Hist. Nat. 3: 234. 1834.
T.: *B. incisifolia* Bertero ex Cambessèdes.

(H) *Bridgesia* W. J. Hooker, Bot. Misc. 2: 222. 1831 [Comp.].
T.: *B. echinopsoides* W. J. Hooker.

4733 **Thouinia** Poiteau, Ann. Mus. Natl. Hist. Nat. 3: 70. 1804.
T.: *T. simplicifolia* Poiteau (*typ. cons.*).

(H) *Thouinia* Thunberg ex Linnaeus f., Suppl. Pl. 9, 89. 1782 [Ol.].
T.: *T. nutans* Linnaeus f.

4747 **Zollingeria** Kurz, J. Asiat. Soc. Bengal 41(2): 303. 1872.
T.: *Z. macrocarpa* Kurz.

(H) *Zollingeria* C. H. Schultz-Bip., Flora 37: 274. 1854 [Comp.].
T.: *Z. scandens* C. H. Schultz-Bip.

4753 **Pancovia** Willdenow, Sp. Pl. 2: 285. 1799.
T.: *P. bijuga* Willdenow.

(H) *Pancovia* Heister ex Fabricius, Enum. 64. 1759 [Ros.].
≡ *Comarum* Linnaeus 1753.

4767 **Schleichera** Willdenow, Sp. Pl. 4: 1096. 1806.
T.: *S. trijuga* Willdenow.

(=) *Cussambium* Lamarck, Encycl. 2: 230. 1786.
T.: *C. spinosum* F. Hamilton (Mem. Wern. Nat. Hist. Soc. 5: 357. 1826).

4820 **Mischocarpus** Blume, Bijdr. 238. 1825.
T.: *M. sundaicus* Blume.

(=) *Pedicellia* Loureiro, Fl. Cochinch. 641, 655. 1790.
T.: *P. oppositifolia* Loureiro.

RHAMNACEAE

4862 **Condalia** Cavanilles, Anales Hist. Nat. 1: 39. 1799.
T.: *C. microphylla* Cavanilles.

(H) *Condalia* Ruiz et Pavón, Prodr. 11. 1794 [Rub.].
T.: *C. repens* Ruiz et Pavón.

4868 **Berchemia** A.-P. de Candolle, Prodr. 2: 22. 1825.
T.: *Rhamnus volubilis* Linnaeus f. [= *B. scandens* (J. Hill) K. Koch].

(≡) *Oenoplea* A. Michaux ex R. A. Hedwig, Gen. Pl. 1: 151. 1806.

4874 **Scutia** (Commerson ex A.-P. de Candolle) A. T. Brongniart, Ann. Sci. Nat. h(Paris) 10: 362. 1827.
T.: *S. indica* A. T. Brongniart, nom. illeg. (*Rhamnus circumscissus* Linnaeus f., *S. circumscissa* (Linnaeus f.) Radlkofer).

(=) *Adolia* Lamarck, Encycl. 1: 44. 1783.
T.: non designatus.

4882 **Colubrina** L. C. Richard ex A. T. Brongniart, Ann. Sci. Nat. (Paris) 10: 368. 1827.
T.: *C. ferruginosa* A. T. Brongniart (*Rhamnus colubrinus* N. J. Jacquin) (*typ. cons.*).

4899 **Colletia** Commerson ex A. L. Jussieu, Gen. Pl. 380. 1789.
T.: *C. spinosa* Lamarck (Tabl. Encycl. 2: 91. 1798).

(H) *Colletia* Scopoli, Intr. 207. 1777 [Ulm.].
T.: *Rhamnus iguanaeus* N. J. Jacquin (Enum. Syst. Pl. 16. 1760).

4905 **Helinus** E. Meyer ex Endlicher, Gen. Pl. 1102. 1840.
T.: *H. ovatus* E. Meyer, nom. illeg. (*Rhamnus mystacinus* W. Aiton, *H. mystacinus* (W. Aiton) E. Meyer ex Steudel).

(≡) *Mystacinus* Rafinesque, Sylva Tell. 30. 1838.

4910 **Ampelocissus** J. E. Planchon, Vigne Am. 8: 371. 1884.
T.: *A. latifolia* (Roxburgh) J. E. Planchon (*Vitis latifolia* Roxburgh) (*typ. cons.*).

(=) *Botria* Loureiro, Fl. Cochinch. 153. 1790.
T.: *B. africana* Loureiro.

4915 **Parthenocissus** J. E. Planchon in A. de Candolle et C. de Candolle, Monogr. Phan. 5: 447. 1887.
T.: *P. quinquefolia* (Linnaeus) J. E. Planchon (*Hedera quinquefolia* Linnaeus) (*typ. cons.*).

4918a **Cayratia** A. L. Jussieu, Dict. Sci. Nat. 10: 103. 1818.
T.: *Columella pedata* Loureiro [= *C. pedata* (Lamarck) Gagnepain (*Cissus pedata* Lamarck)] (etiam vide 7897).

(=) *Lagenula* Loureiro, Fl. Cochinch. 65, 88. 1790.
T.: *L. pedata* Loureiro.

4919 **Leea** D. van Royen ex Linnaeus, Syst. Nat. ed. 12. 2: 627. 1767; Mant. Pl. 17, 124. 1767.
T.: *L. aequata* Linnaeus (*typ. cons.*).

(=) *Nalagu* Adanson, Fam. Pl. 2: 445, 581. 1763.
T.: *Leea asiatica* (Linnaeus) Ridsdale (*Phytolacca asiatica* Linnaeus).

ELAEOCARPACEAE

4927 **Aristotelia** L'Héritier, Stirp. Nov. 31. 1785 vel 1786 ("1784").
T.: *A. macqui* L'Héritier.

(H) *Aristotela* Adanson, Fam. Pl. 2: 125. 1763 [Comp.].
≡ *Othonna* Linnaeus 1753.

TILIACEAE

4938 **Berrya** Roxburgh, Pl. Corom. 3: 60. 1820 ("*Berria*") (*orth. cons.*).
T.: *B. ammonilla* Roxburgh.

(=) *Espera* Willdenow, Ges. Naturf. Freunde Berlin Neue Schriften 3: 450. 1801.
T.: *E. cordifolia* Willdenow.

4943 **Brownlowia** Roxburgh, Pl. Corom. 3: 61. 1820.
T.: *B. elata* Roxburgh.

(=) *Glabraria* Linnaeus, Mant. Pl. 156, 276. 1771.
T.: *G. tersa* Linnaeus.

4948 **Ancistrocarpus** Oliver, J. Linn. Soc., Bot. 9: 173. 1865 ("1867").
T.: *A. brevispinosus* Oliver (*typ. cons.*).

(H) *Ancistrocarpus* Kunth in Humboldt, Bonpland et Kunth, Nova Gen. Sp. 2: ed. fol. 149, ed. qu. 186. 1817 [Phytolacc.].
T.: *A. maypurensis* Kunth.

4957 **Sparmannia** Linnaeus f., Suppl. Pl. 41, (265, 462, "*Sparrmannia* "). 1782.
T.: *S. africana* Linnaeus f.

(H) *Sparmannia* Buchoz, Pl. Nouv. Découv. 3, t. 1. 1779 [Scrophular.].
≡ *Rehmannia* Fischer et C. A. Meyer 1825 (*nom. cons*) (7592).

4959 **Luehea** Willdenow, Ges. Naturf. Freunde Berlin Neue Schriften **3**: 410. 1801 ("*Lühea* ").
T.: *L. speciosa* Willdenow.

(H) *Luehea* F. W. Schmidt, Neue Selt. Pfl. 23. 1793 ("*Lühea* ") [Verben.].
T.: *L. ericoides* F. W. Schmidt.

4960 **Mollia** C. F. P. Martius, Nov. Gen. Sp. Pl. **1**: 96. 1826 ("1824").
T.: *M. speciosa* C. F. P. Martius.

(H) *Mollia* J. F. Gmelin, Syst. Nat. **2**: 303, 420. 1791 [Myrt.].
T.: *M. imbricata* (J. Gaertner) J. F. Gmelin (*Jungia imbricata* J. Gaertner).

MALVACEAE

4995 **Malvastrum** A. Gray, Mem. Amer. Acad. Arts ser. 2. 4: 21. 1849.
T.: *M. wrightii* A. Gray [= *M. aurantiacum* (Scheele) Walpers, *Malva aurantiaca* Scheele].

(=) *Malveopsis* K. B. Presl. Abh. Königl. Böhm. Ges. Wiss. ser. 5. **3**: 449. 1845.
T.: *M. anomala* (Link et Otto) K. B. Presl (*Malva anomala* Link et Otto).

5007 **Pavonia** Cavanilles, Diss. **2**: [App. 2]. 1786; **3**: 132. 1787.
T.: *P. paniculata* Cavanilles.

(=) *Lass* Adanson, Fam. Pl. **2**: 400. 1763.
T.: *Hibiscus spinifex* Linnaeus (Syst. Nat. ed. 10. 1149. 1759).

(=) *Malache* B. C. Vogel in Trew, Pl. Sel. Pinx. Ehret 50. 1772.
T.: *M. scabra* B. C. Vogel.

*5008a **Peltaea** (K. B. Presl) Standley, Contr. U.S. Natl. Herb. **18**: 113. 1916.
T.: *P. ovata* (K. B. Presl) Standley (*Malachra ovata* K. B. Presl).

(=) *Peltostegia* Turczaninow, Bull. Soc. Imp. Naturalistes Moscou **31**: 223. 1858.
T.: *P. parviflora* Turczaninow.

5013 **Hibiscus** Linnaeus, Sp. Pl. 693. 1753.
T.: *H. syriacus* Linnaeus (*typ. cons.*).

5015 **Kosteletzkya** K. B. Presl, Reliq. Haenk. **2**: 130. 1835.
T.: *K. hastata* K. B. Presl.

(=) *Thorntonia* H. G. L. Reichenbach, Consp. 202. 1828.
T.: *Hibiscus pentaspermus* A.-P. de Candolle.

5018 **Thespesia** Solander ex Correa, Ann. Mu. Natl. Hist. Nat. **9**: 290. 1807.
T.: *T. populnea* (Linnaeus) Solander ex Correa (*Hibiscus populneus* Linnaeus).

(≡) *Bupariti* Duhamel, Semis Plantat. Arbr. add. 5. 1760.

TRIPLOCHITONACEAE

5022a **Triplochiton** K. Schumann, Bot. Jahrb. Syst. **28**: 330. 1900.
T.: *T. scleroxylon* K. Schumann.

(H) *Triplochiton* Alefeld, Österr. Bot. Z. **13**: 13. 1863 [Bombac./Malv.].
T.: non designatus.

BOMBACACEAE

5024 **Bombax** Linnaeus, Sp. Pl. 511. 1753.
T.: *B. ceiba* Linnaeus (typus: Rheede,
Hort. Malab. 3: 61. *t. 52*) (*typ. cons.*).

5035 **Bernoullia** Oliver, Hooker's Icon. Pl. (H) *Bernullia* Rafinesque, Aut. Bot. 173. 1840
12: *t. 1169, 1170.* 1873. [Ros.].
T.: *B. flammea* Oliver. T.: non designatus.

5036 **Cumingia** Vidal, Phan. Cuming. Philip. (H) *Cummingia* D. Don in Sweet, Brit. Fl.
211. 1885. Gard. 3: *t. 257.* 1828 [Haemodor.].
T.: *C. philippinensis* Vidal. T.: *C. campanulata* (Lindley) D. Don
h(*Conanthera campanulata* Lindley).

5040 **Neesia** Blume, Nova Acta Phys.-Med. (H) *Neesia* K. Sprengel, Anleit. ed. 2. 2(2):
Acad. Caes. Leop.-Carol. Nat. Cur. 547. 1818 [Comp.].
17(1): 83. 1835. T.: non designatus.
T.: *N. altissima* (Blume) Blume (*Esen-beckia altissima* Blume).

STERCULIACEAE

5053 **Dombeya** Cavanilles, Diss. 2: [App. 1]. (H) *Dombeya* L'Héritier, Stirp. Nov. 33. 1785
1786; 3: 121. 1787. vel 1786 ("1784") [Bignon.].
T.: *D. palmata* Cavanilles. ≡ *Tourrettia* Fougeroux 1787 (*nom.
cons.*) (7766).
(=) *Assonia* Cavanilles, Diss. 2: [App. 2].
1786; 3: 120. 1787.
T.: *A. populnea* Cavanilles.

5060 **Rulingia** R. Brown, Bot. Mag. 48: *t.* (H) *Ruelingia* Ehrhart, Neues Mag. Aertzte
2191. 1820. 6: 297. 1784 [Portulac.].
T.: *R. pannosa* R. Brown. ≡ *Anacampseros* Linnaeus 1758 (vide
2412).

5062 **Byttneria** Loefling, Iter Hispan. 313. (H) *Butneria* Duhamel, Traité Arbr. Arbust.
1758. 1: 113. 1755 [Calycanth.].
T.: *B. scabra* Linnaeus (Syst. Nat. ed. T.: non designatus.
10. 939. 1759).

5075 **Seringia** J. Gay, Mém. Mus. Hist. Nat. (H) *Seringia* K. Sprengel, Anleit. ed. 2. 2(2):
7: 442. 1821. 694. 1818 [Celastr.].
T.: *S. platyphylla* J. Gay, nom. illeg. ≡ *Ptelidium* Du Petit-Thouars 1805.
(*Lasiopetalum arborescens* W. Aiton, *S.
arborescens* (W. Aiton) Druce).

5080 **Pterospermum** Schreber, Gen. 2: 461.
1791.
T.: *P. suberifolium* (Linnaeus) Willde-now (Sp. Pl. 3: 728. 1800) (*Pentapetes
suberifolia* Linnaeus).

5091 **Cola** Schott et Endlicher, Melet. Bot. 33. 1832.
T.: *C. acuminata* (Palisot de Beauvois) Schott et Endlicher (*Sterculia acuminata* Palisot de Beauvois) (*typ. cons.*).

(=) *Bichea* Stokes, Bot. Mater. Med. 2: 564. 1812.
T.: *B. solitaria* Stokes.

DILLENIACEAE

5109 **Saurauia** Willdenow, Ges. Naturf. Freunde Berlin Neue Schriften 3: *t. 4.* 1801 ("*Saurauja*") (*orth. cons.*).
T.: *S. excelsa* Willdenow.

OCHNACEAE

5113 **Ouratea** Aublet, Hist. Pl. Guiane 397. 1775.
T.: *O. guianensis* Aublet.

THEACEAE

5144 **Bonnetia** C. F. P. Martius, Nova Gen. Sp. Pl. 1: 114. 1826 ("1824").
T.: *B.' anceps* C. F. P. Martius (*typ. cons.*).

(H) *Bonnetia* Schreber, Gen. 1: 363. 1789 [The.].
≡ *Mahurea* Aublet 1775.
(=) *Kieseria* C. G. Nees in Wied-Neuwied, Reise Bras. 2: 338. 1821.
T.: *K. stricta* C. G. Nees.

5148 **Gordonia** J. Ellis, Philos. Trans. 60: 518, 520. 1771.
T.: *G. lasianthus* (Linnaeus) J. Ellis (*Hypericum lasianthus* Linnaeus) (etiam vide 8412).

5149 **Laplacea** Kunth in Humboldt, Bonpland et Kunth, Nova Gen. Sp. 5: ed. fol. 161, ed. qu. 207. 1822.
T.: *L. speciosa* Kunth.

5153 **Ternstroemia** Mutis et Linnaeus f., Suppl. Pl. 39, 264. 1782.
T.: *T. meridionalis* Mutis ex Linnaeus f.

(=) *Mokof* Adanson, Fam. Pl. 2: 501. 1763.
T.: non designatus ["Kaempf. Amoen. t. 774"].
(=) *Taonabo* Aublet, Hist. Pl. Guiane 569. 1775.
T.: *T. dentata* Aublet.

236

5155 **Anneslea** Wallich, Pl. Asiat. Rar. **1**: 5. 1829.
T.: *A. fragrans* Wallich.

(H) *Anneslia* R. A. Salisbury, Parad. Lond. *t. 64*. 1807 [Legum.].
T.: *A. falcifolia* R. A. Salisbury.

5157a **Cleyera** Thunberg, Nova Gen. Pl. 68. 1783.
T.: *C. japonica* Thunberg.

(H) *Cleyera* Adanson, Fam. Pl. **2**: 224. 1763 [Logan.].
≡ *Polypremum* Linnaeus 1753.

5157b **Freziera** Willdenow, Sp. Pl. **2**: 1179. 1799 vel 1800.
T.: *F. undulata* (Swartz) Willdenow (*Eroteum undulatum* Swartz) (*typ. cons.*).

(≡) *Eroteum* Swartz, Prodr. *5*, 85. 1788.
(=) *Lettsomia* Ruiz et Pavón, Prodr. 77. 1794.
T.: non designatus.

GUTTIFERAE (CLUSIACEAE)

5171 **Vismia** Vandelli, Fl. Lusit. Bras. 51. 1788.
T.: *V. cayennensis* (N. J. Jacquin) Persoon (Syn. Pl. **2**: 86. 1806) (*Hypericum cayennense* N. J. Jacquin) (*typ. cons.*).

(=) *Caopia* Adanson, Fam. Pl. **2**: 448, 531. 1763.
T.: non designatus ["Marcg. 131."].

5195 **Balboa** Planchon et Triana, Ann. Sci. Nat. Bot. ser. 4. **13**: 315. 1860; **14**: 252. 1860.
T.: *B. membranacea* Planchon et Triana.

(H) *Balboa* Liebmann ex Didrichsen, Vidensk. Meddel. Dansk Naturhist. Foren, Kjøbenhavn **1853**: 106. 1853 [Legum.].
T.: *B. diversifolia* Liebmann ex Didrichsen.

5205 **Platonia** C. F. P. Martius, Nova Gen. Sp. Pl. **3**: 168. 1832 ("1829").
T.: *P. insignis* C. F. P. Martius, nom. illeg. (*Moronobea esculenta* Arruda da Camara, *P. esculenta* (Arruda da Camara) Rickett et Stafleu).

(H) *Platonia* Rafinesque, Caratt. Nuovi Gen. Sp. Sicilia 73. 1810 [Cist.].
≡ *Helianthemum* P. Miller 1754.

DIPTEROCARPACEAE

5214 **Doona** Thwaites, Hooker's J. Bot. Kew Gard. Misc. **3**: *t. 12*. 1851.
T.: *D. zeylanica* Thwaites.

(=) *Caryolobis* J. Gaertner, Fruct. Sem. Pl. **1**: 215. 1788.
T.: *C. indica* J. Gaertner.

5215 **Hopea** Roxburgh, Pl. Corom. **3**: 7, *t. 210*. 1811.
T.: *H. odorata* Roxburgh.

(H) *Hopea* Garden ex Linnaeus, Syst. Nat. ed. 12. 509. 1767; Mant. Pl. 14, 105. 1767 [Symploc.].
T.: *H. tinctoria* Linnaeus.

5221 **Pierrea** F. Heim, Bull. Mens. Soc. Linn. Paris **1891**: 958. 1891.
T.: *P. pachycarpa* F. Heim.

(H) *Pierrea* Hance, J. Bot. **15**: 339. 1877 [Flacourt.].
T.: *P. dictyoneura* Hance.

COCHLOSPERMACEAE

5250 **Cochlospermum** Kunth in Humboldt,
Bonpland et Kunth, Nov. Gen. Sp. 5: ed.
fol. 231, ed. qu. 297. 1822.
T.: *Bombax gossypium* Linnaeus, nom.
illeg. (*Bombax religiosum* Linnaeus
("*religiosa*"), *C. religiosum* (Linnaeus)
Alston).

CANELLACEAE

5254 **Canella** P. Browne, Civ. Nat. Hist.
Jamaica 275. 1756.
T.: *C. winteriana* (Linnaeus) J. Gaertner (Fruct. Sem. Pl. 1: 373. 1788) (*Laurus winteriana* Linnaeus).

5256 **Warburgia** Engler, Pflanzenw. Ost-   (=) *Chibaca* Bertoloni, Mem. Reale Accad.
Afrikas C: 276. 1895.   Sci. Ist. Bologna 4: 545. 1853.
T.: *W. stuhlmannii* Engler.   T.: *C. salutaris* Bertoloni.

VIOLACEAE

5259 **Amphirrhox** K. Sprengel, Syst. Veg.
4(2): 51, 99. 1827.
T.: *A. longifolia* (A. Saint-Hilaire) K.
Sprengel (*Spathularia longifolia* A.
Saint-Hilaire).

5262 **Rinorea** Aublet, Hist. Pl. Guiane 235.   (=) *Conohoria* Aublet, Hist. Pl. Guiane 239.
1775.   1775.
T.: *R. guianensis* Aublet.   T.: *C. flavescens* Aublet.

5271 **Hybanthus** N. J. Jacquin, Enum. Syst.
Pl. 2, 17. 1760.
T.: *H. havanensis* N. J. Jacquin.

FLACOURTIACEAE

5278 **Erythrospermum** Lamarck, Tabl. En-   (=) *Pectinea* J. Gaertner, Fruct. Sem. Pl. 2:
cycl. 2(1): 407, *t. 274.* 1792.   136, *t. 111, fig. 3.* 1790.
T.: *E. pyrifolium* Lamarck (*typ. cons.*).   T.: *P. zeylanica* J. Gaertner.

5304 **Scolopia** Schreber, Gen. 1: 335. 1789. (=) *Aembilla* Adanson, Fam. Pl. 2: 448.
T.: *S. pusilla* (J. Gaertner) Willdenow   1763.
(Sp. Pl. 2: 981. 1799 vel 1800) (*Limonia*   T.: non designatus.
pusilla J. Gaertner).

5311 **Byrsanthus** Guillemin in Delessert, Icon. Sel. Pl. 3: 30, *t. 52*. 1838 ("1837").
T.: *B. brownii* Guillemin.

(H) *Byrsanthes* K. B. Presl, Prodr. Monogr. Lobel. 41. 1836 [Campanul.].
T.: *B. humboldtiana* K. B. Presl, nom. illeg. (*Lobelia nivea* Willdenow).

5320 **Xylosma** G. Forster, Fl. Ins. Austr. 72. 1786.
T.: *X. orbiculata* (J. R. Forster et G. Forster) G. Forster (*Myroxylon orbiculatum* J. R. Forster et G. Forster) (*typ. cons.*) (vide etiam 3584).

5331 **Idesia** Maximowicz, Bull. Acad. Imp. Sci. Saint-Pétersbourg ser. 3. **10**: 485. 1866.
T.: *I. polycarpa* Maximowicz.

(H) *Idesia* Scopoli, Introd. Hist. Nat. 199. 1777 [Verben.].
≡ *Ropourea* Aublet 1775.

5334 **Lunania** W. J. Hooker, London J. Bot. 3:/16/08317. 1844.
T.: *L. racemosa* W. J. Hooker.

(H) *Lunanea* A.-P. de Candolle, Prodr. **2**: 92. 1825 [Stercul.].
≡ *Bichea* Stokes 1812 (etiam vide 5091).

5337 **Samyda** N. J. Jacquin, Enum. Syst. Pl. 4, 21. 1760.
T.: *S. dodecandra* N. J. Jacquin (*typ. cons.*).

(H) *Samyda* Linnaeus, Sp. Pl. 443. 1753 [Mel.].
T.: *S. guidonia* Linnaeus.

5338 **Laetia** Loefling ex Linnaeus, Syst. Nat. ed. 10. 1074, 1373. 1759.
T.: *L. americana* Linnaeus.

5341 **Ryania** Vahl, Ecl. **1**: 51, *t. 9*. 1796.
T.: *R. speciosa* Vahl.

(=) *Patrisa* L. C. Richard, Actes Soc. Hist. Nat. Paris **1**: 110. 1792.
T.: *P. pyrifera* L. C. Richard.

5353 **Tetralix** Grisebach, Cat. Pl. Cub. 8. 1866.
T.: *T. brachypetalus* Grisebach.

(H) *Tetralix* Zinn, Catal. 202. 1757 [Eric.].
T.: *Erica herbacea* Linnaeus (Sp. Pl. 352. 1753).

LOASACEAE

5384 **Eucnide** Zuccarini, Del. Sem. Hort. Bot. Monac. [4]. 1844.
T.: *E. bartonioides* Zuccarini.

(=) *Microsperma* W. J. Hooker, Icon. Pl. **3**: *pl. 234*. 1839.
T.: *M. lobatum* W. J. Hooker ("*lobata*").

5392 **Blumenbachia** H. A. Schrader, Gött. Gel. Anz. **1825**: 1705. 1825.
T.: *B. insignis* H. A. Schrader.

(H) *Blumenbachia* Koeler, Descr. Gram. 28. 1802 [Gram.].
T.: *B. halepensis* (Linnaeus) Koeler (*Holcus halepensis* Linnaeus).

239

ANCISTROCLADACEAE

5400 **Ancistrocladus** Wallich, Num. List. n. 1052. 1829.
T.: *A. hamatus* (Vahl) Gilg (in Engler & Prantl, Nat. Pflanzenfam. 3(6): 276. 1895) (*Wormia hamata* Vahl).

(=) *Bembix* Loureiro, Fl. Cochinch. 259, 282. 1790.
T.: *B. tectoria* Loureiro.

CACTACEAE

5409 **Melocactus** Link et Otto, Verh. Vereins Beförd. Gartenbaues Königl. Preuss. Staaten 3: 417. 1827.
T.: *M. communis* Link et Otto (*Cactus melocactus* Linnaeus) (*typ. cons.*).

(H) *Melocactus* Boehmer in Ludwig, Defin. Gen. Pl. ed. 3. 79. 1760 [Cact.].
≡ *Cactus* Linnaeus 1753 (*nom. rej.* sub 5411).

5411 **Mammillaria** Haworth, Syn. Pl. Succ. 177. 1812.
T.: *M. simplex* Haworth, nom. illeg. (*Cactus mammillaris* Linnaeus, *M. mammillaris* (Linnaeus) H. Karsten) (*typ. cons.*) (etiam vide 5409).

(H) *Mammillaria* Stackhouse, Mém. Soc. Imp. Naturalistes Moscou 2: 55, 74. 1809 [Rhodoph.].
T.: non designatus.
(≡) *Cactus* Linnaeus, Sp. Pl. 466. 1753.

5416 **Rhipsalis** J. Gaertner, Fruct. Sem. Pl. 1: 137. 1788.
T.: *R. cassutha* J. Gaertner.

(=) *Hariota* Adanson, Fam. Pl. 2: 243, 520. 1763.
T.: non designatus ["Plum. ic. 197. f. 2."].

OLINIACEAE

5428 **Olinia** Thunberg, Arch. Bot. (Leipzig) 2(1): 4. 1799.
T.: *O. cymosa* (Linnaeus f.) Thunberg (*Sideroxylon cymosum* Linnaeus f.).

(=) *Plectronia* Linnaeus, Syst. Nat. ed. 12. 138, 183. 1767; Mant. 6, 52. 1767.
T.: *P. ventosa* Linnaeus.

THYMELAEACEAE

5430 **Aquilaria** Lamarck, Encycl. 1: 49. 1783.
T.: *A. malaccensis* Lamarck.

(=) *Agallochum* Lamarck, Encycl. 1: 48. 1783.
T.: non designatus.

5436 **Struthiola** Linnaeus, Syst. Nat. ed. 12. 127. 1767; Mant. Pl. 4, 41. 1767.
T.: *S. virgata* Linnaeus (*typ. cons.*).

(=) *Belvala* Adanson, Fam. Pl. 2: 285. 1763.
T.: *Passerina dodecandra* Linnaeus (Syst. Nat. ed. 10. 1004. 1759).

5446 **Wikstroemia** Endlicher, Prodr. Fl. Nor-
folk. 47. 1833 ("*Wickstroemia*") (*orth.
cons.*).
T.: *W. australis* Endlicher.

(H) *Wikstroemia* H. A. Schrader, Gött. Gel.
Anz. **1821**: 710. 1821 [The.].
T.: *W. fruticosa* H. A. Schrader.

(=) *Capura* Linnaeus, Mant. Pl. **2**: 149, 225.
1771.
T.: *C. purpurata* Linnaeus.

5453 **Thymelaea** P. Miller, Gard. Dict. Abr.
ed. 4. 1754.
T.: *T. sanamunda* Allioni (Fl. Ped. **1**:
132. 1785) (*Daphne thymelaea* Linnaeus) (*typ. cons.*).

5457 **Ovidia** Meisner in A. de Candolle,
Prodr. **14**: 524. 1857.
T.: *O. pillopillo* (C. Gay) Meisner
(*Daphne pillopillo* C. Gay) (*typ. cons.*).

(H) *Ovidia* Rafinesque, Fl. Tell. **3**: 68. 1837
("1836") [Commelin.].
T.: *O. gracilis* (Ruiz et Pavón) Rafinesque (*Commelina gracilis* Ruiz et Pavón).

5467 **Pimelea** Banks ex Solander in J. Gaert-
ner, Fruct. Sem. Pl. **1**: 186. 1788.
T.: *P. laevigata* J. Gaertner, nom illeg.
(*Banksia prostrata* J. R. Forster et G.
Forster, *P. prostrata* (J. R. Forster et G.
Forster) Willdenow).

5467a **Synandrodaphne** Gilg, Bot. Jahrb.
Syst. **53**: 362. 1915.
T.: *S. paradoxa* Gilg.

(H) *Synandrodaphne* Meisner in A. de Candolle, Prodr. **15**(1): 176. 1864 [Laur.].
T.: non designatus.

ELAEAGNACEAE

5471 **Shepherdia** Nuttall, Gen. N. Amer. Pl.
2: 240. 1818.
T.: *S. canadensis* (Linnaeus) Nuttall
(*Hippophaë canadensis* Linnaeus) (*typ.
cons.*).

LYTHRACEAE

*5486 **Nesaea** Kunth in Humboldt, Bonpland et Kunth, Nov. Gen. Sp. **6**(ed.
fol.): 151. 1823.
T.: *N. triflora* (Linnaeus f.) Kunth
(*Lythrum triflorum* Linnaeus f.).

(H) *Nesaea* Lamouroux, Nouv. Bull. Sci. Soc.
Philom. Paris **3**: 185. 1812 [Chloroph.].
T.: non designatus.

SONNERATIACEAE

5497 **Sonneratia** Linnaeus f., Suppl. Pl. 38, 252. 1782.
T.: *S. acida* Linnaeus f.

(=) *Blatti* Adanson, Fam. Pl. 2: 88, 526. 1763.
T.: non designatus.

LECYTHIDACEAE

5505 **Careya** Roxburgh, Pl. Corom. 3: 13. 1811.
T.: *C. herbacea* Roxburgh (*typ. cons.*)
(vide etiam Knuth in Engler, Pflanzenr. iv.219 (Heft 105): 50. 1939).

5506 **Barringtonia** J. R. Forster et G. Forster, Char. Gen. Pl. 38. 1775.
T.: *B. speciosa* J. R. Forster et G. Forster.

(=) *Huttum* Adanson, Fam. Pl. 2: 88, 616. 1763.
T.: non designatus ["Rumph. 3. t. 114 à 116."].

5510 **Gustavia** Linnaeus, Pl. Surin. 12, 17, 18, *t.* [s.n.]. 1775; Amoen. Acad. 8: 266, *t. 5.* 1785.
T.: *G. augusta* Linnaeus.

(=) *Japarandiba* Adanson, Fam. Pl. 2: 448. 1763.
T.: *Gustavia marcgraviana* Miers (vide Miers, Trans. Linn. Soc. London 30: 183. 1874).

RHIZOPHORACEAE

5525 **Carallia** Roxburgh, Pl. Corom. 3: 8, *t. 211.* 1811.
T.: *C. lucida* Roxburgh.

(=) *Karekandel* Wolf, Gen. Pl. Vocab. Char. Def. 73. 1776.
T.: *Karkandela malabarica* Rafinesque (vide R. Ross, Acta Bot. Neerl. 15: 158. 1966).

(=) *Barraldeia* Du Petit-Thouars, Gen. Nov. Madagasc. 24. 1806.
T.: non designatus.

5528 **Weihea** K. Sprengel, Syst. Veg. 2: 559, 594. 1825.
T.: *W. madagascarensis* K. Sprengel.

(≡) *Richaeia* Du Petit-Thouars, Gen. Nov. Madagasc. 25. 1806.

COMBRETACEAE

5538 **Combretum** Loefling, Iter Hispan. 308. 1758.
T.: *C. fruticosum* (Loefling) Stuntz (U.S.D.A. Bur. Plant Industr. Invent. Seeds 31: 86. 1914) (*Gaura fruticosa* Loefling) (vide Linnaeus, Syst. Nat. ed. 10. 999. 1759; Loefling, Iter Hispan. 248. 1758).

(=) *Grislea* Linnaeus, Sp. Pl. 348. 1753.
T.: *G. secunda* Linnaeus.

5543 **Bucida** Linnaeus, Syst. Nat. ed. 10. 1025, 1368. 1759.
T.: *B. buceras* Linnaeus.

(≡) *Buceras* P. Browne, Civ. Nat. Hist. Jamaica 221, *t. 23, fig. 1.* 1756.

5544 **Terminalia** Linnaeus, Syst. Nat. ed. 12. 674 ("638"). 1767; Mant. Pl. 21, 128. 1767.
T.: *T. catappa* Linnaeus.

(≡) *Adamaram* Adanson, Fam. Pl. **2**: (23), 445, 513. 1763 (vide Exell in Regnum Veg. **100**: 26. 1979).

5545 **Buchenavia** Eichler, Flora **49**: 164. 1866.
T.: *B. capitata* (Vahl) Eichler (*Bucida capitata* Vahl).

(=) *Pamea* Aublet, Hist. Pl. Guiane 946. *t. 359.* 1775.
T.: *P. guianensis* Aublet.

MYRTACEAE

5557 **Myrteola** O. Berg, Linnaea **27**: 348. 1856.
T.: *M. microphylla* (Humboldt et Bonpland) O. Berg (*Myrtus microphylla* Humboldt et Bonpland) (*typ. cons.*).

(≡) *Amyrsia* Rafinesque, Sylva Tell. 106. 1838.
(=) *Cluacena* Rafinesque, Sylva Tell. 104. 1838.
LT.: *C. vaccinioides* (Kunth) Rafinesque (*Myrtus vaccinioides* Kunth) (vide McVaugh, Taxon **5**: 139. 1956).

5575 **Calyptranthes** Swartz, Prodr. 5, 79. 1788.
T.: *C. chytraculia* (Linnaeus) Swartz (*Myrtus chytraculia* Linnaeus) (*typ. cons.*).

(≡) *Chytraculia* P. Browne, Civ. Nat. Hist. Jamaica 239. 1756.

5582 **Jambosa** Adanson, Fam. Pl. **2**: 88, 564. 1763 ("*Jambos*") (*orth. cons.*).
T.: *J. vulgaris* A.-P. de Candolle, nom. illeg. (*Eugenia jambos* Linnaeus, *J. jambos* (Linnaeus) Millspaugh) (*typ. cons.*) (etiam vide 5583).

5583 **Syzygium** J. Gaertner, Fruct. Sem. Pl. 1: 166, *t. 33.* 1788.
T.: *S. caryophyllaeum* J. Gaertner (*typ. cons.*).

(H) *Suzygium* P. Browne, Civ. Nat. Hist. Jamaica 240. 1756 [Myrt.].
T.: *Myrtus zuzygium* Linnaeus (Syst. Nat. ed. 10. 1056. 1759).
(=) *Caryophyllus* Linnaeus, Sp. Pl. 515. 1753.
T.: *C. aromaticus* Linnaeus.
(=) *Jambosa* Adanson, Fam. Pl. **2**: 88, 564. 1763 (*nom. cons.*) (5582).

5585 **Piliocalyx** A. T. Brongniart et Gris, Bull. Soc. Bot. France **12**: 185. 1865.
T.: *P. robustus* A. T. Brongniart et Gris.

(H) *Pileocalyx* Gasparrini, Ann. Sci. Nat. Bot. ser. 3. **9**: 220. 1848 [Cucurbit.].
≡ *Mellonia* Gasparrini 1847.

5588 **Metrosideros** Banks ex J. Gaertner, Fruct. Sem. Pl. 1: 170, *t. 34, fig. 10*. 1788.
T.: *M. spectabilis* Solander ex J. Gaertner (*typ. cons.*).

(=) *Nani* Adanson, Fam. Pl. 2: 88, 581. 1763.
T.: *Metrosideros vera* Lindley.

5594 **Xanthostemon** F. Mueller, Hooker's J. Bot. Kew Gard. Misc. 9: 17. 1857.
T.: *X. paradoxus* F. Mueller.

(=) *Nani* Adanson, Fam. Pl. 2: 88, 581. 1763.
T.: *Metrosideros vera* Lindley.

5599 **Leptospermum** J. R. Forster et G. Forster, Char. Gen. Pl. 36. 1775.
T.: *L. scoparium* J. R. Forster et G. Forster (*typ. cons.*).

5600 **Agonis** (A.-P. de Candolle) Sweet, Hort. Brit. ed. 2. 209. 1830.
T.: *A. flexuosa* (Willdenow) Sweet (*Metrosideros flexuosa* Willdenow) (*typ. cons.*).

5601 **Kunzea** Reichenbach, Consp. 175. 1828.
T.: *K. capitata* (J. E. Smith) Heynhold (*Metrosideros capitata* J. E. Smith).

(=) *Tillospermum* R. A. Salisbury, Monthly Rev. 75: 74. 1814.
T.: *Leptospermum ambiguum* J. E. Smith.

5603 **Melaleuca** Linnaeus, Syst. Nat. ed. 12. 509. 1767; Mant. Pl. 14, 105. 1767.
T.: *M. leucadendra* (Linnaeus) Linnaeus (*Myrtus leucadendra* Linnaeus) (*typ. cons.*).

(≡) *Kajuputi* Adanson, Fam. Pl. 2: 84, 530. 1763.

5621 **Thryptomene** Endlicher, Ann. Wiener Mus. Naturgesch. 2: 192. 1839 ("Dec 1838").
T.: *T. australis* Endlicher.

(=) *Gomphotis* Rafinesque, Sylva Tell. 103. 1838.
T.: *G. saxicola* (A. Cunningham ex W. J. Hooker) Rafinesque (*Baeckea saxicola* A. Cunningham ex W. J. Hooker).

5625 **Verticordia** A.-P. de Candolle, Prodr. 3: 208. 1828.
T.: *V. fontanesii* A.-P. de Candolle, nom. illeg. (*Chamaelaucium plumosum* Desfontaines, *V. plumosa* (Desfontaines) Druce) (*typ. cons.*).

MELASTOMATACEAE

5632 **Pterolepis** (A.-P. de Candolle) Miquel, Comm. Phytogr. 2: 72. 1839 ("1840").
T.: *P. parnassiifolia* (A.-P. de Candolle) Triana (*Osbeckia parnassiifolia* A.-P. de Candolle) (*typ. cons.*).

(H) *Pterolepis* H. A. Schrader, Gött. Gel. Anz. 1821: 2071. 1821 [Cyper.].
T.: *P. scirpioides* H. A. Schrader.

5648 **Microlepis** (A.-P. de Candolle) Miquel, Comm. Phytogr. 2: 71. 1839 ("1840").
T.: *M. oleifolia* (A.-P. de Candolle) Triana (Trans. Linn. Soc. London 28: 36. 1871) (*Osbeckia oleifolia* A.-P. de Candolle, "*oleaefolia*") (*typ. cons.*).

5659 **Dissotis** Bentham in W. J. Hooker, Niger Fl. 346. 1849.
T.: *D. grandiflora* (J. E. Smith) Bentham (*Osbeckia grandiflora* J. E. Smith).

(≡) *Hedusa* Rafinesque, Sylva Tell. 101. 1838.

5665 **Monochaetum** (A.-P. de Candolle) Naudin, Ann. Sci. Nat. Bot. ser. 3. 4: 48. 1845.
T.: *M. candolleanum* Naudin, nom. illeg. (*Arthrostemma calcaratum* A.-P. de Candolle, *M. calcaratum* (A.-P. de Candolle) Triana) (*typ. cons.*).

(=) *Ephynes* Rafinesque, Sylva Tell. 101. 1838.
T.: *E. bonplandii* (Humboldt et Bonpland) Rafinesque (*Rhexia bonplandii* Humboldt et Bonpland).

5669 **Cambessedesia** A.-P. de Candolle, Prodr. 3: 110. 1828.
T.: *C. hilariana* (Kunth) A.-P. de Candolle (*Rhexia hilariana* Kunth) (*typ. cons.*).

(H) *Cambessedea* Kunth, Ann. Sci. Nat. (Paris) 2: 336. 1824 [Anacard.].
T.: *Mangifera axillaris* Desrousseaux (in Lamarck, Encycl. 3: 697. 1789).

5676 **Rhynchanthera** A.-P. de Candolle, Prodr. 3: 106. 1828.
T.: *R. grandiflora* (Aublet) A.-P. de Candolle (*Melastoma grandiflorum* Aublet) (*typ. cons.*).

(H) *Rynchanthera* Blume, Tab. Pl. Jav. Orchid. *t. 78.* 1825 [Orchid.].
T.: *R. paniculata* Blume.

5692 **Meriania** Swartz, Fl. Ind. Occ. 1: *t. 15.* 1797.
T.: *M. leucantha* (Swartz) Swartz (*Rhexia leucantha* Swartz) (*typ. cons.*).

(H) *Meriana* Trew, Pl. Sel. Pinx. Ehret 11, *t. 40.* 1754 [Irid.].
≡ *Watsonia* P. Miller 1759 (*nom. cons.*) (1315).

5708 **Bertolonia** Raddi, Mem. Mat. Fis. Soc. Ital. Sci. Modena, Parte Mem. Fis. 18: 384, *t. 5, fig. 3.* 1820.
T.: *B. nymphaeifolia* Raddi.

(H) *Bertolonia* Spin, Jard. St. Sébast. 24. 1809 [Myopor.].
T.: *B. glandulosa* Spin.

5729 **Sonerila** Roxburgh, Fl. Ind. 1: 180. 1820.
T.: *S. maculata* Roxburgh.

5759 **Miconia** Ruiz et Pavón, Prodr. 60. 1794.
T.: *M. triplinervis* Ruiz et Pavón (Syst. 104. 1798) (*typ. cons.*).

(=) *Leonicenia* Scopoli, Intr. 212. 1777.
T.: *Fothergilla mirabilis* Aublet (Hist. Pl. Guiane 440. 1775).

5768 **Bellucia** Necker ex Rafinesque, Sylva Tell. 92. 1838.
T.: *B. nervosa* Rafinesque, nom. illeg. (*Blakea quinquenervia* Aublet) [= *Bellucia grossularioides* (Linnaeus) Triana, *Melastoma grossularioides* Linnaeus].

(H) *Belluccia* Adanson, Fam. Pl. 2: 344, 525. 1763 [Rut.].
≡ *Ptelea* Linnaeus 1753.
(≡) *Apatitia* Desvaux ex W. Hamilton, Prodr. Pl. Ind. Occid. 42. 1825.

5777a **Astronidium** A. Gray, U. S. Explor. Exped. Bot. Phan. [15] 1: 581. 1854.
T.: *A. parviflorum* A. Gray.

(=) *Lomanodia* Rafinesque, Sylva Tell. 97. 1838.
LT.: *L. glabra* (G. Forster) Rafinesque (*Melastoma glabrum* G. Forster) (vide Veldkamp, Taxon 32: 134. 1983).

ONAGRACEAE

5826 **Diplandra** W. J. Hooker et Arnott, Bot. Beechey Voy. 291. 1838.
T.: *D. lopezioides* W. J. Hooker et Arnott.

(H) *Diplandra* Bertero, Mercurio Chileno 13: 612. 1829 [Hydrocharit.].
T.: *D. potamogeton* Bertero.

ARALIACEAE

5852 **Schefflera** J. R. Forster et G. Forster, Char. Gen. Pl. 23. 1775.
T.: *S. digitata* J. R. Forster et G. Forster.

(=) *Sciadophyllum* P. Browne, Civ. Nat. Hist. Jamaica 190. 1756 ("*Sciodaphyllum*").
T.: *S. brownii* K. Sprengel.

UMBELLIFERAE (APIACEAE)

5938 **Anthriscus** Persoon, Syn. Pl. 1: 320. 1805.
T.: *A. vulgaris* Persoon 1805, non Bernhardi 1800 (*Scandix anthriscus* Linnaeus, *A. caucalis* Marschall von Bieberstein) (*typ. cons.*).

(H) *Anthriscus* Bernhardi, Syst. Verz. 1: 113, 168. 1800 [Umbell.].
T.: *A. vulgaris* Bernhardi (*Tordylium anthriscus* Linnaeus).
(=) *Cerefolium* Fabricius, Enum. 36. 1759.
T.: *Scandix cerefolium* Linnaeus (Sp. Pl. 257. 1753).

*5941 **Osmorhiza** Rafinesque, Amer. Monthly Mag. & Crit. Rev. 4: 192. 1819.
T.: *O. claytonii* (Michaux) C. B. Clarke (*Myrrhis claytonii* Michaux).

(≡) *Uraspermum* Nuttall, Gen. Amer. Pl. 192. 1818.

5956 **Bifora** G. F. Hoffmann, Gen. Pl. Umbell. ed. 2. 191. 1816.
T.: *B. dicocca* G. F. Hoffmann, nom. illeg. (*Coriandrum testiculatum* Linnaeus, *B. testiculata* (Linnaeus) K. Sprengel).

246

5964 **Scaligeria** A.-P. de Candolle, Coll. Mém. **5**: 70. 1829.
T.: *S. microcarpa* A.-P. de Candolle.

(H) *Scaligera* Adanson, Fam. Pl. **2**: 323. 1763 [Legum.].
≡ *Aspalathus* Linnaeus 1753.

5977 **Tauschia** Schlechtendal, Linnaea **9**: 607. 1835 ("1834").
T.: *T. nudicaulis* Schlechtendal.

(H) *Tauschia* Preissler, Flora **11**: 44. 1828 [Eric.].
T.: *T. hederifolia* Preissler.

5990 **Lichtensteinia** Chamisso et Schlechtendal, Linnaea **1**: 394. 1826.
T.: *L. lacera* Chamisso et Schlechtendal (*typ. cons.*).

(H) *Lichtensteinia* Willdenow, Ges. Naturf. Freunde Berlin Mag. **2**: 19. 1808 [Lil.].
T.: non designatus.

5992 **Heteromorpha** Chamisso et Schlechtendal, Linnaea **1**: 385. 1826.
T.: *H. arborescens* (Thunberg) Chamisso et Schlechtendal (*Bupleurum arborescens* Thunberg).

(H) *Heteromorpha* Cassini, Bull. Sci. Soc. Philom. Paris **1817**: 12. 1817 [Comp.].
≡ *Heterolepis* Cassini 1820 (*nom. cons.*) (9057).

5998 **Trinia** G. F. Hoffmann, Gen. Pl. Umbell. 92. 1814.
T.: *T. glaberrima* G. F. Hoffmann, nom. illeg. (*Seseli pumilum* Linnaeus, *T. pumila* (Linnaeus) H. G. L. Reichenbach) (*typ. cons.*).

6014 **Trachyspermum** Link, Enum. Hort. Berol. Alt. **1**: 267. 1821.
T.: *T. copticum* (Linnaeus) Link (*Ammi copticum* Linnaeus).

(≡) *Ammios* Moench, Methodus 99. 1794.

6015 **Cryptotaenia** A.-P. de Candolle, Coll. Mém. **5**: 42. 1829.
T.: *C. canadensis* (Linnaeus) A.-P. de Candolle (*Sison canadense* Linnaeus).

(≡) *Deringa* Adanson, Fam. Pl. **2**: 498. 1763.

6018 **Falcaria** Fabricius, Enum. 34. 1759.
T.: *F. vulgaris* Bernhardi (Syst. Verz. 176. 1800) (*Sium falcaria* Linnaeus).

6045 **Polemannia** Ecklon et Zeyher, Enum. Pl. Afric. Austral. 347. 1837.
T.: *P. grossulariifolia* Ecklon et Zeyher.

(H) *Polemannia* K. Bergius ex Schlechtendal, Linnaea **1**: 250. 1826 [Lil.].
T.: *P. hyacinthiflora* K. Bergius ex Schlechtendal.

6052a **Libanotis** Haller ex Zinn, Cat. Pl. Hort. Goett. 226. 1757.
T.: *L. montana* Crantz (*Athamanta libanotis* Linnaeus).

(H) *Libanotis* J. Hill, Brit. Herb. 420. 1756.
T.: *Peucedanum cervaria* (Linnaeus) Lapeyrouse (*Selinum cervaria* Linnaeus) (vide Rauschert, Taxon **31**: 755. 1982).

6058 **Schulzia** K. Sprengel, Pl. Umbell. Prodr. 30. 1813.
T.: *S. crinita* (Pallas) K. Sprengel (*Sison crinitum* Pallas).

(H) *Shultzia* Rafinesque, Med. Repos. ser. 2. **5**: 356. 1808 [Gentian.].
T.: *S. obolarioides* Rafinesque.

6064 **Kundmannia** Scopoli, Intr. 116. 1777.
T.: *K. sicula* (Linnaeus) A.-P. de Candolle (Prodr. 4: 143. 1830) (*Sium sicu-
lum* Linnaeus).

(≡) *Arduina* Adanson, Fam. Pl. 2: 499. 1763.

6070 **Selinum** Linnaeus, Sp. Pl. ed. 2. 350.
1762.
T.: *S. carvifolia* (Linnaeus) Linnaeus
(*Seseli carvifolia* Linnaeus) (*typ. cons.*).

(H) *Selinum* Linnaeus, Sp. Pl. 244. 1753
[Umbell.].
LT.: *S. sylvestre* Linnaeus (vide Hitch-
cock, Prop. Brit. Bot. 139. 1929; Regnum
Veg. 8: 261. 1956).

6083 **Levisticum** J. Hill, Brit. Herb. (fasc. 42)
423. 1756.
T.: *L. officinale* W. D. J. Koch (*Ligusticum levisticum* Linnaeus).

(H) *Levisticum* J. Hill, Brit. Herb. (fasc. 41)
410. 1756 [Umbell.].
≡ *Ligusticum* Linnaeus 1753.

6099 **Bonannia** Gussone, Fl. Sicul. Syn. 1:
355. 1842.
T.: *B. resinifera* Gussone, nom. illeg.
(*Ferula nudicaulis* K. Sprengel, *B. nudicaulis* (K. Sprengel) Rickett et Stafleu).

(H) *Bonannia* Rafinesque, Specchio 1: 115.
1814 [Sapind.].
T.: *B. nitida* Rafinesque.

CORNACEAE

6154 **Alangium** Lamarck, Encycl. 1: 174.
1783.
T.: *A. decapetalum* Lamarck (*typ.
cons.*).

(≡) *Angolam* Adanson, Fam. Pl. 2: 85, 518.
1763.
(=) *Kara-angolam* Adanson, Fam. Pl. 2: 84,
532. 1763.
T.: *Alangium hexapetalum* Lamarck
(Encycl. 1: 175. 1783).

6156 **Curtisia** W. Aiton, Hortus Kew. 1: 162
Aug-Oct 1789.
T.: *C. faginea* W. Aiton, nom. illeg. (*Sideroxylon dentatum* N. L. Burman, *C.
dentatum* (N. L. Burman) C. A. Smith).

(H) *Curtisia* Schreber, Gen. Pl. 199. Apr 1789
[Rut.].
T.: *C. schreberi* J. F. Gmelin.

6157 **Helwingia** Willdenow, Sp. Pl. 4: 716.
1806.
T.: *H. rusciflora* Willdenow, nom. illeg.
(*Osyris japonica* Thunberg, *H. japonica*
(Thunberg) F. G. Dietrich).

(H) *Helvingia* Adanson, Fam. Pl. 2: 345, 553.
1763 [Flacourt.].
≡ *Thamnia* P. Browne 1756 (*nom. rej.* sub
3284).

ERICACEAE

6189 **Loiseleuria** Desvaux, J. Bot Agric. 1:
35. 1813.
T.: *L. procumbens* (Linnaeus) Desvaux
(*Azalea procumbens* Linnaeus).

(≡) *Azalea* Linnaeus, Sp. Pl. 150. 1753.

248

6191 **Rhodothamnus** H. G. L. Reichenbach in Mössler, Handb. ed. 2. **1**: 667, 688. 1827.
T.: *R. chamaecistus* (Linnaeus) H. G. L. Reichenbach (*Rhododendron chamaecistus* Linnaeus).

6195 **Daboecia** D. Don, Edinburgh New Philos. J. **17**: 160. 1834.
T.: *D. polifolia* D. Don, nom. illeg. (*Vaccinium cantabricum* Hudson, *D. cantabrica* (Hudson) K. Koch).

6200 **Lyonia** Nuttall, Gen. N. Amer. Pl. **1**: 266. 1818.
T.: *L. ferruginea* (Walter) Nuttall (*Andromeda ferruginea* Walter).

(H) *Lyonia* Rafinesque, Med. Repos. ser. 2. **5**: 353. 1808 [Polygon.].
≡ *Polygonella* A. Michaux 1803.

6200a **Chamaedaphne** Moench, Methodus 457. 1794.
T.: *C. calyculata* (Linnaeus) Moench (*Andromeda calyculata* Linnaeus).

(H) *Chamaedaphne* Mitchell, Diss. Brev. Bot. Zool. 44. 1769 [Rub.].
≡ *Mitchella* Linnaeus 1753.

6208 **Pernettya** Gaudichaud, Ann. Sci. Nat. (Paris) **5**: 102. 1825 ("*Pernettia*") (*orth. cons.*).
T.: *P. empetrifolia* (Lamarck) Gaudichaud (*Andromeda empetrifolia* Lamarck).

(H) *Pernettya* Scopoli, Intr. 150 ("*Pernetya*"), index 1777 [Campanul.].
≡ *Canarina* Linnaeus 1771 (*nom. cons.*) (8656).

6212 **Arctostaphylos** Adanson, Fam. Pl. **2**: 165, 520. 1763.
T.: *A. uva-ursi* (Linnaeus) K. Sprengel (*Arbutus uva-ursi* Linnaeus) (*typ. cons.*).

(≡) *Uva-ursi* Duhamel, Traité Arbr. Arbust. **2**: 371. 1755.

6215 **Gaylussacia** Kunth in Humboldt, Bonpland et Kunth, Nova Gen. Sp. **3**: ed. fol. 215, ed. qu. 275. 1819.
T.: *G. buxifolia* Kunth.

6232 **Cavendishia** Lindley, Edward's Bot. Reg. **21**: sub. *t. 1791*. 1835.
T.: *C. nobilis* Lindley.

(H) *Cavendishia* S. F. Gray, Nat. Arr. Brit. Pl. **1**: 689. 1821 [Hepat.].
≡ *Antoiria* Raddi 1818.
(=) *Chupalon* Adanson, Fam. Pl. **2**: 164, 538. 1763.
T.: non designatus ["Chupalulones, Nieremb."].

EPACRIDACEAE

6251 **Lebetanthus** Endlicher, Gen. Pl. 1411 ("*Lebethanthus*"), 1458. 1841 (*orth. cons.*).
T.: *L. americanus* (W. J. Hooker) Endlicher (*Prionitis americana* W. J. Hooker) (*typ. cons.*).

(≡) *Allodape* Endlicher, Gen. Pl. 749. 1839.

6254 **Richea** R. Brown, Prodr. 555. 1810.
T.: *R. dracophylla* R. Brown.

(H) *Richea* Labillardière, Voy. Rech. Pérouse
1: 186, *t. 16.* 1800 [Comp.].
T.: *R. glauca* Labillardière.

(=) *Cystanthe* R. Brown, Prodr. 555. 1810.
T.: *C. sprengelioides* R. Brown.

6260 **Epacris** Cavanilles, Icon. 4: 25, *t. 344.* 1797.
T.: *E. longiflora* Cavanilles (*typ. cons.*).

(H) *Epacris* J. R. Forster et G. Forster, Char. Gen. Pl. 10. 1775 [Epacr.].
LT.: *E. longifolia* J. R. Forster et G. Forster (vide Regnum Veg. 8: 262. 1956).

6262a **Leucopogon** R. Brown, Prodr. 541. 1810.
T.: *L. lanceolatus* R. Brown, nom. illeg. (*Styphelia parviflora* H. Andrews, *L. parviflorus* (H. Andrews) Lindley).

(=) *Perojoa* Cavanilles, Icon. 4: 29. 1797.
T.: *P. microphylla* Cavanilles.

DIAPENSIACEAE

6275 **Shortia** Torrey et A. Gray, Amer. J. Sci. Arts 42: 48. 1842.
T.: *S. galacifolia* Torrey et A. Gray.

(H) *Shortia* Rafinesque, Aut. Bot. 16. 1840 [Cruc.].
T.: *S. dentata* Rafinesque (*Arabis dentata* Torrey et A. Gray 1838, non Clairville 1811).

6277 **Galax** Sims, Bot. Mag. 20: *t. 754.* 1804.
T.: *G. urceolata* (Poiret) Brummitt (*Pyrola urceolata* Poiret).

(H) *Galax* Linnaeus, Sp. Pl. 200. 1753 [Hydrophyll.].
T.: *G. aphylla* Linnaeus.

THEOPHRASTACEAE

6282 **Jacquinia** Linnaeus, Fl. Jamaica 27. 1760 ("*Jaquinia*") (*orth. cons.*).
T.: *J. ruscifolia* N. J. Jacquin (Enum. Syst. Pl. 15. 1760).

MYRSINACEAE

6285 **Ardisia** Swartz, Prodr. 3, 48. 1788.
T.: *A. tinifolia* Swartz (*typ. cons.*).

(=) *Katoutheka* Adanson, Fam. Pl. 2: 159, 534. 1763.
T.: *Psychotria dalzellii* J. D. Hooker (vide Ridsdale in Manilal, Bot. Hist. Hort. Malab. 136. 1980).

(=) *Vedela* Adanson, Fam. Pl. 2: 502. 1763.
T.: non designatus ["*Viscioides* Plum. M.S. vol. 6. t. 100."].

(=) *Icacorea* Aublet, Hist. Pl. Guiane, Suppl. 1. 1775.
T.: *I. guianensis* Aublet.

(=) *Bladhia* Thunberg, Nova Gen. Pl. 6. 1781.
T.: *B. japonica* Thunberg.

6288 **Heberdenia** Banks ex A. de Candolle, Ann. Sci. Nat. Bot. ser. 2. **16**: 79. 1841.
T.: *H. excelsa* Banks ex A. de Candolle (Prodr. **8**: 106. 1844), nom. illeg. (*Anguillaria bahamensis* J. Gaertner, *H. bahamensis* (J. Gaertner) Sprague) (etiam vide 974).

6291 **Labisia** Lindley, Edward's Bot. Reg. **31**: ad *t. 48.* 1845.
T.: *L. pothoina* Lindley.

(=) *Angiopetalum* Reinwardt, Syll. Pl. **2**: 7. 1825 vel 1826.
T.: *A. punctatum* Reinwardt.

6301 **Cybianthus** C. F. P. Martius, Nova Gen. Sp. Pl. **3**: 87. 1831 ("1829").
T.: *C. penduliflorus* C. F. P. Martius (*typ. cons.*).

(=) *Peckia* Vellozo, Fl. Flum. 51. 1825.
T.: non designatus.

6304 **Wallenia** Swartz, Prodr. 2, 31. 1788.
T.: *W. laurifolia* Swartz.

6310 **Embelia** N. L. Burman, Fl. Indica 62. 1768.
T.: *E. ribes* N. L. Burman.

(≡) *Ghesaembilla* Adanson, Fam. Pl. **2**: 449. 1763.

(=) *Pattara* Adanson, Fam. Pl. **2**: 447, 588. 1763.
T.: *Ardisia tseriam-cottam* Roemer et Schultes (Syst. Veg. **4**: 518. 1819).

PRIMULACEAE

6318 **Douglasia** Lindley, Quart. J. Sci. Lit. Arts **1827**: 385. 1827.
T.: *D. nivalis* Lindley.

(H) *Douglassia* P. Miller, Gard. Dict. Abr. ed. 4. 1754 [Verben.].
≡ *Volkameria* Linnaeus 1753.

(=) *Vitaliana* Sesler in Donati [Auszug Natur-Gesch. Adriat. Meers 66. 1753]; Essai Hist. Nat. Mer Adriat. 69. 1758.
T.: *V. primuliflora* Bertoloni (Fl. Ital. **2**: 368. 1835) (*Primula vitaliana* Linnaeus).

PLUMBAGINACEAE

6348 **Acantholimon** Boissier, Diagn. Pl. Orient. **7**: 69. Jul-Oct 1846.
T.: *A. glumaceum* (Jaubert et Spach) Boissier (*Statice glumacea* Jaubert et Spach).

(≡) *Armeriastrum* (Jaubert et Spach) Lindley, Veg. Kingd. 641. Jan-Mai 1846.

6350 **Armeria** Willdenow, Enum. Pl. Hort. Berol. 333. 1809.
T.: *A. vulgaris* Willdenow (*Statice armeria* Linnaeus).

(≡) *Statice* Linnaeus, Sp. Pl. 274. 1753 (vide P. Miller, Gard. Dict. Abr. ed. 4. 1754; Hitchcock, Prop. Brit. Bot. 143. 1929).

6351 **Limonium** P. Miller, Gard. Dict. Abr. ed. 4. 1754.
T.: *L. vulgare* P. Miller (*Statice limonium* Linnaeus) (*typ. cons.*).

SAPOTACEAE

6365 **Labatia** Swartz, Prodr. 2: 32. 1788.
T.: *L. sessiliflora* Swartz.

(H) *Labatia* Scopoli, Intr. 197. 1777 [Aquifol.].
≡ *Macoucoua* Aublet 1775.

6368a **Planchonella** Pierre, Not. Bot. Sapot. 34. 1890.
T.: *P. obovata* (R. Brown) Pierre (*Sersalisia obovata* R. Brown) (*typ. cons.*).

(=) *Hormogyne* A. de Candolle, Prodr. 8: 176. 1844.
T.: *H. cotinifolia* A. de Candolle.

6370 **Argania** Roemer et J. A. Schultes, Syst. Veg. 4: xlvi, 502. 1819.
T.: *A. sideroxylon* Roemer et J. A. Schultes, nom. illeg. (*Sideroxylon spinosum* Linnaeus, *A. spinosa* (Linnaeus) Skeels).

6373 **Dipholis** A. de Candolle, Prodr. 8: 188. 1844.
T.: *D. salicifolia* (Linnaeus) A. de Candolle (*Achras salicifolia* Linnaeus).

(=) *Spondogona* Rafinesque, Sylva Tell. 35. 1838.
T.: *S. nitida* Rafinesque, nom. illeg. (*Bumelia pentagona* Swartz).

6374 **Bumelia** Swartz, Prodr. 3, 49. 1788.
T.: *B. retusa* Swartz (*typ. cons.*).

(=) *Robertia* Scopoli, Intr. 154. 1777.
T.: *Sideroxylon decandrum* Linnaeus (Mant. 48. 1767).

6382 **Niemeyera** F. Mueller, Fragm. 7: 114. 1870.
T.: *N. prunifera* (F. Mueller) F. Mueller (*Chrysophyllum pruniferum* F. Mueller).

(H) *Niemeyera* F. Mueller, Fragm. 6: 96. 1867 [Orchid.].
T.: *N. stylidioides* F. Mueller.

6384 **Cryptogyne** J. D. Hooker in Bentham et J. D. Hooker, Gen. Pl. 2: 656. 1876.
T.: *C. gerrardiana* J. D. Hooker.

(H) *Cryptogyne* Cassini, Dict. Sci. Nat. 50: 491, 493, 498. 1827 [Comp.].
T.: *C. absinthioides* Cassini.

6386a **Manilkara** Adanson, Fam. Pl. 2: 166, 574. 1763.
T.: *M. kauki* (Linnaeus) Dubard (Ann. Inst. Bot.-Géol. Colon. Marseille ser. 3. 3: 9. 1915) (*Mimusops kauki* Linnaeus) (*typ. cons.*).

(=) *Achras* Linnaeus, Sp. Pl. 1190. 1753.
T.: *A. zapota* Linnaeus.

EBENACEAE

6408 **Brachynema** Bentham, Trans. Linn. Soc. London 22: 126. 1857.
T.: *B. ramiflorum* Bentham.

(H) *Brachynema* Griffith, Not. Pl. Asiat. 4: 176. 1854 [Verben.].
T.: *B. ferrugineum* Griffith.

252

STYRACACEAE

6410 **Halesia** J. Ellis ex Linnaeus, Syst. Nat. ed. 10. 1044, 1369. 1759.
T.: *H. carolina* Linnaeus.

(H) *Halesia* P. Browne, Civ. Nat. Hist. Jamaica 205. 1756 [Rub.].
T.: *Guettarda argentea* Lamarck (Encycl. 3: 54. 1789).

OLEACEAE

6421 **Forsythia** Vahl, Enum. 1: 39. 1804.
T.: *F. suspensa* (Thunberg) Vahl (*Ligustrum suspensum* Thunberg).

(H) *Forsythia* Walter, Fl. Carol 153. 1788 [Saxifrag.].
T.: *F. scandens* Walter.

6422 **Schrebera** Roxburgh, Pl. Corom. 2: 1, *t. 101.* 1799.
T.: *S. swietenioides* Roxburgh.

(H) *Schrebera* Linnaeus, Sp. Pl. ed. 2. 1662. 1763 [Convolvul.].
T.: *S. schinoides* Linnaeus.

6428 **Linociera** Swartz ex Schreber, Gen. 2: 784. 1791.
T.: *L. ligustrina* (Swartz) Swartz (*Thouinia ligustrina* Swartz).

(=) *Mayepea* Aublet, Hist. Pl. Guiane 81. 1775.
T.: *M. guianensis* Aublet.

(=) *Ceranthus* Schreber, Gen. 14. 1789.
T.: *C. schreberi* J. F. Gmelin (Syst. Nat. 2: 26. 1791).

LOGANIACEAE

6450 **Logania** R. Brown, Prodr. 454. 1810.
T.: *L. floribunda* R. Brown, nom. illeg. (*Euosma albiflora* H. Andrews, *L. albiflora* (H. Andrews) Druce) (*typ. cons.*).

(H) *Loghania* Scopoli, Intr. 236. 1777 [Marcgrav.].
= *Souroubea* Aublet 1775.

(≡) *Euosma* H. Andrews, Bot. Repos. 8: *t. 520.* 1808.

6468 **Peltanthera** Bentham in Bentham et J. D. Hooker, Gen. Pl. 2: 797. 1876.
T.: *P. floribunda* Bentham.

(H) *Peltanthera* Roth, Nov. Pl. Sp. 132. 1821 [Apocyn.].
T.: *P. solanacea* Roth.

GENTIANACEAE

6483 **Belmontia** E. Meyer, Comm. Pl. Afr. Austr. 183. 1838 ("1837").
T.: *B. cordata* E. Meyer, nom. illeg. (*Sebaea cordata* Roemer et J. A. Schultes, nom. illeg., *Gentiana exacoides* Linnaeus, *Belmontia exacoides* (Linnaeus) Druce) (*typ. cons.*.

(≡) *Parrasia* Rafinesque, Fl. Tell. 3: 78. 1837 ("1836").

253

6484 **Enicostema** Blume, Bijdr. 848. 1826.
T.: *E. littorale* Blume.

6501 **Bartonia** Muhlenberg ex Willdenow,
Ges. Naturf. Freunde Berlin Neue
Schriften **3**: 444. 1801.
T.: *B. tenella* Willdenow [= *B. virginica* (Linnaeus) Britton, Sterns et Poggenburg, *Sagina virginica* Linnaeus].

6504 **Orphium** E. Meyer, Comm. Pl. Afr.
Austr. 181. 1838 ("1837").
T.: *O. frutescens* (Linnaeus) E. Meyer
(*Chironia frutescens* Linnaeus).

6509a **Gentianella** Moench, Methodus 284.
1794.
T.: *G. tetrandra* Moench, nom. illeg.
(*Gentiana campestris* Linnaeus, *Gentianella campestris* (Linnaeus) Börner).

(=) *Amarella* Gilibert, Fl. Lit. Inch. **1**: 36.
1782.
LT.: *Gentiana amarella* Linnaeus (vide
Rauschert, Taxon **25**: 192. 1976).

6513 **Halenia** Borkhausen, Arch. Bot. (Leipzig) **1**(1): 25. 1796.
T.: *H. sibirica* Borkhausen, nom. illeg.
(*Swertia corniculata* Linnaeus, *H. corniculata* (Linnaeus) Cornaz).

6526 **Schultesia** C. F. P. Martius, Nova Gen.
Sp. Pl. **2**: 103. 1827.
T.: *S. crenuliflora* C. F. P. Martius (*typ. cons.*).

(H) *Schultesia* K. Sprengel, Pugill. **2**: 17. 1815
[Gram.].
T.: *S. petraea* (Thunberg) K. Sprengel
(*Chloris petraea* Thunberg).

6544 **Villarsia** Ventenat, Choix *t. 9.* 1803.
T.: *V. ovata* (Linnaeus f.) Ventenat
(*Menyanthes ovata* Linnaeus f.) (*typ. cons.*).

(H) *Villarsia* J. F. Gmelin, Syst. Nat. **2**: 306,
447. 1791 [Gentian.].
T.: *V. aquatica* J. F. Gmelin.

APOCYNACEAE

6559 **Carissa** Linnaeus, Syst. Nat. ed. 12. 189.
1767; Mant. Pl. 7, 52. 1767.
T.: *C. carandas* Linnaeus.

(≡) *Carandas* Adanson, Fam. Pl. **2**: 171, 532.
1763.

6562 **Landolphia** Palisot de Beauvois, Fl.
Oware **1**: 54. 1805.
T.: *L. owariensi* Palisot de Beauvois.

(=) *Pacouria* Aublet, Hist. Pl. Guiane 268.
1775.
T.: *P. guianensis* Aublet.
(=) *Vahea* Lamarck, Tabl. Encycl. **1**: *t. 169.*
1792.
T.: *V. gummifera* Lamarck (Tabl. Encycl. **2**: 292. 1819).

6564 **Willughbeia** Roxburgh, Pl. Corom. **3**:
77, *t. 280.* 1820.
T.: *W. edulis* Roxburgh.

(H) *Willughbeja* Scopoli ex Schreber, Gen. **1**:
162. 1789, nom. illeg. [Apocyn.].
T.: non designatus.

254

6566 **Urnularia** Stapf, Hooker's Icon. Pl. 28: t. 2711. 1901.
T.: *U. beccariana* (O. Kuntze) Stapf (*Ancylocladus beccarianus* O. Kuntze).

(H) *Urnularia* P. Karsten, Enum. Fung. Lapponia 209. 1866 [Fungi].
T.: *U. boreella* P. Karsten.

6583 **Alstonia** R. Brown, Asclepiadeae 64. 1810.
T.: *A. scholaris* (Linnaeus) R. Brown (*Echites scholaris* Linnaeus) (*typ. cons.*).

(H) *Alstonia* Scopoli, Intr. 198. 1777 [Apocyn.].
≡ *Pacouria* Aublet 1775 (vide 6562).

6588 **Aspidosperma** C. F. P. Martius et Zuccarini, Flora 7(1) (Beil. 4): 135. 1824.
T.: *A. tomentosum* C. F. P. Martius et Zuccarini (*typ. cons.*).

(=) *Coutinia* Vellozo, Quinogr. Portug. 166. 1799.
T.: *C. illustris* Vellozo.

(=) *Macaglia* L. C. Richard ex Vahl, Skr. Naturhist.-Selsk. 6: 107. 1810.
LT.: *M. alba* Vahl (vide Woodson, Ann. Missouri Bot. Gard. 38: 136. 1951).

6616 **Alyxia** Banks ex R. Brown, Prodr. 469. 1810.
T.: *A. spicata* R. Brown (*typ. cons.*).

(≡) *Gynopogon* J. R. Forster et G. Forster, Char. Gen. Pl. 18. 1775.

6626 **Kopsia** Blume, Catalogus 12. 1823.
T.: *K. arborea* Blume.

(H) *Kopsia* Dumortier, Comment. Bot. 16. 1822 [Orobanch.].
T.: *K. ramosa* (Linnaeus) Dumortier (*Orobanche ramosa* Linnaeus).

6632 **Thevetia** Linnaeus, Opera Var. 212. 1758.
T.: *T. ahouai* (Linnaeus) A. de Candolle (Prodr. 8: 345. 1844) (*Cerbera ahouai* Linnaeus) (*typ. cons.*).

(≡) *Ahouai* P. Miller, Gard. Dict. Abr. ed. 4. 1754.

6639 **Urceola** Roxburgh, Asiat. Res. 5: 169. 1799.
T.: *U. elastica* Roxburgh.

(H) *Urceola* Vandelli, Fl. Lusit. Bras. 8. 1788 [Spermatoph.].
T.: non designatus.

6670 **Spirolobium** Baillon, Bull. Mens. Soc. Linn. Paris 1: 773. 1889.
T.: *S. cambodianum* Baillon.

(H) *Spirolobium* Orbigny, Voy. Amér. Mér. 8 (Atlas) Bot. (1): t. 13. ?1839 ("1847") [Legum.].
T.: *S. australe* Orbigny.

6677 **Chonemorpha** G. Don, Gen. Hist. 4: 76. 1837.
T.: *C. macrophylla* G. Don, nom. illeg. (*Echites fragrans* Moon, *C. fragrans* (Moon) Alston).

(≡) *Belutta-kaka* Adanson, Fam. Pl. 2: 172, 525. 1763.

6683 **Ichnocarpus** R. Brown, Asclepiadeae 50. 1810.
T.: *I. frutescens* (Linnaeus) W. T. Aiton (Hort. Kew. ed. 2. 2: 69. 1811) (*Apocynum frutescens* Linnaeus).

6691 **Parsonsia** R. Brown, Asclepiadeae 53. 1810.
T.: *P. capsularis* (G. Forster) R. Brown ex Endlicher (Ann. Wiener Mus. Naturgesch. **1**: 175. 1836) (*Periploca capsularis* G. Forster) (*typ. cons.*).

(H) *Parsonsia* P. Browne, Civ. Nat. Hist. Jamaica 199. 1756 [Lythr.].
T.: *P. herbacea* Jaume Saint-Hilaire (Expos. Fam. Nat. **2**: 178. 1805) (*Lythrum parsonsia* Linnaeus).

6702 **Prestonia** R. Brown, Asclepiadeae 58. 1810.
T.: *P. tomentosa* R. Brown.

(H) *Prestonia* Scopoli, Intr. 218. 1777 [Malv.].
≡ *Lass* Adanson 1763.

ASCLEPIADACEAE

6726 **Camptocarpus** Decaisne in A. de Candolle, Prodr. **8**: 493. 1844.
T.: *C. mauritianus* (Lamarck) Decaisne (*Cynanchum mauritianum* Lamarck) (*typ. cons.*).

(H) *Camptocarpus* K. Koch, Linnaea **17**: 304. 1843 [Boragin.].
≡ *Oskampia* Moench 1794.

6772 **Schubertia** C. F. P. Martius, Nov. Gen. Sp. Pl. **1**: 55, *t. 33.* 1824.
T.: *S. multiflora* C. F. P. Martius (*typ. cons.*).

(H) *Schubertia* Mirbel, Nouv. Bull. Sci. Soc. Philom. Paris **3**: 123. 1812 [Pin.].
≡ *Taxodium* L. C. Richard 1810.

6857 **Oxypetalum** R. Brown, Asclepiadeae 30. 1810.
T.: *O. banksii* J. A. Schultes (in Roemer et J. A. Schultes, Syst. Veg. **6**: 91. 1820).

(=) *Gothofreda* Ventenat, Choix *t. 60.* 1808.
T.: *G. cordifolia* Ventenat.

*6870 **Brachystelma** R. Brown, Bot. Mag. **49**: t. 2343. 1822.
T.: *B. tuberosum* (Meerburg) R. Brown ex Sims ("*tuberosa*") (*Stapelia tuberosa* Meerburg).

(=) *Microstemma* R. Brown, Prodr. 459. 1810.
T.: *M. tuberosum* R. Brown.

6889 **Pectinaria** Haworth, Suppl. Pl. Succ. 14. 1819.
T.: *P. articulata* (W. Aiton) Haworth (*Stapelia articulata* W. Aiton).

(H) *Pectinaria* Bernhardi, Syst. Verz. **1**: 113. 1800 [Umbell.].
≡ *Scandix* Linnaeus 1753 (vide Hitchcock, Prop. Brit. Bot. 141. 1929).

6914 **Dregea** E. Meyer, Comm. Pl. Afr. Austr. 199. 1838.
T.: *D. floribunda* E. Meyer.

(H) *Dregea* Ecklon et Zeyher, Enum. Pl. Afric. Austral. 350. 1837 [Umbell.].
T.: non designatus.

CONVOLVULACEAE

6979 **Bonamia** Du Petit-Thouars, Hist. Vég. Iles France 33, *t. 8.* 1804.
T.: *B. madagascariensis* Poiret (in Lamarck, Encycl. Suppl. **1**: 677. 1810) (*typ. cons.*).

6994 **Calystegia** R. Brown, Prodr. 483. 1810. (≡) *Volvulus* Medikus, Philos. Bot. 2: 42.
T.: *C. sepium* (Linnaeus) R. Brown 1791.
(*Convolvulus sepium* Linnaeus) (*typ.
cons.*).

6997 **Merremia** Dennstedt ex Endlicher, (=) *Operculina* Silva Manso, Enum. Subst.
Gen. Pl. 1: 1403. 1841. Bras. 16. 1837.
T.: *M. hederacea* (N. L. Burman) H. T.: *O. turpethum* (Linnaeus) Silva Man-
Hallier (*Evolvulus hederaceus* N. L. so (*Convolvulus turpethum* Linnaeus).
Burman). (=) *Camonea* Rafinesque, Fl. Tell. 4: 81. 1838
("1836").
T.: *C. bifida* (Vahl) Rafinesque (*Convol-
vulus bifidus* Vahl).

7003 **Ipomoea** Linnaeus, Sp. Pl. 159. 1753.
T.: *I. pes-tigridis* Linnaeus (*typ. cons.*).

7003a **Pharbitis** Choisy, Mém. Soc. Phys. (≡) *Convolvuloides* Moench, Methodus 451.
Genève 6: 438. 1833. 1794.
T.: *P. hispida* Choisy, nom. illeg. (*Con-* (=) *Diatremis* Rafinesque, Ann. Gén. Sci.
volvulus purpureus Linnaeus, *P. purpu-* Phys. 8: 271. 1821.
rea (Linnaeus) J. O. Voigt) (*typ. cons.*). T.: *Convolvulus nil* Linnaeus.
(=) *Diatrema* Rafinesque, Herb. Raf. 80.
1833 med.
T.: *D. trichocarpa* Rafinesque, nom. illeg. (*Convolvulus carolinus* Linnaeus).

HYDROPHYLLACEAE

7022 **Nemophila** Nuttall, J. Acad. Nat. Sci. (=) *Viticella* Mitchell, Diss. Brev. Bot. Zool.
Philadelphia 2(1): 179. 1822 (med.?); 42. 1769.
Nuttall in W. Barton, Fl. N. Amer. 2: 71. T.: non designatus.
Jul-Dec 1822.
T.: *N. phacelioides* Nuttall.

7023 **Ellisia** Linnaeus, Sp. Pl. ed. 2. 1662. (H) *Ellisia* P. Browne, Civ. Nat. Hist. Jamaica
1763. 262. 1756 [Verben.].
T.: *E. nyctelea* (Linnaeus) Linnaeus T.: *E. acuta* Linnaeus (Syst. Nat. ed. 10.
(*Ipomoea nyctelea* Linnaeus). 1121. 1759).

7029 **Hesperochiron** S. Watson, U.S. Geol. (=) *Capnorea* Rafinesque, Fl. Tell. 3: 74.
Expl. 40th Par. 5: 281. 1871. 1837 ("1836").
T.: *H. californicus* (Bentham) S. Wat- T.: *C. nana* (Lindley) Rafinesque (*Nico-*
son (*Ourisia californica* Bentham). *tiana nana* Lindley).

7033 **Nama** Linnaeus, Syst. Nat. ed. 10. 950. (H) *Nama* Linnaeus, Sp. Pl. 226. 1753 [Hy-
1759. drophyll.].
T.: *N. jamaicensis* Linnaeus (*typ. cons.*; T.: *N. zeylanica* Linnaeus ("*zaylanica* ").
etiam vide Choisy in A. de Candolle,
Prodr. 10: 182. 1846).

7035 **Wigandia** Kunth in Humboldt, Bonpland et Kunth, Nova Gen. Sp. **3**: ed.
fol. 98, ed. qu. 126. 1819.
T.: *W. caracasana* Kunth (*typ. cons.*).

7037 **Hydrolea** Linnaeus, Sp. Pl. ed. 2. 328.
1762.
T.: *H. spinosa* Linnaeus.

BORAGINACEAE

7042 **Bourreria** P. Browne, Civ. Nat. Hist.
Jamaica 168, 492 ("*Beureria* "). 1756.
T.: *B. baccata* Rafinesque (*Cordia
bourreria* Linnaeus).

(H) *Beureria* Ehret, Pl. Papil. Rar. *t. 13.* 1755
[Calycanth.].
T.: non designatus.

7056 **Trichodesma** R. Brown, Prodr. 496.
1810.
T.: *T. zeylanicum* (N. L. Burman) R.
Brown (*Borago zeylanica* N. L. Burman) (*typ. cons.*).

(=) *Borraginoides* Boehmer in Ludwig, Defin. Gen. Pl. ed. 3. 18. 1760.
T.: *Borago indica* Linnaeus (Sp. Pl. 137.
1753).

7082 **Amsinckia** J. G. C. Lehmann, Sem.
Hort. Bot. Hamburg. **1831**: 3, 7. 1831.
T.: *A. lycopsoides* Lehmann.

7097 **Alkanna** Tausch, Flora **7**: 234. 1824.
T.: *A. tinctoria* Tausch [= *Lithospermum tinctorium* Linnaeus] (*typ. cons.*).

(H) *Alkanna* Adanson, Fam. Pl. **2**: 444, 514.
1763 [Lythr.].
≡ *Lawsonia* Linnaeus 1753.

7102 **Mertensia** Roth, Catalecta **1**: 34. 1797.
T.: *M. pulmonarioides* Roth.

(=) *Pneumaria* J. Hill, Veg. Syst. **7**: 40. 1764.
T.: non designatus.

7124 **Rochelia** H. G. L. Reichenbach, Flora
7: 243. 1824.
T.: *R. saccharata* H. G. L. Reichenbach,
nom. illeg. (*Lithospermum dispermum*
Linnaeus f., *R. disperma* (Linnaeus f.)
R. Wettstein).

(H) *Rochelia* Roemer et J. A. Schultes, Syst.
Veg. **4**: xi, 108. 1819 [Boragin.].
≡ *Lappula* Gilibert 1792.

7124a **Vaupelia** A. Brand, Repert. Spec.
Nov. Regni Veg. **13**: 82. 1914.
T.: *V. barbata* (Vaupel) A. Brand
(*Trichodesma barbatum* Vaupel).

(H) *Vaupellia* Grisebach, Fl. Brit. W. I. 460.
1861 [Gesner.].
T.: *V. calycina* Grisebach.

VERBENACEAE

7139 **Urbania** R. Philippi, Verz. Antofagasta
Pfl. 60. 1891.
T.: *U. pappigera* R. Philippi (*typ.
cons.*).

(H) *Urbania* Vatke, Österr. Bot. Z. **25**: 10.
1875 [Scrophular.].
T.: *U. lyperiifolia* Vatke.

7148 **Bouchea** Chamisso, Linnaea 7: 252. 1832.
T.: *B. pseudogervao* (A. Saint-Hilaire) Chamisso (*Verbena pseudogervao* A. Saint-Hilaire) (*typ. cons.*).

*7148a **Chacasanum** E. Meyer, Comm. Pl. Afr. Austr. 275. 1838.
T.: *C. cernuum* (Linnaeus) E. Meyer (*Buchnera cernua* Linnaeus) (*typ. cons.*).

(=) *Plexipus* Rafinesque, Fl. Tellur. 2: 104. 1837.
T.: *P. cuneifolius* (Linnaeus f.) Rafinesque (*Buchnera cuneifolia* Linnaeus f.).

7151 **Stachytarpheta** Vahl, Enum. 1: 205. 1804.
T.: *S. jamaicensis* (Linnaeus) Vahl (*Verbena jamaicensis* Linnaeus) (*typ. cons.*).

(≡) *Valerianoides* Medikus, Philos. Bot. 1: 177. 1789.
(=) *Vermicularia* Moench, Suppl. Meth. 150. 1802.
T.: non designatus.

7156 **Amasonia** Linnaeus f., Suppl. Pl. 48, 294. 1782.
T.: *A. erecta* Linnaeus f.

(=) *Taligalea* Aublet, Hist. Pl. Guiane 625. 1775.
T.: *T. campestris* Aublet.

7157 **Casselia** C. G. Nees et C. F. P. Martius, Nova Acta Phys.-Med. Acad. Caes. Leop.-Carol. Nat. Cur. 11: 73. 1823.
T.: *C. serrata* C. G. Nees et C. F. P. Martius (*typ. cons.*).

(H) *Casselia* Dumortier, Comment. Bot. 21. 1822 [Boragin.].
≡ *Mertensia* Roth 1797 (*nom. cons.*) (7102).

7181 **Tectona** Linnaeus f., Suppl. Pl. 20 ("*Tektona*"), 151. 1782.
T.: *T. grandis* Linnaeus f.

(≡) *Theka* Adanson, Fam. Pl. 2: 445. 1763.

7182a **Xerocarpa** Lam, Verben. Malay. Archip. 98. 1919.
T.: *X. avicenniifoliola* Lam.

(H) *Xerocarpa* (G. Don) Spach, Hist. Nat. Vég. Phan. 9: 583. 1840 [Gooden.].
T.: non designatus.

7185 **Premna** Linnaeus, Mant. Pl. 154, 252. 1771.
T.: *P. serratifolia* Linnaeus (*typ. cons.*).

(=) *Appella* Adanson, Fam. Pl. 2: 84, 519. 1763.
T.: non designatus ["H.M. 1. t. 53"].

LABIATAE (LAMIACEAE)

7227 **Stenogyne** Bentham, Edward's Bot. Reg. 15: sub t. 1292. 1830.
T.: *S. rugosa* Bentham (*typ. cons.*).

7249 **Glechoma** Linnaeus, Sp. Pl. 578. 1753. ("*Glecoma*") (*orth. cons.*).
T.: *G. hederacea* Linnaeus.

7250 **Dracocephalum** Linnaeus, Sp. Pl. 594. 1753.
T.: *D. moldavica* Linnaeus (*typ. cons.*).

7299 **Sphacele** Bentham, Edward's Bot. Reg. 15: sub *t. 1289.* 1829.
T.: *S. lindleyi* Bentham, nom. illeg. (*Stachys salviae* Lindley, *Sphacele salviae* (Lindley) Briquet) (*typ. cons.*).

7305 **Micromeria** Bentham, Edward's Bot. Reg. 15: sub *t. 1282.* 1829.
T.: *M. juliana* (Linnaeus) Bentham ex H. G. L. Reichenbach (Fl. Germ. Excurs. 311. 1831) (*Satureja juliana* Linnaeus).

7306 **Saccocalyx** Cosson et Durieu, Ann. Sci. Nat. Bot. ser. 3. 20: 80. 1853.
T.: *S. satureioides* Cosson et Durieu.

7312 **Amaracus** Gleditsch, Syst. Pl. Stamin. Situ 189. 1764.
T.: *A. dictamnus* (Linnaeus) Bentham (*Origanum dictamnus* Linnaeus) (*typ. cons.*).

7314 **Majorana** P. Miller, Gard. Dict. Abr. ed. 4. 1754.
T.: *M. hortensis* Moench (Meth. 406. 1794) (*Origanum majorana* Linnaeus) (*typ. cons.*).

7317 **Pycnanthemum** A. Michaux, Fl. Bor.-Amer. 2: 7. 1803.
T.: *P. incanum* (Linnaeus) A. Michaux (*Clinopodium incanum* Linnaeus) (*typ. cons.*).

7321 **Bystropogon** L'Héritier, Sert. Angl. 19. 1789 ("1788").
T.: *B. plumosus* L'Héritier ("*plumosum*") (*typ. cons.*).

7327 **Cunila** Linnaeus, Syst. Nat. ed. 10. 1359. 1759.
T.: *C. mariana* Linnaeus, nom. illeg. (*Satureja origanoides* Linnaeus; *C. origanoides* (Linnaeus) N. L. Britton).

7342 **Hyptis** N. J. Jacquin, Collectanea 1: 101, 103. 1787 ("1786").
T.: *H. capitata* N. J. Jacquin (*typ. cons.*).

(=) *Alguelaguen* Adanson, Fam. Pl. 2: 505. 1763.
T.: non designatus ["Feuill, t. 1."].

(=) *Phytoxis* Molina, Sag. Stor. Nat. Chili ed. 2. 145. 1810.
T.: *P. sideritifolia* Molina.

(=) *Xenopoma* Willdenow, Ges. Naturf. Freunde Berlin Mag. 5: 399. 1811.
T.: *X. obovatum* Willdenow.

(=) *Zygis* Desvaux ex W. Hamilton, Prodr. Pl. Ind. Occid. 40. 1825.
T.: *Z. aromatica* W. Hamilton.

(H) *Amaracus* J. Hill, Brit. Herb. 381. 1756 [Lab.].
≡ *Majorana* P. Miller 1754 (*nom. cons.*) (7314).

(=) *Hofmannia* Heister ex Fabricius, Enum. 61. 1759.
T.: *Origanum sipyleum* Linnaeus (Sp. Pl. 589. 1753).

(=) *Furera* Adanson, Fam. Pl. 2: 193, 560. 1763.
T.: *Satureja virginiana* Linnaeus (Sp. Pl. 567. 1753).

(H) *Cunila* Linnaeus ex P. Miller, Gard. Dict. Abr. ed. 4. 1754.
LT.: *Sideritis romana* Linnaeus (vide Reveal et Strachan, Taxon 29: 333. 1980).

(=) *Mesosphaerum* P. Browne, Civ. Nat. Hist. Jamaica 257. 1756.
LT.: *M. suaveolens* (Linnaeus) O. Kuntze (*Ballota suaveolens* Linnaeus) (vide O. Kuntze, Revis. Gen. Pl. 2: 525. 1891).

(=) *Condea* Adanson, Fam. Pl. 2: 504. 1763.
LT.: *Satureja americana* Poiret (vide O. Kuntze, Revis. Gen. Pl. 2: 524. 1891).

7346 **Alvesia** Welwitsch, Trans. Linn. Soc. London 27: 55, *t. 19.* 1869.
T.: *A. rosmarinifolia* Welwitsch.

(H) *Alvesia* Welwitsch, Anais Cons. Ultramar. Parte Não Off. **1858**: 587. 1859 [Legum.].
T.: *A. bauhinioides* Welwitsch.

7350 **Plectranthus** L'Héritier, Stirp. Nov. 84 verso. 1788.
T.: *P. fruticosus* L'Héritier (*typ. cons.*).

SOLANACEAE

7377 **Nicandra** Adanson, Fam. Pl. 2: 219, 582. 1763.
T.: *N. physalodes* (Linnaeus) J. Gaertner (Fruct. Sem. Pl. 2: 237. 1791) (*Atropa physalodes* Linnaeus) (*typ. cons.*).

(≡) *Physalodes* Boehmer in Ludwig, Defin. Gen. Pl. ed. 3. 41. 1760.

7380 **Dunalia** Kunth in Humboldt, Bonpland et Kunth, Nova Gen. Sp. 3: ed. fol. 43, ed. qu. 55, *t. 194.* 1818.
T.: *D. solanacea* Kunth.

(H) *Dunalia* K. Sprengel, Pugill. 2: 25. 1815 [Rub.].
≡ *Lucya* A.-P. de Candolle (*nom. cons.*) (8140).

7382 **Iochroma** Bentham, Edward's Bot. Reg. 31: *t. 20.* 1845.
T.: *I. tubulosum* Bentham, nom. illeg. (*Habrothamnus cyaneus* Lindley, *I. cyaneum* (Lindley) M. L. Green) (*typ. cons.*).

(=) *Diplukion* Rafinesque, Sylva Tell. 53. 1838.
T.: non designatus.

(=) *Valteta* Rafinesque, Sylva Tell. 53. 1838.
T.: non designatus.

7388 **Hebecladus** Miers, London J. Bot. 4: 321. 1845.
T.: *H. umbellatus* (Ruiz et Pavón) Miers (*Atropa umbellata* Ruiz et Pavón) (*typ. cons.*).

(≡) *Kokabus* Rafinesque, Sylva Tell. 55. 1838.

(=) *Ulticona* Rafinesque, Sylva Tell. 55. 1838.
T.: non designatus.

(=) *Kukolis* Rafinesque, Sylva Tell. 55. 1838.
T.: *K. bicolor* Rafinesque.

7392 **Triguera** Cavanilles, Diss. 2 [append.]: i, *t. A.* 1786.
T.: *T. ambrosiaca* Cavanilles (*typ. cons.*).

(H) *Triguera* Cavanilles, Diss. 1: 41. 1785 [Bombac.].
T.: *T. acerifolia* Cavanilles.

7393 **Scopolia** N. J. Jacquin, Observ. Bot. 1: 32. 1764 ("*Scopola*") (*orth. cons.*).
T.: *S. carniolica* N. J. Jacquin.

(H) *Scopolia* Adanson, Fam. Pl. 2: 419. Jul-Aug 1763 [Cruc.].
≡ *Ricotia* Linnaeus, Aug 1763 (*nom. cons.*) (2968).

7398 **Athenaea** Sendtner in C. F. P. Martius, Fl. Bras. 10: 133. 1846.
T.: *A. picta* (C. F. P. Martius) Sendtner (*Witheringia picta* C. F. P. Martius) (*typ. cons.*).

(H) *Athenaea* Adanson, Fam. Pl. 2: 121. 1763 [Comp.].
≡ *Struchium* P. Browne 1756.

(=) *Deprea* Rafinesque, Sylva Tell. 57. 1838.
T.: non designatus.

7400 **Withania** Pauquy, Diss. Bellad. 14.
1825.
T.: *W. frutescens* (Linnaeus) Pauquy
(*Atropa frutescens* Linnaeus) (*typ. cons.*).

7407a **Lycianthes** (Dunal) Hassler, Annuaire
Conserv. Jard. Bot. Genève **20**: 180.
1917.
T.: *L. lycioides* (Linnaeus) Hassler (*Solanum lycioides* Linnaeus).

(≡) *Otilix* Rafinesque, Med. Fl. **2**: 87. 1830.
(=) *Parascopolia* Baillon, Hist. Pl. **9**: 338.
1888.
T.: *P. acapulcensis* Baillon.

7414 **Solandra** Swartz, Kongl. Vetensk.
Acad. Nya Handl. **8**: 300. 1787.
T.: *S. grandiflora* Swartz.

(H) *Solandra* Linnaeus, Syst. Nat. ed. 10.
1269, 1380. 1759 [Umbell.].
T.: *S. capensis* Linnaeus.

GOETZEACEAE

7421 **Goetzea** Wydler, Linnaea **5**: 423, *t. 8.*
1830.
T.: *G. elegans* Wydler.

(H) *Goetzea* H. G. L. Reichenbach, Consp.
150. 1828 [Legum.].
≡ *Rothia* Persoon 1807 (*nom. cons.*)
(3659).

SCROPHULARIACEAE

7467 **Aptosimum** Burchell ex Bentham, Edward's Bot. Reg. **22**: sub *t. 1882.* 1836.
T.: *A. depressum* Burchell ex Bentham
[= *A. procumbens* (J. G. C. Lehmann)
Steudel, *Ohlendorffia procumbens* J. G.
C. Lehmann].

(=) *Ohlendorffia* J. G. C. Lehmann, Sem.
Hort. Bot. Hamburg. **1835**: 7. 1835.
T.: *O. procumbens* J. G. C. Lehmann.

7472 **Hemimeris** Linnaeus f., Suppl. Pl. 45,
280. 1782.
T.: *H. montana* Linnaeus f. (*typ. cons.*).

(H) *Hemimeris* Linnaeus, Pl. Rar. Afr. 8.
1760 [Scrophular.].
T.: *H. bonae-spei* Linnaeus.

7474 **Calceolaria** Linnaeus, Kongl. Vetensk.
Acad. Handl. **31**: 286. 1770.
T.: *C. pinnata* Linnaeus.

(H) *Calceolaria* Loefling, Iter Hispan. 183.
1758 [Viol.].
T.: non designatus.

7485 **Anarrhinum** Desfontaines, Fl. Atlant.
2: 51. 1798.
T.: *A. pedatum* Desfontaines (*typ. cons.*).

(=) *Simbuleta* Forsskål, Fl. Aegypt.-Arab.
115. 1775.
T.: *S. forskaohlii* J. F. Gmelin ("*Forskåhlii*") (Syst. Nat. **2**: 242. 1791).

7510 **Tetranema** Bentham, Bot. Reg. **29**: *t.
52.* 1843.
T.: *T. mexicanum* Bentham.

262

7517 **Manulea** Linnaeus, Syst. Nat. ed. 12. 419. Oct. 1767; Mant. Pl. 12, 88. Oct 1767.
T.: *M. cheiranthus* (Linnaeus) Linnaeus (*Lobelia cheiranthus* Linnaeus).

(≡) *Nemia* P. J. Bergius, Descr. Pl. Cap. 160, 162. Sep 1767.

7518 **Chaenostoma** Bentham, Companion Bot. Mag. **1**: 374. 1836.
T.: *C. aethiopicum* (Linnaeus) Bentham (*Buchnera aethiopica* Linnaeus) (*typ. cons.*).

(=) *Palmstruckia* Retzius, Obs. Bot. Pugill. 15. 1810.
T.: *Manulea foetida* (Andrews) Persoon (*Buchnera foetida* Andrews).

7523 **Zaluzianskya** F. W. Schmidt, Neue Selt. Pfl. 11. 1793.
T.: *Z. villosa* F. W. Schmidt.

(H) *Zaluzianskia* Necker, Hist. & Commentat. Acad. Elect. Sci. Theod.-Palat. **3** (Phys.): 303. 1775 [Pteridoph.].
≡ *Marsilea* Linnaeus 1753 (vide Christensen, Index Fil. lvii. 1906).

7532 **Limnophila** R. Brown, Prodr. 442. 1810.
T.: *L. gratioloides* R. Brown, nom. illeg. (*Hottonia indica* Linnaeus, *L. indica* (Linnaeus) Druce).

(≡) *Hydropityon* C. F. Gaertner, Suppl. Carp. 19. 1805.
(=) *Ambuli* Adanson, Fam. Pl. **2**: 208. 1763.
T.: *A. aromatica* Lamarck (Encycl. **1**: 128. 1783).
(=) *Diceros* Loureiro, Fl. Cochinch. 381. 1790.
T.: *D. cochinchinensis* Loureiro.

7534 **Stemodia** Linnaeus, Syst. Nat. ed. 10. 1118, 1374. 1759.
T.: *S. maritima* Linnaeus.

(≡) *Stemodiacra* P. Browne, Civ. Nat. Hist. Jamaica 261. 1756.

7546 **Bacopa** Aublet, Hist. Pl. Guiane 128. 1775.
T.: *B. aquatica* Aublet.

(=) *Moniera* P. Browne, Civ. Nat. Hist. Jamaica 269. 1756.
T.: non designatus.
(=) *Brami* Adanson, Fam. Pl. **2**: 208. 1763.
T.: *B. indica* Lamarck (Encycl. **1**: 459. 1785).

7549 **Micranthemum** A. Michaux, Fl. Bor.-Amer. **1**: 10. 1803.
T.: *M. orbiculatum* A. Michaux, nom. illeg. (*Globifera umbrosa* J. F. Gmelin, *M. umbrosum* (J. F. Gmelin) Blake).

(≡) *Globifera* J. F. Gmelin, Syst. Nat. **2**: 32. 1791.

7556 **Glossostigma** Wight et Arnott, Nova Acta Phys.-Med. Acad. Caes. Leop.-Carol. Nat. Cur. **18**: 355. 1836.
T.: *G. spathulatum* Arnott, nom. illeg. (*Limosella diandra* Linnaeus, *G. diandrum* (Linnaeus) O. Kuntze).

(≡) *Peltimela* Rafinesque, Atlantic J. **1**: 199. 1833.

7559 **Artanema** D. Don in Sweet, Brit. Fl. Gard. ser. 2. 3: *t. 234.* 1834.
T.: *A. fimbriatum* (W. J. Hooker ex R. Graham) D. Don (*Torenia fimbriata* W. J. Hooker ex R. Graham).

(=) *Bahel* Adanson, Fam. Pl. 2: 210. 1763.
T.: *Columnea longifolia* Linnaeus.

7592 **Rehmannia** Liboschitz ex Fischer et C. A. Meyer, Index Sem. Hort. Petrop. 1: 36. 1835.
T.: *R. sinensis* (Buchoz) Liboschitz et Fischer ex C. A. Meyer ("*chinensis-*") (*Sparmannia sinensis* Buchoz) (etiam vide 4957).

7602 **Seymeria** Pursh, Fl. Amer. Sept. 2: 736. 1814.
T.: *S. tenuifolia* Pursh, nom. illeg. (*Afzelia cassioides* J. F. Gmelin *S. cassioides* (J. F. Gmelin) Blake) (*typ. cons.*) (etiam vide 3509).

7604a **Agalinis** Rafinesque, New Fl. 2: 61. 1837.
T.: *A. palustris* Rafinesque, nom. illeg. (*Gerardia purpurea* Linnaeus, *A. purpurea* (Linnaeus) Pennell) (*typ. cons.*).

(=) *Virgularia* Ruiz et Pavón, Prodr. 92. 1794.
LT.: *V. lanceolata* Ruiz et Pavón (vide D'Arcy, Taxon 28: 419-420. 1979).
(=) *Chytra* C. F. Gaertner, Suppl. Carp. 184, t. 214. 1807.
T.: *C. anomala* C. F. Gaertner.
(=) *Tomanthera* Rafinesque, New Fl. 2: 65. 1837.
LT.: *T. lanceolata* Rafinesque (vide D'Arcy, Taxon 28: 419-420. 1979).

7632 **Cordylanthus** Nuttall ex Bentham in A. de Candolle, Prodr. 10: 597. 1846.
T.: *C. filifolius* Nuttall ex Bentham, nom. illeg. (*Adenostegia rigida* Bentham, *C. rigidus* (Bentham) Jepson).

(≡) *Adenostegia* Bentham in Lindley, Nat. Syst. Bot. ed. 2. 445. 1836.

7645 **Bartsia** Linnaeus, Sp. Pl. 602. 1753.
T.: *B. alpina* Linnaeus (*typ. cons.*).

7649 **Rhynchocorys** Grisebach, Spic. Fl. Rumel. 2: 12. 1844.
T.: *R. elephas* (Linnaeus) Grisebach (*Rhinanthus elephas* Linnaeus).

(≡) *Elephas* P. Miller, Gard. Dict. Abr. ed. 4. 1754.

7650 **Lamourouxia** Kunth in Humboldt, Bonpland et Kunth, Nov. Gen. Sp. 2: ed. fol. 269, ed. qu. 335. 1818 ("1817").
T.: *L. multifida* Kunth.

(H) *Lamourouxia* C. A. Agardh, Syn. Alg. Scand. xiv. 1817 [Rhodoph.].
T.: *L. elegans* (Lamouroux) C. A. Agardh (*Claudea elegans* Lamouroux).

7665 **Anemopaegma** C. F. P. Martius ex Meisner, Pl. Vasc. Gen. **1**: 300. 1840; **2**: 208. 1840 ("*Anemopaegmia*") (*orth. cons.*).
T.: *A. mirandum* (Chamisso) A.-P. de Candolle (*Bignonia miranda* Chamisso).

(=) *Cupulissa* Rafinesque, Fl. Tell. **2**: 57. 1837.
T.: *C. grandifolia* (N. J. Jacquin) Rafinesque (*Bignonia grandiflora* N. J. Jacquin).

(=) *Platolaria* Rafinesque, Sylva Tell. 78. 1838.
T.: *P. flavescens* Rafinesque, nom. illeg. (*Bignonia orbiculata* N. J. Jacquin).

7668 **Cuspidaria** A.-P. de Candolle, Biblioth. Universelle Genève ser. 2. **17**: 125. Sep 1838; Rev. Bignon. 9. Oct 1838.
T.: *C. pterocarpa* (Chamisso) A.-P. de Candolle (Prodr. **9**: 178. 1845) (*Bignonia pterocarpa* Chamisso) (*typ. cons.*).

(H) *Cuspidaria* (A.-P. de Candolle) Besser, Enum. Pl. **2**: 104. 1822 [Cruc.].
T.: *C. biebersteinii* Andrzejowski ex Besser, nom. illeg. (*Cheiranthus cuspidatus* Marschall von Bieberstein).

7673 **Haplolophium** Chamisso, Linnaea **7**: 556. 1832 ("*Aplolophium*") (*orth. cons.*).
T.: *H. bracteatum* Chamisso.

7679 **Phaedranthus** Miers, Proc. Roy. Hort. Soc. London ser. 2. **3**: 182. 1863.
T.: *P. lindleyanus* Miers.

(=) *Sererea* Rafinesque, Sylva Tell. 107. 1838.
T.: *S. heterophylla* Rafinesque, nom. illeg. (*Bignonia heterophylla* Willdenow, nom. illeg., *Bignonia kerere* Aublet).

7697 **Lundia** A.-P. de Candolle, Biblioth. Universelle Genève ser. 2. **17**: 127. Sep 1838; Rev. Bignon. 11. Oct 1838.
T.: *L. glabra* A.-P. de Candolle (Prodr. **9**: 180. 1845) (*typ. cons.*).

(H) *Lundia* H. C. F. Schumacher, Beskr. Guin. Pl. **2**: 5, [231]. 1827 [Flacourt.].
T.: *L. monacantha* H. C. F. Schumacher.

7705 **Bignonia** Linnaeus, Sp. Pl. 622. 1753.
T.: *B. capreolata* Linnaeus (*typ. cons.*).

7714 **Campsis** Loureiro, Fl. Cochinch. 377. 1790.
T.: *C. adrepens* Loureiro.

(=) *Notjo* Adanson, Fam. Pl. **2**: 226, 582. 1763.
T.: non designatus.

7741 **Dolichandrone** (Fenzl) Seemann, Ann. Mag. Nat. Hist. ser. 3. **10**: 31. 1862.
T.: *D. spathacea* (Linnaeus f.) K. Schumann (*Bignonia spathacea* Linnaeus f.).

(≡) *Pongelia* Rafinesque, Sylva Tell. 78. 1838.

7757 **Enallagma** (Miers) Baillon, Hist. Pl. **10**: 54. 1888.
T.: *E. cucurbitina* (Linnaeus) Baillon ex K. Schumann (in Engler et Prantl, Nat. Pflanzenfam. 4(3b): 247. 1895) (*Crescentia cucurbitina* Linnaeus).

(≡) *Dendrosicus* Rafinesque, Sylva Tell. 80. 1838.

7760 **Colea** Bojer ex Meisner, Pl. Vasc. Gen. 1: 301. 1840; 2: 210. 1840.
T.: *C. colei* (Bojer ex W. J. Hooker) M. L. Green (*Bignonia colei* Bojer ex W. J. Hooker) (*typ. cons.*).

(≡) *Odisca* Rafinesque, Sylva Tell. 80. 1838.
(=) *Uloma* Rafinesque, Fl. Tell. 2: 62. 1837 ("1836").
T.: *U. telfairiae* (Bojer) Rafinesque (*Bignonia telfairiae* Bojer).

7766 **Tourrettia** Fougeroux, Mém. Acad. Sci. (Paris) 1784: 205. 1787 ("*Tourretia*") (*orth. cons.*).
T.: *T. lappacea* (L'Héritier) Willdenow (Sp. Pl. 3: 263. 1800) (*Dombeya lappacea* L'Héritier) (etiam vide 5053).

OROBANCHACEAE

7792 **Epifagus** Nuttall, Gen. N. Amer. Pl. 2: 60. 1818.
T.: *E. americana* Nuttall, nom. illeg. (*Orobanche virginiana* Linnaeus, *E. virginiana* (Linnaeus) Barton).

GESNERIACEAE

7800 **Ramonda** L. C. Richard in Persoon, Syn. Pl. 1: 216. 1805.
T.: *R. pyrenaica* Persoon, nom. illeg. (*Verbascum myconi* Linnaeus, *R. myconi* (Linnaeus) H. G. L. Reichenbach).

(H) *Ramondia* Mirbel in A.-P. de Candolle, Bull. Sci. Soc. Philom. Paris 2: 179. 1801 [Pteridoph.].
T.: *R. flexuosa* (Linnaeus) Mirbel (*Ophioglossum flexuosum* Linnaeus).

7808 **Oreocharis** Bentham in Bentham et J. D. Hooker, Gen. Pl. 2: 1021. 1876.
T.: *O. benthamii* C. B. Clarke (in A. de Candolle et C. de Candolle, Monogr. Phan. 5: 63. 1883) (*Didymocarpus oreocharis* Hance).

(H) *Oreocharis* (Decaisne) Lindley, Veg. Kingd. 656. 1846 [Boragin.].
T.: non designatus.

7809 **Didissandra** C. B. Clarke in A. de Candolle et C. de Candolle, Monogr. Phan. 5: 65. 1883.
T.: *D. elongata* (Jack) C. B. Clarke (*Didymocarpus elongatus* Jack) (*typ. cons.*).

(=) *Ellobum* Blume, Bijdr. 746. 1826.
T.: *E. montanum* Blume.

7810 **Didymocarpus** Wallich, Edinburgh Philos. J. 1: 378. 1819.
T.: *D. primulifolius* D. Don (Fl. Nepal. 123. 1825) (*typ. cons.*).

(=) *Henckelia* K. Sprengel, Anleit. ed. 2. 2: 402. 1817.
T.: *H. incana* (Vahl) K. Sprengel (*Roettlera incana* Vahl).

7824 **Aeschynanthus** Jack, Trans. Linn. Soc. London **14**: 42, *t. 2, f. 3*. 1823 ("1825"). T.: *A. volubilis* Jack (*typ. cons.*).

(=) *Trichosporum* D. Don, Edinburgh Philos. J. **7**: 84. 1822. T.: non designatus.

*7833 **Rhynchoglossum** Blume, Bijdr. Fl. Ned. Ind. 741. 1826 ("*Rhinchoglossum*") (*orth. cons.*). T.: *R. obliquum* Blume.

7835 **Acanthonema** J. D. Hooker, Bot. Mag. **88**: *t. 5339*. 1862. T.: *A. strigosum* J. D. Hooker.

(H) *Acanthonema* J. G. Agardh, Öfvers. Förh. Kongl. Svenska Vetensk.-Akad. **3**: 104. 1846 [Rhodoph.]. T.: *A. montagnei* J. G. Agardh, nom. illeg. (*Conferva oxyclada* Montagne, *C. aculeata* Montagne 1839, non Suhr 1834).

7853 **Mitraria** Cavanilles, Anales Ci. Nat. **3**: 230. 1801. T.: *M. coccinea* Cavanilles.

(H) *Mitraria* J. F. Gmelin, Syst. Nat. **2**: 771, 799. 1791 [Barrington.]. ≡ *Commersona* Sonnerat 1776, non *Commersonia* J. R. Forster et G. Forster 1775.

7854 **Sarmienta** Ruiz et Pavón, Prodr. 4. 1794. T.: *S. repens* Ruiz et Pavón, nom. illeg. (*S. scandens* (J. D. Brandis) Persoon, *Urceolaria scandens* J. D. Brandis).

(≡) *Urceolaria* J. D. Brandis in Molina, Vers. Naturgesch. Chili 133. 1786.

7860 **Alloplectus** C. F. P. Martius, Nov. Gen. Sp. Pl. **3**: 53. 1829. T.: *A. sparsiflorus* C. F. P. Martius (*typ. cons.*).

(=) *Crantzia* Scopoli, Intr. 173. 1777. T.: *Besleria cristata* Linnaeus. (=) *Vireya* Rafinesque, Specchio **1**: 194. 1814. T.: *V. sanguinolenta* Rafinesque.

7866 **Codonanthe** (C. F. P. Martius) Hanstein, Linnaea **26**: 209. 1854 ("1853"). T.: *C. gracilis* (C. F. P. Martius) Hanstein (*Hypocyrta gracilis* C. F. P. Martius) (*typ. cons.*).

(H) *Codonanthus* G. Don, Gen. Hist. **4**: 166. 1838 [Logan.]. T.: *C. africana* G. Don.

7874 **Achimenes** Persoon, Syn. Pl. **2**: 164. 1806. T.: *A. coccinea* (Scopoli) Persoon (*Buchnera coccinea* Scopoli) (*typ. cons.*).

(H) *Achimenes* P. Browne, Civ. Nat. Hist. Jamaica 270. 1756 [Gesner.]. T.: non designatus [*A. Major, herbacea, subhirsuta, oblique assurgens* P. Browne].

7878 **Seemannia** Regel, Gartenflora **4**: 183, *t. 126*. 1855. T.: *S. ternifolia* Regel.

267

7887a **Rechsteineria** Regel, Flora **31**: 247. 1848.
T.: *R. allagophylla* (C. F. P. Martius) Regel (*Gesneria allagophylla* C. F. P. Martius).

(≡) *Alagophyla* Rafinesque, Fl. Tell. **2**: 33. 1837.

(=) *Megapleilis* Rafinesque, Fl. Tell. **2**: 57. 1837.
T.: *M. tuberosa* Rafinesque, nom. illeg. (*Gesneria bulbosa* Ker-Gawler).

(=) *Styrosinia* Rafinesque, Fl. Tell. **2**: 97. 1837.
T.: *S. coccinea* Rafinesque, nom. illeg. (*Gesneria aggregata* Ker-Gawler).

(=) *Tulisma* Rafinesque, Fl. Tell. **2**: 98. 1837.
T.: *T. verticillata* (W. J. Hooker) Rafinesque (*Gesneria verticillata* W. J. Hooker).

COLUMELLIACEAE

7897 **Columellia** Ruiz et Pavón, Prodr. 3, *t. 1*. 1794.
T.: *C. oblonga* Ruiz et Pavón (*typ. cons.*).

(H) *Columella* Loureiro, Fl. Cochinch. 85. 1790 [Vit.].
≡ *Cayratia* A. L. Jussieu 1818 (*nom. cons.*) (4918a).

LENTIBULARIACEAE

7900 **Polypompholyx** J. G. C. Lehmann, Nov. Stirp. Pug. **8**: 48. 1844.
T.: *P. tenella* J. G. C. Lehmann (*Utricularia tenella* R. Brown) (*typ. cons.*).

(=) *Cosmiza* Rafinesque, Fl. Tell. **4**: 110. 1838 ("1836").
T.: *C. coccinea* Rafinesque, nom. illeg. (*Utricularia multifida* R. Brown).

ACANTHACEAE

7908 **Elytraria** A. Michaux, Fl. Bor.-Amer. **1**: 8. 1803.
T.: *E. virgata* A. Michaux, nom. illeg. (*Tubiflora caroliniensis* J. F. Gmelin, *E. caroliniensis* (J. F. Gmelin) Persoon).

(≡) *Tubiflora* J. F. Gmelin, Syst. Nat. **2**: 27. 1791.

7914 **Thunbergia** Retzius, Physiogr. Sälsk. Handl. **1**(3): 163. 1780 ("1776").
T.: *T. capensis* Retzius.

(H) *Thunbergia* Montin, Kongl. Vetensk. Acad. Handl. **34**: 288, *t. 11*. 1773 [Rub.].
T.: *T. florida* Montin ex Retzius (Physiogr. Sälsk. Handl. **1**(3): 163. 1780) (*T. capensis* Montin ex Linnaeus f. 1782, non Retzius 1780).

7932 **Phaulopsis** Willdenow, Sp. Pl. **3**: 342. 1800 ("*Phaylopsis*") (*orth. cons.*).
T.: *P. parviflora* Willdenow, nom. illeg. (*Micranthus oppositifolius* J. C. Wendland, *P. oppositifolia* (J. C. Wendland) Lindau) (etiam vide 1313).

7972 **Crabbea** W. H. Harvey, London J. Bot. 1: 27. 1842.
T.: *C. hirsuta* W. H. Harvey.

(H) *Crabbea* W. H. Harvey, Gen. S. Afr. Pl. 276. 1838 [Acanth.].
T.: *C. pungens* W. H. Harvey.

7990 **Stenandrium** C. G. Nees in Lindley, Nat. Syst. Bot. ed. 2. 444. 1836.
T.: *S. mandioccanum* C. G. Nees.

(=) *Gerardia* Linnaeus, Sp. Pl. 610. 1753.
T.: *G. tuberosa* Linnaeus.

8014 **Carlowrightia** A. Gray, Proc. Amer. Acad. Arts 13: 364. 1878.
T.: *C. linearifolia* (Torrey) A. Gray (*Schaueria linearifolia* Torrey).

(=) *Cardiacanthus* C. G. Nees et Schauer in A. de Candolle, Prodr. 11: 331. 1847.
T.: *C. neesianus* Schauer ex C. G. Nees.

8028 **Tetramerium** C. G. Nees in Bentham, Bot. Voy. Sulphur 147. 1846.
T.: *T. polystachyum* C. G. Nees.

(H) *Tetramerium* C. F. Gaertner, Suppl. Carp. 90. 1805 [Rub.].
T.: *T. odoratissimum* C. F. Gaertner, nom. illeg. (*Ixora americana* Linnaeus).

(=) *Henrya* C. G. Nees in Bentham, Bot. Voy. Sulphur *t. 49.* 1845; 148. 1846.
T.: *H. insularis* C. G. Nees.

8031 **Dicliptera** A. L. Jussieu, Ann. Mus. Natl. Hist. Nat. 9: 267. 1807.
T.: *D. chinensis* (Linnaeus) A. L. Jussieu (*Justicia chinensis* Linnaeus) (*typ. cons.*).

(≡) *Diapedium* C. König, Ann. Bot. (König & Sims) 2: 189. 1805 ("1806").

8037 **Odontonema** C. G. Nees, Linnaea 16: 300. post Jun 1842.
T.: [Garden specimen without date or collector] (GZU) [= *O. rubrum* (Vahl) Kuntze (*Justicia rubra* Vahl)] (*typ. cons.*).

(H) *Odontonema* C. G. Nees ex Endlicher, Gen. Pl. Suppl. 2: 63 Mar-Jun 1842 [Acanth.].
LT.: *Justicia lucida* Andrews (vide Baum et Reveal, Taxon 29: 336. 1980).

8039 **Mackaya** W. H. Harvey, Thes. Cap. 1: 8, *t. 13.* 1859.
T.: *M. bella* W. H. Harvey.

(H) *Mackaia* S. F. Gray, Nat. Arr. Brit. Pl. 1: 391. 1821 [Phaeoph.].
T.: non designatus.

8042 **Schaueria** C. G. Nees, Ind. Sem. Hort. Ratisb. 1838; Linnaea 13 (litt.): 119. 1839.
T.: *S. calycotricha* (Link et Otto) C. G. Nees (*Justicia calycotricha* Link et Otto).

(H) *Schauera* C. G. Nees in Lindley, Nat. Syst. Bot. ed. 2. 202. 1836 [Laur.].
≡ *Endlicheria* C. G. Nees 1833 (*nom. cons.*) (2811a).

(=) *Flavicoma* Rafinesque, Fl. Tell. 4: 63. 1838 med. ("1836").
T.: non designatus.

8069 **Fittonia** E. Coemans, Fl. Serres Jard. Eur. 15: 185. 1865.
T.: *F. verschaffeltii* (Lemaire) Van Houtte (*Gymnostachyum verschaffeltii* Lemaire).

(=) *Adelaster* Lindley ex Veitch, Gard. Chron. 1861: 499. 1861.
T.: *A. albivenis* Lindley ex Veitch.

8079 **Isoglossa** Örsted, Vidensk. Meddel. Dansk Naturhist. Foren. Kjøbenhavn **1854**: 155. 1854.
T.: *I. origanoides* (C. G. Nees) Lindau (*Rhytiglossa origanoides* C. G. Nees).

(≡) *Rhytiglossa* C. G. Nees ex Lindley, Nat. Syst. Bot. ed. 2. 285, 444. 1836.

8096 **Anisotes** C. G. Nees in A. de Candolle, Prodr. **11**: 424. 1847.
T.: *A. trisulcus* (Forsskål) C. G. Nees (*Dianthera trisulca* Forsskål).

(H) *Anisotes* Lindley ex Meisner, Pl. Vasc. Gen. **1**: 117. 1838; **2**: 84. 1838 [Lythr.].
T.: *A. hilariana* Meisner, nom. illeg. (*Lythrum anomalum* A. Saint-Hilaire).

(≡) *Calasias* Rafinesque, Fl. Tell. **4**: 64. 1838 ("1836").

8097 **Jacobinia** C. G. Nees ex Moricand, Pl. Nouv. Amér. 156, *t. 92.* 1846.
T.: *J. lepida* C. G. Nees ex Moricand.

8100 **Trichocalyx** I. B. Balfour, Proc. Roy. Soc. Edinburgh **12**: 87. 1884.
T.: *T. obovatus* I. B. Balfour (*typ. cons.*).

(H) *Trichocalyx* Schauer, Nova Acta Phys.-Med. Acad. Caes. Leop.-Carol. Nat. Cur. **19** (suppl. 2): 86. 1841 [Myrt.].
≡ *Calytrix* Labillardière 1806.

RUBIACEAE

8126 **Bikkia** Reinwardt, Syll. Pl. **2**: 8. 1825 vel 1826.
T.: *B. tetrandra* (Linnaeus f.) A. Gray (*Portlandia tetrandra* Linnaeus f.).

8130 **Lerchea** Linnaeus, Mant. Pl. **2**: 155, 256. 1771.
T.: *L. longicauda* Linnaeus.

(H) *Lerchia* Haller ex Zinn, Catal. 30. 1757 [Chenopod.].
T.: non designatus.

8136 **Kohautia** Chamisso et Schlechtendal, Linnaea **4**: 156. 1829.
T.: *K. senegalensis* Chamisso et Schlechtendal.

(=) *Duvaucellia* S. Bowdich in T. Bowdich, Exc. Madeira 259. 1825.
T.: *D. tenuis* S. Bowdich.

8140 **Lucya** A.-P. de Candolle, Prodr. **4**: 434. 1830.
T.: *L. tuberosa* A.-P. de Candolle, nom. illeg. (*Peplis tetrandra* Linnaeus, *L. tetrandra* (Linnaeus) K. Schumann) (etiam vide 7380).

8158 **Cruckshanksia** W. J. Hooker et Arnott, Bot. Misc. **3**: 361. 1833.
T.: *C. hymenodon* W. J. Hooker et Arnott.

(H) *Cruckshanksia* W. J. Hooker, Bot. Misc. **2**: 211. 1831 [Geran.].
T.: *C. cistiflora* W. J. Hooker.

8162 **Payera** Baillon, Bull. Mens. Soc. Linn. Paris **1**: 178. 1878.
T.: *P. conspicua* Baillon.

(H) *Payeria* Baillon, Adansonia **1**: 50. 1860 [Mel.].
T.: *P. excelsa* Baillon.

270

8181 **Wendlandia** Bartling ex A.-P. de Candolle, Prodr. 4: 411. 1830.
T.: *W. paniculata* (Roxburgh) A.-P. de Candolle (*Rondeletia paniculata* Roxburgh) (*typ. cons.*).

(H) *Wendlandia* Willdenow, Sp. Pl. 2: 275. 1799 [Menisperm.].
≡ *Androphylax* Wendland 1798 (*nom. rej.* sub 2570).

8183 **Augusta** Pohl, Pl. Bras. 2: 1. 1828 vel 1829; Flora 12: 118. 1829.
T.: *A. lanceolata* Pohl (*typ. cons.*) [= *A. longifolia* (K. Sprengel) Rehder, *Ucriana longifolia* K. Sprengel].

(H) *Augusta* Leandro, Denkschr. Königl. Akad. Wiss. München, Cl. Math. Phys. 7: 235. 1821 [Comp.].
T.: non designatus.

8197 **Hymenodictyon** Wallich in Roxburgh, Fl. Ind. 2: 148. 1824.
T.: *H. excelsum* (Roxburgh) A.-P. de Candolle (Prodr. 4: 358. 1824) (*Cinchona excelsa* Roxburgh).

(=) *Benteca* Adanson, Fam. Pl. 2: 166, 525. 1763.
T.: *B. rheedei* Roemer et J. A. Schultes (Syst. Veg. 4: 706. 1819).

8204 **Manettia** Mutis ex Linnaeus, Mant. Pl. 2: 553, 558. 1771.
T.: *M. reclinata* Linnaeus.

(H) *Manettia* Boehmer in Ludwig, Defin. Gen. Pl. ed. 3. 99. 1760 [Scrophular.].
≡ *Selago* Linnaeus 1753.
(=) *Lygistum* P. Browne, Civ. Nat. Hist. Jamaica 142. 1756.
T.: *Petesia lygistum* Linnaeus (Syst. Nat. ed. 10. 894. 1759).

8209 **Cosmibuena** Ruiz et Pavón, Fl. Peruv. Chil. 3: 2. 1802.
T.: *C. obtusifolia* Ruiz et Pavón, nom. illeg. (*Cinchona grandiflora* Ruiz et Pavón, *Cosmibuena grandiflora* (Ruiz et Pavón) Rusby) (*typ. cons.*).

(H) *Cosmibuena* Ruiz et Pavón, Prodr. 10. 1794 [Ros.].
LT.: *Hirtella cosmibuena* Lamarck (vide Regnum Veg. 8: 271. 1956).

8215 **Schizocalyx** Weddell, Ann. Sci. Nat. Bot. ser. 4. 1: 73. 1854.
T.: *S. bracteosus* Weddell.

(H) *Schizocalyx* Scheele, Flora 26: 568, 575. 1843 [Lab.].
T.: non designatus.

8227 **Mitragyna** Korthals, Observ. Naucl. Ind. 19. 1839.
T.: *M. parvifolia* (Roxburgh) Korthals (*Nauclea parvifolia* Roxburgh) (*typ. cons.*).

(H) *Mitragyne* R. Brown, Prodr. 452. 1810 [Logan./Gentian.].
≡ *Mitrasacme* Labillardière 1804.
(=) *Mamboga* Blanco, Fl. Filip. 140. 1837.
T.: *M. capitata* Blanco.

8228 **Uncaria** Schreber, Gen. 1: 125. 1789.
T.: *U. guianensis* (Aublet) J. F. Gmelin (Syst. Nat. 2: 370. 1791) (*Ourouparia guianensis* Aublet).

(≡) *Ourouparia* Aublet, Hist. Pl. Guiane 177. 1775.

8237 **Acranthera** Arnott ex Meisner, Pl. Vasc. Gen. 1: 162. 1838; 2: 115. 1838.
T.: *A. ceylanica* Arnott ex Meisner.

(=) *Psilobium* Jack, Malayan Misc. 2(7): 84. 1822.
T.: *P. nutans* Jack.

8241 **Schradera** Vahl, Ecl. 1: 35, t. 5. 1796.
T.: *S. involucrata* (Swartz) K. Schumann (in C. F. P. Martius, Fl. Bras. 6(6): 295. 1889) (*Fuchsia involucrata* Swartz).

(H) *Schraderia* Heister ex Medikus, Philos. Bot. 2: 40. 1791 [Lab.].
≡ *Arischrada* Pobedimova 1972.

271

8244 **Coptophyllum** Korthals, Ned. Kruidk. Arch. **2**: 161. 1850 ("1851").
T.: *C. bracteatum* Korthals.

(H) *Coptophyllum* G. Gardner, London J. Bot. **1**: 133. 1842 [Pteridoph.].
T.: *C. buniifolium* G. Gardner.

8250 **Coccocypselum** P. Browne, Civ. Nat. Hist. Jamaica 144. 1756 ("*Coccocipsilum*") (*orth. cons.*).
T.: *C. repens* Swartz (*typ. cons.*).

(=) *Sicelium* P. Browne, Civ. Nat. Hist. Jamaica 144. 1756.
T.: non designatus.

8265 **Pentagonia** Bentham, Bot. Voy. Sulphur *t. 39*. 1844; 105. 1845.
T.: *P. macrophylla* Bentham.

(H) *Pentagonia* Heister ex Fabricius, Enum. ed. 2. 336. 1763 sero [Solan.].
≡ *Nicandra* Adanson, Jul-Aug 1763 (*nom. cons.*) (7377).

8285 **Gardenia** J. Ellis, Philos. Trans. **51**(2): 935, *t. 23*. 1761.
T.: *G. jasminoides* J. Ellis.

(H) *Gardenia* J. Colden, Essays Observ. Phys. Lit. Soc. Edinb. **2**: 2. 1756 [Gutt.].
T.: non designatus.

8296 **Villaria** Rolfe, J. Linn. Soc., Bot. **21**: 311. 1884.
T.: *V. philippinensis* Rolfe.

(H) *Vilaria* Guettard, Mém. Minéral. Dauphiné **1**: clxx. 1779 [Comp.].
T.: *V. subacaulis* Guettard.

8312 **Zuccarinia** Blume, Bijdr. 1006. 1826 vel 1827.
T.: *Z. macrophylla* Blume.

(H) *Zuccarinia* Märklin, Ann. Wetterauischen Ges. Gesammte Naturk. **2**: 252. 1811 [Spermatoph.].
T.: *Z. verbenacea* Märklin.

8316 **Duroia** Linnaeus f., Suppl. 30, 209. 1782.
T.: *D. eriopila* Linnaeus f.

(=) *Pubeta* Linnaeus, Pl. Surin. 16. 1775.
T.: non designatus.

8353 **Mesoptera** J. D. Hooker in Bentham et J. D. Hooker, Gen. Pl. **2**: 130. 1873.
T.: *M. maingayi* J. D. Hooker.

(H) *Mesoptera* Rafinesque, Herb. Raf. 73. 1833 [Orchid.].
≡ *Liparis* L. C. Richard 1818 (*nom. cons.*) (1556).

8357 **Cuviera** A.-P. de Candolle, Ann. Mus. Natl. Hist. Nat. **9**: 222. 1807.
T.: *C. acutiflora* A.-P. de Candolle.

(H) *Cuviera* Koeler, Descr. Gram. 328 ("382"). 1802 [Gram.].
T.: *C. europaea* (Linnaeus) Koeler (*Elymus europaeus* Linnaeus).

8365 **Timonius** A.-P. de Candolle, Prodr. **4**: 461. 1830.
T.: *T. rumphii* A.-P. de Candolle, nom. illeg. (*Erithalis timon* K. Sprengel, *T. timon* (K. Sprengel) Merrill) (*typ. cons.*).

(=) *Porocarpus* J. Gaertner, Fruct. Sem. Pl. **2**: 473. 1791.
T.: *P. helminthotheca* J. Gaertner.

(=) *Polyphragmon* Desfontaines. Mém. Mus. Hist. Nat. **6**: 5. 1820.
T.: *P. sericeum* Desfontaines.

(=) *Helospora* Jack, Trans. Linn. Soc. London **14**: 127. 1823 ("1825").
T.: *H. flavescens* Jack.

(=) *Burneya* Chamisso et Schlechtendal, Linnaea **4**: 188. 1829.
T.: *B. forsteri* Chamisso et Schlechtendal, nom. illeg. (*Erithalis polygama* G. Forster).

8366 **Chomelia** N. J. Jacquin, Enum. Syst. Pl. 1, 12. 1760.
T.: *C. spinosa* N. J. Jacquin.

(H) *Chomelia* Linnaeus, Opera Var. 210. 1758 [Rub.].
LT.: *Rondeletia asiatica* Linnaeus (vide Dandy, Taxon **18**: 470. 1969).

8388 **Psilanthus** J. D. Hooker, Hooker's Icon. Pl. **12**: 28, *t. 1129*. Apr 1873; in Bentham et J. D. Hooker, Gen. Pl. **2**: 115. Apr 1873.
T.: *P. mannii* J. D. Hooker.

(H) *Psilanthus* (A.-P. de Candolle) A. L. Jussieu ex M. J. Roemer, Fam. Nat. Syn. Monogr. **2**: 132, 198. 1846 [Passiflor.].
T.: *P. viridiflorus* (Cavanilles) M. J. Roemer (*Passiflora viridiflora* Cavanilles).

8397 **Trichostachys** J. D. Hooker in Bentham et J. D. Hooker, Gen. Pl. **2**: 128. 1873.
T.: *T. longifolia* Hiern (in Oliver, Fl. Trop. Afr. **3**: 227. 1877).

(H) *Trichostachys* Welwitsch, Syn. Mad. Drog. Med. Angola 19. 1862 [Prot.].
T.: *T. speciosa* Welwitsch.

8399 **Psychotria** Linnaeus, Syst. Nat. ed. 10. 929, 1364. 1759.
T.: *P. asiatica* Linnaeus.

(=) *Psychotrophum* P. Browne, Civ. Nat. Hist. Jamaica 160. 1756.
T.: non designatus.

(=) *Myrstiphyllum* P. Browne, Civ. Nat. Hist. Jamaica 152. 1756.
T.: non designatus.

8410 **Geophila** D. Don, Prodr. Fl. Nepal. 136. 1825.
T.: *G. reniformis* D. Don, nom. illeg. (*Psychotria herbacea* N. J. Jacquin, *G. herbacea* (N. J. Jacquin) K. Schumann).

(H) *Geophila* Bergeret, Fl. Basses-Pyrénées **2**: 184. 1803 [Lil.].
T.: *G. pyrenaica* Bergeret.

8411 **Cephaëlis** Swartz, Prodr. 3, 45. 1788.
T.: *C. muscosa* (N. J. Jacquin) Swartz (*Morinda muscosa* N. J. Jacquin) (*typ. cons.*).

(=) *Evea* Aublet, Hist. Pl. Guiane 100. 1775.
T.: *E. guianensis* Aublet.

(=) *Carapichea* Aublet, Hist. Pl. Guiane 167. 1775.
T.: *C. guianensis* Aublet.

(=) *Tapogomea* Aublet, Hist. Pl. Guiane 157. 1775.
T.: *T. violacea* Aublet.

8412 **Lasianthus** Jack, Trans. Linn. Soc. London **14**: 125. 1823 ("1825").
T.: *L. cyanocarpus* Jack (*typ. cons.*).

(H) *Lasianthus* Adanson, Fam. Pl. **2**: 398, 568. 1763 [The.].
≡ *Gordonia* J. Ellis 1771 (*nom. cons.*) (5148).

(=) *Dasus* Loureiro, Fl. Cochinch. 141. 1790.
T.: *D. verticillata* Loureiro.

8428 **Gaertnera** Lamarck, Tabl. Encycl. 1(2): 272, *t. 167*. 1792.
T.: *G. vaginata* Lamarck.

(H) *Gaertnera* Schreber, Gen. Pl. 1: 290. Apr 1789 [Malpigh.].
≡ *Hiptage* J. Gaertner 1790 (*nom. cons.*) (4208).

(H) *Gaertneria* Medikus, Philos. Bot. 1: 45. Apr 1789 [Comp.].
≡ *Franseria* Cavanilles 1793 (*nom. cons.*) (9147).

8430 **Paederia** Linnaeus, Syst. Nat. ed. 12. 189. 1767; Mant. Pl. 7, 52. 1767.
T.: *P. foetida* Linnaeus.

(≡) *Daun-contu* Adanson, Fam. Pl. 2: 146. 1763.

(=) *Hondbessen* Adanson, Fam. Pl. 2: 158. 1763.
T.: *Paederia valli-kara* A. L. Jussieu.

8445 **Nertera** Banks et Solander ex J. Gaertner, Fruct. Sem. Pl. 1: 124. 1788.
T.: *N. depressa* J. Gaertner.

(=) *Gomozia* Mutis ex Linnaeus f., Suppl. Pl. 17, 129. 1782.
T.: *G. granadensis* Linnaeus f.

8473 **Borreria** G. F. W. Meyer, Prim. Fl. Esseq. 79. 1818.
T.: *B. suaveolens* G. F. W. Meyer (*typ. cons.*).

(H) *Borrera* Acharius, Lichenogr. Universalis 93, 496. 1810. [Fungi: Lich.].
T.: non designatus.

(=) *Tardavel* Adanson, Fam. 2: 145. 1763.
T.: non designatus ["H.M. 9. t. 76."].

8473a **Robynsia** Hutchinson in Hutchinson et Dalziel, Fl. W. Trop. Afr. 2: 108. 1931.
T.: *R. glabrata* Hutchinson.

(H) *Robynsia* Drapiez in Lemaire, Hortic. Universel 2: 127, 231. 1841 [Amaryllid.].
T.: *R. geminiflora* Drapiez.

8485 **Asperula** Linnaeus, Sp. Pl. 103. 1753.
T.: *A. arvensis* Linnaeus (*typ. cons.*).

VALERIANACEAE

8530 **Fedia** J. Gaertner, Fruct. Sem. Pl. 2: 36. 1790.
T.: *F. cornucopiae* (Linnaeus) J. Gaertner (*Valeriana cornucopiae* Linnaeus) (*typ. cons.*).

(H) *Fedia* Adanson, Fam. Pl. 2: 152. 1763 [Valerian.].
T.: *Valeriana ruthenica* Willdenow.

8535 **Patrinia** A. L. Jussieu, Ann. Mus. Natl. Hist. Nat. 10: 311. 1807.
T.: *P. sibirica* (Linnaeus) A. L. Jussieu (*Valeriana sibirica* Linnaeus) (*typ. cons.*).

DIPSACACEAE

8541 **Cephalaria** H. A. Schrader ex Roemer et J. A. Schultes, Syst. Veg. 3: 1, 43. 1818.
T.: *C. alpina* (Linnaeus) Roemer et J. A. Schultes (*Scabiosa alpina* Linnaeus) (*typ. cons.*).

(=) *Lepicephalus* Lagasca, Gen. Sp. Pl. 7. 1816.
T.: non designatus.

CUCURBITACEAE

8596 **Ecballium** A. Richard, Dict. Class. Hist. Nat. 6: 19. 1824.
T.: *E. elaterium* (Linnaeus) A. Richard (*Momordica elaterium* Linnaeus).

(≡) *Elaterium* P. Miller, Gard. Dict. Abr. ed. 4. 1754.

8598 **Citrullus** H. A. Schrader in Ecklon et Zeyher, Enum. Pl. Afric. Austral. 279. 1836.
T.: *C. vulgaris* H. A. Schrader (*Cucurbita citrullus* Linnaeus) (*typ. cons.*).

(≡) *Anguria* P. Miller, Gard. Dict. Abr. ed. 4. 1754.
(=) *Colocynthis* P. Miller, Gard. Dict. Abr. ed. 4. 1754.
T.: non designatus.

8627 **Cayaponia** Silva Manso, Enum. Subst. Braz. 31. 1836.
T.: *C. diffusa* Silva Manso (*typ. cons.*).

8629 **Echinocystis** Torrey et A. Gray, Fl. N. Amer. 1: 542. 1840.
T.: *E. lobata* (A. Michaux) Torrey et A. Gray (*Sicyos lobata* A. Michaux).

(=) *Micrampelis* Rafinesque, Med. Repos. ser. 2. 5: 350. 1808.
T.: *M. echinata* Rafinesque.

8636 **Sechium** P. Browne, Civ. Nat. Hist. Jamaica 355. 1756.
T.: *S. edule* (N. J. Jacquin) Swartz (*Sicyos edulis* N. J. Jacquin) (*typ. cons.*).

CAMPANULACEAE

8651 **Michauxia** L'Héritier, Michauxia. 1788.
T.: *M. campanuloides* L'Héritier (*typ. cons.*).

8656 **Canarina** Linnaeus, Mant. Pl. 2: 148 ("*Canaria*"), 225 ("*Canaria*"), 588. 1771.
T.: *C. campanula* Linnaeus, nom. illeg. (*Campanula canariensis* Linnaeus, *Canarina canariensis* (Linnaeus) Vatke).

(≡) *Mindium* Adanson, Fam. Pl. 2: 134. 1763 (vide Rafinesque, Fl. Tell. 2: 78. 1837).

*8663 **Prismatocarpus** L'Héritier, Sert. Angl. 1. 1789.
T.: *P. paniculatus* L'Héritier (*typ. cons.*).

8668 **Wahlenbergia** H. A. Schrader ex Roth, Nov. Pl. Sp. 399. 1821.
T.: *W. elongata* (Willdenow) H. A. Schrader ex Roth (*Campanula elongata* Willdenow) [= *Campanula capensis* Linnaeus, *W. capensis* (Linnaeus) A. de Candolle].

(=) *Cervicina* Delile, Descr. Egypte, Hist. Nat. 7. 1813.
T.: *C. campanuloides* Delile.

8680 **Sphenoclea** J. Gaertner, Fruct. Sem. Pl.
1: 113. 1788.
T.: *S. zeylanica* J. Gaertner.

8706 **Downingia** Torrey, Rep. Explor. Rail- (≡) *Bolelia* Rafinesque, Atlantic J. 1: 120.
road Pacif. Ocean 4(1, 4): 116. 1857. 1832.
T.: *D. elegans* (Lindley) Torrey (*Clin-
tonia elegans* Lindley).

GOODENIACEAE

8716 **Scaevola** Linnaeus, Mant. Pl. 2: 145.
1771.
T.: *S. lobelia* Murray (Syst. Veg. 178.
1774), nom illeg. (*Lobelia plumieri* Linnaeus, *S. plumieri* (Linnaeus) Vahl).

STYLIDIACEAE

8724 **Stylidium** Swartz ex Willdenow, Sp. Pl. (H) *Stylidium* Loureiro, Fl. Cochinch. 220.
4: 7, 146. 1805. 1790 [Alang.].
T.: *S. graminifolium* Swartz (*typ.* T.: *S. chinense* Loureiro.
cons.).

COMPOSITAE (ASTERACEAE)

8751 **Vernonia** Schreber, Gen. 2: 541. 1791.
T.: *V. noveboracensis* (Linnaeus) Willdenow (Sp. Pl. 3: 1632. 1803) (*Serratula
noveboracensis* Linnaeus) (*typ. cons.*).

8761 **Piptolepis** C. H. Schultz-Bip., Pollichia (H) *Piptolepis* Bentham, Pl. Hartw. 29. 1840
20-21: 380. 1863. [Ol.].
T.: *P. ericoides* (Lamarck) C. H. T.: *P. phillyreoides* Bentham.
Schultz-Bip. (*Conyza ericoides* La-
marck).

8772 **Soaresia** C. H. Schultz-Bip., Pollichia (H) *Soaresia* Allemão, Trab. Soc. Vellosia-
20-21: 376. 1863. na (Bibliot. Guanabara) 72. 1851 [Mor.].
T.: *S. velutina* C. H. Schultz-Bip.
T.: *S. nitida* Allema = to.

8775a **Pseudelephantopus** Rohr, Skr. Naturhist.-Selsk. 2(1): 214. 1792. ("*Pseudo-
Elephantopus*") (*orth. cons.*).
T.: *P. spicatus* (Aublet) C. F. Baker
(Trans. Acad. Sci. St. Louis 12: 45, 54,
56. 1902) (*Elephantopus spicatus* Aublet).

8808 **Brachyandra** R. Philippi, Fl. Atac. 34. 1860.
T.: *B. macrogyne* R. Philippi.

(H) *Brachyandra* Naudin, Ann. Sci. Nat. Bot. ser. 3. **2**: 143. 1844 [Melastomat.].
T.: *B. perpusilla* Naudin.

8818 **Mikania** Willdenow, Sp. Pl. **3**: 1742. 1803.
T.: *M. scandens* (Linnaeus) Willdenow (*Eupatorium scandens* Linnaeus) (*typ. cons.*).

8823 **Brickellia** S. Elliott, Sketch Bot. S. Carolina **2**: 290. 1823.
T.: *B. cordifolia* S. Elliott.

(H) *Brickellia* Rafinesque, Med. Repos. ser. 2. **5**: 353. 1808 [Polemon.].
≡ *Ipomopsis* A. Michaux 1803.
(=) *Kuhnia* Linnaeus, Sp. Pl. ed. 2. 1662. 1763.
T.: *K. eupatorioides* Linnaeus.
(=) *Coleosanthus* Cassini, Bull. Sci. Soc. Philom. Paris **1817**: 67. 1817.
T.: *C. cavanillesii* Cassini.

8826 **Liatris** J. Gaertner ex Schreber, Gen. **2**: 542. 1791.
T.: *L. squarrosa* (Linnaeus) A. Michaux (Fl. Bor.-Amer. **2**: 92. 1803) (*Serratula squarrosa* Linnaeus) (*typ cons.*).

(≡) *Lacinaria* J. Hill, Veg. Syst. **4**: 49. 1762.

8840 **Bradburia** Torrey et A. Gray, Fl. N. Amer. **2**: 250. 1842.
T.: *B. hirtella* Torrey et A. Gray.

(H) *Bradburya* Rafinesque, Fl. Ludov. 104. 1817 [Legum.].
LT.: *B. scandens* Rafinesque (vide Regnum Veg. **100**: 232. 1979).

8843 **Chiliophyllum** R. Philippi, Linnaea **33**: 132. 1864.
T.: *C. densifolium* R. Philippi.

(H) *Chiliophyllum* A.-P. de Candolle, Prodr. **5**: 554. 1836 [Comp.].
T.: *C. globosum* (Ortega) A.-P. de Candolle (*Anthemis globosa* Ortega).

8844 **Chrysopsis** (Nuttall) S. Elliott, Sketch Bot. S. Carolina **2**: 333. 1823.
T.: *C. mariana* (Linnaeus) S. Elliott (*Inula mariana* Linnaeus) (*typ cons.*).

(≡) *Diplogon* Rafinesque, Amer. Monthly Mag. & Crit. Rev. **4**: 195. 1819.

8852 **Haplopappus** Cassini, Dict. Sci. Nat. **56**: 168. 1828 ("*Aplopappus*") (*orth. cons.*).
T.: *H. glutinosus* Cassini.

8855 **Bigelowia** A.-P. de Candolle, Prodr. **5**: 329. 1836.
T.: *B. nudata* (A. Michaux) A.-P. de Candolle (*Chrysocoma nudata* A. Michaux) (*typ. cons.*).

(H) *Bigelowia* Rafinesque, Amer. Monthly Mag. & Crit. Rev. **1**: 442. 1817 [Caryophyll.].
T.: *B. montana* Rafinesque.

8862 **Pteronia** Linnaeus, Sp. Pl. ed. 2. 1176. 1763.
T.: *P. camphorata* Linnaeus.

(=) *Pterophorus* Boehmer in Ludwig, Defin. Gen. Pl. ed. 3. 165. 1760.
T.: non designatus ["Vaill. Act. Paris. 1719."].

8887 **Amellus** Linnaeus, Syst. Nat. ed. 10. 1225, 1377. 1759.
T.: *A. lychnites* Linnaeus (*typ. cons.*).

(H) *Amellus* P. Browne, Civ. Nat. Hist. Jamaica 317. 1756 [Comp.].
T.: *Santolina amellus* Linnaeus.

8898 **Callistephus** Cassini, Dict. Sci. Nat. 37: 491. 1825.
T.: *C. chinensis* (Linnaeus) C. G. Nees (Gen. Sp. Aster. 222. 1833) (*Aster chinensis* Linnaeus).

(≡) *Callistemma* Cassini, Bull. Sci. Soc. Philom. Paris 1817: 32. 1817.

8909 **Celmisia** Cassini, Dict. Sci. Nat. 37: 259. 1825.
T.: *C. longifolia* Cassini (*typ. cons.*).

(H) *Celmisia* Cassini, Bull. Sci. Soc. Philom. Paris 1817: 32. Feb 1817 [Comp.].
T.: *C. rotundifolia* Cassini (Dict. Sci. Nat. 7: 356. Mai 1817).

8916 **Olearia** Moench, Suppl. Meth. 254. 1802.
T.: *O. dentata* Moench, nom. illeg. (*Aster tomentosus* J. C. Wendland, *O. htomentosa* (J. C. Wendland) A.-P. de Candolle).

(=) *Shawia* J. R. Forster et G. Forster, Char. Gen. Pl. 48. 1775.
T.: *S. paniculata* J. R. Forster et G. Forster.

8918 **Sommerfeltia** Lessing, Syn. Gen. Compos. 189. 1832.
T.: *S. spinulosa* (K. Sprengel) Lessing (*Conyza spinulosa* K. Sprengel).

(H) *Sommerfeltia* Flörke ex Sommerfelt, Kongel. Norske Videnskabersselsk. Skr. 19de Aarhundr. 2(2): 60. 1827 [Fungi: Lich.].
T.: *S. arctica* Flörke.

8919 **Felicia** Cassini, Bull. Sci. Soc. Philom. Paris 1818: 165. 1818.
T.: *F. tenella* (Linnaeus) C. G. Nees (Gen. Sp. Aster. 208. 1833) (*Aster tenellus* Linnaeus).

(=) *Detris* Adanson, Fam. Pl. 2: 131, 549. 1763.
T.: non designatus.

8926 **Conyza** Lessing, Syn. Gen. Compos. 203. 1832.
T.: *C. chilensis* K. Sprengel (Novi Prov. 14. 1819) (*typ. cons.*).

(H) *Conyza* Linnaeus, Sp. Pl. 861. 1753 [Comp.].
LT.: *C. squarrosa* Linnaeus (vide M. L. Green, Prop. Brit. Bot. 181. 1929; Regnum Veg. 8: 275. 1956).
(=) *Eschenbachia* Moench, Methodus 573. 1794.
T.: *E. globosa* Moench, nom. illeg. (*Erigeron aegyptiacus* Linnaeus).
(=) *Dimorphanthes* Cassini, Bull. Sci. Soc. Philom. Paris 1818: 30. 1818.
T.: non designatus.
(=) *Laennecia* Cassini, Dict. Sci. Nat. 25: 91. 1822.
T.: *L. gnaphalioides* (Kunth) Cassini (*Conyza gnaphalioides* Kunth).

8939 **Blumea** A.-P. de Candolle, Arch. Bot. (Paris) 2: 514. 1833.
T.: *B. balsamifera* (Linnaeus) A.-P. de Candolle (Prodr. 5: 447. 1836) (*Conyza balsamifera* Linnaeus) (*typ. cons.*).

(H) *Blumia* C. G. Nees, Flora 8: 152. 1825 [Magnol.].
T.: *B. candollei* (Blume) C. G. Nees (*Talauma candollei* Blume, "*candollii*").

(=) *Placus* Loureiro, Fl. Cochinch. 496. 1790.
LT.: *P. tomentosus* Loureiro (vide Merrill, Trans. Amer. Philos. Soc. ser. 2. 24: 387. 1935).

8969 **Filago** Linnaeus, Sp. Pl. 927, 1199, add. post indicem. 1753.
T.: *F. pyramidata* Linnaeus (*typ. cons.*).

8994 **Cassinia** R. Brown, Observ. Compositae 126. 1817.
T.: *C. aculeata* (Labillardière) R. Brown (*Calea aculeata* Labillardière) (*typ. cons.*).

(H) *Cassinia* R. Brown in W. T. Aiton, Hortus Kew. ed. 2. 5: 184. 1813 [Comp.].
T.: *C. aurea* R. Brown.

9006 **Helichrysum** P. Miller, Gard. Dict. Abr. ed. 4. 1754 ("*Elichrysum*") (*orth. cons.*).
T.: *H. orientale* (Linnaeus) J. Gaertner (*Gnaphalium orientale* Linnaeus) (*typ. cons.*).

9009 **Podotheca** Cassini, Dict. Sci. Nat. 23: 561. 1822.
T.: *P. angustifolia* (Labillardière) Lessing (*Podosperma angustifolium* Labillardière).

(≡) *Podosperma* Labillardière, Nov. Holl. Pl. 2: 35, *t. 177.* 1806.

9028 **Angianthus** J. C. Wendland, Coll. Pl. 2: 31, *t. 48.* 1810.
T.: *A. tomentosus* J. C. Wendland.

(=) *Siloxerus* Labillardière, Nov. Holl. Pl. 2: 57, *t. 209.* 1806.
T.: *S. humifusus* Labillardière.

9039 **Disparago** J. Gaertner, Fruct. Sem. Pl. 2: 463. 1791.
T.: *D. ericoides* (P. J. Bergius) J. Gaertner (*Stoebe ericoides* P. J. Bergius).

9050 **Relhania** L'Héritier, Sert. Angl. 22. 1789.
T.: *R. paleacea* (Linnaeus) L'Héritier (*Leyssera paleacea* Linnaeus).

(=) *Osmites* Linnaeus, Sp. Pl. ed. 2. 1285. 1763.
LT.: *O. bellidiastrum* Linnaeus (vide Bremer, Taxon 28: 412. 1979).

9054 **Podolepis** Labillardière, Nov. Holl. Pl. 2: 56, *t. 208.* 1806.
T.: *P. rugata* Labillardière.

9057 **Heterolepis** Cassini, Bull. Sci. Soc. Philom. Paris **1820**: 26. 1820.
T.: *Arnica inuloides* Vahl (Symb. Bot. **2**: 91. 1791 [= *H. aliena* (Linnaeus f.) Druce, *Oedera aliena* Linnaeus f.] (etiam vide 5992).

9059 **Printzia** Cassini, Dict. Sci. Nat. **37**: 463, 488. 1825.
T.: *P. cernua* (P. J. Bergius) Druce (*Inula cernua* P. J. Bergius).

9065 **Iphiona** Cassini, Bull. Sci. Soc. Philom. Paris **1817**: 153. 1817.
T.: *I. dubia* Cassini, nom. illeg. (*Conyza pungens* Lamarck) [= *I. mucronata* (Forsskål) Ascherson et Schweinfurth] (*typ. cons.*).

9091 **Pallenis** (Cassini) Cassini, Dict. Sci. Nat. **23**: 566. 1822.
T.: *P. spinosa* (Linnaeus) Cassini (*Buphthalmum spinosum* Linnaeus).

9101 **Lagascea** Cavanilles, Anales Ci. Nat. **6**: 331. 1803 ("*Lagasca*") (*orth. cons.*).
T.: *L. mollis* Cavanilles.

(=) *Nocca* Cavanilles, Icon. **3**: 12. 1795.
T.: *N. rigida* Cavanilles.

9147 **Franseria** Cavanilles, Icon. **2**: 78. 1793.
T.: *F. ambrosioides* Cavanilles, nom. illeg. (*Ambrosia arborescens* P. Miller 1768) (etiam vide 8428).

9150 **Podanthus** Lagasca, Gen. Sp. Pl. 24. 1816.
T.: *P. ovatifolius* Lagasca.

(H) *Podanthes* Haworth, Syn. Pl. Succ. 32. 1812 [Asclepiad.].
LT.: *P. pulchra* Haworth (vide Mansfeld, Bull. Misc. Inform. **1935**: 451. 1935).

9155 **Zinnia** Linnaeus, Syst. Nat. ed. 10. 1221, 1377. Mai-Jun 1759.
T.: *Z. peruviana* (Linnaeus) Linnaeus (*Chrysogonum peruvianum* Linnaeus).

(≡) *Crassina* Scepin, De Acido Veg. 42. 1758.
(=) *Lepia* J. Hill, Exot. Bot. *t. 29.* Feb-Sep 1759.
T.: non designatus.

9157 **Heliopsis** Persoon, Syn. Pl. **2**: 473. 1807.
T.: *H. helianthoides* (Linnaeus) Sweet (Hort. Brit. **2**: 487. 1826) (*Buphthalmum helianthoides* Linnaeus) (*typ. cons.*).

9166 **Eclipta** Linnaeus, Mant. Pl. **2**: 157, 286. 1771.
T.: *E. erecta* Linnaeus, nom. illeg. (*Verbesina alba* Linnaeus, *E. alba* (Linnaeus) Hasskarl) (*typ. cons.*).

(≡) *Eupatoriophalacron* P. Miller, Gard. Dict. Abr. ed. 4. 1754.

280

9168 **Selloa** Kunth in Humboldt, Bonpland et Kunth, Nova Gen. Sp. 4: ed. fol. 208, ed. qu. 265. 1820.
T.: *S. plantaginea* Kunth.

(H) *Selloa* K. Sprengel, Novi Prov. 36. 1819 [Comp.].
T.: *S. glutinosa* K. Sprengel.

9192 **Wedelia** N. J. Jacquin, Enum. Syst. Pl. 8, 28. 1760.
T.: *W. fruticosa* N. J. Jacquin.

(H) *Wedelia* Loefling, Iter Hispan. 180. 1758 [Nyctagin.].
≡ *Allionia* Linnaeus 1759 (*nom. cons.*) (2348).

9208 **Salmea** A.-P. de Candolle, Cat. Pl. Horti Monsp. 140. 1813.
T.: *S. scandens* (Linnaeus) A.-P. de Candolle (*Bidens scandens* Linnaeus) (*typ. cons.*).

(H) *Salmia* Cavanilles, Icon. 3: 24. 1795 [Lil.].
T.: *S. spicata* Cavanilles.

9215 **Actinomeris** Nuttall, Gen. N. Amer. Pl. 2: 181. 1818.
T.: *A squarrosa* Nuttall, nom. illeg. (*Coreopsis alternifolia* Linnaeus, *A. alternifolia* (Linnaeus) A.-P. de Candolle) (*typ. cons.*).

(≡) *Ridan* Adanson, Fam. Pl. 2: 130, 598. 1763.

9222 **Guizotia** Cassini, Dict. Sci. Nat. 59: 237, 247, 248. 1829.
T.: *G. abyssinica* (Linnaeus f.) Cassini (*Polymnia abyssinica* Linnaeus f.).

9224 **Synedrella** J. Gaertner, Fruct. Sem. Pl. 2: 456. 1791.
T.: *S. nodiflora* (Linnaeus) J. Gaertner (*Verbesina nodiflora* Linnaeus).

(≡) *Ucacou* Adanson, Fam. Pl. 2: 131, 615. 1763.

9241 **Balduina** Nuttall, Gen. N. Amer. Pl. 2: 175. 1818.
T.: *B. uniflora* Nuttall (*typ. cons.*).

(=) *Mnesiteon* Rafinesque, Fl. Ludov. 67. 1817.
T.: non designatus.

9247 **Marshallia** Schreber, Gen. 2: 810. 1791.
T.: *M. obovata* (Walter) Beadle et Boynton (*Athanasia obovata* Walter) (*typ. cons.*).

9258 **Layia** W. J. Hooker et Arnott ex A.-P. de Candolle, Prodr. 7: 294. 1838.
T.: *L. gaillardioides* W. J. Hooker et Arnott) A.-P. de Candolle (*Tridax gaillardioides* W. J. Hooker et Arnott, "*galardioides*").

(H) *Layia* W. J. Hooker et Arnott, Bot. Beechey Voy. 182. 1833 ("1841") [Legum.].
T.: *L. emarginata* W. J. Hooker et Arnott.

(=) *Blepharipappus* W. J. Hooker, Fl. Bor.-Amer. 1: 316. 1833 ("1840").
LT.: *B. scaber* W. J. Hooker (vide Arnott in Lindley, Nat. Syst. Bot. ed. 2. 443. 1836).

9285 **Villanova** Lagasca, Gen. Sp. Pl. 31. 1816.
T.: *V. alternifolia* Lagasca (*typ. cons.*).

(H) *Villanova* Ortega, Nov. Pl. Descr. Dec. 47. 1797 [Comp.].
T.: *V. bipinnatifida* Ortega.
(=) *Unxia* Linnaeus f., Suppl. Pl. 56, 368. 1782.
T.: *U. camphorata* Linnaeus f.

9289 **Thymopsis** Bentham in Bentham et J. D. Hooker, Gen. Pl. **2**: 407. 1873.
T.: *T. wrightii* Bentham, nom. illeg. (*Tetranthus thymoides* Grisebach, *T. thymoides* (Grisebach) Urban).

(H) *Thymopsis* Jaubert et Spach, Ill. Pl. Orient. **1**: 72. 1842 [Gutt.].
T.: *T. aspera* Jaubert et Spach.

9291 **Schkuhria** Roth, Catalecta **1**: 116. 1797.
T.: *S. abrotanoides* Roth.

(H) *Sckuhria* Moench, Methodus 566. 1794 [Comp.].
T.: *S. dichotoma* Moench, nom. illeg. (*Siegesbeckia flosculosa* L'Héritier).

9322 **Oedera** Linnaeus, Mant. Pl. **2**: 159, 291. 1771.
T.: *O. prolifera* Linnaeus, nom illeg. (*Buphthalmum capense* Linnaeus, *O. capensis* (Linnaeus) Druce.

(H) *Oedera* Crantz, Duab. Drac. Arbor. xxx. 1768 [Lil.].
T.: *O. dragonalis* Crantz.

9365 **Peyrousea** A.-P. de Candolle, Prodr. **6**: 76. 1838.
T.: *P. oxylepis* A.-P. de Candolle, nom. illeg. (*Cotula umbellata* Linnaeus f., *P. umbellata* (Linnaeus f.) Fourcade) (*typ. cons.*).

(H) *Peyrousia* Poiret, Dict. Sci. Nat. **39**: 363. 1826 [Irid.].
≡ *Lapeirousia* Pourret 1788.

9382 **Robinsonia** A.-P. de Candolle, Arch. Bot. (Paris) **2**: 333. 1833.
T.: *R. gayana* Decaisne (Ann. Sci. Nat. Bot. ser. 2. **1**: 28. 1834) (*typ. cons.*).

(H) *Robinsonia* Scopoli, Intr. 218. 1777 [Quiin.].
≡ *Touroulia* Aublet 1775.

9405 **Gynura** Cassini, Dict. Sci. Nat. **34**: 391. 1825.
T.: *G. auriculata* Cassini (Opusc. Phytol. **3**: 100. 1834) (*typ. cons.*).

(=) *Crassocephalum* Moench, Methodus 516. 1794.
T.: *C. cernuum* Moench, nom. illeg. (*Senecio rubens* B. Jussieu ex N. J. Jacquin, *C. rubens* (B. Jussieu ex N. J. Jacquin) S. Moore).

9412 **Ligularia** Cassini, Bull. Sci. Soc. Philom. Paris **1816**: 198. 1816.
T.: *L. sibirica* (Linnaeus) Cassini (Dict. Sci. Nat. **26**: 402. 1823) (*Othonna sibirica* Linnaeus).

(H) *Ligularia* H. A. Duval, Pl. Succ. Horto Alencon. 11. 1809 [Saxifrag.].
≡ *Sekika* Medikus 1791.
(=) *Senecillis* J. Gaertner, Fruct. Sem. Pl. **2**: 453. 1791.
T.: non designatus.

9425 **Dimorphotheca** Moench, Methodus 585. 1794.
T.: *D. pluvialis* (Linnaeus) Moench (*Calendula pluvialis* Linnaeus) (*typ. cons.*).

9428 **Tripteris** Lessing, Linnaea 6: 95. 1831.
T.: *T. arborescens* (N. J. Jacquin) Lessing (*Calendula arborescens* N. J. Jacquin) (*typ. cons.*).

9431 **Ursinia** J. Gaertner, Fruct. Sem. Pl. 2: 462. 1791.
T.: *U. paradoxa* (Linnaeus) J. Gaertner (*Arctotis paradoxa* Linnaeus).

9434 **Gazania** J. Gaertner, Fruct. Sem. Pl. 2: 451. 1791.
T.: *G. rigens* (Linnaeus) J. Gaertner (*Othonna rigens* Linnaeus).

(=) *Meridiana* J. Hill, Veg. Syst. 2: 121**. 1761.
T.: *M. tesselata* J. Hill (Hort. Kew. 26. 1768).

9438 **Berkheya** Ehrhart, Neues Mag. Aertzte 6: 303. 1784.
T.: *B. fruticosa* (Linnaeus) Ehrhart (*Atractylis fruticosa* Linnaeus).

(≡) *Crocodilodes* Adanson, Fam. Pl. 2: 127, 545. 1763.

9439 **Didelta** L'Héritier, Stirp. Nov. 55. 1786.
T.: *D. tetragoniifolia* L'Héritier.

(=) *Breteuillia* Buchoz, Grand Jard. t. 62. 1785.
T.: *B. trianensis* Buchoz.

9446 **Siebera** J. Gay, Mém. Soc. Hist. Nat. Paris 3: 344. 1827.
T.: *S. pungens* (Lamarck) A.-P. de Candolle (Prodr. 6: 531. 1838) (*Xeranthemum pungens* Lamarck).

(H) *Sieberia* K. Sprengel, Anleit. ed. 2. 2(1): 282. 1817 [Orchid.].
T.: non designatus.

9457 **Saussurea** A.-P. de Candolle, Ann. Mus. Natl. Hist. Nat. 16: 156, 196. 1810.
T.: *S. alpina* (Linnaeus) A.-P. de Candolle (*Serratula alpina* Linnaeus) (*typ. cons.*).

(H) *Saussuria* Moench, Methodus 388. 1794 [Lab.].
T.: *S. pinnatifida* Moench, nom. illeg. (*Nepeta multifida* Linnaeus).

9464 **Silybum** Adanson, Fam. Pl. 2: 116, 605. 1763.
T.: *S. marianum* (Linnaeus) J. Gaertner (Fruct. Sem. Pl. 2: 378. 1791) (*Carduus marianus* Linnaeus) (*typ. cons.*).

(≡) *Mariana* J. Hill, Veg. Syst. 4: 19. 1762.

9466 **Galactites** Moench, Methodus 558. 1794.
T.: *G. tomentosa* Moench (*Centaurea galactites* Linnaeus).

9476 **Amberboa** (Persoon) Lessing, Syn. Gen. Compos. 8. 1832.
T.: *A. moschata* (Linnaeus) A.-P. de Candolle (Prodr. 6: 560. 1837) (*Centaurea moschata* Linnaeus) (*typ. cons.*).

(=) *Amberboi* Adanson, Fam. Pl. 2: 117, 516. 1763.
T.: *Centaurea lippii* Linnaeus.
(=) *Chryseis* Cassini, Bull. Sci. Soc. Philom. Paris 1817: 33. 1817.
T.: *C. odorata* Cassini, nom. illeg. (*Centaurea amberboi* P. Miller).
(=) *Lacellia* Viviani, Fl. Lib. Sp. 58. 1824.
T.: *L. libyca* Viviani.

9479 **Cnicus** Linnaeus, Sp. Pl. 826. 1753.
T.: *C. benedictus* Linnaeus (*typ. cons.*).

9483 **Moquinia** A.-P. de Candolle, Prodr. 7: 22. 1838.
T.: *M. racemosa* (K. Sprengel) A.-P. de Candolle (*Conyza racemosa* K. Sprengel) (*typ. cons.*).

(=) *Moquinia* A. Sprengel, Tent. Suppl. 9. 1828.
T.: *M. rubra* A. Sprengel.

9490 **Stifftia** Mikan, Del. Fl. Faun. Bras. *t. 1.* 1820.
T.: *S. chrysantha* Mikan.

9511 **Schlechtendalia** Lessing, Linnaea 5: 242. 1830.
T.: *S. luzulifolia* Lessing.

(H) *Schlechtendalia* Willdenow, Sp. Pl. 3: 2125. 1803 [Comp.].
T.: *S. glandulosa* (Cavanilles) Willdenow (*Willdenowa glandulosa* Cavanilles).

9528 **Gerbera** Linnaeus, Opera Var. 247. 1758.
T.: *G. linnaei* Cassini (Dict. Sci. Nat. 18: 460. 1821) (*Arnica gerbera* Linnaeus) (*typ. cons.*).

9529 **Chaptalia** Ventenat, Jard. Cels *t. 61.* 1802.
T.: *C. tomentosa* Ventenat.

9545 **Moscharia** Ruiz et Pavón, Prodr. 103. 1794.
T.: *M. pinnatifida* Ruiz et Pavón (Syst. 186. 1798).

(H) *Moscharia* Forsskål, Fl. Aegypt.-Arab. lxxiv, 158. 1775 [Lab.].
T.: *M. asperifolia* Forsskål.

9560 **Krigia** Schreber, Gen. 2: 532. 1791.
T.: *K. virginica* (Linnaeus) Willdenow (Sp. Pl. 3: 1618. 1803) (*Tragopogon virginicus* Linnaeus) (*typ. cons.*).

9566 **Rhagadiolus** Jussieu, Gen. Pl. 168. 1789.
T.: *R. edulis* Gaertner (Fruct. Sem. Pl. 2: 354. 1791) (*Lapsana rhagadiolus* Linnaeus).

(H) *Rhagadiolus* Zinn, Cat. Pl. Hort. Gotting. 436. 1757.
LT.: *Hyoseris hedypnois* Linnaeus (vide Meikle, Taxon 29: 159. 1980).

9574 **Leontodon** Linnaeus, Sp. Pl. 798. 1753.
T.: *L. hispidus* Linnaeus (*typ. cons.*).

9576 **Stephanomeria** Nuttall, Trans. Amer. Philos. Soc. ser. 2. 7: 427. 1841.
T.: *S. minor* (W. J. Hooker) Nuttall (*Lygodesmia minor* W. J. Hooker) (*typ. cons.*).

(=) *Ptiloria* Rafinesque, Atlantic J. 1: 145. 1832.
T.: non designatus.

9578 **Rafinesquia** Nuttall, Trans. Amer. Philos. Soc. ser. 2. 7: 429. 1841.
T.: *R. californica* Nuttall.

(H) *Rafinesquia* Rafinesque, Fl. Tell. **2**: 96. 1837 ("1836"), nom. altern. [Legum.].
≡ *Hosackia* Bentham ex Lindley 1829.

9581 **Podospermum** A.-P. de Candolle in Lamarck et A.-P. de Candolle, Fl. Franç. ed. 3. 4: 61. 1805.
T.: *P. laciniatum* (Linnaeus) A.-P. de Candolle (*Scorzonera laciniata* Linnaeus) (*typ. cons.*).

(≡) *Arachnospermum* F. W. Schmidt, Samml. Phys.-Ökon. Aufsätze 1: 274. 1795.

9592 **Taraxacum** Wiggers, Prim. Fl. Holsat. 56. 1780.
T.: *T. officinale* Wiggers (*Leontodon taraxacum* Linnaeus) (*typ. cons.*).

(H) *Taraxacum* Zinn, Catal. 425. 1757 [Comp.].
≡ *Leontodon* Linnaeus 1753.

9604 **Pyrrhopappus** A.-P. de Candolle, Prodr. 7: 144. 1838.
T.: *P. carolinianus* (Walter) A.-P. de Candolle (*Leontodon carolinianus* Walter) (*typ. cons.*).

9604a **Thorelia** Gagnepain, Notul. Syst. (Paris) 4: 18. 1920.
T.: *T. montana* Gagnepain.

(H) *Thorelia* Hance, J. Bot. **15**: 268. 1877 [Spermatoph.].
T.: *T. deglupta* Hance.

[HETEROPYXIDACEAE]

9712 **Heteropyxis** Harvey, Thes. Cap. **2**: 18. 1863.
T.: *H. natalensis* Harvey.

(H) *Heteropyxis* Griffith, Not. Pl. Asiat. **4**: 524. 1854 [Bombac.].
T.: *Boschia griffithii* Masters.

XV. FOSSIL PLANTS

SPHENOPHYLLACEAE

Sphenophyllum A. T. Brongniart, Prodr. Hist. Vég. Foss. 68. 1828.
T.: *S. emarginatum* (A. T. Brongniart) A. T. Brongniart (*Sphenophyllites emarginatus* A. T. Brongniart).

(≡) *Sphenophyllites* A. T. Brongniart, Mém. Mus. Hist. Nat. 8: 209 ("*Sphoenophyllites*"), 234, t. 13[2], f. 8. 1822.
(=) *Rotularia* Sternberg, Versuch Fl. Vorwelt 1(2): 33. 1821.
T.: *R. marsileifolia* Sternberg ("*marsiliaefolia*").

CALAMITACEAE

Asterophyllites A. T. Brongniart, Prodr. Hist. Vég. Foss. 159. 1828.
T.: *A. equisetiformis* (Sternberg) A. T. Brongniart (*Bornia equisetiformis* Sternberg).

(H) *Asterophyllites* A. T. Brongniart, Mém. Mus. Hist. Nat. 8: 210. 1822 [Foss.].
T.: *A. radiatus* A. T. Brongniart.

(≡) *Bornia* Sternberg, Versuch Fl. Vorwelt 1 (Tentamen): xxviii. 1825.

(=) *Bechera* Sternberg, Versuch Fl. Vorwelt 1 (Tentamen): xxx. 1825.
T.: *B. ceratophylloides* Sternberg.

(=) *Brukmannia* Sternberg, Versuch Fl. Vorwelt 1 (Tentamen): xxix, *t. 19, f. 1-2, t. 45, f. 2, t. 58, f. 1.* 1825.
T.: *B. tenuifolia* (Sternberg) Sternberg (*Schlotheimia tenuifolia* Sternberg).

Calamites A. T. Brongniart, Hist. Vég. Foss. 1: 121. 1828.
T.: *C. suckowii* A. T. Brongniart (*typ. cons.*).

(H) *Calamitis* Sternberg, Versuch Fl. Vorwelt 1(1): 22. 1820 [Foss.].
T.: *C. pseudobambusia* Sternberg.

MEDULLOSACEAE

Dolerotheca Halle, Kongl. Svenska Vetenskapsakad. Handl. ser. 3. 12: 42, *t. 9, t. 10, f. 1-2.* 1933.
T.: *D. fertilis* (Renault) Halle (*Dolerophyllum fertile* Renault).

(=) *Discostachys* Grand'Eury, Géol. Paléontol. Bassin Houiller Gard *t. 8, fig. 2.* 1890.
T.: *D. cebennensis* Grand'Eury.

MEGALOPTERIDACEAE

Megalopteris (Dawson) E. B. Andrews, Rep. Geol. Surv. Ohio 2(2): 415. 1875.
T.: *M. dawsonii* (Hartt) E. B. Andrews (*Neuropteris dawsonii* Hartt).

(=) *Cannophyllites* A. T. Brongniart, Prodr. Hist. Vég. Foss. 130. 1828.
T.: *C. virletii* A. T. Brongniart.

GLOSSOPTERIDACEAE

Glossopteris A. T. Brongniart, Prodr. Hist. Vég. Foss. 54. 1828.
T.: *G. browniana* A. T. Brongniart (Hist. Vég. Foss. 1: 222. 1831).

(H) *Glossopteris* Rafinesque, Anal. Nat. Tabl. Univ. 205. 1815 [Pteridoph.].
≡ *Phyllitis* J. Hill 1757.

GINKGOACEAE

Baiera C. F. W. Braun, Beitr. Petrefacten-Kunde 6: 20. 1843.
T.: *B. muensteriana* (K. B. Presl) Heer (*Sphaerococcites muensterianus* K. B. Presl).

(H) *Bajera* Sternberg, Versuch Fl. Vorwelt 1 (Tentamen): 28. 1825 [Foss.: Calamit.].
T.: *B. scanica* Sternberg.

Cardiocarpus A. T. Brongniart, Rech. Graines Foss. Silic. 20, *t. A 1-2.* 1880.
T.: *C. drupaceus* A. T. Brongniart.

Cordaianthus Grand'Eury, Mém. Acad. Roy. Sci. Inst. France **24**: 227, *t. 26, f. 4-12.* 1877.
T.: *C. gemmifer* Grand'Eury.

Cordaites Unger, Gen. Sp. Pl. Foss. 277. 1850.
T.: *C. borassifolia* (Sternberg) Unger (*Flabellaria borassifolia* Sternberg).

(H) *Cardiocarpus* Reinwardt, Syll. Pl. Nov. **2**: 14. 1826 [Simaroub.].
T.: *C. amarus* Reinwardt.

(=) *Botryoconus* Göppert, Paleontographica **12**: 152, *t. 21, f. 1.* 1864.
T.: *B. goldenbergii* Göppert.

(≡) *Neozamia* Pomel, Bull. Soc. Géol. France ser. 2. **3**: 655. 1846.

APPENDIX IIIB

In the following list the **nomina conservanda** are arranged in alphabetical sequence. The same abbreviations and conventions as in Appendix IIIA (see p. 111) are used.

Names listed in this Appendix fall under the special provisions of Art. 14.2 and 14.4 (see also Art. 69.3).

A listed name may have been published either as the name of a new taxon or as a combination based on an earlier, legitimate basionym. The effect may be quite different. For example, conservation of the name *Lycopersicon esculentum* against *Lycopersicon lycopersicum* does not preclude the use of the homotypic *Solanum lycopersicum*, as this is not a "combination based on a rejected name" (Art. 14.7).

Lycopersicon esculentum P. Miller, Gard. Dict. ed. 8. 1768 (*Solanum lycopersicum* Linnaeus, Sp. Pl. 185. 1753) [Solan.].
LT.: "9. *Lycopersicon*", Herb. Linnaeus no. 248.16 (**LINN**) (vide Terrell et al., Taxon **32**: 310-312. 1983).

(≡) *Lycopersicon lycopersicum* (Linnaeus) H. Karsten, Deut. Fl. 966. 1882 ("*Lycopersicum lycopersicum*").

Triticum aestivum Linnaeus, Sp. Pl. 85. 1753 [Gram.].
LT.: "*Triticum* 3", Herb. Clifford no. 24 (**BM**) (vide Bowden, Canad. J. Bot. **37**: 674. 1959).

(=) *Triticum hybernum* Linnaeus, Sp. Pl. 86. 1753.
LT.: "*Triticum* 2", Herb. Clifford no. 24 (**BM**) (vide Hanelt, Schultze-Motel et Jarvis, Taxon **32**: 492. 1983).

APPENDIX IV

The names printed in **bold-face** type, and all combinations based on these names, are ruled as rejected under Art. 69.1, and none is to be used.

These names are neither illegitimate nor invalid under Art. 6.

Later homonyms of a rejected name, and names illegitimate because of inclusion of the type of a rejected name, have no conserved status and are not to be used.

The rejected names are arranged in alphabetical sequence.

*   Rejection approved by the General Committee, and authorized under Art. 15 pending final decision by the next Congress.

Anthospermum ciliare Linnaeus, Sp. Pl. ed. 2. 1512. 1763 [Rubi.].
LT.: Herb. Linnaeus no. 1233.4 (**LINN**) (vide Brummitt, Taxon **36**: 73-74. 1987).

Arthonia lurida Acharius, Lichenogr. Universalis 143. 1810 [Fungi].
T.: "Helvetia", Schleicher (**UPS**).

Bromus purgans Linnaeus, Sp. Pl. 76. 1753 [Gram.].
LT.: P. Kalm, Herb. Linnaeus no. 93.11 (**LINN**) (vide Hitchcock, Contr. U. S. Natl. Herb. **12**: 122. 1908).

*Buchnera euphrasioides Vahl, Symb. Bot. **3**: 81. 1794 (*Buchnera euphrasioides* (Vahl) Bentham, Companion Bot. Mag. **1**: 364. 1836) [Scrophular.].
LT.: Ghana, König, Vahl Herb. microf. **5**: I. 2, 3 (**C**) (vide Hepper, Taxon **35**: 390. 1986).

Crataegus oxyacantha Linnaeus, Sp. Pl. 477. 1753 [Ros.].
LT.: Herb. Linnaeus no. 643.12 (**LINN**) (vide Dandy, Bot. Soc. Exch. Club Brit. Isles Rep. **12**: 867-868. 1946).

*Helotium Tode, Fungi Mecklenb. Sel. **1**: 22. 1790 : Fries, Syst. Mycol. **3**, index: 94. 1832 [Fungi].
LT.: *H. glabrum* Tode : Fries (vide Donk, Beih. Nova Hedwigia **5**: 123. 1962).

Justicia verticillaris Linnaeus fil., Suppl. Pl. 85. 1782 (*Hypoëstes verticillaris* (Linnaeus fil.) Solander ex Roemer et Schultes, Syst. Veg. **1**: 140. 1817) [Acanth.].
LT.: Herb. Thunberg no. 427 (**UPS**) (vide Brummitt et al., Taxon **32**: 658-659. 1983).

Lecidea synothea Acharius, Kongl. Vetensk. Acad. Nya Handl. **29**: 236. 1808 [Fungi].
LT.: Suecia, 1807, Swartz (**BM**) (vide Cannon et Hawksworth, Taxon **32**: 479. 1983).

Lichen jubatus Linnaeus, Sp. Pl. 1155. 1753 [Fungi].
LT.: Herb. Linnaeus no. 1273.281 (**LINN**) (vide Hawksworth, Taxon **19**: 238-239. 1970).

Lupinus hirsutus Linnaeus, Sp. Pl. 721. 1753 [Legum.].
LT.: "*Lupinus hirsutus* L. – 1015 ns. Roy. prodr. 367", Herb. van Royen (**L** no. 908.119-125) (vide Lee et Gladstones, Taxon **28**: 618-620. 1979).

Melianthus minor Linnaeus, Sp. Pl. 639. 1753 [Melianth.].
LT: C. Commelijn, Hort. Med. Amstel. Pl. *t. 4.* 1706 (vide Wijnands, Taxon **34**: 314. 1985).

Musa humilis Aublet, Hist. Pl. Guiane 931. 1775 (*Heliconia humilis* (Aublet) Jacquin, Pl. Hort. Schoenbr. **1**: 23. 1797) [Mus.].
LT.: French Guiana, Aublet (**BM**) (vide Andersson, Taxon **33**: 524. 1984).

Peziza tax. infragen. **Phialea** Persoon, Mycol. Eur. **1**: 276. 1822 (*Peziza* ser. *Phialea* (Persoon : Fries) Fries, Syst. Mycol. **2**(1): 116. 1822; *Phialea* (Persoon : Fries) Gillet, Champ. France Discomyc. 93. 1881-1882).
T.: *Peziza phiala* Vahl : Fries.

Phacidium musae Léveillé, Ann. Sci. Nat. Bot. ser. 3. **5**: 303. 1846 [Fungi].
T.: M. Bonpland (**PC**, Herb. Amér. Equat.).

Rotala decussata A.-P. de Candolle, Prodr. **3**: 76. 1828 [Lythr.].
T.: Endeavour River, Queensland, Australia, R. Brown (**G-DC**).

Scabiosa papposa Linnaeus, Sp. Pl. 101. 1753 (*Pterocephalus papposus* (Linnaeus) Coulter, Mém. Dipsac. 31. 1823) [Dipsac.].
LT.: "14 *Scabiosa*", Herb. van Royen (**L** no. 902.125-731) (vide Meikle, Taxon **31**: 542. 1982).

Solanum indicum Linnaeus, Sp. Pl. 187. 1753 [Solan.].
LT.: Ceylon, Hermann 94, Herb. Hermann **3**: 16 (**BM**) (vide Hepper, Bot. J. Linn. Soc. **76**: 288. 1978).

Solanum sodomeum Linnaeus, Sp. Pl. 187. 1753 [Solan.].
LT.: Ceylon, Hermann 95, Herb. Hermann **3**: 30 (**BM**) (vide Hepper, Bot. J. Linn. Soc. **76**: 290. 1978).

Solanum verbascifolium Linnaeus, Sp. Pl. 184. 1753 [Solan.].
LT.: "*verbascifolium* 1", Herb. Linnaeus no. 248.1 (**LINN**) (vide Roe, Taxon **17**: 177. 1968).

Stilbum cinnabarinum Montagne, Ann. Sci. Nat. Bot. ser. 2. **8**: 360. 1837 [Fungi].
T.: Cuba, M. Ramon, Herb. Montagne (**PC**).

Stipa columbiana Macoun, Cat. Canad. Pl. **2**: 101. 1888 [Gram.].
LT.: Yale, B.C., May 17, 1875, Macoun (**CAN** no. 9899) (vide Hitchcock, Contr. U. S. Natl. Herb. **24**: 253. 1925).

INDEX TO APPENDIX IIIA

In this Index, the Roman numerals refer to the sections i-xiii preceding the Spermatophyta and to section xv following them. In sections i-xiii the conserved generic names are listed alphabetically.

The Arabic numerals refer to the section xiv Spermatophyta, where the names are arranged according to the Dalla Torre & Harms system.

Names printed in roman type refer to conserved names (left columns).

Names printed in *italics* refer to rejected names (right columns). In the sections i-xiii and xv the italicised names will be found under the heading of the corresponding conserved name (mentioned in each entry). For example: "*Acetabulum* ix (Acetabularia)" means that the rejected name *Acetabulum* will be found in section ix (Chlorophyceae) under the conserved name *Acetabularia*.

Allionia 2348
Allodape 6251
Alloplectus 7860
Allosorus xiii (Cheilanthes)
Alocasia 752
Aloidella xi (Aloina)
Aloina xi
Alpinia 1328
Alstonia 6583
Alternaria x
Alvesia 7346
Alysicarpus 3810
Alyxia 6616
Amanita x
Amanitopsis x
Amaracus 7312
Amarella 6509a
Amaryllis 1176
Amasonia 7156
Amberboa 9476
Amberboi 9476
Amblostima 1006
Amblyodon xi
Ambuli 7532
Amellus 8887
Amerimnon 3821
Amianthium 955
Ammios 6014
Amomum 1344
Amorphophallus 723
Ampelocissus 4910
Amphibia ii (Bostrychia)
Amphicarpaea 3860
Amphidium xi
Amphinomia 3657
Amphirrhox 5259
Amphisphaeria x
Amphithrix i (Homoethrix)
Amsinckia 7082
Amyrsia 5557
Anabaena i
Anacampseros 2412
Anacolia xi
Anadyomene ix
Anarrhinum 7485
Anastomaria x (Gyrodon)
Ancistrocarpus 4948
Ancistrocladus 5400
Andira 3841
Androgyne 1714
Androphylax 2570
Androstachys 4299a
Anecochilus 1500
Anemia xiii
Anemopaegma 7665

Anepsa 957
Angianthus 9028
Angiopetalum 6291
Angiopoma x (Drechslera)
Angiopteris xiii
Angolam 6154
Anguillaria 974
Anguria 8598
Anictangium xi
 (Anoectangium)
Aniotum 3848
Anisomeridium x
Anisonema viii
Anisotes 8096
Anneslea 5155
Anneslia 5155
Annulina ix (Cladophora)
Anodontium xi (Drummondia)
Anoectangium xi
Anoectochilus 1500
Anthophysa iv
Anthriscus 5938
Antiaris 1956
Anzia x
Apalatoa 3495
Apatitia 5768
Aphananthe 1904
Aphanochaete ix
Aphanothece 1
Aphoma 975
Apios 3874
Apona ii (Lemanea)
Aponogeton 65
Aposphaeria x
Appella 7185
Aptosimum 7467
Apuleia 3532
Apuleja 3532
Aquilaria 5430
Arabidopsis 2999
Arachnites 1386
Arachnitis 1386
Arachnodiscus iii
 (Arachnoidiscus)
Arachnoidiscus iii
Arachnospermum 9581
Araiostegia xiii
Araliopsis 4073
Arceuthobium 2091
Arctostaphylos 6212
Ardisia 6285
Arduina 6064
Aremonia 3377
Arenga 575
Areschougia ii

Argania 6370
Argolasia 1236
Argyrolobium 3673
Aristotela 4927
Aristotelia 4927
Armeria 6350
Armeriastrum 6348
Armoracia 2965a
Aronia 3338a
Arrhenopterum xi
 (Aulacomnium)
Artanema 7559
Arthonia x
Artocarpus 1946
Artotrogus x (Pythium)
Aschersonia x
Aschistodon IX (Ditrichum)
Ascidium x (Ocellularia)
Ascolepis 454
Ascophora x (Rhizopus)
Ascophyllum v
Asperula 8485
Aspidosperma 6588
Aspidostigma 4085
Assonia 5053
Astasia viii
Astelia 1111
Asteristion x (Phaeotrema)
Asterophyllites xv
Astrocaryum 668
Astronidium 5777a
Atalantia 4096
Atamosco 1181
Athenaea 7398
Atherurus 787
Atractylocarpus xi
Atrichum xi
Atropis 384
Atylus 2026
Audouinella ii
Augea 3967
Augia 3967
Augusta 8183
Aulacia 4089
Aulacodiscus iii
Aulacomnium xi
Auricula iii
Avoira 668
Aytonia xii (Plagiochasma)
Azalea 6189

Babiana 1310
Bacciferv (Cystoseira)
Bacopa 7546
Bahel 7559

Baiera xv
Baillouviana ii (Dasya)
Baitaria 2407
Balanites 3980
Balanoplis 1891a
Balbisia 3932
Balboa 5195
Balduina 9241
Balsamea 4151
Bambos 424
Bambusa 424
Bambusina ix
Banisteria 4226
Banksia 2068
Barbarea 2961
Barbula xi
Barclaya 2515
Barosma 4036
Barraldeia 5525
Barringtonia 5506
Bartonia 6501
Bartramia xi
Bartramidula xi
Bartsia 7645
Baryxylum 3561
Basilaea 1088
Basteria 2663
Bathelium x (Trypethelium)
Baumgartia 2570
Baxtera 1044
Baxteria 1044
Bazzania xii
Bechera xv (Asterophyllites)
Belamcanda 1285
Belis 31
Bellevalia 1093
Belluccia 5768
Bellucia 5768
Belmontia 6483
Belou 4099
Belutta-Kaka 6677
Belvala 5436
Bembix 5400
Benjamina 4063
Benteca 8197
Benzoin 2821
Berchemia 4868
Bergena 3182
Bergenia 3182
Bergera 4090
Berkheya 9438
Berlinia 3516
Bernardia 4397
Berniera 2804
Bernieria 2804

Bernoullia 5035
Bernullia 5035
Berrya 4938
Bertolonia 5708
Bessera 1055
Beureria 7042
Beverna 1310
Biarum 784
Bichatia i (Gloeocapsa)
Bichea 5091
Bifida ii (Rhodophyllis)
Bifora 5956
Bigelowia 8855
Bignonia 7705
Bihai 1321
Bikkia 8126
Bikukulla 2856
Bilimbiospora x
    (Leptophaeria)
Billia 4722
Billya 4722
Bistella 3201
Bivonaea 2902
Bivonea 2902
Bladhia 6285
Blandfordia 1021
Blasteniospora x (Xanthoria)
Blatti 5497
Blepharipappus 9258
Bletilla 1533
Blossevillea v (Cystophora)
Blumea 8939
Blumenbachia 5392
Blumia 8939
Blysmus 468a
Bobartia 1284
Boehmia x (Leptoglossum)
Boenninghausenia 4011
Boenninghausia 4011
Boldu 2759
Bolelia 8706
Boletus x
Bombax 5024
Bonamia 6979
Bonannia 6099
Bonaveria 3694
Bonnetia 5144
Bornia xv (Asterophyllites)
Borraginoides 7056
Borrera 8473
Borreria 8473
Borrichius ii (Dudresnaya)
Boscia 3106
Bostrychia ii
Botor 3914

Botria 4910
Botrydiopsis vi
Botryocladia ii
Botryoconus xv
    (Cordaianthus)
Botryonipha x (Stilbella)
Botryophora 4516
Bouchea 7148
Bougainvillea 2350
Bourreria 7042
Bouteloua 295
Bowiea 1011
Boykiana 3185
Boykinia 3185
Brachtia 1751
Brachyandra 8808
Brachynema 6408
Brachysteleum xi
    (Ptychomitrium)
Brachystelma 6870
Bradburia 8840
Bradburya 8840
Brami 7546
Brassavola 1619
Braunea 2577
Brebissonia iii
Breteuillia 9439
Brexia 3225
Breynia 4303
Brickellia 8823
Bridgesia 4730
Brigantiaea x (Lopadium)
Brodiaea 1053
Brosimum 1957
Broussonetia 1923
Brownea 3524
Brownlowia 4943
Brucea 4120
Brukmannia xv
    (Asterophyllites)
Brunia 3292
Bryantea 2797
Bryoxiphium xi
Bucco 4037
Bucephalon 1917
Buceras 5543
Buchenavia 5545
Buchloë 308
Bucida 5543
Buckleya 2109
Buda 2450
Buekia 1328
Bulbine 985
Bulbophyllum 1705
Bulbostylis 471a

Bullardia x (Melanogaster)
Bumelia 6374
Bupariti 5018
Buraeavia 4331
Burcardia 968
Burchardia 968
Bureava 4331
Burneya 8365
Bursa-pastoris 2986
Bursera 4150
Burtonia 3629
Butea 3876
Butneria 5062
Byrsanthes 5311
Byrsanthus 5311
Byssus ix (Trentepohlia)
Bystropogon 7321
Byttneria 5062

Cacara 3908
Cactus 5411
Cailliea 3452
Cajan 3892
Cajanus 3892
Calacinum 2208
Calamites xv
Calamitis xv (Calamites)
Calandrinia 2407
Calanthe 1631
Calasias 8096
Calceolaria 7474
Caldesiella x (Tomentella)
Calesiam 4563
Caliphruria 1196
Calliandra 3444
Calliblepharis ii
Callista 1694
Callistachys 3624
Callistemma 8898
Callistephus 8898
Callixene 1146
Calodendrum 4035
Caloplaca x
Calopogon 1534
Calorophus 815
Calucechinus 1889
Calusparassus 1889
Calvatia x
Calycanthus 2663
Calypogeia xii
Calypso 1559
Calyptranthes 5575
Calystegia 6994
Camassia 1087
Cambessedea 5669

Cambessedesia 5669
Cammarum 2528
Camoensia 3589
Camonea 6997
Camphora 2782
Campnosperma 4578
Campsis 7714
Camptocarpus 6726
Campulosus 286
Campylobasidum x
  (Septobasidium)
Campylus 2583
Cananga 2684
Canarina 8656
Canavali 3891
Canavalia 3891
Candida x
Candollea 8724
Canella 5254
Cannophyllites xv
  (Megalopteris)
Cansjera 2124
Cantuffa 3553
Caopia 5171
Capnoides 2858
Capnorchis 2856
Capnorea 7029
Capriola 282
Capsella 2986
Capura 5446
Carallia 5525
Carandas 6559
Carapichea 8411
Cardaminum 2965
Cardiacanthus 8014
Cardiocarpus xv
Careya 5505
Carissa 6559
Carlowrightia 8014
Carpolepidum xii (Plagiochila)
Carpomitra v
Carrichtera 2936
Carrodorus iv (Hydrurus)
Carya 1882
Caryolobis 5214
Caryophyllus 5583
Cassebeera xiii (Doryopteris)
Casselia 7157
Cassinia 8994
Castanopsis 1891a
Castela 4118
Castelia 4118
Catenella ii
Catevala 1029
Catha 4627

Catharinea xi (Atrichum)
Cavendishia 6232
Cayaponia 8627
Caylusea 3122
Cayratia 4918a
Cebatha 2570
Cecropia 1971
Cedrus 23
Celmisia 8909
Centritractus vi
Centrophorum 134c
Centrosema 3858
Cephaëlis 8411
Cephalaria 8541
Cephalotus 3176
Cephaloziella xii
Ceraia 1694
Ceramianthemum ii
  (Gracilaria)
Ceramion ii (Ceramium)
Ceramium ii
Cerania x (Thamnolia)
Ceranthus 6428
Cerataulina iii
Ceratophora x (Gloeophyllum)
Cerefolium 5938
Cervicina 8668
Cesius xii (Gymnomitrion)
Ceterac xiii (Ceterach)
Ceterach xiii
Cetraria x
Ceuthospora x
Chacasanum 7148a
Chaenostoma 7518
Chaetangium ii (Suhria)
Chaetocarpus 4467
Chaetomorpha ix
Chamaedaphne 6200a
Chamaedorea 594
Chamissoa 2297
Chaos ii (Porphyridium)
Chaptalia 9529
Chasmone 3673
Cheilanthes xiii
Chibaca 5256
Chiliophyllum 8843
Chiloscyphus xii
Chimonanthus 2663a
Chionographis 951
Chlamydomonas ix
Chlamysporum 992
Chlorea x (Letharia)
Chlorociboria x
Chlorococcum ix
Chlorogalum 1007

Chloromonas ix
Chloronitum ix
   (Chaetomorpha)
Chloronotus xi (Crossidium)
Chloroxylon 4065
Chloroxylum 4065
Choaspis ix (Sirogonium)
Chomelia 8366
Chondria ii
Chondropsis x
Chondrospora x (Anzia)
Chonemorpha 6677
Chordaria v
Chorispermum 3051
Chorispora 3051
Chrosperma 955
Chrozophora 4355
Chryseis 9476
Chrysopogon 134c
Chrysopsis 8844
Chrysothrix x
Chupalon 6232
Chylocladia ii
Chytra 7604a
Chytraculia 5575
Chytraphora v (Carpomitra)
Cieca 4349
Ciliaria ii (Calliblepharis)
Cinnamomum 2782
Circinnus 3693
Cirrhopetalum 1704
Cistella x
Citrullus 8598
Cladaria x (Ramaria)
Cladaria 1569
Cladonia x
Cladophora ix
Cladophoropsis ix
Clarisia 1937
Clathrospermum 2691a
Clathrospora x (Pleospora)
Clavaria x
Clavatula ii (Catenella)
Cleyera 5157a
Clianthus 3753
Clisosporium x
   (Coniothyrium)
Clompanus 3834
Cluacena 5557
Cluzella iv (Hydrurus)
Cnicus 9479
Coccochloris i (Aphanothece)
Coccocypselum 8250
Coccoloba 2209
Cocculus 2570

Cochlospermum 5250
Codiaeum 4454
Codiolum ix (Urospora)
Codonanthe 7866
Codonanthus 7866
Coelopyrum 4578
Coeomurus x (Uromyces)
Coilotapalus 1971
Cola 5091
Colea 7760
Coleanthus 228
Coleochaete ix
Coleosanthus 8823
Collea 1488
Collema x
Colletia 4899
Collybia x
Colocasia 755
Colocynthis 8598
Colophermum v (Ectocarpus)
Colubrina 4882
Columella 7897
Columellia 7897
Coluteastrum 3756
Combretum 5538
Commiphora 4151
Compsoa 967
Condalia 4862
Condea 7342
Conferva ix (Cladophora)
Conia x (Lepraria)
Conicephala xii
   (Conocephalum)
Coniocarpon x (Arthonia)
Coniogramme xiii
Coniothyrium x
Conjugata ix (Spirogyra)
Conocephalum xii
Conocybe x
Conohoria 5262
Convolvuloides 7003a
Conyza 8926
Copaiba 3490
Copaifera 3490
Copaiva 3490
Coprinarius x (Panaeolus)
Coptophyllum 8244
Cordaianthus xv
Cordaites xv
Cordula 1393
Cordyceps x
Cordylanthus 7632
Cordyline 1108
Corinophoros x (Peccania)
Cornea ii (Gelidium)

Coronopus 2884
Correa 4031
Correia 4031
Cortaderia 329
Cortinarius x
Corycarpus 356
Corydalis 2858
Corynephorus 269
Corynomorpha ii
Cosmibuena 8209
Cosmiza 7900
Coublandia 3837
Coumarouna 3845
Courondi 4662
Coutinia 6588
Crabbea 7972
Cracca 3745
Crantzia 7860
Crassina 9155
Crassocephalum 9405
Craterella x (Craterellus)
Craterellus x
Crepidopus x (Pleurotus)
Crocodilodes 9438
Crocynia x
Crossidium xi
Cruckshanksia 8158
Crudia 3495
Crumenula ix (Lepocinclis)
Crypsis 221
Cryptanthus 846
Cryptogyne 6384
Cryptopleura ii
Cryptotaenia 6015
Cryptothecia x
Ctenium 286
Ctenodus ii (Phacelocarpus)
Cudrania 1942
Culcasia 690
Cumingia 5036
Cummingia 5036
Cunila 7327
Cunninghamia 31
Cunonia 3275
Cunto 4079
Cupulissa 7665
Curcuma 1351
Curtisia 6156
Cuspidaria 7668
Cussambium 4767
Cuviera 8357
Cyanotis 904
Cyanotris 1087
Cyathula 2312
Cybele 2066

Elaphrium 4150
Elaterium 8596
Elatostema 1988
Elayuna 3528
Elephas 7949
Eleutherine 1292
Ellimia 3126
Ellisia 7023
Ellisius ii (Heterosiphonia)
Ellobum 7809
Elutheria 4190
Elytraria 7908
Elytrospermum 468b
Embelia 6310
Emex 2194
Enallagma 7757
Enargea 1146
Encentrus 4627
Enchidium 4449
Encoelia x
Endespermum 4470
Endlichera 2811a
Endlicheria 2811a
Endophis x (Leptorhaphis)
Endosigma iii (Pleurosigma)
Endospermum 4470
Enicostema 6484
Enneastemon 2691a
Entada 3468
Enteromorpha ix
Epacris 6260
Ephedrosphaera x (Nectria)
Ephemerella xi
Ephemerum xi
Ephippium 1704
Ephynes 5665
Epibaterium 2570
Epicostorus 3316
Epidendrum 1614
Epidermidophyton x
   (Epidermophyton)
Epidermophyton x
Epifagus 7792
Epineuron ii (Vidalia)
Epipactis 1482
Epiphylla ii (Phyllophora)
Eranthis 2528
Erebinthus 3718
Erica 1697
Ericaria v (Cystoseira)
Eriocladus x (Lachnocladium)
Eriosema 3898
Erophila 2989a
Eroteum 5157b
Erporkis 1516

Erythrospermum 5278
Erythrotrichia ii
Eschenbachia 8926
Espera 4938
Etlingera 1344
Eucharis 1196
Euclidium 3038
Eucnide 5384
Eucomis 1088
Eulophia 1648
Eulophus 1648
Euosma 6450
Eupatoriophalacron 9166
Euphyllodium iii (Podocystis)
Eupodiscus iii
Euriosma 3898
Euscaphis 4667
Eusideroxylon 2793
Eustichium xi (Bryoxiphium)
Euterpe 631
Evea 8411
Exilaria iii (Licmophora)
Exocarpos 2097
Eysenhardtia 3708
Eystathes 4281

Fagara 3991
Fagaster 1889
Fagopyrum 2202
Falcaria 6018
Falcata 3860
Fasciata (Fascia) v (Petalonia)
Fastigiaria ii (Furcellaria)
Fedia 8530
Felicia 8919
Ferolia 1957
Ficinia 465
Filago 8969
Filaspora x (Rhabdospora)
Filix xiii (Dryopteris)
Fimbriaria ii (Odonthalia)
Fimbristylis 471
Fittonia 8069
Flavicoma 8042
Flemingia 3899
Forsythia 6421
Franseria 9147
Freesia 1316
Freziera 5157b
Frustulia iii
Funckia 1111
Funicularius v (Himanthalia)
Furcellaria ii
Furera 7317
Fuscaria ii (Rhodomela)
Fusidium x (Cylindrocarpon)

Gabura x (Collema)
Gaertnera 8428
Gaertneria 8428
Galactites 9466
Galax 6277
Galearia 4455
Galedupa 3836
Gamoscyphus xi
   (Heteroscyphus)
Gansblum 2989a
Gardenia 8285
Gasparrinia x (Caloplaca)
Gastrochilus 1822
Gastroclonium ii
Gausapia x (Septobasidium)
Gautieria x
Gaylussacia 6215
Gazania 9434
Gelidium ii
Gelona x (Pleurotus)
Genosiris 1289
Gentianella 6509a
Geophila 8410
Gerardia 7990
Gerbera 9528
Ghesaembilla 6310
Gigalobium 3468
Giganthemum 3589
Glabraria 4943
Glandulifolia 4038
Glechoma 7249
Gleichenia xiii
Glenospora x (Septobasidium)
Globifera 7549
Glochidion 4302
Gloeocapsa i
Gloeococcus ix
Gloeophyllum x
Gloiococcus x (Gloeococcus)
Glossopteris xv
Glossostigma 7556
Glossula 1403a
Glyceria 383
Glycine 3864
Glycosmis 4087
Glycycarpus 4600
Glyphocarpa xi (Anacolia)
Glyphocarpa xi (Bartramidula)
Goetzea 7421
Goldbachia 2923
Gomozia 8445
Gomphonema iii
Gomphotis 5621
Gongolaria v (Cystoseira)
Gongrosira ix

Goniotrichum ii
  (Erythrotrichia)
Gordonia 5148
Gothofreda 6857
Gracilaria ii
Grammita ii (Polysiphonia)
Granularius v (Dictyopteris)
Graphorkis 1648a
Graphorkis 1648
Grateloupella ii (Polysiphonia)
Grevillea 2045
Grislea 5538
Grona 3807
Guaiabara 2209
Guarea 4190
Guatteria 2679
Guizotia 9222
Gustavia 5510
Gymnocephalus xi
  (Aulacomnium)
Gymnoderma x
Gymnogrammitis xiii
  (Araiostegia)
Gymnomitrion xii
Gymnopus x (Collybia)
Gymnoscyphus xii
  (Solenostoma)
Gymnosporia 4627
Gymnostomum xi
Gymnozyga ix (Bambusina)
Gynandropsis 3087
Gynocephalum 4712
Gynopogon 6616
Gynura 9405
Gyrocephalus x (Gyromitra)
Gyrodinium vii
Gyrodon x
Gyromitra x
Gyrosigma iii
Gyrosigma iii (Pleurosigma)
Gyroweisia xi

Haematococcus ix
Haenkea 4038
Halenia 6513
Halesia 6410
Halidrys v
Halimeda ix
Hantzschia iii
Hapale 756
Hapaline 756
Haplohymenium xi
Haplolophium 7673
Haplomitrium xii
Haplopappus 8852

Haplophyllum 4012a
Hariota 5416
Harissona 4117
Harrisonia 4117
Hartogia 4037
Haworthia 1029
Hebecladus 7388
Heberdenia 6288
Hebokia 4667
Hedusa 5659
Hedwigia xi
Heisteria 2147
Heleophylax 468b
Helichrysum 9006
Helicodiceros 779
Heliconia 1321
Helinus 4905
Heliopsis 9157
Helleborine 1482
Helminthocladia ii
Helminthora ii
Helminthosporium x
Helodium xi
Heloniopsis 952
Helosis 2163
Helospora 8365
Helvingia 6157
Helwingia 6157
Hemiaulus iii
Hemichlaena 465
Hemieva 3187
Hemimeris 7472
Hemiptychus iii
  (Arachnoidiscus)
Hemisphaeria x (Daldinia)
Henckelia 7810
Hendersonia x (Stagonospora)
Henrya 8028
Hepetis 878
Herbacea v (Desmarestia)
Hermesias 3524
Hermupoa 3103
Herposteiron ix
  (Aphanochaete)
Hesperochiron 7029
Hessea 1166
Heteranthera 924
Heteranthus 272
Heterolepis 9057
Heteromorpha 5992
Heteropteris 4226
Heteropyxis 9712
Heteroscyphus xi
Heterosiphonia ii
Hexalepis 891

Hexonix 952
Heydia 2103
Hibiscus 5013
Hicorius 1882
Hierochloë 206
Hierochontis 3038
Hildenbrandia ii
Himanthalia v
Himantoglossum 1399
Hippeastrum 1208
Hippoperdon x (Calvatia)
Hippurina v (Desmarestia)
Hiptage 4208
Hirneola x
Hoffmannseggia 3557
Hofmannia 7312
Hohenbuehelia x (Pleurotus)
Hoiriri 861
Holcus 257
Holigarna 4604
Holodiscus 3332
Holomitrium xi
Holothrix 1408
Homa + i' d 784
Homalocenchrus 194
Homoeocladia iii (Nitzschia)
Homoeothrix i
Hondbessen 8430
Hookera xi (Hookeria)
Hookeria xi
Hopea 5215
Hormiscia ix (Urospora)
Hormogyne 6368a
Hormosira v
Hosta 1018
Houttuynia 1857
Hugueninia 2997
Humboldtia 3518
Humboltia 3518
Humida ix (Prasiola)
Humiria 3953
Huttum 5506
Hyalina v (Desmarestia)
Hybanthus 5271
Hydnum x
Hydrodictyon ix
Hydrolapatha ii (Delesseria)
Hydrolea 7037
Hydrophora x (Mucor)
Hydropisphaera x (Nectria)
Hydropityon 7532
Hydrostemma 2515
Hydrurus iv
Hygroamblystegium xi
Hylogyne 2062

Hymenocarpos 3693
Hymenochaete x
Hymenodictyon 8197
Hypaelyptum 452
Hyperbaena 2611
Hypholoma x
Hypnum xi
Hypocistis 2180
Hypoderma x
Hypodiscus 816
Hypolaena 815
Hypopeltis xiii (Polystichum)
Hypoxylon x
Hyptis 7342

Icacorea 6285
Ichnocarpus 6683
Ichthyomethia 3839
Idesia 5331
Ilicioides 4615
Ilmu 1261
Imhofia 1175
Inocarpus 3848
Inochorion ii (Rhodophyllis)
Iochroma 7382
Ioxylon 1918
Iphigenia 975
Iphiona 9065
Ipo 1956
Ipomoea 7003
Iresine 2339
Iria 471
Iridaea ii
Iridea ii (Iridaea)
Iridorkis 1558
Isidium x (Pertusaria)
Isoglossa 8079
Isopogon 2026
Ithyphallus x (Mutinus)
Ixia 1302

Jacobinia 8097
Jacquinia 6282
Jambolifera 4079
Jambosa 5582
Jambosa 5583
Jamesia 3209
Japarandiba 5510
Jimensia 1533
Johnsonia 1037
Josephia 2069
Julocroton 4349
Juncoides 937

Kajuputi 5603
Kalawael 3424a
Kaliformis ii (Chylocladia)
Kara-angolam 6154
Karekandel 5525
Karkinetron 2208
Karstenia x
Katoutheka 6285
Katou-Tsjeroe 4604
Kennedia 3868
Kernera 2908
Kerstingiella 3910a
Kieseria 5144
Killinga 462
Knightia 2064
Kniphofia 1024
Kohautia 8136
Kokabus 7388
Kokera 2297
Kolman x (Collema)
Kopsia 6626
Kosteletzkya 5015
Kozola 952
Krempelhuberia x
  (Pseudographis)
Krigia 9560
Kuhnia 8823
Kuhnistera 3710
Kukolis 7388
Kundmannia 6064
Kunzea 5601
Kyllinga 462

Labatia 6365
Labisia 6291
Lacellia 9476
Lachnanthes 1161
Lachnocladium x
Lacinaria 8826
Lactarius x
Laelia 1617
Laennecia 8926
Laetia 5338
Laetinaevia x
Lagascea 9101
Lagenula 4918a
Lamarckia 374
Laminaria v
Lamourouxia 7650
Lampranthus 2405a
Lanaria 1236
Landolphia 6562
Langermannia x (Calvatia)
Lannea 4563
Laothoë 1007

Laplacea 5149
Laportea 1980
Larrea 3973
Laschia x (Hirneola)
Lasianthus 8412
Lass 5007
Laurelia 2775
Laurencia ii
Laxmannia 1032
Layia 9258
Leaeba 2570
Lebetanthus 6251
Lecanactis x
Leda ix (Zygogonium)
Leea 4919
Leersia 194
Leiotheca xi (Drummondia)
Lejeunea xii
Lemanea ii
Lembidium xii
Lenormanda ii (Lenormandia)
Lenormandia ii
Lens 3853
Leobordia 3657
Leonicenia 5759
Leontodon 9574
Leontopetaloides 1248
Leopoldia 1095a
Leopoldia 1208
Leperzia 1211
Lepia 9155
Lepicaulon 985a
Lepicephalus 8541
Lepidanthus 816
Lepidocarpus 2035
Lepidopilum xi
Lepidostemon 3022
Lepidozia xii
Lepocinclis viii
Lepra x (Pertusaria)
Lepraria x
Leproncus x (Pertusaria)
Leptaxis 3196
Leptocarpus 808
Leptodon xi
Leptoglossum x
Leptohymenium xi
  (Platygyrium)
Leptorhaphis x
Leptorkis 1556
Leptospermum 5599
Leptosphaeria x
Leptostomum xi
Lerchea 8130
Lerchia 8130

299

Lessertia 3756
Letharia x
Lettsomia 5157b
Leucadendron 2037
Leucocarpum 4623
Leucoloma xi
Leucopogon 6262a
Leucospermum 2036
Levisticum 6083
Liatris 8826
Libanotis 6052a
Libertia 1283
Lichen x (Parmelia)
Lichina x
Lichtensteinia 5990
Licmophora iii
Ligularia 9412
Liliastrum 982
Limnanthes 4542
Limnophila 7532
Limodorum 1483
Limonium 6351
Lindera 2821
Lindleya 3328
Linkia 2023
Linociera 6428
Liparis 1556
Lipocarpha 452
Lippius xii (Saccogyna)
Lissochilus 1648
Listera 1494
Lithophragma 3197
Lithothamnion ii
Lithothamnium ii
    (Lithothamnion)
Litsea 2798
Lloydia 1077
Lobularia 3013
Locandi 4109
Logania 6450
Loghania 6450
Loiseleuria 6189
Lomanodia 5777a
Lomatia 2063
Lonchocarpus 3834
Lonchostoma 3286
Lopadium x
Lophanthera 4247
Lophidium xiii (Schizaea)
Lophiodon xi (Ditrichum)
Lophiostoma x
Lophodermium x
Lopholejeunea xii
Lopho-lejeunea xii
    (Lopholejeunea)

Loranthus 2074
Lorea v (Himanthalia)
Lotononis 3657
Lotophyllus 3673
Loudetia 278a
Lourea 3899
Lucernaria ix (Zygnema)
Lucya 8140
Ludovia 682
Luehea 4959
Lunanea 5334
Lunania 5334
Lundia 7697
Luzula 937
Luzuriaga 1146
Lycianthes 7407a
Lycopodioides xiii
    (Selaginella)
Lyginia 800
Lygistum 8204
Lygodium xiii
Lygos 3675a
Lyngbyea i (Lyngbya)
Lyngbya 1
Lyonia 6200
Lysanthe 2045
Lysigonium iii (Melosira)

Macaglia 6588
Machaerium 3823
Mackaia 8039
Mackaya 8039
Maclura 1918
Macrocalyx 7023
Macrodon xi (Leucoloma)
Macrolobium 3517
Macroplodia x (Sphaeropsis)
Macrosporium x (Alternaria)
Macrotyloma 3910a
Mahonia 2566
Maianthemum 1119
Majorana 7314
Malache 5007
Malapoenna 2798
Malcolmia 3032
Malnaregam 4096
Malvastrum 4995
Malveopsis 4995
Mamboga 8227
Mammillaria 5411
Mancoa 2973
Manettia 8204
Manilkara 6386a
Manisuris 127
Manuela 7517

Mappia 4693
Marasmius x
Marchesinia xii
Mariana 9464
Mariscus 459
Marshallia 9247
Martensia ii
Martinellius xii (Radula)
Martinezia 612, 631
Mastigophora xii
Matteuccia xiii
Matthiola 3042
Maximiliana 660
Maximilianea 660
Mayepea 6428
Medicago 3688
Meesia xi
Megalangium xi
    (Acidodontium)
Megalopteris xv
Megapleilis 7887a
Megathecium x (Melanospora)
Megotigea 779
Meibomia 3807
Meistera 1344
Melaleuca 5603
Melancranis 465
Melanogaster x
Melanoleuca x
Melanospora x
Melocactus 5409
Melosira iii
Membranifolia ii
    (Phyllophora)
Meratia 2663a
Meriana 5692
Meriania 5692
Meridiana 9434
Merkia xi (Pellia)
Merremia 6997
Mertensia 7102
Mesembryanthemum 2405
Mesoptera 8353
Mesosphaerum 7342
Metasequoia 32a
Metrosideros 5588
Metroxylon 565
Metzleria xi (Atractylocarpus)
Michauxia 8651
Miconia 5759
Micrampelis 8629
Micrandra 4435
Micranthemum 7549
Micranthus 1313
Microchaete 1

Microcystis 1
Microlepis 5648
Micromelum 4089
Micromeria 7305
Micromphale x (Marasmius)
Microsperma 5384
Microspora ix
Microstemma 6870
Microstylis 1553
Microthelia x
 (Anisomeridium)
Microtropis 4621
Mikania 8818
Millettia 3720
Milligania 1112
Miltonia 1778
Mindium 8656
Miquelia 4713
Mischocarpus 4820
Mison x (Phellinus)
Mitragyna 8227
Mitragyne 8227
Mitraria 7853
Mittenothamnium xi
Mnesiteon 9241
Mniobryum xi
Mnium xi
Moenchia 2432
Mokof 5153
Mollia 4960
Mondo 1140
Moniera 7546
Monilia x
Moniliformia v (Hormosira)
Monochaetum 5665
Monotris 1408
Monstera 700
Montrichardia 730
Moorea 329
Moquinia 9483
Moraea 1265
Morenia 594
Moscharia 9545
Mougeotia ix
Mucor x
Mucuna 3877
Muehlenbeckia 2208
Muellera 3837
Muellera 3834
Muelleriella xi
Muraltia 4278
Murdannia 899a
Murraya 4090
Musaefolia v (Alaria)
Mutinus x

Mycobonia x
Mycoporum x
Mylia xii
Myridium x (Laetinaevia)
Myrinia xi
Myriophylla ii (Botryocladia)
Myriostigma x (Cryptothecia)
Myristica 2750
Myroxylon 3584
Myrstiphyllum 8399
Myrteola 5557
Myrtopsis 4020
Mystacinus 4905
Myxolibertella x (Phomopsis)
Myxonema ix (Stigeoclonium)

Nageia 13
Nalagu 4919
Nama 7033
Nani 5588, 5594
Naravelia 2542
Nardia xii
Naregamia 4172
Naron 1265a
Narthecium 944
Nasturtium 2965
Naudinia 4060
Nazia 143
Neckera xi
Neckeria xi (Neckera)
Nectandra 2790
Nectria x
Needhamia 3718
Neesia 5040
Nelanaregam 4172
Nemastoma ii
Nemia 7517
Nemopanthus 4615
Nemophila 7022
Neolitsea 2797
Neottia 1495
Neozamia xv (Cordaites)
Nephroia 2570
Nereidea ii (Plocamium)
Nerine 1175
Nertera 8445
Nervilia 1468
Nesaea 5486
Neslia 2988
Nestronia 2109
Neurocarpus v (Dictyopteris)
Nialel 4189
Nicandra 7377
Nicolaia 1337a
Nidularia x

Niemeyera 6382
Nigredo x (Uromyces)
Nissolia 3784
Nissolius 3823
Nitophyllum ii
Nitzschia iii
Nocca 9101
Nodularia i
Nodularius v (Ascophyllum)
Nodulosphaeria x
Nodulosphaeria x
 (Leptosphaeria)
Nomochloa 468a
Nothofagus 1889
Nothopegia 4600
Nothoscordum 1050
Notjo 7714
Nunnezharia 594
Nuphar 2514
Nyctophylax 1332
Nymphaea 2513
Nymphozanthus 2514

Oberonia 1558
Obsitila 985a
Ocellularia x
Odisca 7760
Odonthalia ii
Odontia x (Tomentella)
Odontonema 8037
Oedera 9322
Oenoplea 4868
Oeonia 1834
Ohlendorffia 7467
Oidium x
Olearia 8916
Oligomeris 3126
Olinia 5428
Omphalandria 4472
Omphalea 4472
Oncidium 1779
Opa 3339
Opegrapha x
Operculina 6997
Ophiocytium vi
Ophiopogon 1140
Ophthalmidium x (Porina)
Oplismenus 169
Opospermum v (Elachista)
Orbignya 657
Orchiastrum 1490
Orectospermum 3500
Oreocharis 7808
Oreodoxa 612, 631
Ormocarpum 3792

Ormosia 3597
Ornithopteris xiii (Anemia)
Orphium 6504
Orthopixis xi (Aulacomnium)
Orthopixis xi (Leptostomum)
Orthopogon 169
Orthothecium xi
Oscularia 2405a
Osmites 9050
Osmorhiza 5941
Osmundea ii (Laurencia)
Ostrya 1885
Otilix 7407a
Ouratea 5113
Ouret 2317
Ourouparia 8228
Outea 3517
Ovidia 5457
Oxylobium 3624
Oxypetalum 6857
Oxytria 1006
Oxytropis 3767

Pachyrhizus 3908
Pacouria 6562
Padina v
Paederia 8430
Paepalanthus 830
Palisota 894
Pallassia 4035
Pallavicinia xii
Pallenis 9091
Palma-Filix 7
Palmstruckia 7518
Paludana 1344
Pamea 5545
Panaeolus x
Pancovia 4753
Panel 4087
Panicastrella 320
Panisea 1714
Pantocsekia iii
Panus x
Paphiopedilum 1393a
Papillaria xi
Papyracea ii (Cryptopleura)
Papyrius 1923
Paradisea 982
Parapetalifera 4036
Paraphysorma x (Staurothele)
Parasacharomyces x (Candida)
Parascopolia 7407a
Parendomyces x (Candida)
Parmelia x
Parrasia 6483

Parsonsia 6691
Parthenocissus 4915
Patella x (Scutellinia)
Patagonium 3800
Patersonia 1289
Patrinia 8535
Patrisa 5341
Pattara 6310
Pavonia 5007
Payera 8162
Payeria 8162
Peccania x
Peckia 6301
Pectinaria 6889
Pectinea 5278
Pedicellaria 3087
Pedicellia 4820
Pedilanthus 4501
Pelae 4281
Pelexia 1488
Pellaea xiii
Pellia xii
Pellionia 1987
Peltaea 5008a
Peltandra 747
Peltanthera 6468
Peltigera x
Peltimela 7556
Peltogyne 3500
Peltophorum 3561
Peltostegia 5008a
Pentaceras 3998
Pentaceros 3998
Pentagonia 8265
Pentapodiscus iii
    (Aulacodiscus)
Peribotryon x (Chrysothrix)
Pericampylus 2568
Peridermium x
Peripherostoma x (Daldinia)
Peristylus 1403a
Pernettya 6208
Pernetya 6208
Perojoa 6262a
Peronia iii
Persea 2783
Persoonia 2023
Pertusaria x
Perytis 1893
Petalonia v
Petalostemon 3710
Petermannia 1258
Pettera 3676
Petteria 3676
Peumus 2759

Peyrousea 9365
Pezicula x
Phacelocarpus ii
Phacidium x
Phacus viii
Phaedranthus 7679
Phaeotrema x
Pharbitis 7003a
Pharomitrium xi
    (Pterygoneurum)
Phaseoloides 3722
Phaulopsis 7932
Phellinus x
Phibalis x
Phillipsia x
Philodendron 739
Phlyctis x
Pholiota x
Pholiotella x (Conocybe)
Pholiotina x (Conocybe)
Phoma x
Phomopsis x
Phragmipedium 1393b
Phrynium 1368
Phucagrostis 60
Phyllachora x
Phyllactidium ix (Coleochaete)
Phyllaurea 4454
Phyllocladus 15
Phyllodes 1368
Phyllona ii (Porphyra)
Phyllophora ii
Phyllorkis 1705
Phyllostachys 417
Phyllosticta x
Physalodes 7377
Physconia x
Physedium xi (Ephemerella)
Physocara 3316
Physocarpus 3316
Phytocrene 4712
Phytoxis 7299
Piaropus 921
Pickeringia 3619
Picramnia 4131
Picrodendron 4134
Pierrea 5221
Pigafetta 567
Pilea 1984
Pileocalyx 5585
Piliocalyx 5585
Piliostigma 3528
Pimelea 5467
Pinellia 787
Pinnularia iii

Piptochaetium 212
Piptolepis 8761
Piratinera 1957
Piscidia 3839
Pistolochia 2858
Pitcairnia 878
Pithecellobium 3441
Pittosporum 3252
Placodion x (Peltigera)
Placus 8939
Plagiochasma xii
Plagiochila xii
Planchonella 6368a
Plaso 3876
Platanthera 1410
Plathymenia 3466
Platisphaera x (Lophiostoma)
Platolaria 7665
Platonia 5205
Platychloris ix (Chloromonas)
Platygyrium xi
Platylepis 1516
Platylophus 3269
Platyphyllum x (Cetraria)
Plaubelia xi (Trichostomum)
Plectranthus 7350
Plectronia 5428
Plenckia 4637
Pleonosporium ii
Pleospora x
Pleurendotria 3197
Pleuridium xi
Pleurochaete xi (Tortella)
Pleurolobus 3807
Pleuropus x (Pleurotus)
Pleurosigma iii
Pleurospa 730
Pleurotus x
Pleurozium xi
Plexipus 7148a
Plocamium ii
Plocaria ii (Gracilaria)
Plumaria ii
Plumaria ii (Plilota)
Pneumaria 7102
Podalyria 3621
Podanthes 9150
Podanthus 9150
Podocarpus 13
Podocarpus 15
Podocratera x (Tholurna)
Podocystis iii
Podolepis 9054
Podopogon 212
Podosperma 9009

Podospermum 9581
Podospora x
Podotheca 9009
Pogomesia 910
Poiretia 3789
Polemannia 6045
Polia 2455
Polichia 2467
Pollichia 2467
Pollinia 134c
Polyacanthus 4627
Polyblastia x
Polycarpaea 2455
Polychroa 1987
Polycystis x (Urocystis)
Polyedrium vi (Tetraëdriella)
Polygonastrum 1118
Polygonum 2201
Polyneura ii
Polyphragmon 8365
Polypompholyx 7900
Polyschidea v (Saccorhiza)
Polysiphonia ii
Polystachya 1565
Polystichum xiii
Pongam 3720, 3836
Pongamia 3836
Pongamia 3720
Pongelia 7741
Pongelion 4124
Porina x
Porocarpus 8365
Porphyra ii
Porphyridium ii
Porphyrostromium ii
   (Erythrotrichia)
Posidonia 57
Possira 3574
Prasiola x
Premna 7185
Prestoea 612
Prestonia 6702
Printzia 9059
Prionitis ii
Prismatocarpus 8663
Prismatoma ii
   (Corynomorpha)
Pritchardia 542
Protea 2035
Protium 4137
Protococcus ix
   (Chlamydomonas)
Psammospora x (Melanoleuca)
Pselium 2568
Pseudelephantopus 8775a

Pseudo-brasilium 4131
Pseudo-fumaria 2858
Pseudographis x
Pseudolarix 25
Pseudomonilia x (Candida)
Psilanthus 8388
Psilobium 8237
Psophocarpus 3914
Psora x
Psychotria 8399
Psychotrophum 8399
Pteretis xiii (Matteuccia)
Pteridium xiii
Pterigynandrum xi
   (Platygyrium)
Pterocarpus 3828
Pterococcus 4421
Pterogonium xi (Platygyrium)
Pterolepis 5632
Pterolobium 3553
Pteronia 8862
Pterophorus 8862
Pterophyllus x (Pleurotus)
Pteropsis xiii (Drymoglossum)
Pterospermum 5080
Pterostylis 1449
Pterota 3991
Pterygoneurum xi
Ptilochaeta 4234
Ptiloria 9576
Ptilota ii
Ptychomitrium xi
Ptyxostoma 3286
Pubeta 8316
Puccinellia 384
Pucciniola x (Uromyces)
Pulina x (Lepraria)
Pulveraria x (Chrysothrix)
Pungamia 3836
Pupal 2314
Pupalia 2314
Pycnanthemum 7317
Pycnoporus x
Pygmaea x (Lichina)
Pyrenacantha 4709
Pyrenodesmia x (Caloplaca)
Pyrenula x
Pyrrhopappus 9604
Pythion 723
Pythium x
Pyxidicula iii (Rhopalodia)

Quinata 3823
Quinchamalium 2120

Racodium x
Raddetes x (Conocybe)
Radula xii
Rafinesquia 9578
Ramalina viiia
Ramaria x
Ramonda 7800
Ramondia 7800
Raphanis 2965a
Raphanozon x (Telamonia)
Rapistrum 2956
Razoumofskya 2091
Reboulia xii
Rechsteineria 7887a
Rehmannia 7592
Reichardia 3553
Reineckea 1129
Reineria 3718
Relhania 9050
Renealmia 1331
Restio 804
Resupinatus x (Pleurotus)
Retama 3657a
Reticula ix (Hydrodictyon)
Reussia 923
Rhabdonema iii
Rhabdospora x
Rhagadiolus 9566
Rhaphiolepis 3339
Rhaphis 134c
Rhipidium x
Rhipsalis 5416
Rhizopus x
Rhizosolenia iii
Rhodomela ii
Rhodophyllis ii
Rhodopis 3871
Rhodopsis 3871
Rhodothamnus 6191
Rhodymenia ii
Rhopalodia iii
Rhynchanthera 5676
Rhynchocorys 7649
Rhynchoglossum 7833
Rhynchosia 3897
Rhynchospora 492
Rhyssopteris 4222
Rhytiglossa 8079
Riccardia xii
Richaeia 5528
Richea 6254
Ricotia 2968
Ridan 9215
Riedelia 1332
Rinorea 5262

Rivularia i
Robertia 6374
Robillarda x
Robinsonia 9382
Robynsia 8473a
Roccella x
Rochea 3171
Rochelia 7124
Romulea 1261
Rothia 3659
Rottboelia 127
Rottboellia 127
Rotularia xv (Sphenophyllum)
Rourea 3424
Ruelingia 5060
Rulingia 5060
Rutstroemia x
Ryania 5341
Rymandra 2064
Rynchanthera 5676

Saccharina v (Laminaria)
Saccidium 1408
Saccocalyx 7306
Saccogyna xii
Saccolabium 1822
Saccorhiza v
Sagotia 4452
Saguerus 575
Sagus 565
Salacia 4662
Salgada 2793
Salken 3838
Salmea 9208
Salmia 9208
Salomonia 4277
Samadera 4109
Samyda 5337
Sanseverinia 1110
Sansevieria 1110
Santalodes 3424a
Santaloides 3424a
Sarcanthus 1824
Sarcoderma ii (Porphyridium)
Sarcodum 3753
Sargassum v
Sargentia 4074
Sarmienta 7854
Sarothamnus 3682a
Satyrium 1430
Saurauia 5109
Saussurea 9457
Saussuria 9457
Savastana 206
Scaevola 8716

Scaligera 5964
Scaligeria 5964
Scalius xii (Haplomitrium)
Scalptrum iii (Gyrosigma)
Scalptrum iii (Pleurosigma)
Scandalida 3699
Scapania xii
Schauera 8042
Schaueria 8042
Schefflera 5852
Schelhameria 962
Schelhammera 962
Schisandra 2656
Schizaea xiii
Schizocalyx 8215
Schizonotus 3323
Schizothecium x (Podospora)
Schkuhria 9291
Schlechtendalia 9511
Schleichera 4767
Schlotheimia xv
(Asterophyllites)
Schmidtia 312
Schoenodum 808
Schoenolirion 1006
Schoenoplectus 468b
Schotia 3506
Schouwia 2940
Schradera 8241
Schraderia 8241
Schranckia 3448
Schrankia 3448
Schrebera 6422
Schubertia 6772
Schultesia 6526
Schultzia 6058
Schulzia 6058
Sciadophyllum 5852
Sckuhria 9291
Sclerodontium xi (Leucoloma)
Scleropyrum 2103
Sclerotinia x
Scolochloa 381
Scolopia 5304
Scopolia 7393
Scopularia 1408
Scurrula 2074
Scutarius ii (Nitophyllum)
Scutellinia x
Scutia 4874
Scytophyllum 4627
Scytosiphon v
Scytosiphon v (Dictyosyphon)
Sechium 8636
Securidaca 4275

Securigera 3694
Securinega 4297
Sedoidea ii (Gastroclonium)
Seemannia 7878
Segestria x (Porina)
Selaginella xiii
Selaginoides xiii (Selaginella)
Selinum 6070
Selloa 9168
Senecillis 9412
Septaria x (Septoria)
Septobasidium x
Septoria x
Sequoia 32
Serapias 1397
Serda x (Gloeophyllum)
Sererea 7679
Seringia 5075
Serpentinaria ix (Mougeotia)
Sertularia ix (Halimeda)
Sesban 3747
Sesbania 3747
Sesia x (Gloeophyllum)
Setaria 171
Seymeria 7602
Shawia 8916
Shepherdia 5471
Shortia 6275
Shutereia 3863
Shuteria 3863
Sicelium 8250
Siebera 9446
Sieberia 9446
Sieglingia 280
Sigmatella iii (Nitzschia)
Siliquarius v (Halidrys)
Siloxerus 9028
Silybum 9464
Simarouba 4111
Simaruba 4111, 4150
Simbuleta 7485
Simethis 987
Siphonychia 2477
Siraitos 951
Sirogonium ix
Sitodium 1946
Skimmia 4083
Smilacina 1118
Smithia 3796
Soaresia 8772
Soja 3864
Solandra 7414
Solenostoma xii
Solori 3838
Somion x (Spongipellis)

Sommerfeltia 8918
Sonerila 5729
Sonneratia 5497
Sophia 2997
Soranthe 2028
Sorbaria 3323
Sordaria x
Sorghum 134b
Sorgum 134b
Soria 3038
Sorocephalus 2028
Sparmannia 4957
Spathelia 4066
Spergularia 2450
Spermatochnus v
Sphacele 7299
Sphaerella ix
    (Chlamydomonas)
Sphaeria x (Hypoxylon)
Sphaeriopsis x
    (Amphisphaeria)
Sphaerophorus x
Sphaeropsis x
Sphaerotheca x
Sphaerozosma ix
Sphenoclea 8680
Sphenomeris xiii
Sphenophyllites xv
    (Sphenophyllum)
Sphenophyllum xv
Sphinctocystis iii
    (Cymatopleura)
Spiranthes 1490
Spirhymenia ii (Vidalia)
Spirodinium vii (Gyrodinium)
Spirodiscus vi (Ophiocytium)
Spirogyra ix
Spirolobium 6670
Spirostylis 2078
Splachnon ix (Enteromorpha)
Spondogona 6373
Spongipellis x
Spongocladia ix
    (Cladophoropsis)
Spongopsis ix (Chaetomorpha)
Sporodictyon x (Polyblastia)
Stachyanthus 4715
Stachygynandrum xiii
    (Selaginella)
Stachytarpheta 7151
Stagonospora x
Statice 6350
Stauroptera iii (Pinnularia)
Staurothele x
Steganotropis 3858

Stelis 1587
Stellandria 2656
Stellorkis 1468
Stemodia 7534
Stemodiacra 7534
Stenandrium 7990
Stenanthium 957
Stenocarpus 2066
Stenogyne 7227
Stenoloma xiii (Sphenomeris)
Stenophyllus 471a
Stephanomeria 9576
Stereocaulon x
Stereococcus ix (Gongrosira)
Steriphoma 3103
Stifftia 9490
Stigeoclonium ix
Stilbella x
Stilophora v
Stimegas 1393
Stizolobium 3877
Streptylis 899a
Struthanthus 2078
Struthiola 5436
Struvea ix
Stylidium 8724
Styllaria iii (Licmophora)
Stylurus 2045
Styrosinia 7887a
Suaeda 2261
Suhria ii
Suksdorfia 3187
Sulitra 3756
Sutherlandia 3754
Suzygium 5583
Swartzia 3574
Sweetia 3582
Sychnogonia x (Thelopsis)
Symphyglossum 1834a
Symphyoglossum 1834a
Symplocarpus 708
Symplocia x (Crocynia)
Synandrodaphne 5467a
Synedrella 9224
Syringidium iii (Cerataulina)
Syringodea 1260
Syringospora x (Candida)
Syzygium 5583

Tacca 1248
Taetsia 1108
Taligalea 7156
Talinum 2406
Taonabo 5153
Tapeinanthus 2074a

Tapeinochilos 1360
Tapinanthus 2074a
Tapinothrix i (Homoeothrix)
Tapogomea 8411
Taralea 3845
Taraxacum 9592
Tardavel 8473
Tariri 4131
Tauschia 5977
Teclea 4085
Tectona 7181
Tekel 1283
Telamonia x
Telopea 2062
Tema 166
Tephrosia 3718
Terminalia 5544
Terminalis 1108
Ternstroemia 5153
Tessella iii (Rhabdonema)
Tetradonta ix (Chloromonas)
Tetraëdriella vi
Tetragonolobus 3699
Tetralix 5353
Tetramerium 8028
Tetranema 7510
Tetrapodiscus iii
    (Aulacodiscus)
Thamnea 3284
Thamnia 3284
Thamnium x (Roccella)
Thamnolia x
Theka 7181
Thelopsis x
Thelypteris xiii
Theodora 3506
Thespesia 5018
Thevetia 6632
Tholurna x
Thomsonia 723
Thorelia 9604a
Thorntonia 5015
Thouinia 4733
Thryallis 4244
Thryptomene 5621
Thunbergia 7914
Thymelaea 5453
Thymopsis 9289
Thysanotus 992
Tiliacora 2577
Tillospermum 5601
Timmia xi
Timonius 8365
Tinantia 910
Tinospora 2583

Tissa 2450
Tithymaloides 4501
Tithymalus 4498a
Tittmannia 3285
Tobira 3252
Toddalia 4077
Tolmiea 3196
Toluifera 3584
Tomanthera 7604a
Tomentella x
Torresia 206
Torreya 17
Tortella xi
Tortula xi
Touchiroa 3495
Toulichiba 3597
Tounatea 3574
Tournesol 4355
Tourrettia 7766
Tovara 3081
Tovaria 3081
Trachyandra 985a
Trachylejeunea xii
Trachyspermum 6014
Tragus 143
Tremella x
Trentepohlia ix
Treubia xii
Triceros 4666
Trichilia 4195
Trichocalyx 8100
Trichocolea xii
Trichodesma 7056
Trichodesmium i
Tricholoma x
Trichosporum 7824
Trichostachys 8397
Trichostomum xi
Trichostomum xi (Ditrichum)
Tricondylus 2063
Tricyrtis 967
Trigoniastrum 4264
Trigonostemon 4449
Triguera 7392
Trinia 5998
Triplochiton 5022a
Tripodiscus iii (Aulacodiscus)
Tripteris 9428
Trochera 201
Trombetta x (Craterellus)
Trophis 1917
Trypethelium x
Tryphia 1408
Tsjeru-caniram 2124
Tubercularia x

Tubiflora 7908
Tuburcina x (Urocystis)
Tulbaghia 1047
Tulisma 7887a
Tumboa 48
Turpinia 4666

Ucacou 9224
Ugena xiii (Lygodium)
Uloma 7760
Ulticona 7388
Ulva ix
Uncaria 8228
Unifolium 1119
Unxia 9285
Uraspermum 5941
Urbania 7139
Urceola 6639
Urceolaria 7854
Urceolina 1211
Urnularia 6566
Urocystis x
Uromyces x
Uropedium 1393b
Urospora ix
Ursinia 9431
Urticastrum 1980
Uva-ursi 6212

Vaginarius x (Amanitopsis)
Vaginata x (Amanitopsis)
Vagnera 1118
Vahea 6562
Vahlia 3201
Valerianoides 7151
Vallota 1178
Valota 1178
Valteta 7382
Vanieria 1942
Variolaria x (Pertusaria)
Vaupelia 7124a
Vaupellia 7124a
Vedela 6285
Veitchia 639
Venana 3225
Ventenata 272
Ventenatia 272
Venturia x
Vermicularia 7151
Vernonia 8751
Verrucaria x
Versipellis x (Xerocomus)
Vertebrata ii (Polysiphonia)
Verticordia 5625
Vibo 2194

INDEX

A separate index to Appendix IIIA (Nomina generica conservanda et rejicienda) will be found on pp. 291-307.

The references are not to pages but to the Articles, Recommendations, etc., of the Code, as follows:

Pre. = Preamble; Roman numerals = Principles; Arabic numerals = Articles; Arabic numerals followed by letters = Recommendations; Ex. = Examples; Div. = Division; H. = App. I (Hybrids); App. II-IV = Appendix II-IV.

Other abbreviations include: etiam v. = etiam vide (see also); v. = vide (see).

-a, 73B.1(a, b), 73C.1(a, c).
ä (= ae), 73.6.
å (= ao), 73.6.
Abbreviation (author's name), 46A; – (personal name), 73B.Note 1.
Abies balsamea, 23.Ex.4.
Absence of a Rule, Pre.9.
Abstracting journals, 29A.
Abutilon album, – *flore flavo.* 23.Ex.9.
Acaena anserinifolia, – ×*anserovina*, – *ovina*, H.10.Ex. 3.
Acanthococos (*-coccus*), 64.Ex.9.
Acanthoeca (*-ica*), 64.Ex.7.
Acceptance of name (valid publ.), 34.1(a), 34.2, 45.1.
-aceae, 18.1, 61A.1.
Acer pseudoplatanus, 73.Ex.13.
Aceras, 76.2(c).
-achne, 76.2(b).
Acosmus, 34.Ex.4.
Adenanthera bicolor, 7.Ex.3.
Adiantum capillus-veneris, 23.Ex.1.
Adjectival form of word (for epithet), 21.2, 21B, 23A, 73.8, 73D.
Adjective (agreement of epithet), 21.2, 23.5, 24.2; – (ordinal; not admissible as epithet), 23.6(b); – (plural, as epithet), 21.2, 21B; – (used as a noun to coin a generic name), 20A.1(f); – (used as substantive), 18.1, 19.1.

Adonis, 76.1.
Adoption (epithet of illegitimate name), 72.Note 1, 72A.
-ae, 73.10, 73C.1(a), 73G.1(a).
a̋e, ae, æ, 73.6.
ae (= ä), 73.6.
Aegilops speltoides, – *squarrosa*, H.3.Ex.3.
Aesculus sect. *Aesculus*, – sect. *Calothyrsus*, – sect. *Macrothyrsus*, – sect. *Pavia*, – *hippocastanum*, 52.Ex.2.
Aextoxicaceae, 18.Ex.1.
×*Aextoxicon*, 18.Ex.1.
Agaricus, – tribus *Pholiota*, 33.Ex.13; – *albus corticis*, 23.Ex.10; – *octogesimus nonus*, 23.Ex.8.
Agathopyllum, – *neesianum*, 68.Ex.1.
Agati, 76.Ex.4.
Agreement (grammatical, between name and epithet), 21.2, 23.5, 24.2, 32.5.
×*Agroelymus*, 40.Ex.7, H.8.Ex.1.
×*Agrohordeum*, H.8.Ex.1, H.9.Ex.1.
×*Agropogon*, H.3.Ex.1, H.6.Ex.1; – *littoralis*, H.3.Ex.1.
Agropyron, 40.Ex.7, H.8.Ex.1, H.9.Ex.1.
Agrostis, H.2.Ex.1, H.6.Ex.1; – *radiata*, 63.Ex.11; – *stolonifera*, H.2.Ex.1.
Albizia(*-izzia*), 73H.Ex.1.
albomarginatus, 73G.1(b).
-ales, 17.1.

Aspidium berteroanum, 73C.1(c).
aster (*asteris*), 75.Ex.1.
Aster, 18.5, 19.Ex.4, 40.Ex.4; – *novae-angliae*, 73.Ex.14.
Asteraceae, 18.5, 19.Ex.4.
x*Asterago*, 40.Ex.4.
Astereae (-*inae*, -*oideae*), 19.Ex.4.
Asterostemma, 64.Ex.5.
Astragalus (*Cycloglottis*), – (*Phaca*), 21A.Ex.1; – *cariensis*, 64.Ex.4; – *contortuplicatus*, 21A.Ex.1; – *matthewsiae* (-*ii*), 73.Ex.18; – *rhizanthus*, 64.Ex.4.
Astrostemma, 64.Ex.5.
astrum (*astri*), 75.Ex.1.
Atherospermataceae (-*spermeae*, -*spermaceae*), 18.Ex.5.
Atriplex, 23.Ex.6, 76.Ex.1; – "*nova*", 23.Ex.6.
Atropa bella-donna, 23.Ex.1.
atropurpureus, 73G.1(b).
auct. non (use in citation), 50D.
augusti, 73C.2.
Author's name, v. Citation of author's name.
Authors of proposals, Div. III.4.
Autograph (eff. publ.), 29.2, 29.3.
Automatic establishment (names), v. Autonyms.
Automatic typification (names), 10.5, 16.1, 17.1.
Autonyms (automatically established names), 19.3, 19.4, 22.1, 22.2, 26, 46.1, 68.2; – (of conserved name), 14.9; – (definition), 6.8; – (epithet in new comb.), 57.Note 1; – (priority), 57.3; – (type), 7.21; – (valid publication), 32.1, 32.6, 33.1.
Available (printed matter), 30.1; – (zool. nom.), 45.4, 45.Ex.4.
Avowed substitute (nomen novum), 7.11, 33.2, 33.Note 1, 72.1(b), 72.Note 1; – (for basionym without rank), 35.2; – (formal error), 33.Ex.8-9, 59.6.

Backcross, H.4.1.
Bacteria (nomenclature), Pre.7, I, 14.4(footnote), 65.Note 1; etiam v. Groups not covered by this Code.
Baeothryon, 20.Ex.10.
balansana (-*num*, -*nus*), 73C.1(c).
Bartlingia, 50C.Ex.1.
Bartramia, 20.Ex.1.
Basidiomycetes (pleomorphic), 59.
Basionym (author citation), 49; – (definition), 33.2; – (indication required), 33.2; – (superfluous name), 63.3; – (use in typification), 7.12; – (without rank), 35.2.
beatricis, 73C.2.

Behen behen, – *vulgaris*, 55.Ex.8.
Beiträge Naturkunde (Ehrhart), 20.Ex.10.
Belladonna, H.6.Ex.2.
Berberis, 14.Ex.2.
Bertol. (abbreviation), 46A.3.
Bibliographic error (not invalidating publ.), 33.2.
Bibliographic references, v. Citation.
Bigeneric hybrids, H.6.
Binary name or combination (infrageneric), 21.1; – (infraspecific), 24.4; – (specific), 23.1; – (homonym or tautonym), 55.1(a).
Binary nomenclature (not consistently employed), 23.6(c).
Binding decision (on homonyms), 64.3(footnote); – (on Code), Div. III.4.
Biverbal (epithets), 23.1; – (generic names), 20.3.
Blandfordia grandiflora, 63.Ex.6.
Blephilia, – *ciliata*, 33.Ex.2.
Blue-green algae, Pre.7; etiam v. Algae.
Boletellus, 76A.Ex.1.
Boletus, 76A.Ex.1; – *aereus carne lutea*, 23.Ex.10; – *piperatus*, 50E.Ex.2; – *testaceus scaber*, 23.Ex.10; – *vicesimus sextus*, 23.Ex.8.
Bornet & Flahault (Révision Nostocacées), 13.1(e).
Botanical nomenclature (independent of zoological nom.), I.
Bouchea, 73B.Ex.1.
Bougainvillea, 73.Ex.4.
Br., R. (abbreviation), 46A.5.
Braddleya (*Bradlea*, *Bradleja*), 64.Ex.6.
Brassavola, H.6.Ex.6.
Brassica, 18.5; – *campestris*, H.9.Ex.3; – *nigra*, 23.Ex.4.
Brassicaceae, 18.5.
brauniarum, 73C.1(b).
Brazzeia, 34.Ex.7.
brienianus, 73C.4(b).
British Desmidieae (Ralfs), 13.1(e).
Bromeliineae, 17.Ex.2.
Bromus iaponicus (*japonicus*), 73.Ex.8.
Brosimum, 34.Ex.9.
Bryophyta (nomenclature committee), Div. III.2; – (starting points), 13.1(b, c).
Bulbostylis, 14.Ex.6.
Bureau of Nomenclature, Div. III.3.

Cacalia napaeifolia (*napeaefolia*), 73.Ex.12.
Cainito, 63.Ex.1.
Calamintha calamintha, – *officinalis*, 55.Ex.7.
Calandrinia polyandra, 72.Ex.2.
Calicium debile, 46.Ex.5.

Dendromecon, 76.2(b).
-dendron, 76.2(c).
Dendrosicus, 14.Ex.5.
Dentaria, 57.Ex.2.
Descriptio generico-specifica, 42.1.
Description (accompanying name), 32, 32A, 32B, 32E, 36, 36A, 41, 42; – (author), 46.2; – (combined generic and specific), 42; – (generic), 41.2, 42; – (Latin), 36, 36A, 39; – (monotypic genus), 42; – (prelinnaean), 32A; – (reference to), 32, 32A, 36, 38, 39, 41, 42.1; – (requirement valid publ.) 32.1(c), 36, 41, 42.
Descriptive phrase (instead of epithet), 23.Ex.9.
Designation (unitary, of species), 20.4(b).
Desmidiaceae (starting point), 13.1(e).
Desmodium griffithianum, 73C.1(d).
Desmostachya (Desmostachys), 64.Ex.9.
De Wild. (abbreviation), 46A.1.
Diacritical signs, 46B.2, 73.6, 73A.2, 73B.1(d), 73C.3.
Diaeresis, 73.6.
Diagnosis (definition), 32.2; etiam v. Description.
Diagnostic characters (change), 47, 47A; – (change, retention of name), 51.
Dianthus monspessulanus, 23.Ex.1.
Dicera, – dentata, 52.Ex.1.
Didymopanax gleasonii, 46C.Ex.1.
Different senses (name used in), v. Name (excluding type).
Different spelling (Linnaean generic names), 13.4.
Digitalis grandiflora, – mertonensis, – purpurea, H.3.Ex.3.
Dillenia, 73B.1(c).
Dionysia – sect. Ariadne, – sect. Dionysiopsis, 54.Ex.2.
Diospyros, 76.1.
Dipterocarpus, 76.Ex.2.
Direct reference (valid publ.), 32.3, 33.2, 33.3, 45.1.
Discordant elements (type), 7B.5, 9.2, [70].
Distribution of printed matter (eff. publ.), 29.1.
Divisio(n) 3.1, 4.1; – (name), 16, 16A.1, 16A.4; – (priority), 11.4, 16.Note 1; – (typification), 10.5, 16.
Division of a genus (retention or choice of name), 52; – (gender of new names), 76A.
Division of a species (retention or choice of name), 53.1.
Division of infraspecific taxa (retention or choice of name), 53.2.

Division of taxa (retention or choice of name), 51-53.
Doubt (superfluous name), 63.Note 1; – (taxonomic, and valid publ.), 34.2.
Doubtful consequences, Pre.9.
Dracunculus, – vulgaris, 55.Ex.6.
Drimys, 18.Ex.3.
Drypeteae (-inae), 61A.Ex.1.
dubuyssonii, 73C.4(c).
Duplicate, 7.6, 7B.1.
Duration, 32.Ex.3.
Durvillaea, 64.Ex.9, 73B.Ex.1.
Dussia martinicensis, 6.Ex.1.

e (transcription of è, é or ê), 73.6.
-e, 73C.1(a).
-ea, 73B.1(a).
-eae, 19.2, 19.6, 61A.1.
Earlier homonyms, v. Homonyms.
Earlier names, v. Priority.
Eccilia, 57.Ex.6.
Echinocarpus, 57.Ex.1.
Echium lycopsis, – altera species, 7.Ex.4.
Eclipta alba, – erecta, – prostrata, 57.Ex.5.
Economic importance, 14.2.
Ectocarpus mucronatus, 33.Ex.3.
Editorial committee, Div.III.2.
Effective publication, 6.1, 29-31; – (conditional for valid publ.), 32.1; – (date), 30, 30A; – (descriptions on exsiccata), 31; etiam v. First author.
Egeria, 32.Ex.1.
Ehrhart (Beiträge Naturk., Phytophylacium), 20.Ex.10.
Element (type), 7.2-9, 7.16, 7B.5, 9, 10A.
Elementa Botanica (Necker), 20.Ex.11.
Elenchus Fungorum (Fries), 13.1(d).
Elenchus Veg. (Kramer), 23.Ex.11.
xElyhordeum, H.8.Ex.1.
xElymopyrum, 40.Ex.7.
xElymotriticum, H.8.Ex.1.
Elymus, 20A.1(j), 40.Ex.7, H.3.Ex.2, H.8.Ex.1; – europaeus, 63.Ex.12; – farctus, – subsp. boreoatlanticus, – xlaxus, – repens, H.5.Ex.1.
Embelia sarasiniorum, 23.Ex.1.
emendavit (emend.) 47A, 48.Ex.1.
Enallagma, 14.Ex.5.
Enantioblastae, 17.Ex.1.
Enargea, 14.Ex.4.
Ending, v. Termination.
Englera, 73B.Ex.1.
Englerastrum, 73B.Ex.1.
Englerella, 73B.Ex.1.
-ensis 73D.

Entoloma, 57.Ex.6.

Enumeration (ordinal adjective, not epithet), 23.6(b).

-eos, 73G.1(a).

Ephemeral publications, 29A.

Epiphyllum, H.6.Ex.1 & Ex.4.

Epithet, v. Final, Infrageneric, Specific, etc.

Equisetum palustre var. *americanum*, – f. *fluitans*, 6.Ex.2.

-er (personal names ending in), 73B.1(b), 73C.1.

Erica, 19.Ex.3, 19A.Ex.1, H.9.Ex.2; – *cinerea*, H.9.Ex.2.

Ericaceae, 19.Ex.3, 19A.Ex. 1.

×*Ericalluna*, – *bealei*, H.9.Ex.2.

Ericeae (*-oideae*), 19.Ex.3.

Erigeron, 76.Ex.1.

Erroneous application (epithet, on transfer), 55.2, 56.2.

Error (bibliographic, valid publ.), 33.2; – (typographic or orthographic), 73.

Eryngium nsect. *Alpestria*, – sect. *Alpina*, – sect. *Campestria*, H.9.Ex.1; – *amorginum*, 73D.Ex.1.

Erysimum hieraciifolium var. *longisiliquum*, 64.Ex.16.

Erythrina, – *micropteryx*, – *poeppigiana*, 34.Ex.6.

-es, 73G.1(a).

Eschweilera (*Eschweileria*), 64.Ex.5.

Established custom, Pre.9.

et, 46C.1.

et al., 46C.2.

Etching (eff. publ.), 29.3.

Etymology (name or epithet), V, 64.Ex.10, 73I.

Eu-, 21.3.

Euanthe, – *sanderiana*, H.8.Ex.2.

Euastrum binale, 46.Ex.4.

Eucalyptus, 76.Ex.1.

Eugenia laurina, 50A.Ex.1.

Eulophus, – *peucedanoides*, 33.Ex.2.

Eunotia gibbosa, 44.Ex.2.

Eupenicillium brefeldianum, 59.Ex.1.

Euphony of names, Pre.1.

Euphorbia subg. *Esula*, 22.Ex.4; – sect. *Tithymalus*, – subsect. *Tenellae*, 21.Ex.1; – *amygdaloides*, – *characias*, – subsp. *wulfenii*, – ×*cornubiensis*, H.5.Ex.2; – *esula*, 22.Ex.4; – *jaroslavii* 34.Ex.10; – ×*martini*, – subsp. *cornubiensis*, H.5.Ex.2; – *peplis*, 64.Ex.9; – *peplus*, 22.Ex.4, 64.Ex.9; – *wulfenii*, H.5.Ex.2; – *yaroslavii*, 34.Ex.10.

ex (use in citation), 46.3, 50A.2.

Examples (use in Code), Pre.3.

exclusa specie (excl. sp.), 47A.

exclusa varietate (excl. var.), 47A.

Exclusion of type, 47, 48, 63.2, 72.Ex.3.

excluso genere (excl. gen.), 47A.

Excoecaria, 57.Ex.4.

Exercitia Phytol. (Gilibert), 23.Ex.11.

Exsiccata (eff. publ.), 31.

Faba, 18.5.

Fabaceae, 18.5, 19.7.

Faboideae, 19.7.

Fagus, 76.Ex.1; – *silvatica* (*sylvatica*), 73.Ex.1.

Family(-ia), 3, 4.1; – (alternative names), 10.4, 11.1, 18.6; – (change of rank), 61; – (conservation), 14, 15, 15A, App.II; – (name), 18; – (type of name), 10.4.

Fancy names (cultivars), 28.Note 2.

Farinosae, 17.Ex.1.

Fascicles, v. Works appearing in parts.

fedtschenkoi, 73C.1(a).

Female or male (signs), v. Signs; – (type specimens), 7.5; etiam v. Gender.

Ficus exasperata, – *irumuensis*, 50D.Ex.1; – *neoëbudarum*, 73.Ex.13; – *stortophylla*, 50D.Ex.1.

Figure, 32D; – (algae), 39, 39A; – (fossil pl.), 7.18, 7B.6, 38; – (generic description), 41.Note 2, 42.2; – (non-vascular plants), 42.2, 44.2; – (monotypic genus), 42.2; – (original material), 7.4 (footnote); – (scale), 32D.3; – (as specific or infrasp. descr.), 44; – (as type), 7.3, 7.5, 7.8, 7.9, 8.4, 9.3, 10.3, 37.3, 37.5; – (valid publication), 32.Note 1, 42.2, 44.1.

Filago, 20.Ex.1.

Final epithet, 11.3, 26, 27, 56.1, 57.1; – (definition), 11.3 (footnote); – (of autonyms), 57.Note 1.

Final vote (nomencl. proposals), Div.III.4.

First author (to choose gender), 76.3; – (to designate type), 8; – (to select from simultaneous homonyms), 64.5; – (to select from orthographic variants), 75.2; – (to unite taxa), 57.2.

Flora Aegypt.-Arab. (Forsskål), 23.Ex.5.

Flora Bor.-Amer. (Michaux), 30.Ex.1.

Flora Carol. (Walter), 20.Ex.8.

Flora Vorwelt (Sternberg), 13.1(f).

Flora Europae (Gandoger), 33.Ex.12.

Flora Lituan. (Gilibert), 23.Ex.11.

Floras (publ. new taxa in), 45A.

Flore Kouy Tchéou (Léveillé), 29.Ex.3.

Flower colour, 32.Ex.3.

Folium (inadmissible as generic name), 20.Ex.5.

Form(a), 4.1; – (epithet), 24, 24A-B, 26, 26A.3, 27; – (special), 4.Note 3; etiam v. Infraspecific name or epithet.

Formae speciales, 4.Note 3.

Form-genera, 3.3-4, 7.19, 11.1.

Form-taxa, 3.Note 1, 11.1, 25, 59.3, 59.5.

Formula (hybrids), 23.6(d), H; – (definition), H.2.1, H.10.3.

Forsskål (Flora Aegypt.-Arab.), 23.Ex.5.

Fossil plants, Pre.7; – (not assignable to family), 3.3; – (Committee), Div.III.2; – (definition) Pre. (footnote 2), 13.3; – (more than one correct name), 11.1; – (no Latin diagnosis required), 36.1; – (starting point), 13.1(f); – (union with taxon of non-fossil plants), 58; – (type), 7.18-19, 7B.6, 9.4; – (valid publ.), 36.1, 38.

Fr. (abbreviation), 46A.2.

Fries (Elenchus Fungorum), 13.1(d); – (Systema Mycologicum), 13.1(d), 33.5; etiam v. Sanctioned names.

Fucales, 17.Ex.2.

Fuirena, 20.Ex.9; – *umbellata*, 43.Ex.4.

Fumaria bulbosa var. *solida*, 49.Ex.5; – *densiflora*, 40.Ex.3; – *gussonei*, 23.Ex.1; – *officinalis*, – ×*salmonii*, 40.Ex.3; – *solida*, 49.Ex.5.

Fungal components (lichens), 13.1(d).

Fungi, Pre.7; – (classes), 16A.3; – (Committee), Div.III.2; – (divisions), 16A.1; – (epithets), 73H; – (International Code of Nomenclature of Bacteria), 65.Note 1; – (living culture), 9A; – (more than one correct name), 11.1, 59; – (parasitic), 4.Note 3, 32E, 73H; – (pleomorphic), 13.6, 59, 59A, 62; – (starting points), 13.1(d); – (subclasses), 16A.3; – (subdivisions), 16A.2; – (typification), 7.19, 7.20, 59.

Fungi Exsicc. Suec. (Lundell et Nannfeldt), 31.Ex.1.

Future acceptance (provisional name), 34.1, H.9.Note 2.

Gaertner f. (abbreviation), 46A.4.

Galium tricorne, – *tricornutum*, 63.Ex.7.

Gandoger (Flora Europae), 33.Ex.12.

Garden origin (names, citation), 28, 46.3.

Gardener's Dictionary (Miller), 23.Ex.11, 33.Ex.1.

-*gaster*, 76.2(b).

Gasteromycetes (starting point), 13.1(d).

×*Gaulnettya*, 40.Ex.5.

Gaultheria, 40.Ex.5.

×*Gaulthettya*, 40.Ex.5.

Geaster, 75.Ex.1.

Geastrum, 75.Ex.1; – (-*rvm*), – *hygrometricum* (-*cvm*), 73.Ex.7.

Gender (conservation of), 14.10; – (of generic names), 20A.1(i), 76, 76A; etiam v. Agreement, Sex.

Genera Plantarum (Linnaeus), 13.4.

General Committee, 14.11, 15, 15A, 64.3 (footnote), 64.Ex.10, 69.2, Div.III.1 (footnote), Div.III.2 & 4.

Generic name, 20, 20A; – (choice), 57A; – (conservation), 7.17, 14, 15, 15A, 73.Note 1 & 2, App. III; – (consisting of two words), 20.3; – (etymology), 73I; – (form), 20, 20A; – (gender), 20A.1(i), 76, 76A; – (hybrid), H.6-9; – (hyphens in), 73.Note 2; – (Linnaean), 13.4, 41.Note 1, [74]; – (orthogr. variants), 75; – (derived from personal name), 20A.1(i), 73B; – (rejection), 14, App. III; – (repetition), 22; – (retention on division of genus), 52; – (type), 7.17, 10.1-3, 10A; – (used as epithets), 73.Ex.11, 73F; – (valid publ.), 41.2, 42.

Generico-specific description, 42.

Genitive (base of family name), 18.1; – (in epithet), 23A, 73.Ex.12, 73.10, 73C.

Gentiana lutea, – *tenella* var. *occidentalis*, 6.Ex.2.

genuinus, 7B.4, 24.3.

Genus, 3, 4.1; – (change of rank), 49, 60-61A; – (division), 52; – (name v. Generic name); – (form of name), 20, 20A; – (monotypic), 42; – (form-), 3.3, 7.19; – (subdivision of v. Subdivision of a genus); – (type of name), 10.1-3.

Geographical names (in plant names), 23A, 73.7, 73D.

Geranium robertianum, 23.Ex.1.

Gerardia, 73B.Ex.1.

Gerardiina, 64.Ex.9.

Gerrardina, 64.Ex.9.

Gesneria donklarii, 46.Ex.3.

Giffordia, – *mucronata*, 33.Ex.3.

Gigartina cordata var. *splendens*, 7.Ex.2.

Gilibert (Exercitia Phytol., Flora Lituan.), 23.Ex.11.

Ginkgo, 18.Ex.2.

Ginkgoaceae, 18.Ex.2.

glazioui, 73C.1(a).

Gleditschia, (*Gleditsia*), 73.Ex.9.

Globba brachycarpa, – *trachycarpa*, 73.Ex.2.

Globularia cordifolia, 47A.Ex.1.

Gloeosporium balsameae, 23.Ex.4.

Gloriosa, 20.Ex.1.

Gluta benghas, – *renghas*, 45.Ex.3, 73.Ex.3.

Gmelin, C. C., – J. F., – J. G., – S. G. (abbreviations), 46A.4.

Gomont (Monogr. Oscillariées), 13.1(e).

317

Magnolia foetida, – grandiflora, – virginiana var. foetida, 60.Ex.2.

Mahonia, 14.Ex.2.

Mail vote (preliminary), Div.III.4.

Male or female (signs), v. Signs; – (type specimens), 7.5; etiam v. Gender.

Malpighia, 22.Ex.1-2; – subg. Homoiostylis, – subg. Malpighia, 22.Ex.1; – sect. Apyrae, – sect. Malpighia, 22.Ex.2; – glabra, 22.Ex.1.

Maltea, H.6.Ex.5.

Malvastrum bicuspidatum subsp. (var.) tumidum, 34.Ex.11.

Malvineae, 17.Ex.2.

Manihot, 20.Ex.1, 76.Ex.5; – gossypiifolia, 76.Ex.5.

Manuscript (eff. publ.), 29.

Manuscript name (citation), 50A.

Martia, 73B.Ex.1.

martii, 73C.2, 73C.4(e).

Martiusia, 73B.Ex.1.

Material (handwritten, eff. publ.), 29.2-3; – (living culture, from type), 9A; – (original, typification), 7.4-5, 7A, 7B.2, 8.1.

Mazocarpon, 3.Ex.2.

Mechanical method (typification), 7B.3, 8.1(c), 8.Ex.1.

-mecon, 76.2(b).

Medicago orbicularis, – polymorpha var. orbicularis, 49.Ex.1.

Melilotus 73G.1(c); 76.1.

Meliola, 33.Ex.7; – albiziae, – albizziae, 73H.Ex.1.

Meliosma, 73G.1(c).

Melissa calamintha, 55.Ex.7.

Mentha, 62.Ex.1; – aquatica, H.2.Ex.1, H.11.Ex.3; – arvensis, H.2.Ex.1; – ×piperita, – nsubsp. piperita, – nsubsp. pyramidalis, H.11.Ex.3; – f. hirsuta, H.12.Ex.1; – ×smithiana, H.3.Ex.1; – spicata, H.2.Ex.1; – subsp. spicata, – subsp. tomentosa, H.11.Ex.3.

Mention (subordinate taxa, not valid publ.), 34.1(d).

Mesembryanthemum (Mesembrianthemum), 73.Ex.1; – sect. Minima, 34.Ex.3.

Mespilodaphne mauritiana, 68.Ex.1.

Mespilus, – arbutifolia var. nigra, 33.Ex.4.

Metallic etching (eff. publ.), 29.3.

Metasequoia, – disticha, – glyptostroboides, 58.Ex.2.

Michx. (abbreviation), 46A.3.

Michaux (Flora Bor.-Amer.), 30.Ex.1.

Microfilm (eff. publ.), 29.1.

Micromeria benthamii, – ×benthamineolens, – pineolens, H.10.Ex.3.

Micropteryx poeppigiana, 34.Ex.6.

Microspecies (Gandoger), 33.Ex.12.

Miller (Gardener's Dictionary), 23.Ex.11, 33.Ex.1.

Mimosa cineraria, – cinerea, 64.Ex.15.

Minthe, 62.Ex.1.

Minuartia, – stricta, 55.Ex.2.

Misapplication (new comb.) 48.Note 1, 55.2, 56.2.

Misapplied name (citation), 50D; – (new name for), 33.Note 1.

Misidentifications (citation), 50D.

Misipus, 52.Ex.1.

Misplaced term (rank), 33.4-5.

Misspelled name (citation), 50F; – (correction), 73.

Mixture (type), 7.6 (footnote), 9.2.

Modification of the Code, Pre.6, Div.III.

-monado-, 16.2.

Monarda ciliata, 33.Ex.2.

Monochaete (Monochaetum), 64.Ex.9.

Monogr. Iconogr. Oedogoniaceen (Hirn), 13.1(e).

Monogr. Oscillariées (Gomont), 13.1(e).

Monotropeae (-oideae) 19A.Ex.1.

Monotypic genus (valid publ. name), 42.

Monstrosity, [71].

Montia parvifolia, – subsp. flagellaris, – subsp. parvifolia, 25.Ex.1.

Morphological technical term (as name), 20.2.

Mouriri subg. Pericrene, 6.Ex.2.

Mucor chrysospermus, 59.Ex.5.

Müll. Arg. (abbreviation), 46A.4.

Multigeneric hybrids, H.6.3-4.

Multiplication sign (hybrids), H.1-H.3A.

munronis, 73C.2.

Musci (starting point), 13.1(b).

mutatis characteribus (mut. char.), 47A.

-myces, 76.2(a).

-mycetes (-idae), 16A.3.

Mycographie Suisse (Secretan), 23.Ex.10.

Mycosphaerella aleuritidis, 59.Ex.3.

-mycota, 16A.1.

-mycotina, 16A.2.

Myogalum boucheanum, 34.Ex.5.

Myosotis, 51.Ex.1, 73G.1(b).

Myrcia laevis, – lucida, 7.Ex.1.

Myrtus serratus, 50A.Ex.1.

Myxomycetes (starting point), 13.1(d).

n- (= notho-), H.3.1.

-n-, 73C.1(c).

ñ (transcription of), 73.6.

Name(s) v. Authors, Species, Genus, etc.; –
(definition), 6.6; – (excluding type), 48, 69;
– (likely to be confused), 64.3; – (ultimately
based on generic names), 7.1, 10.5, 16.
napaulensis, 64.Ex.8.
Nasturtium, 14.Ex.3; – *nasturtium-aquaticum*,
23.Ex.3.
Natural order (ordo naturalis, instead of family), 18.2.
nec (use in citation), 50C.
Necker (Elementa Botanica), 20.Ex.11.
Nelumbo, 18.Ex.2.
Nelumbonaceae, 18.Ex.2.
-nema, 76.2(c).
-nemato-, 16.2.
Neotype, 7.4, 7.9-10, 7C, 8, 9.1; – (definition),
7.9.
nepalensis, 64.Ex.8.
Nepeta ×*faassenii*, 40.Ex.1.
Neves-armondia, 20.Ex.7.
New combination, v. Infrageneric name, Speci-
fic epithet, Infraspecific epithet.
New name (chosen when no legitimate one
available), 72; – (substitute, typification),
7.11-12; – (without rank), 35.2; etiam v.
Avowed substitute.
Newspapers (eff. publ.), 29.4.
nidus-avis, 73G.1(b).
nipalensis, 64.Ex.8.
Nixus (for ordo), 17.2.
nobis, nob., 46D.
Nolanea, 57.Ex.6.
nom. cons. (abbreviation, used in citation),
50E.1.
nom. nud. (abbreviation), 50B.
Nomen conservandum, v. Conserved name,
Nom. cons.
Nomen novum, v. Avowed substitute.
Nomen nudum (citation), 50B.
Nomenclator Botanicus (Steudel), 33.Ex.1.
Nomenclatural synonym, 14.4.
Nomenclatural type, v. Type, Holotype, Iso-
type, Lectotype, Neotype, Syntype.
Nomenclaturally superfluous name, 63.
Nomenclature (contrary to Rules), Pre.8; –
(independent of zoological nom.), I.
Nomenclature Committees, Div.III.2; etiam v.
Committees (nomenclature).
Nomenclature Section. Div.III.
Nomina conservanda, v. Conserved name.
Nomina familiarum conservanda, App.II.
Nomina generica conservanda et rejicienda,
App.III.
Nomina rejicienda, 14, 15, 15A, App.III.
Nomina utique rejicienda, 69, App.IV.

non (use in citation), 50C.
Non-fossil plants, Pre.7, 9.3; – (definition), Pre.
(footnote 2), 13.3; – (starting points), 13.1;
– (union with fossil plants), 58.
Non-vascular plants, (illustration with analysis), 44.2; – (type), 9.1; – (valid publ., monotypic genera), 42.2.
Nostocaceae (starting point), 13.1(e).
Nothomorph, H.12.2.
Nothotaxon, 3.2, 4.3, 50, H.1, H.3; etiam
v. Hybrid.

-o-, 73G.1(a). ö (= oe), 73.6.
ø (= oe), 73.6.
Object of the Rules and Recommendations,
Pre.4-5.
Objective synonym, 14.4 (footnote).
obrienii, 73C.4(b).
-odes, 76.4.
-odon, 76.2(a).
Odontoglossum, H.6.Ex.7.
oe (= ö, ø), 73.6.
Oedogoniaceae (starting point), 13.1(e).
Oenothera depressa, – ×*hoelscheri*, – *rubri-
caulis*, – ×*wienii*, H.4.Ex.1.
Offset (eff. publ.), 29.3.
-oideae, 19.1, 19.6, 61A.1.
-oides, 76.4.
okellyi, 73C.4(b).
Oncidium, H.6.Ex.7.
Oplopanax, 76.2(a).
-opsida, 16A.3.
Orbea, 8.Ex.3.
×*Orchicoeloglossum mixtum*, H.11.Ex.2.
Orchis, 76.1; – *fuchsii*, H.11.Ex.2; – *rotundifolia*, 46.Ex.3.
Order, 3.1, 4.1, 17; – (name), 17, 17A; – ("nat-
ural"), 18.2; – (priority), 11.4, 16.Note 1; –
(typification), 10.5, 16.1.
Order of ranks, 3-5, 33.4-5.
Ordinal adjective (as epithet), 23.6(b).
Ordo, v. Order.
Original material (typification), 7.4-5, 7A,
7B.2, 8.1; – (rediscovery), 8.1.
Original spelling, 45.2, 73, 73B.1(d), 73C.3,
73E; – (citation), 50F; – (of author's name),
46B.
originalis, 24.3, 26.Ex.1.
originarius, 24.3.
Ormocarpum, 76.Ex.2.
Ornithogalum, – *boucheanum*, – *undulatum*,
34.Ex.5.
Orobanche artemisiae, – *artemisiepiphyta*, –
columbariae, – *columbarihaerens*, – *rapum*,
– *sarothamnophyta*, 62.Ex.1.

Orontiaceae (sectio), 33.Ex.10.
Orthographic error, 45.2, 73, 75.
Orthographic variants, 75; – (citation), 50F; – (conservation), 14.10; – (correction), 73, 75.3; – (definition), 75.Note 1; – (derived from host name), 73H; – (Linnaean), 13.4.
Orthography, 73-75; – (conserved), 14.10; – (original, citation), 50F.
-os, 73G.1(a).
Osbeckia, 64.Ex.2.
-osma, 76.2(b),
Ostrya virginiana, 73D.Ex.1.
Ottoa, 73B.1(a).
-ous, 73G.1(a).

Page reference (defined), 33.2 (footnote).
Pagination (separates), 45C.
Palaeobotany, v. Fossil plants.
Palmae, 18.5.
-panax, 76.2(a).
Panax nossibiensis, 44.Ex.1.
Papaver rhoeas, 23.Ex.1.
Papilionaceae, 18.5, 19.7, 51.2.
Papilionoideae, 19.7.
Parasites, 4.Note 3, 32E, 73H.
Paratype, 7.8, 7B.1.
Parentage (hybrid), v. hybrid (parentage).
Parentheses (author citation), 49; – (correct name of type element), 10A; – (original status), 50; – (subdivision of genus), 21A.
Parietales, 17.Ex.1.
Particle, 46A.1, 73C.4.
Parts, v. Works appearing in parts.
Pecopteris, 3.Ex.1.
Peltophorum (Peltophorus), 64.Ex.9.
Penicillium brefeldianum, 59.Ex.1.
Peperomia san-felipensis, 73.Ex.14.
Peponia (Peponium), 64.Ex.9.
Pereskia opuntiaeflora (opuntiiflora), 73.Ex.11.
Peridermium balsameum, 23.Ex.4.
Periodicals (popular), 29A.
Permanence of text (eff. publ.), 29A.
Pernettya, 40.Ex.5.
Personal names (in plant names), 20A.1(i), 23A, 73.7, 73B-C, H.6.3-4, H.6A.
Persoon (Synopsis Method. Fung.), 13.1(d); etiam v. Sanctioned names.
Petalodinium, 45.Ex.5.
Petrefactenkunde (Schlotheim), 13.1(f).
Petrophiloides, 58.Ex.1.
Petrosimonia brachiata, – oppositifolia, 62.Ex.4.
Peyrousea, 73B.Ex.1.
Peziza corticalis, 33.Ex.6.

Phaelypea, 42.Ex.3.
Phaeocephalum, 20.Ex.10.
xPhilageria, H.9.Ex.1.
Philesia, H.9.Ex.1.
Philgamia, – hibbertioides, 42.Ex.4.
Phippsia, H.6.Ex.5.
Phlox, – divaricata subsp. divaricata, – subsp. laphamii, H.3.Ex.3; – drummondii 'Sternenzauber', 28.Ex.1; – pilosa subsp. ozarkana, H.3.Ex.3.
Phlyctidia, – andensis, – boliviensis, – brasiliensis, – hampeana, – ludoviciensis, – sorediiformis, 43.Ex.2.
Phlyctis andensis, – boliviensis, – brasiliensis, – sorediiformis, 43.Ex.2.
Phoenicosperma, 57.Ex.1.
Pholiota, 33.Ex.13.
Phoradendron (-dendrum), 73.Ex.1.
Phrase (descriptive, instead of epithet), 23. Ex.9.
-phyceae, 16.2, 16A.3.
-phycidae, 16A.3.
Phyllachora annonicola (anonicola), 73H.Ex.1.
Phyllanthus, 47A.Ex.1.
Phyllerpa prolifera var. firma, 24.Ex.4.
Physospermum, 29.Ex.1.
Physconia, 10.Ex.1.
-phyta, 16.2, 16A.1.
Phyteuma, 76.Ex.1.
-phytina, 16A.2.
Phytophylacium (Ehrhart), 20.Ex.10.
Picea abies, – excelsa, 63.Ex.4.
Pinus abies, – excelsa, 63.Ex.4; – mertensiana, 55.Ex.9.
Piptolepis, – phillyreoides, 42.Ex.2.
Piratinera, 34.Ex.9.
Pirus, – mairei, 50F.Ex.1.
Pisocarpium, 76.Ex.2.
Planera aquatica, 43.Ex.3.
Plantae Veron. (Séguier), 41.Ex.2.
Platycarya, 58.Ex.1.
Plectranthus, – fruticosus, – punctatus, 7.Ex.5.
Pleomorphic fungi, 13.6, 59, 59A, 62.1; etiam v. Anamorph, Holomorphic fungi.
Pleonasm (spec. name), 23B.1(e).
Pleuripetalum (Pleuropetalum), 64.Ex.5.
Plumbaginaceae, 18.Ex.1.
Plumbago, 18.Ex.1.
Plural adjective (used as substantive), 18.1, 19.1, 21.2, 21B.
Poa, 18.5, 19.Ex.2.
Poaceae, 18.5, 19.Ex.2.
Poëae, 19.Ex.2.
-pogon, 76.2(a).
poikilantha (-anthes), 64.Ex.8.

Polifolia, 73.Ex.11.
polyanthemos (*-anthemus*), 64.Ex.8.
Polycarpaea, 76.Ex.2.
Polycarpon, 76.Ex.2.
Polycnemum oppositifolium. 62.Ex.4.
Polygamous species (holotype), 9.Ex.1.
Polygonales, 17.Ex.2.
Polygonum pensylvanicum, 73D.Ex.1.
Polyploid, H.3.Ex.1.
Polypodium, – *australe*, – ×*font-queri*, – ×*shi-vasiae*, – *vulgare*, 63.Ex.13; – nsubsp. *man-toniae*, H.3.Ex.1; – subsp. *prionodes*, H.2.Ex.1; – subsp. *vulgare*, 63.Ex.13, H.2.Ex.1.
Polypogon, H.2.Ex.1, H.6.Ex.1; – *monspelien-sis*, H.2.Ex.1.
Pooideae, 19.Ex.2.
Popular periodicals, 29A.
Populus ×*canadensis* var. *marylandica*, – var. *serotina*, H.12.Ex.1.
Porella, – *pinnata*, 13.Ex.1.
porsildiorum, 23A.1.
Position (correct name), IV, 6.5, 11.1.
Potamogeton, 18.Ex.1.
Potamogetonaceae, 18.Ex.1.
Potentilla atrosanguinea, – *atrosanguinea-pe-data*, – *pedata*, H.10.Ex.1.
×*Potinara*, H.6.Ex.6.
p. p. (pro parte), 47A.
Precision (achieved by citation), 46-50F.
Prefix, 4.1, 16A.2, 73B.Note 1, 73C.4.
Preliminary mail vote, Div.III.4.
Pre-linnean, v. Pre-starting-point.
Preparation (type), 9.1-2.
Preservation of original material, 7A; – (place), 8.4, 37.5.
President International Association Plant Tax-onomy, Div.III.2.
President Nomenclature Section, Div.III.3.
Pre-starting-point (authors, citation), 46.3; – (publications, validation by), 7.15, 32A.
Primula sect. *Dionysiopsis*, 54.Ex.2.
Principles, Pre.2, Div.I; – (priority and typifica-tion, not above family), 10.5, 11.4.
Printed matter (effective publication), 29-31.
Priority, III, IV; – (above family), 11.4, 16.Note 1, 16B; – (and conservation), 14.1; – (auto-nyms), 57.3; – (inside rank), 61; – (legiti-mate names taken into consideration), 45.3; – (limitation), 11.2-3, 13-15, 57.1, 58-59; – (names of pleomorphic fungi), 59.4; – (names without rank), 35.2; – (outside rank), 60; – (of publication), III; – (typifica-tion), 8.

pro hybr., 50.
pro parte (p.p.), 47A.
pro. sp., 50.
pro syn. (use in citation), 50A.
Prodromus (Swartz), 63.Ex.10.
Prokaryotic groups, Pre.7, I (footnote); etiam v. Bacteria.
Proposals for amendment of the Code, Div. III.4.
Proposals to conserve names, 14.11, 15, 15A.
Protea, 50E.Ex.1.
Protodiniferidae, 45.Ex.7.
Protologue, 7B.3-4, 7C, 8, 10.2-3; – (defini-tion), 8(footnote).
Provisional approval (conserved or rejected name), 15, 15A.
Provisional name, 34.1(b).
×*Pseudadenia*, H.6.Ex.3.
Pseudocompounds, 73G.1(b).
Pseudoditrichaceae, 41.Ex.1.
Pseudoditrichum, – *mirabile*, 41.Ex.1.
Pseudorchis, H.6.Ex.3.
Pseudo-salvinia, 73.Ex.15.
Psilotum truncatum, 63.Ex.9.
Pteridium aquilinum subsp. (var.) *caudatum*, 26A.Ex.2.
Pteridophyta (Committee), Div.III.2; – (start-ing point), 13.1(a).
Pteris caudata, 26A.Ex.2.
pteroides (*-oideus*), 64.Ex.8.
Ptilostemon, 55.Ex.4; – sect. *Cassinia*, – nsect. ×*Platon*, – sect. *Platyrhaphium*, – nsect. ×*Plinia*, – sect. *Ptilostemon*, H.7.Ex.1; – *chamaepeuce*, – *muticus*, 55.Ex.4.
Public meeting (eff. publ.), 29.1.
Publication, v. Valid, Effective, etc.
Publishing author, 34A, 46, 46C-D.
Puccinellia, H.6.Ex.5.
×*Pucciphippsia*, H.6.Ex.5.
Pulsatilla montana subsp. *australis*, – subsp. *dacica*, – var. *serbica*, 49.Ex.6.
Purpose (of giving name), Pre.1; etiam v. Ob-ject.
Pyroleae (*-oideae*), 19A.Ex.1.
Pyrus, 33.Ex.4; – *calleryana*, – *mairei*, 50F.Ex.1.

Quercus alba, – ×*deamii*, – *macrocarpa*, – *muehlenbergii*, H.10.Ex.4.
Question mark (valid publ.), 34.2.
Quisqualis, 20.Ex.7.
Quotation marks (original spelling), 50F; – (superfluous name), 63.Note 1.

Radicula, 20.Ex.2.
Radiola, – *linoides*, – *radiola*, 72.Ex.1.

Radix (inadmissible as generic name), 20.Ex.5.
Ralfs (British Desmidieae), 13.1(e).
Rank(s), 1-5; – (change of), 51, 60, 61, 61A; – (change, citation), 49; – (correct name), IV, 6.5, 11; – (hybrids), 3.2, 4.3, 50, H.1.1, H.3.1, H.5, H.12; – (indication), 21.1, 21A, 24.1, 35; – (misplaced term), 33.4, 33.5; – (principal), 3; – (priority inside rank), 60, 61; – (simultaneous publ. at different), 34.3; – (supplementary), 4.2; – (valid publ.), 35.
Rankless names, 35.2.
Raphidomonas, 16.Ex.1.
Raphidophyceae, 16.Ex.1.
Rapporteur-général, Div.III.2, Div.III.3.
Ravenelia cubensis, 59.Ex.2.
Ravensara, 68.Ex.1.
Reasons for changing a name, Pre.8.
Recent v. Non-fossil.
Recommendations, Pre.5.
Recorder, Div.III.3.
Rediscovery of original material (typification), 8.1.
Reference (direct or indirect, valid publ.), 32.1(c), 32.3-4, 41; – (full and direct), 33.2, 33.3, 45.1; – (page), 33.2 (footnote); – (to illustration), 38, 39, 39A; – (to Latin diagnosis), 36; – (to pre-starting point descr.), 7.15, 32A; – (validating family or generic name), 41; etiam v. Citation.
Regnum (kingdom), 3.1, 4.1.
Reihe (for order), 17.2.
Rejected names, 14, 69, App.III-IV.
Rejection of names, 15.1, 15A, 62-72; – (homonyms), 64; – (inappropriate names, etc.), 62.1; – (names based on discordant elements), [70]; – (names based on monstrosity), [71]; – (name used in sense excluding type), 69; etiam v. Nomina rejicienda.
Relative order of ranks, 5; etiam v. Misplaced term.
Relevant rule (absence), Pre.9.
Remodelling of taxa (retention of name), 51-53.
remyi, 73C.4(d).
Renanthera, H.8.Ex.2.
Repetition of epithet, 22, 22A, 26, 26A, 27.
Repetition of generic name (as epithet), 22.
Replacement (of illegitimate names), 72.
Reprints, v. Separates.
Requirements for publication, v. Effective, Valid.
Residue (lectotypification), 7B.5 & 7.
Restoration of a rejected name, 14.6-7.

Retention of a name or epithet, 51-56; – (change of circumscription), 51; – (change of rank) 60A; – (division of taxon), 51-53; – (excluding type), 48.Note 2; – (generic), 52; – (infrageneric), 54-56; – (specific or infraspecific), 53, 55, 56; – (type excluded), 48.Note 2; – (union of taxa), 57; – (with altered termination), 61A.1.
Retention of original orthography, 73, 73E, 75.
Retroactive rules, VI.
Revision Nostocacées (Bornet & Flahault), 13.1(e).
Rhamnus, 76.Ex.1; – subg. (sect.) *Pseudofrangula*, – *alnifolia*, 22A.Ex.1.
Rhaptopetalaceae, 34.Ex.7.
Rhaptopetalum, 34.Ex.7.
Rheedia kappleri, 9.Ex.1.
Rheum ×*cultorum*, 40.Ex.2.
Rhizoctonia microsclerotia, 59.Ex.4.
Rhododendreae (-*oideae*), 19.Ex.3.
Rhododendron, 19.Ex.3, 20.Ex.1, 22.Ex.3; – subg. (sect.) *Anthodendron*, – subg. *Pentanthera*, – *luteum*, 22.Ex.3.
Rhodomenia (*Rhodymenia*), 14.Ex.8.
Rhodophyllaceae, 18.Ex.1.
Rhodophyllidaceae, 18.Ex.1.
Rhodophyllis, 18.Ex.1.
Rhodophyllus, 18.Ex.1, 57.Ex.6.
Rhodora, 19.Ex.3.
Rhodoreae, 19.Ex.3.
Rhynchostylis, H.6.Ex.7.
Rich. (abbreviation), 46A.2.
Richardia, 62.Ex.5.
Richardsonia, 62.Ex.5.
richardsonis, 73C.2.
Ricinocarpos sect. *Anomodiscus*, 21.Ex.1.
×*Rodrettiopsis*, H.6.Ex.7.
Rodriguezia, H.6.Ex.7.
Romanization (author's name), 46B.
Rorippa, 14.Ex.3, 20.Ex.2.
Rosa, 18.Ex.1, 19.Ex.1, 20.Ex.1, 46.Ex.1; – *canina*, H.3.Ex.3; – *gallica*, – var. *eriostyla*, – var. *gallica*, 46.Ex.1; – *glutinosa* var. *leioclada*, – *jundzillii* var. *leioclada*, 24.Ex.5; – *pissardii* (-*di*, -*ti*), 73.Ex.16; – *toddiae* (-*ii*), 73.Ex.17; – *webbiana*, 73C.1(d).
Rosaceae, 18.Ex.1, 19.Ex.1, 46.Ex.1.
Roseae (-*oideae*), 19.Ex.1.
Rubia, 64.Ex.9.
Rubus, 64.Ex.9; – *amnicola*, 23.Ex.4; – *quebecensis*, 73D.Ex.1.
Rules (definition, object), Pre.; – (retroactive), VI.
-*rum*, 73C.1.

Species, 2-4; – (as hybrid), 50; – (division of), 53.1; – (regarded as sum of its subordinate taxa), 25; etiam v. Specific name or epithet.

Species Muscorum (Hedwig), 13.1(b).

Species naturales (Necker), 20.Ex.11.

Species Plantarum (Linnaeus), v. Linnaeus.

Species Plantarum (Willdenow), 30.Ex.1.

Specific name or epithet, 23-23B; – (based on host plant), 73H; – (change of rank), 61; – (conservation), 14, 15, 15A; – (erroneous application on transfer), 55.2; – (etymology), 73I; – (formation), 23B; – (hybrids), H.10; – (not illegitimate under illeg. generic name), 68; – (initial letter), 73.2, 73F; – (multiverbal), 23.1; – (orthography, homonyms), 64.3; – (from personal name), 73C; – (repetition), 26, 26A, 27; – (retention on change of rank), 61A.3; – (retention on division), 53; – (transfer to another genus), 55; – (type), 9; – (valid publication), 42-44.

Specimen (basis of illustration), 32D.2, 39A; – (type), 7-10A, 37.5.

Spelling, v. Orthography, Orthographic variants.

Spergula stricta, 55.Ex.2.

Spermatites, 3.Ex.1.

Spermatophyta (Committee), Div.III.2; – (starting point), 13.1(a).

Sphagnaceae (starting point), 13.1(c).

Spina (inadmissible as generic name), 20.Ex.5.

Spiritus asper (transcription), 73A.2.

Spondias mombin, 23.Ex.1.

ß (= ss), 73.4.

St. (in epithet), 73C.4(d).

St.-Hil. (abbrev.), 46A.5.

Stability of nomenclature, Pre.1, 14.2.

Stachys, 76.1; – *ambigua*, 50.Ex.1; – *palustris* subsp. (var.) *pilosa*, 26A.Ex.1.

Standard species, 8.Ex.2.

Stapelia, – *hirsuta*, – *variegata*, 8.Ex.3.

Starting-point, 13, 32.1(a); – (citation prest.-p.), 46.3.

stat. nov., 7.12.

Status (name under Code), 6.6, 12; – (hybrids, alteration), 50.

steenisii, 73C.4(e).

-*stemon*, 76.2(a).

Stenocarpus, 76.Ex.2.

Sternberg (Flora Vorwelt), 13.1(f).

Steudel (Nomenclator Botanicus), 33.Ex.1.

Stickman (Herb. Amboinense), 34.Ex.2.

-*stigma*, 76.2(c).

Stigmaria, 3.Ex.1.

Stillingia, 57.Ex.4.

-*stoma*, 76.2(c).

strassenii, 73C.4(e).

Stratigraphic relations (fossils), 13.3.

Strychnos, 76.1.

Suaeda, – *baccata*, – *vera*, 43.Ex.1.

Sub- (prefix), 4.1.

Subclass(is), 4.1; – (name), 16.1, 16A.3, 16A.4; – (priority), 11.4, 16.Note 1; – (type of name), 10.5.

Subdivisio(n), 4.1; – (name), 16.1, 16A.2, 16A.4; – (priority), 11.4, 16.Note 1; – (type of name), 10.5.

Subdivision of a family (definition), 4.Note 1; – (name), 19, 19A; – (type of name), 10.4, 19.

Subdivision of a genus, 10, 21-22; – (definition), 4.Note 1; – (choice of name), 54; – (homonyms), 64.4; – (hybrid), 40.Note 1, H.7, H.9; – (name or epithet), 21-22A, 40.Note 1, H.7, H.9; – (transfer), 54; – (type of name), 7B.7, 10, 22, 22A; – (valid publ.), 43.

Subfamily (-ia), 4.1; – (name), 19.1, 19A; etiam v. Subdivision of a family.

Subforma, 4.1; etiam v. Infraspecific name or epithet.

Subfossil plants (union with taxon of recent plants), 58.

Subgenus, 4.1; – (change of rank), 61A.2; – (epithet), 21, 21A-B, 22, 22A; – (hybrid), H.7, H.9; – (name derived from personal name), 73B; – (name), 21, 21A-B; – (type-, epithet), 22, 22A.2; etiam v. Subdivision of a genus.

Subjective synonym, 14.4(footnote).

Suborder, 4.1; – (name), 16.1, 17; – (priority), 11.4, 16.Note 1; – (typification), 10.5.

Subordinate ranks, v. Hierarchy.

Subordinate taxa (mention not valid publ.), 34.1(d); – (species sum of), 25.

Subregnum, 4.1.

Subsectio(n), 4.1; etiam v. Subdivision of a genus.

Subseries, 4.1; etiam v. Subdivision of a genus.

Subspecies, 4.1; – (change of rank), 61, 61A.3-4; – (choice of name on division), 53.2; – (epithet), 24, 24A-B, 26, 26A, 27; – (valid publ.), 43; etiam v. Infraspecific name or epithet.

Substitute name, v. Avowed substitute.

Subtribe(-us), 4.1; – (name), 19, 19A.

Subvariety(-as), 4.1; – etiam v. Infraspecific name or epithet.

Suffix, 16A.2, 73B.Note 1; etiam v. Termination.

Sum of subordinate taxa (species, etc.), 25.

326

– (familia), 10.4; – (figure), 9.3; – (forma), 9, 26, 26A.3, 27; – (genus), 10; – (hybrid), H.9.Note 1, H.10.Note 1; – (inclusion), 63.1-2; – (indication by means of epithet), 22.4, 22.5, 22A, 24.3; – (indication, new taxa), 37, 37A; – (infraspecific taxa), 9, 26-27; – (material, heterogeneous), 7B.5, 9.2; – (material, living), 9.5, 9A; – (method), II, 7-10; – (nomen novum), 7.11; – (obligate choice on division of taxa), 52, 53; – (place of preservation), 37.5; – (pleomorphic fungi), 7.Note 2, 59, 59A; – (previous descr.), 7.15, 7.16, 7.Note 1; – (section), 22, 22A; – (species), 9, 26, 26A; – (starting point after 1753), 7.14; – (subdivision of a family), 10.4; – (subdivision of a genus), 10, 22, 22A; – (subspecies), 9, 26-27; – (superfluous name), 7.13, 63; – (taxonomic position), 13.2; – (varietas), 9, 26-27.

Typescripts (eff. publ.), 29.1.
Typical element of taxon not type, 7.2.
typicus, 7B.4, 24.3.
Typification, v. Type.
Typographic error, 73.1, 75.
Typography, 46B.2, 73.2, 73.4-6, 73F.

ü (= ue), 73.6.
Ubochea, 73B.Ex.1.
ue (= ü), 73.6.
Uffenbachia, 73.Ex.6.
Ulmus racemosa, 50C.Ex.1.
Umbelliferae, 18.5.
Union of biverbal epithets, 23.1.
Union of taxa, 51.1(b), 57-58; – (and conserved names), 14.5; – (non-fossil and fossil plants), 58.
Unitary designation of species, 20.4(b).
Unpublished names, 23B.1(i), 34A.
Upper-case letter, v. Capital letter.
Uredinales (starting point), 13.1(d).
Uredo cubensis, 59.Ex.2; – *pustulata*, 73.Ex.7.
Uromyces fabae, 23.Ex.1.
Urtica "dubia?", 23.Ex.5.
Urvillea, 64.Ex.9, 73B.Ex.1.
-us, 73B.1(c), 73G.1(a).
Usage (current), 69.4; – (persistent, excl. type), 69; etiam v. Custom, Tradition.
Ustilaginales, 17.Ex.2, 59.1; – (starting point), 13.1(d).
Utricularia inflexa, – var. *stellaris*, – *stellaris*, – var. *coromandeliana*, – var. *stellaris*, 26.Ex.3.
Uva-ursi, 20.Ex.6.

v (use as vowel), 73.5.
Vaccinieae (*-oideae*), 19A.Ex.1.
Vaillantia (*Valantia*), 73.Ex.9.
Valeriana sect. *Valerianopsis*, 21.Ex.1.
Valid (botanical nomenclature), v. Valid publication.
Valid publication, 6.2, 12, 32-45; – (acceptance by author), 34.1(a), 34.2, 45.1; – (algae), 36, 39, 39A, 45.4; – (alternative names), 34.3, 34. Note 1; – (basionym), 33.2; – (combination), 33.1-2; – (date), 45; – (definition) 6.2; – (family name), 41.1; – (fossil pl.), 36.1, 38; – (future acceptance), 34.1(b); – (generic name), 41.2, 42; – (hybrids), 40, H.9-10; – (illegitimate family name), 18.3; – (by means of illustrations), 41.Note 2, 42.2, 44; – (indication of type), 37; – (infrageneric names), 43; – (Latin diagnosis), 36; – (Linnaean genera), 13.4, 41.Note 1; – (mention subordinate taxa), 34.1(d); – (monotypic genus), 42; – (name not accepted by author), 34.1(a); – (nomen novum), 33.2; – (orthographic variant), 75; – (provisional names), 34.1(b); – (rank to be indicated), 35; – (by reference), 7.15-16, 32.1(c), 32A, 41, 45.1; – (species in monotypic genus), 41.3, 42.1; – (species name), 41.3; – (starting point), 13; – (synonym), 34.1(c); – (in works written in a modern language), 45A.
vanbruntiae, 73C.4(e).
Vanda, H.6.Ex.7, H.8.Ex.2; – *lindleyana*, 73C.1(c).
vanderhoekii, 73C.4(e).
Variant, v. Orthographic variants.
Variants arising in cultivation, 4.Note 2, 28.
Variety(-as), 4.1; – (change of rank), 61, 61A; – (choice of name on division), 53.2; – (epithet), 24, 24A-B, 25-26, 26A, 27; – (valid publ.), 43; – (without rank before 1890), 35.3; etiam v. Infraspecific name or epithet.
x *Vascostylis*, H.6.Ex.7.
vechtii, 73C.4(e).
Verbascum sect. *Aulacosperma*, 64.Ex.12; – *lychnitis*, – *nigro-lychnitis*, H.10.Ex.2; – *nigrum*, 23.Ex.4, H.10.Ex.2; – x*schiedeanum*, H.10.Ex.2.
Verbena hassleriana, 73C.1(d).
Verbesina alba, – *prostrata*, 57.Ex.5.
veridicus, 24.3.
verlotiorum, 73C.1(b).
Vernacular names (in names or epithets), 73.7, 73F; – (gender as generic names), 76.3.
Veronica anagallis-aquatica, 23.Ex.2, 73.Ex.14.
verus, 24.3.
Vexillifera, – *micranthera*, 6.Ex.1.

Vffenbachia, 73.Ex.6.

Viburnum x*bodnantense* "Dawn", 28.Ex.1; – *ternatum*, 46.Ex.2.

Vice-Rapporteur, Div.III.3.

Vicia, 18.5.

Vinca major, 23.Ex.4.

Vincetoxicum, 62.Ex.1.

Viola hirta, 24.Ex.5; – "*qualis*", 23.Ex.5; – *tricolor* var. *hirta*, 24.Ex.5.

vonhausenii, 73C.4(e).

Voting, Div.III.4.

Vowel (connecting), 73.8, 73G.1(a.2); – (use in names), 73.5-6, 73B, 73C.1, 73G.

Vredo pvstvlata, 73.Ex.7.

w (permissible in Latin plant names), 73.4.

Wahlenbergia, 60.Ex.1.

Walter (Flora Carol.), 20.Ex.8.

Waltheria americana, – *indica*, 57.Ex.3.

Webb & Heldreich (Cat. Pl. Hisp. Blanco Lect.), 29.Ex.2.

Weihea, 14.Ex.1.

Willdenow (Species Plantarum), 30.Ex.1.

Wilsonara, H.6.Ex.7.

wilsoniae, 73C.1(b).

Wintera, 18.Ex.3.

Winteraceae, 18.Ex.3.

Wood, fossil, 9.4.

Words (not intended as names), 20.4(a); 23.6(a), 43.Note 1; – (compound), 73G; – (compound, gender), 76.2.

Works appearing in parts (dates of publication), 35.4, 45B-C.

x (vs. x), H.3A.

Xanthoceras, 76.2(c).

Xanthoxylon, 50F.Ex.2.

Xerocomus, 76A.Ex.1.

y (permissible in plant names), 73.4.

Zanthoxylum caribaeum var. *floridanum*, – *cribrosum*, 50F.Ex.2.

zeylanica, 64.Ex.8, 73.Ex.1.

Zoological nomenclature, I; – (International Code of), 14.4(footnote), 45.Ex.4 (footnote) & Ex.5-7.

Zygophyllum billardierei (*billardierii*), 73.Ex.10.